The Ocean Sunfishes

Evolution, Biology and Conservation

Editors

Tierney M. Thys

California Academy of Sciences
San Francisco, CA
USA

Graeme C. Hays

Centre for Integrative Ecology
Deakin University
Geelong
Australia

Jonathan D.R. Houghton

Queen's University Belfast
School of Biological Sciences
Belfast
UK

CRC Press is an imprint of the
Taylor & Francis Group, an **informa** business
A SCIENCE PUBLISHERS BOOK

Cover illustration reproduced by kind courtesy of Mike Johnson.

First edition published 2021
by CRC Press
6000 Broken Sound Parkway NW, Suite 300, Boca Raton, FL 33487-2742

and by CRC Press
2 Park Square, Milton Park, Abingdon, Oxon, OX14 4RN

Library of Congress Cataloging-in-Publication Data

Names: Thys, Tierney M., 1966- editor.
Title: The ocean sunfishes : evolution, biology and conservation / editors, Tierney M. Thys, California Academy of Sciences, San Francisco, CA, Graeme C. Hays, Centre for Integrative Ecology, Deakin University, Geelong, Australia, Jonathan D.R. Houghton, Queen's University Belfast, School of Biological Sciences, Belfast, UK.
Description: Boca Raton : CRC Press, [2020] | Includes bibliographical references and index.
Identifiers: LCCN 2020019431 | ISBN 9780367359744 (hardcover)
Subjects: LCSH: Ocean sunfish--Evolution. | Ocean sunfish--Biology. | Ocean sunfish--Conservation.
Classification: LCC QL638.M64 O24 2020 | DDC 597/.64--dc23
LC record available at https://lccn.loc.gov/2020019431

ISBN: 978-0-367-35974-4 (hbk)

Typeset in Times New Roman
by Radiant Productions

Foreword

As executive director of a public aquarium that's home to more than 500 different species, I probably shouldn't admit that I have a favorite. But I do. It's the ocean sunfish, or *Mola mola.*

As with many marine species, there's much we don't know about these remarkable fishes, despite the fact they've fascinated and enchanted people across the globe for thousands of years. They're a favorite of the Japanese—from their classical art to a contemporary Pokémon character. They were referenced in Roman times by Pliny the Elder, and are respected in Polynesian culture. In California, where Monterey Bay Aquarium is located, they were part of the diet of indigenous people—a conclusion drawn by anthropologists based on abundant *Mola* remains found in 4,000-year-old midden sites on the southern California coast.

Everything about ocean sunfishes is so unlikely, which is perhaps what makes them so oddly endearing. It starts with their half-a-fish body shape and their impressive size (Astonishingly, they increase in size more than 600 million times from their larval state to full maturity, with some individuals growing to weigh 2,300 kg and spanning 3 m from tip to "tail").

It's remarkable to realize they can grow so large when they begin life as tiny plankton, adrift in an ocean filled with hungry mouths. The lucky few that do manage to survive to maturity are the progeny of mothers who can produce hundreds of millions of eggs during their lifespan.

I'm not alone in my fascination for the ocean sunfish. Aquariums in Japan have a long history of featuring *Molidae* in their living exhibits. We sought our colleagues' advice before including ocean sunfish in our Open Sea exhibit when it first opened in 1996. We were (and remain) one of the few aquariums in the United States ever to exhibit *Mola mola.* They have been a consistent favorite with our visitors ever since—although they continue to challenge our animal care team.

Their popularity makes them effective ambassadors for their wild kin in the global ocean. Sharing their story gives us opportunities each day to talk with visitors—two million people a year at the Monterey Bay Aquarium, and three million who connect with us through social media—about how threats to ocean health from climate change, poorly managed fisheries and plastic pollution put ocean sunfish and other marine life at risk.

We've found that the emotional bonds our visitors form with ocean sunfish at the aquarium make them more receptive to learning about threats to ocean health that put wild sunfish at risk—and to ask what they can do to make a difference.

Inspiring people, and connecting them with the story of these impressive fishes, is the impetus for The Ocean Sunfishes: *Evolution, Biology and Conservation.* The book also demonstrates that effective communication begins and ends with rigorous science. You'll find it in abundance in this volume that brings the latest *Molidae* research together in a single place.

Scientific discoveries are advancing at a rapid pace, thanks to new tools and technologies. This book reflects the full scope of what's been learned about these singular fishes. It also highlights questions that science has yet to answer and offers an invitation to new generations of researchers to build on the work of their predecessors. I hope the book inspires students and scientists to keep expanding the knowledge, appreciation and conservation of *Molidae* around the world.

Science and storytelling go hand in hand, and editors Tierney Thys, Graeme Hays and Jonathan Houghton have included an important chapter devoted to the timeless hold ocean sunfish have on our imaginations. I see this every day at the aquarium. Whether *Mola mola* inspire or amuse, entice divers to travel to see them in the wild, prompt TED Talks or disparaging social media diatribes, this is certain: No one is indifferent to these impressive fishes.

I'm confident, based on our experiences at Monterey Bay Aquarium, that we can harness fascination for *Molidae* in ways that will secure a bright future for the living ocean, and for these unlikely charmers. They deserve no less.

November, 2019

Julie Packard
Monterey, California USA

Preface

The oft repeated mantra that "nothing is known about the elusive ocean sunfishes" no longer holds true as evidenced by the wealth of information presented in this book: *The Ocean Sunfishes: Evolution, Biology and Conservation*. When we first began researching these bizarre behemoths, back in the early 1990s in the case of T. Thys, the field of ocean sunfish research was wide open. Very few people, outside of devoted ichthyologists, had ever heard of ocean sunfishes, let alone dedicated substantial time to their study. Decades later, this story has changed dramatically. Interest and explorations into the Molidae as well as many elusive ocean animals, have exploded—a surge fueled, in part, by social media, an insatiably curious, ever-growing human population and crowd-sourced datasets. Each new discovery has been accompanied by an increased scientific appreciation for marine megafauna as individual entities, mobile data-gathering assistants and powerful players in the vast and varied ocean food web.

Our book draws from an impressive worldwide selection of molid researchers with contributors hailing from Australia, Austria, Brazil, Chile, Denmark, Ecuador, Ireland, Italy, Japan, New Zealand, Portugal, Spain, Switzerland, Taiwan, United Kingdom and the USA. This broad geographic distribution of contributors mirrors the circum-global nature of the Molidae themselves who boast a remarkably wide geographic range. Molid sightings span from north of the Arctic Circle off Norway to the Beagle Channel off Ushuaia, Argentina and everywhere in between.

Our book is organized as a journey from the fossil origins of pre-Miocene Molidae through the various aspects of molid life history to the future of molids in an ocean greatly impacted by overfishing and increasing climate pollution. Each chapter ends with a set of remaining questions specific to that area of research. It is our hope that this book will be the go-to resource for anyone with a deep interest in the ocean sunfishes and most importantly as a springboard for future researchers eager to make new discoveries.

The Ocean Sunfishes: Evolution, Biology and Conservation has been a labor of love and a richly collaborative effort. We hope you enjoy reading it as much as we enjoyed putting it together. We extend a special thanks to all the contributors for their excellent research, willingness to join this effort and for being so delightful to work with. We would also like to thank the members of the Molidae family for their limitless capacity to astound, inspire and entice the public and research communities alike to better understand our ocean world. The ocean sunfishes are a wide-eyed reminder that we still have much to learn about our wondrous blue planet.

January 2020

Tierney M. Thys, Carmel California USA
Graeme C. Hays, Geelong Australia
Jonathan D.R. Houghton, Belfast, N. Ireland

Acknowledgments

All the editors would like to thank Tasha Phillips, Tor Mowatt-Larsen, Olivia Daly and Lawrence Eagling for their invaluable and greatly appreciated help in editing, proof-reading and formatting figures. We would also like to thank whole-heartedly the ocean sunfishes themselves for being such inspirational and captivating creatures.

Tierney M. Thys

A big thank you to Graeme Hays and Jonathan Houghton for their willingness, expertise, enthusiasm and encouragement to take on this labor of love and to all the amazing and inspirational collaborators and chapter coauthors. Thanks to Steven A. Wainwright, my graduate school advisor for introducing me to ocean sunfish and for being such a visionary; my family and community of friends; Eugenie Clark, Sylvia Earle, Mark Bertness, Jim Tyler, Bruce Collette, Chuck Baxter and Hong Young Yan for encouraging my tetraodontiform obsession. Huge thanks to National Geographic for steadfast support of my molid research endeavors; Sea Studios for tagging support and media training and; Mike Johnson a wonderful photographer, friend and field assistant who helps me run oceansunfish.org and provided the beautiful cover of the book. Thanks to wonderful friends including Lukas Kubicek for unwavering dedication and sleuthing into all things molidae; Marianne Nyegaard for tireless research and willingness to go the many extra miles; Monterey Bay Aquarium staff, especially John O'Sullivan and Chuck Farwell for teaching me how to tag, along with Heidi Dewar at the SWFC; the CRC editors, especially Raju for answering our constant stream of questions; Inga Potter for invaluable editing help; Rachel Kelly for searching the world for mola larvae; Jamie Watts and Nick Bezio for their beautiful drawings, Emily Nixon and Sarah Matye for corralling the citizen science sightings site; Don Kohr for tirelessly retrieving esoteric sunfish resources for me; supporters of Adopt a Sunfish Project for donating to our research efforts; Bruce Robison for a stack of original sunfish reprints dating back to the 1800s that he hoped I'd put to good use (I have!). Thanks to Alastair Dove for encouraging me to take on this book; Steve Karl, Anna Bass and Todd Streelman for their pioneering molid genetics work; the California Academy of Sciences for hosting me as a research associate. And last but not least, thank you to my family—my wonderful in-laws, amazing sisters and remarkable Ma for constant support and unwavering love of all living creatures and beloved Brett, Marina and Grant, for tolerating my sunfish obsession, joining in and always being so supportive.

Graeme C. Hays

Thanks to Tierney Thys for very kindly inviting me to be part of this adventure and for her endless enthusiasm and support throughout the gestation of this book. Jonathan Houghton and Tom Doyle for introducing me to ocean sunfish during amazing field-trips on the west coast of Ireland and for the long-term collaborations we have enjoyed, over two decades, looking at the roles of jellyfish as food for marine megafauna.

Jonathan D.R. Houghton

Thank you to Tierney Thys for the very generous offer to co-edit this book with her; Graeme Hays for allowing myself and Tom Doyle to go off on a sunfish tangent during our post-doctoral years; Tom

Doyle for his enthusiasm, intellect and shared propensity to speculate wildly about sunfish after a few pints (to the dismay of Graeme Hays). Thanks Graeme for humoring our early intellectual forays into the unknown; Lukas Kubicek for introducing me to the stunning opportunities the Mediterranean Sea offers for sunfish research and all his help in the field.

G&M Williams Fund for a travel bursary to Tierney Thys to visit Queen's University Marine Laboratory, Portaferry for an editorial retreat. My lab at QUB for keeping the science (and humor) coming during my shift to a more sedentary academic lifestyle. My sunfish collaborators over the years (especially Tom Doyle, Chris Harrod & Natasha Phillips) without whom I would lay no claim to knowing anything about ocean sunfish. And last (but so far from least) my family for keeping me afloat all these years.

Dedication

This book is dedicated to:
Dr. Stephen A. Wainwright a visionary scientist, artist and lover of all anatomical oddities

Nature reveals great secrets in her extreme forms.

Contents

Foreword iii
Julie Packard

Preface v
Tierney M. Thys, Graeme C. Hays and *Jonathan D.R. Houghton*

Acknowledgments vi

1. **Evolution and Fossil Record of the Ocean Sunfishes** 1
Giorgio Carnevale, Luca Pellegrino and *James C. Tyler*

2. **Phylogeny, Taxonomy and Size Records of Ocean Sunfishes** 18
Etsuro Sawai, Marianne Nyegaard and *Yusuke Yamanoue*

3. **Genetic Insights Regarding the Taxonomy, Phylogeography and Evolution of Ocean Sunfishes (Molidae: Tetraodontiformes)** 37
Eric J. Caldera, Jonathan L. Whitney, Marianne Nyegaard, Enrique Ostalé-Valriberas, Lukas Kubicek and *Tierney M. Thys*

4. **Overview of the Anatomy of Ocean Sunfishes (Molidae: Tetraodontiformes)** 55
Katherine E. Bemis, James C. Tyler, Eric J. Hilton and *William E. Bemis*

5. **Locomotory Systems and Biomechanics of Ocean Sunfish** 72
Yuuki Y. Watanabe and *John Davenport*

6. **Reproductive Biology of the Ocean Sunfishes** 87
Kristy Forsgren, Richard S. McBride, Toshiyuki Nakatsubo, Tierney M. Thys, Carol D. Carson, Emilee K. Tholke, Lukas Kubicek and *Inga Potter*

7. **Ocean Sunfish Larvae: Detections, Identification and Predation** 105
Tierney M. Thys, Marianne Nyegaard, Jonathan L. Whitney, John P. Ryan, Inga Potter, Toshiyuki Nakatsubo, Marko Freese, Lea M. Hellenbrecht, Rachel Kelly, Katsumi Tsukamoto, Gento Shinohara, Tor Mowatt-Larssen and *Lukas Kubicek*

8. **Movements and Foraging Behavior of Ocean Sunfish** 129
Lara L. Sousa, Itsumi Nakamura and *David W. Sims*

9. **The Diet and Trophic Role of Ocean Sunfishes** 146
Natasha D. Phillips, Edward C. Pope, Chris Harrod and *Jonathan D.R. Houghton*

10. **Parasites of the Ocean Sunfishes** 160
Ana E. Ahuir-Baraja

11. **Biotoxins, Trace Elements, and Microplastics in the Ocean Sunfishes (Molidae)** 186
Miguel Baptista, Cátia Figueiredo, Clara Lopes, Pedro Reis Costa, Jessica Dutton, Douglas H. Adams, Rui Rosa and *Joana Raimundo*

12. Fisheries Interactions, Distribution Modeling and Conservation Issues of the Ocean Sunfishes **216**
Marianne Nyegaard, Salvador García Barcelona, Natasha D. Phillips and *Etsuro Sawai*

13. Sunfish on Display: Husbandry of the Ocean Sunfish *Mola mola* **243**
Michael J. Howard, Toshiyuki Nakatsubo, João Pedro Correia, Hugo Batista, Núria Baylina, Carlos Taura, Kristina Skands Ydesen and *Martin Riis*

14. Ocean Sunfishes and Society **263**
Tierney M. Thys, Marianne Nyegaard and *Lukas Kubicek*

15. Unresolved Questions About Ocean Sunfishes, Molidae—A Family Comprising Some of the World's Largest Teleosts **280**
Graeme C. Hays, Jonathan D.R. Houghton, Tierney M. Thys, Douglas H. Adams, Ana E. Ahuir-Baraja, Jackie Alvarez, Miguel Baptista, Hugo Batista, Núria Baylina, Katherine E. Bemis, William E. Bemis, Eric J. Caldera, Giorgio Carnevale, Carol D. Carson, João Pedro Correia, Pedro Reis Costa, Olivia Daly, John Davenport, Jessica Dutton, Lawrence E. Eagling, Cátia Figueiredo, Kristy Forsgren, Marko Freese, Salvador García-Barcelona, Chris Harrod, Alex Hearn, Lea Hellenbrecht, Eric J. Hilton, Michael J. Howard, Rachel Kelly, Lukas Kubicek, Clara Lopes, Tor Mowatt-Larssen, Richard McBride, Itsumi Nakamura, Toshiyuki Nakatsubo, Emily Nixon, Marianne Nyegaard, Enrique Ostalé-Valriberas, Luca Pellegrino, Natasha D. Phillips, Edward C. Pope, Inga Potter, Joana Raimundo, Martin Riis, Rui Rosa, John P. Ryan, Etsuro Sawai, Gento Shinohara, David W. Sims, Lara L. Sousa, Carlos Taura, Emilee Tholke, Katsumi Tsukamoto, James C. Tyler, Yuuki Y. Watanabe, Kevin C. Weng, Jonathan L. Whitney, Yusuke Yamanoue and Kristina S. Ydesen

Index **297**

Chapter 1

Evolution and Fossil Record of the Ocean Sunfishes

Giorgio Carnevale,[1,*] *Luca Pellegrino*[1] and *James C. Tyler*[2]

Introduction to Molidae

Molidae is a very small family of pelagic tetraodontiform fishes, commonly referred to as ocean sunfishes, which has five currently recognized species arranged in three genera, *Masturus*, *Mola* and *Ranzania* (e.g., Fraser-Brunner 1951, Tyler 1980, Santini and Tyler 2002, Nyegaard et al. 2018). Most of the molid species have global distributions in tropical to temperate seas (Fraser-Brunner 1951). For additional distribution information see Etsuro et al. 2020 [Chapter 2], Caldera et al. 2020 [Chapter 3] and Hays et al. 2020 [Chapter 15]. Species of the genera *Masturus* and *Mola* can reach gigantic sizes, growing to about 3 m in length and weighing more than 2,300 kg (e.g., Gudger 1928, 1937a, Santini and Tyler 2002, Forsgren et al. 2020 [Chapter 6]). *Mola* spp. are also considered one of the most fecund extant vertebrates, with individual females capable of producing up to tens of millions of eggs at one time (e.g., Schmidt 1921, Parenti 2003, Forsgren et al. 2020 [Chapter 6]). Due to their highly unusual morphology (Fig. 1), the species from family Molidae have attracted the attention of naturalists for many centuries (e.g., Steenstrup and Lütken 1898, Fraser-Brunner 1951).

This peculiar body plan is defined by numerous autapomorphic features (Fraser-Brunner 1951, Winterbottom 1974, Tyler 1980, Santini and Tyler 2002), most notably a short and rigid vertebral column and the loss of the swim bladder in adults, axial musculature, ribs, and pelvic and caudal fins. In particular, the posteriorly truncated appearance of these fishes is related to the disappearance of the skeletal components of a typical teleost caudal peduncle supporting a normal caudal fin; rather, these caudal peduncle components are replaced by a very deep and abruptly abbreviated tail region (Fig. 1), named the clavus by Fraser-Brunner (1951). From a skeletal point of view, the clavus is formed by a series of rod-like bones and fin rays situated between the posteriormost rays of the dorsal and anal fins. Johnson and Britz (2005) demonstrated that the caudal fin is completely lost in molids (except possibly for a few elongate rays at the center of the clavus in *Masturus* spp.) and that the skeleton of the clavus is formed by the modified posteriormost elements of the dorsal and anal fins; this hypothesis has also been confirmed by the cutaneous branch innervation pattern of the fin rays of the clavus, which is identical to that of the dorsal and anal fin rays (Nakae and Sasaki 2006). To

[1] Dipartimento di Scienze della Terra, Università degli Studi di Torino, Via Valperga Caluso, 35 I-10125 Torino, Italy.
Email: lu.pellegrino@unito.it
[2] National Museum of Natural History, Smithsonian Institution, Washington, DC 20560-0106, USA.
Email: TYLERJ@si.edu
* Corresponding author: giorgio.carnevale@unito.it

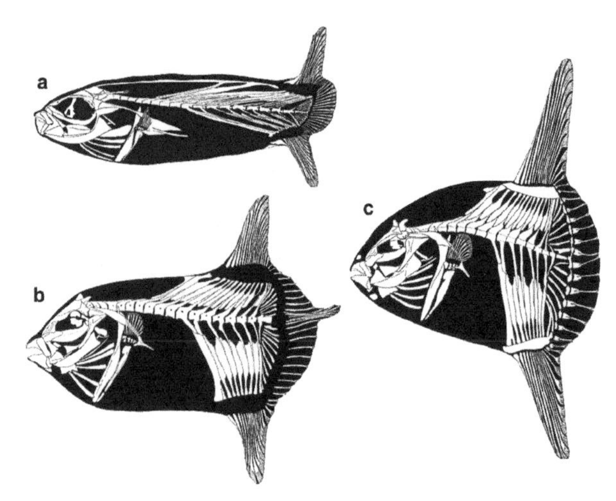

Figure 1. Range of diversity in skeletal morphology in the molids (modified from Tyler 1980). (a) *Ranzania laevis* (Pennant 1776); (b) *Masturus lanceolatus* (Liénard 1840); (c) *Mola mola* (Linnaeus 1758).

compensate for the absence of a true caudal fin and caudal peduncle, molids rely on the dorsal and anal fins for swimming, using their considerably developed muscles that function as lift-generating wings (e.g., Watanabe and Sato 2008, Davenport et al. 2018, Watanabe and Davenport 2020 [Chapter 5]).

Because of their abbreviated body shape and habit of passively floating on the sea surface, the ocean sunfishes were historically regarded as slow-moving surface-dwellers that fed primarily on gelatinous zooplankton (e.g., Fraser-Brunner 1951, Hooper et al. 1973). However, recent studies have revealed that they are highly active animals undergoing considerable horizontal and vertical migrations, and that they have apparent ontogenetic dietary shifts in which young individuals feed benthically on a variety of prey, and adults are also capable of chasing fast-moving midwater prey (e.g., Nakamura and Sato 2014, Phillips et al. 2015, Nyegaard et al. 2017, Sousa et al. 2016, Sousa et al. 2020 [Chapter 8]). In fact, *Masturus* and *Mola* spp. are known to exploit a variety of food items (e.g., algae, seaweed, eelgrass, sponges, hydroids, jellyfish, ctenophores, molluscs, crustaceans, echinoderms, salps, and fishes; e.g., Weems 1985), while *Ranzania* spp. may have a diet that includes seaweeds (Barnard 1927, but perhaps only incidentally) as well as crustaceans and other invertebrates, small fishes, and even squid in a particularly diverse diet (e.g., Nyegaard et al. 2017).

Since the publication of the monumental "Le Régne Animal" by Cuvier (1817), molids have been grouped with pufferfishes (Tetraodontidae) and porcupinefishes (Diodontidae) within the Gymnodontes based on the shared possession of highly modified teeth incorporated into a parrot-like beak, and scales modified into several kinds of prickly spines and variously developed basal plates. Although the composition of gymnodonts has been modified since the time of Cuvier by the insertion of several fossil families or distinctive genera (e.g., Avitoplectidae, Balkariidae, *Ctenoplectus*, Eoplectidae, and Zignoichthyidae) and one extant family (Triodontidae), the existence of a close relationship between molids, pufferfishes, and porcupinefishes has been confirmed by a number of subsequent studies (e.g., Regan 1903, Breder and Clark 1947, Winterbottom 1974, Tyler 1980, Lauder and Liem 1983, Tyler and Sorbini 1996, Santini and Tyler 2003, Santini et al. 2013b, Arcila et al. 2015, Bannikov et al. 2017, Arcila and Tyler 2017).

The Molidae is one of the tetraodontiform lineages with the least known fossil record and it is based mostly upon isolated jaws and dermal plates (Santini and Tyler 2003). According to Tyler and Santini (2002), the scarcity of fossil molids results from the synergistic effect of two factors not conducive to fossilization, namely their pelagic habitat and the weakly ossified and spongy nature of their skeleton. The goal of this chapter is to provide an overview of the currently known fossils belonging to the family Molidae and to briefly discuss their evolutionary and paleoecological significance.

Skeletal Features and Fossilization Potential

The overall skeletal anatomy of the molids has been investigated in great detail by a number of authors (e.g., Gregory and Raven 1934, Gudger 1937b, Raven 1939a, 1939b, Fraser-Brunner 1951, Tyler 1980, Bemis et al. 2020 [Chapter 4]). While an adult molid's most conspicuous skeletal feature is the posterior clavus which replaces the normal caudal fin and peduncle (Fig. 1C), they have a variety of additional remarkable anatomical features including: a large bone in the middle to rear of the interorbital septum that Tyler (1980) identified as a basisphenoid; a parrot-like beak mostly formed by a thick mass of osteodentine; large deep plates of firm cartilage that intervene between the bases of the dorsal- and anal-fin rays and their pterygiophores and their anteriormost larval vertebral element is incorporated ontogenetically into the basioccipital in at least two taxa (e.g., Tyler 1980, Britski et al. 1985, Santini and Tyler 2002, 2003, Britz and Johnson 2005). The molid 'basisphenoid' however, has been recently reinterpreted by Britz and Johnson (2012) as highly modified pterosphenoids because of its ontogenetic origin from the lamina marginalis of the chondrocranium, but this begs the question of what the bilaterally paired bones in the upper rear of the orbit identified by Tyler (1980) as pterosphenoids are. These persistent questions are examples of how much we still have to learn about molid skeletal morphology.

One of the most striking features of the molid skeleton is the considerable reduction of the bony tissue and the concomitant abundance of cartilage (Fraser-Brunner 1951). The bones are weakly ossified and characterized by a delicate fibrous or spongy texture, especially in the species of the genera *Masturus* and *Mola* (e.g., Gregory and Raven 1934, Raven 1939b). The reduced ossification of the bones, however, is associated with the possession of a sort of exoskeleton constituted by a continuous cover of scales, characterized by variably enlarged basal plates (Tyler 1980, Katayama and Matsuura 2016). These usually somewhat polygonal scale plates tend to bear a central tubercle, and are more or less in close contact with each other and bounded to the hypodermis (a thick subcutaneous collagenous tissue layer; labeled capsule in Davenport et al. 2018; Watanabe and Davenport 2020 [Chapter 5]) in *Masturus* and *Mola* spp. but form a robust and thickened carapace in *Ranzania* spp. (e.g., Fraser-Brunner 1951, Tyler 1980, Davenport et al. 2018). The skin of large-sized individuals of some species also exhibits other ossifications with a spongy texture that can reach considerable thickness (Watanabe and Davenport 2020 [Chapter 5]). In particular, a thick ovoid dermal plate (nasal plate) may be found on the snout just above the mouth and an elongate and thick subcylindrical bony plate (jugular plate) can be located on the chin, just below the mouth (e.g., Leriche 1926, Deinse 1953, Tyler 1980, Weems 1985, Carnevale and Godfrey 2018; Fig. 1). Additional dermal plates with high preservation potential can be observed along the posterior edge of the clavus in large-sized individuals of *Mola* spp. These "caudal" plates, called paraxial ossicles by Fraser-Brunner (1951), develop around the bifurcate distal ends of the fin rays and exhibit a different morphology and size based on their position along the clavus (Fraser-Brunner 1951, Tyler 1980), with the largest ones located in its medial portion and the smaller ones at its distal extremities. The anterior face of each of these plates has a longitudinal furrow that allows for the insertion of the membrane that connects the rays of the clavus, whereas their lateral faces are rough and notably rugose (e.g., Leriche 1926, Bemis et al. 2020 [Chapter 4]).

The parrot-like beak is one the most remarkable features of the Molidae, which has been traditionally used to associate them to other gymnodonts within the Tetraodontiformes. The upper beak consists of the fused contralateral premaxillae, whereas the lower beak is formed by the fusion of the two contralateral dentaries. These beak-like jaws are formed by a thick mass of osteodentine that overlies a basal unit of chondroid tissue (e.g., Britski et al. 1985). The biting edges of the beaks are toothless, without any discrete teeth or dental units visible. A triturating surface can be present posterior to the beak in the upper and lower jaws; however, the trituration teeth undergo extensive ontogenetic change (Tyler 1980), gradually decreasing in size and eventually disappearing in large specimens (see Weems 1985, K. Bemis *unpublished data*). There are taxonomically significant differences in the number, size, and morphology of the teeth in the known species, both extant and fossil (Tyler 1980, Weems 1985).

Due to their weakly ossified skeleton, rich in cartilage and characterized by a delicate spongy and fibrous texture, skeletal remains of molids are very rare as fossils, especially those pertaining to the genus *Mola*, while *Masturus* is completely unknown in the record. For this reason, most of the fossils unquestionably referred to in this group consist of isolated jaws and dermal plates.

The Fossil Record of the Molidae

Eomola bimaxillaria: Morphological Features and Phylogenetic Interpretation of a Putative Eocene Ocean Sunfish

Two upper jaws plus a few associated cranial bones from the Middle Eocene (Bartonian; 40-39 Ma) Kuma Horizon cropping out along the Pshekha River, in the vicinity of the Gorny Luch farmstead, Krasnodar Region, North Caucasus (Russia) document the earliest putative fossil evidence of the Molidae (Tyler and Bannikov 1992). The available material (Fig. 2) consists of two specimens, of which the paratype solely consists of an isolated left premaxilla. The holotype (Fig. 2a) comprises the premaxillae, maxillae, quadrates, a presumed interopercle, and a coracoid. These bones exhibit the typical striations indicative of a weakly ossified skeleton. Based on its premaxillary measurements and on those of *Mola mola* specimens of comparable size, a standard length of about 320 mm has been estimated by Tyler and Bannikov (1992) for the holotype of *Eomola bimaxillaria.*

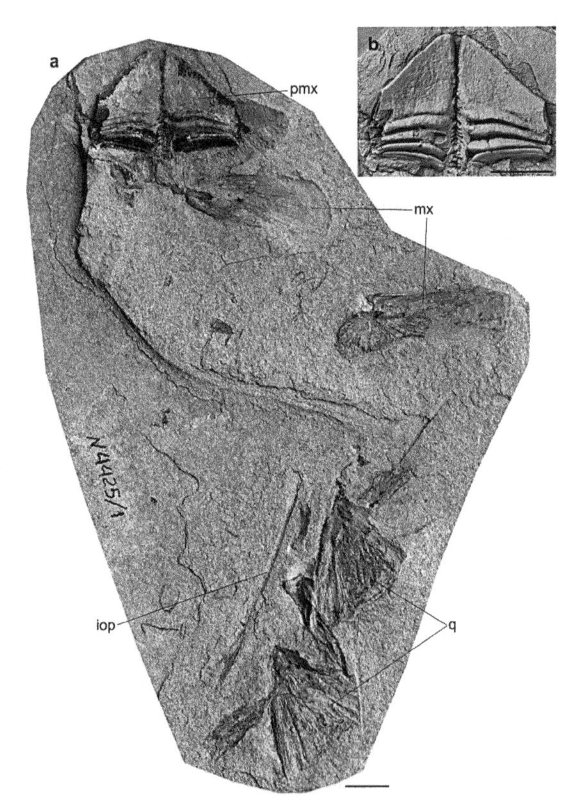

Figure 2. *Eomola bimaxillaria* Tyler and Bannikov, 1992 from the Bartonian Kuma Horizon, Gorny Luch, North Caucasus, Russia. (a) holotype, PIN 4425/1, general view of the premaxillae (pmx), maxillae (mx), quadrates (q) and interopercle (iop); (b) detail of the premaxillae in ventral view (coated with ammonium chloride to enhance the contrast). Scale bars 10 mm. Photos courtesy of Alexandre F. Bannikov.

Unlike in other molids, the premaxillae of *Eomola bimaxillaria* are separated and articulated with each other in the midline through interdigitating processes (as in tetraodontids; versus fused in the midline as in extant molids and diodontids) clearly visible along the medial edges of these approximately triangular bones (Fig. 2b). Each of the two contralateral premaxillae has four laterally elongate trituration teeth. A ridge located anterior to the trituration teeth possibly represents the absorption area of an older set of teeth. There is no evidence of small individual outer teeth along the biting edge of the beak. *Eomola bimaxillaria* also differs from other molid taxa by having a posteriorly bifurcate (vs. simple) interopercle similar to that of triodontids and several tetraodontids and diodontids; this condition, therefore, seems to be plesiomorphic for gymnodonts. Moreover, additional slight differences between *Eomola* and other molids concern the overall shape of the articular head of the maxilla and the relative proportions of some of the bones. Despite these differences, Tyler and Bannikov (1992) assigned this Middle Eocene fossil taxon to the family Molidae based on its possession of two derived features: laterally elongate trituration teeth and weakly ossified bones with a striated external texture.

Eomola is the only ocean sunfish with unfused premaxillae. The possession of unfused premaxillae is also characteristic of some other gymnodonts, including triodontids and tetraodontids, and is regarded as the plesiomorphic condition for the gymnodonts (Santini and Tyler 2003). Based on this feature, Tyler and Bannikov (1992) concluded that *Eomola* is the sister taxon to the group formed by the other molid genera (see also Santini and Tyler 2003).

As a final note, it is worth mentioning that Patterson (1993) erroneously used the name *Eoranzania* in reference to *Eomola*, at that time in the process of being described by Tyler and Bannikov (1992), who had been in consultation with Patterson about it. Thus, the generic name *Eoranzania* must be considered as a *nomen nudum*.

The Neogene Record of the Genus *Mola*

Due to its weakly ossified and delicate skeleton, the fossil record of the genus *Mola* is solely based on isolated beak-like jaws and dermal plates. The earliest putative representative of this genus consists of a partially preserved upper jaw from Argentina reported by Woodward (1901) (see also Gregory and Raven 1934, Romer 1966, Carroll 1988). This fragmentary specimen, assigned by Woodward (1901) to an indeterminate species of *Mola*, was collected from the "Patagonian Formation" in the Chubut Province. The "Patagonian Formation" corresponds to an informal unit that comprises several lithostratigraphic units originated in shallow marine to estuarine biotopes, with ages ranging from the late Oligocene to the early Miocene (e.g., Cuitiño et al. 2017). In the beds of the "Patagonian Formation" exposed in the Chubut Province, abundant marine vertebrate remains are known from the Gaiman Formation (e.g., Cione 1978, Scasso and Castro 1999). The age of the Gaiman Formation has been defined as early Miocene (Aquitanian-Burdigalian, between 23 and 18 Ma; Cuitiño et al. 2012, 2015).

Numerous beak-like toothless jaws, dermal plates, and caudal ossicles from the middle Miocene to Pliocene of Europe and North America have been referred to the extinct species *Mola pileata* (Fig. 3). Material belonging to this species was described for the first time by van Beneden (1881) from the Miocene of the Antwerp region, Belgium, who erroneously referred some isolated molid bones to two different species of the seabream *Pagrus* (*P. pileatus* and *P. torus*). Two years later, van Beneden (1883) created the species *Orthagoriscus* (=*Mola*) *chelonopsis* based on isolated jaws also from the Antwerp region. Leriche (1907) considered the fossils reported by van Beneden (1881, 1883) as synonyms of *Orthagoriscus* (=*Mola*) *pileatus* and reported additional fossils of this species from the middle Miocene of the Rhône Basin. Subsequently, Leriche (1926) described numerous isolated jaws (Figs. 3a-e), dermal plates (Figs. 3f-j), and (caudal) paraxial ossicles (Figs. 3k-l) from the Anversian (middle Miocene) of Kessel, southeast of the city of Antwerp, and correctly interpreted their anatomical identity based on detailed comparative observations on the extant *Mola mola*.

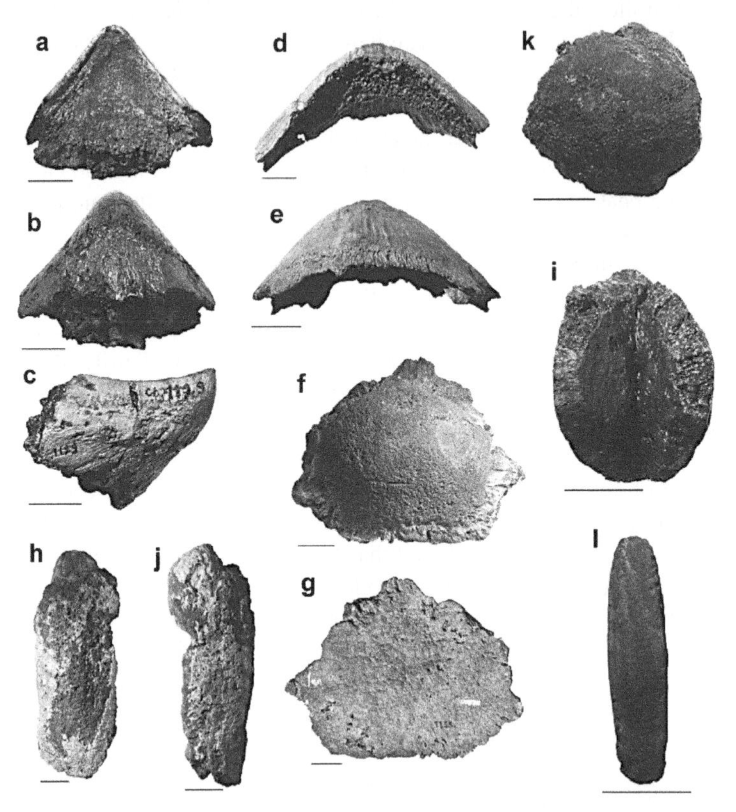

Figure 3. *Mola pileata* (van Beneden, 1881) from the Middle Miocene Antwerp Sands, Kessel, Belgium. (a) IRSNB P1129, premaxillary beak, ventral view; (b) IRSNB P1129, premaxillary beak, dorsal view; (c) IRSNB P1129, premaxillary beak, right lateral view; (d) IRSNB P1128, dentary beak, dorsal view; (e) IRSNB P1128, dentary beak, ventral view; (f) IRSNB P1133, nasal plate, external view; (g) IRSNB P1133, nasal plate, internal view; (h) IRSNB P1137, jugular plate, external view; (j) IRSNB P1137, jugular plate, left lateral view; (k) IRSNB P2960, medial paraxial ossicle, left lateral view; (i) IRSNB P2960, medial paraxial ossicle, anterior view; (l) IRSNB P2961, dorsal/ventral paraxial ossicle, left lateral view. Scale bars 20 mm. Photos courtesy of Olivier Lambert.

The age of this material derived from the Antwerp Sands cropping out in the vicinity of Kessel has been established based on marine palynomorphs as late Langhian to middle Serravallian, between about 15 and 12 Ma (see Lambert and Louwye 2006). Well-preserved jugular plates and caudal (paraxial) ossicles referred to *Mola pileata* have been reported from the middle Miocene mid- to outer-neritic clay of the Eibergen Member of the Breda Formation exposed in the Wiegerink quarry close to Groenlo and in the vicinity of Delden, east Netherlands (Deinse 1953, van den Bosch et al. 1975). King (2016) suggested a possible Serravallian age for the Eibergen Member of the Breda Formation. In North America a single premaxillary beak has been reported from the upper Serravallian (about 12 Ma) bed 19 of the Choptank Formation (see Vogt et al. 2018), Chesapeake Group, Maryland (Weems 1985, Carnevale and Godfrey 2018). Weems (1985) and Purdy et al. (2001) described some premaxillary beaks from the Pliocene (upper Zanclean-Piacenzian; Ward and Gilinsky 1993) Yorktown Formation of North Carolina. According to Weems (1985), *Mola pileata* can be recognized primarily based on features of the premaxillary beak, which is completely toothless and totally lacking the palatal tooth brace, with the toothless shelf anterior to the location of the former tooth position much longer than in *Mola mola*, such that the anteroposterior beak length is greater than the lateral beak width at the level of the back of the shelf.

The existence of indeterminate material (jaws and caudal ossicles) referred to the family Molidae from the middle to late Miocene of southern and central California (Round Mountain Silt of Kern

County, Monterey Formation of Santa Barbara County, Modelo Formation of Ventura County, Puente Formation of Los Angeles County, Topanga Formation of Orange County), which includes some isolated beaks in many ways similar to those characteristic of the genus *Mola* has been announced (although not described and figured) by Hakel and Stewart (2002) (see also Fierstine et al. 2012).

Finally, very abundant remains (paraxial ossicles and beak-like jaws) referred to as *Mola mola* have been collected from at least two archeological sites in the Southern California Channel Islands (Eel Point on San Clemente Island and Little Harbor on Santa Catalina Island) and from a coastal site in Santa Barbara County (Porcasi and Andrews 2001). These archeozoological remains reveal the existence of considerable fishery activity in southern California on this species since about 7000 cal BP or, more generally, during the early, middle, and late Holocene, also demonstrating that this species was exploitable, although with an undetermined fishing technique.

Extinct Representatives of the Genus *Ranzania*

The fossil record of the genus *Ranzania* is restricted to the Miocene and includes at least four extinct species plus some incomplete remains of problematic taxonomic placement (Fig. 4). The relatively more robust skeletal structure, as well as the tough carapace formed by articulated polygonal scales characteristic of *Ranzania* appear to increase its fossilization potential; this is suggested by the occurrence of partially complete articulated skeletal remains (Uyeno and Sakamoto 1994, Carnevale 2007, Carnevale and Santini 2007).

A few relatively small premaxillary beaks referred by Weems (1985) to *Ranzania tenneyorum* from the Langhian (between 16 and 15 Ma) zone 10 of the Plum Point Member of the Calvert Formation (see Vogt et al. 2018), Chesapeake Group, Maryland and Virginia represent the earliest evidence of *Ranzania* in the fossil record (Fig. 4h). These beaks are unique within the genus *Ranzania* in having three or more well-developed pairs of tooth rows emerging on a bony shelf, separated by a sill from the main level of the palatal bracing bone, and a notch clearly exposed just behind the tooth rows (Weems 1985, Carnevale and Godfrey 2018). Weems (1985) suggested that at least one of the premaxillary beaks described by Leriche (1926) from the Antwerp Sands should be assigned to *Ranzania tenneyorum*.

Another species, *Ranzania grahami*, is known based on abundant premaxillary (Fig. 4a) and dentary (Fig. 4b) beaks and dermal (mostly nasal and jugular) plates, including a partially complete and remarkably thick dermal shield formed by irregularly-shaped polygonal plates (Fig. 4c), from the Langhian (about 15 Ma) zones 11 and 12 of the Plum Point Member of the Calvert Formation (see Vogt et al. 2018), Chesapeake Group, Maryland and Virginia (Weems 1985, Carnevale and Godfrey 2018; but see also Berry 1941, Weems 1974). The premaxillary beak of *Ranzania grahami* is massive and lacks a bony shelf on the palatal bracing; usually it is toothless, although some very small and irregularly disposed teeth can be rarely observed (Fig. 4a). The dentary beak has a very robust biting edge and it is usually toothless (Fig. 4b). The nasal and jugular (Fig. 4f-g) plates of this species are very similar to those of the extant *Mola mola* (Weems 1985).

A single incomplete carapace constituted by regular polygonal plates with denticulate margins and with a central large tubercle from the Serravallian (between 13.39 and 12.62 Ma) deposits of the Tufillo Unit, Torricella Peligna, Abruzzo, Italy (Carnevale 2007) represents the holotype and only known specimen of *Ranzania zappai* (Fig. 4d).

The Middle Miocene (probably Serravallian) *Ranzania ogaii* from the Hiranita Formation Chichibumachi Group, Saitama Prefecture, Japan is also known based on a single incomplete specimen (Uyeno and Sakamoto 1994; Fig. 5). The specimen was collected in the city of Chichibu and consists of a partially complete articulated skeleton lacking most of the head and the clavus. Of the head, an unidentifiable bone and a single opercle are recognizable. The axial skeleton is represented by part of the vertebral column, dorsal- and anal-fin pterygiophores, and a left cleithrum. Overall, the morphology of the skeletal elements is generally consistent with that of the extant *Ranzania laevis*. *Ranzania ogaii* differs from its congenerics in having a dermal cover of regular polygonal plates

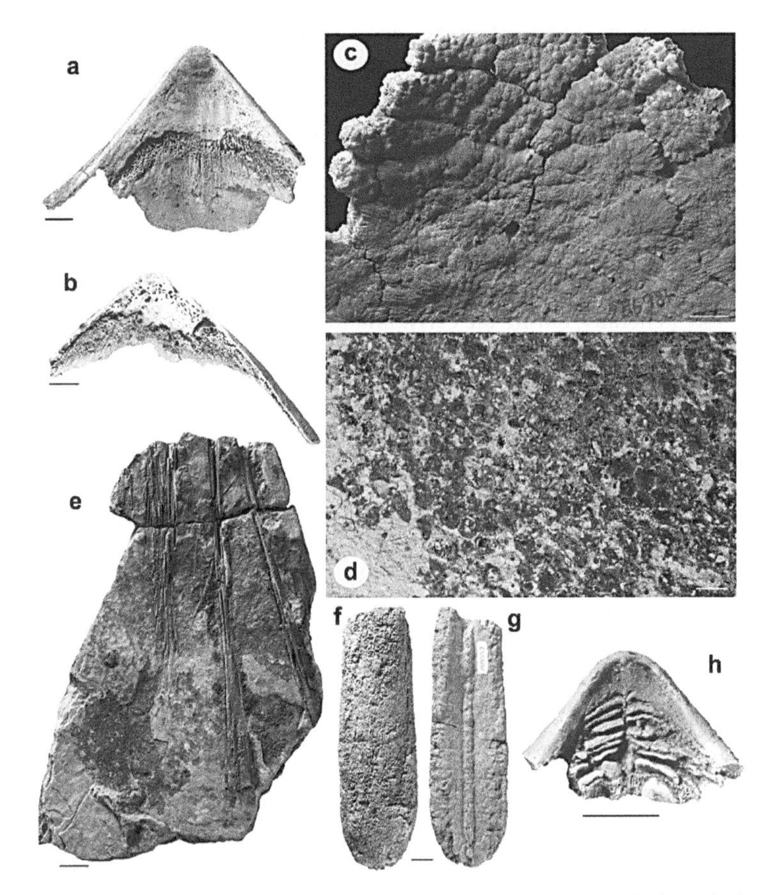

Figure 4. Miocene representatives of the genus *Ranzania*. *Ranzania grahami* Weems, 1985 from the Langhian "zone 11" of the Calvert Formation, Horsehead Cliffs, Westmoreland State Park, Westmoreland County, Virginia; holotype, USNM 186986, (a) premaxillary beak, ventral view, (b) dentary beak, dorsal view, (c) articulated dorsal dermal shield, internal view; scale bars 10 mm. (d) *Ranzania zappai* Carnevale, 2007 from the Serravallian deposits of the Tufillo Unit, Torricella Peligna, Abruzzo, Italy; holotype, MSNUP PEL057, articulated dermal plate (carapace), internal view; scale bar 2 mm. (e) *Ranzania* spp. from the Messinian diatomites of the Chelif Basin, Algeria. MNHN ORA 1777, anal-fin pterygiophores; scale bar 20 mm. *Ranzania grahami* Weems, 1985 from the Langhian "zone 11" of the Calvert Formation, Horsehead Cliffs, Westmoreland State Park, Westmoreland County, Virginia; holotype, USNM 186986, (f) jugular plate, external view, (g) jugular plate, internal view; scale bars 10 mm. (h) *Ranzania tenneyorum* Weems, 1985 from the Langhian "zone 10" of the Calvert Formation, local basal phosphate horizon, Gravett's Mill Pond, King William County, Virginia; holotype, premaxillary beak, ventral view; scale bar 10 mm. Photos (a), (b), (c), (f), (g), (h) courtesy of Stephen J. Godfrey; (d) courtesy of Chiara Sorbini; (e) courtesy of Gäel Clement.

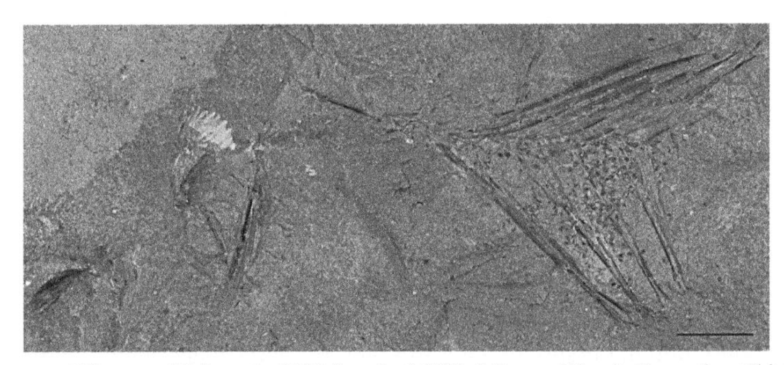

Figure 5. *Ranzania ogaii* Uyeno and Sakamoto, 1994 from the Middle Miocene Hiranita Formation, Chichibumachi Group, Saitama Prefecture, Japan. Holotype, NMSN PV-17186, left lateral view. Scale bar 20 mm. Photo courtesy of Naoki Kohno.

with smooth outer margin and bearing a single central prominent tubercle; certain scales of small specimens of *Ranzania laevis* are very similar to those of *R. ogaii*, but they change during ontogeny with a considerable reduction of the central tubercle (Uyeno and Sakamoto 1994). Carnevale (2007) suggested that the prominent central tubercle in the dermal plates of *Ranzania ogaii* and *R. zappai* could be regarded as the retention of a juvenile trait into adulthood.

The youngest Miocene record of *Ranzania* consists of a single largely incomplete skeleton solely represented by part of the anal-fin pterygiophores complement (Fig. 4e) from the Messinian (6.7–5.95 Ma) of the Chelif Basin, Algeria (Carnevale and Santini 2007). Because of its incompleteness, it is not possible to interpret the taxonomic status of this specimen at the species level.

The Miocene Genus *Austromola*

The molid *Austromola angerhoferi* from the Early Miocene Konservat-Lagerstätte cropping out near Pucking, in the Linz area, Upper Austria certainly represents one of the most spectacular Cenozoic teleost fish (Gregorova et al. 2009). At least three partially complete articulated skeletons (Fig. 6) with a maximum estimated length exceeding 3 m have been found in the laminated pelagic silty clay of the Ebelsberg Formation that originated in the North Alpine Foreland Basin during the early Aquitanian (22.4–22.2 Ma). These deposits accumulated on the northern shelf of the Central Paratethys in a pelagic context characterized by intense upwelling (Grunert et al. 2010).

While the majority or part of the axial skeleton is recognizable in all three specimens, only a small part of the head is recognizable in the holotype and paratype (Gregorova et al. 2009). The overall physiognomy of the body of *Austromola angerhoferi*, including the rounded profile of its pectoral fin, resembles that of *Mola* spp. Like *Ranzania*, *Austromola* possesses ten caudal vertebrae

Figure 6. *Austromola angerohoferi* from the Aquitanian of the Ebelsberg Formation, Pucking, Austria. Paratype, NHMW 2008z0037/0001, left lateral view. Scale bar 1 m. Photo courtesy of Mathias Harzhauser.

(versus nine in *Mola* and eight in *Masturus* spp.). *Austromola* shares with *Mola* and *Ranzania* spp. the possession of a shaft-like posteriormost caudal vertebra. Moreover, the distal ends of the proximal pterygiophores of the clavus lack the bony ossicles characteristic of *Mola* spp. Finally, unlike in other molid genera (see Tyler 1980), the hypaxial proximal pterygiophores of the clavus of *Austromola* are distinctly longer than the epaxial ones, and the rays of the clavus are simple without any apparent trace of distal bifurcation (Fig. 6).

The gigantic specimens of *Austromola angerhoferi* certainly are among the largest Cenozoic fossil teleosts. According to Gregorova et al. (2009) a body length of about 320 cm and a maximum depth of 400 cm can be estimated based on linear extrapolation of body proportions and morphometric parameters of large specimens of *Mola mola*. These authors also hypothesized that *Austromola* represents the sister taxon of the pair formed by *Masturus* and *Mola* spp. and that the clade formed by *Austromola* + (*Masturus* + *Mola*) is the sister clade to *Ranzania*.

Interrelationships of the Molidae

The monophyletic status and relative position of the Molidae within the gymnodonts has been discussed in several papers (e.g., Winterbottom 1974, Tyler 1980, Santini and Tyler 2003, Bannikov et al. 2017). Recent studies involving both extant and extinct taxa concur to indicate the Paleogene family Zignoichthyidae as the sister group to the Molidae (Arcila et al. 2015, Close et al. 2016, Arcila and Tyler 2017). These studies, however, provide a different timing of the separation between these two lineages, with Arcila et al. (2015) and Arcila and Tyler (2017) suggesting a Cretaceous split whereas Close et al. (2016) hypothesized a Paleocene event. Regardless of the precise age for the origin of the molid lineage, the earliest putative evidence of this group in the record dates back to the middle Eocene with *Eomola bimaxillaria*, a species characterized by the plesiomorphic retention of separated premaxillae articulating medially by interdigitating processes (Tyler and Bannikov 1992). *Eomola* has been postulated as the sister-group to the clade formed by the four other genera (Fig. 7), this clade being characterized by, among its other features, the jaws fused along the midline (Tyler and Bannikov 1992, Santini and Tyler 2002, 2003, Close et al. 2016, Arcila and Tyler 2017).

As far as the intrarelationships of the extant genera of the Molidae are concerned, both the morphology-based and molecular-based phylogenetic analyses suggest that *Masturus* and *Mola* spp. are more closely related to each other than to *Ranzania* (e.g., Bonaparte 1841, Gill 1897, Fraser-Brunner 1951, Tyler 1980, Santini and Tyler 2002, Yamanoue et al. 2004, Bass et al. 2005, Alfaro et al. 2007, Yoshita et al. 2009). As mentioned above, the phylogenetic position of the Early Miocene *Austromola* has been analyzed by Gregorova et al. (2009), who hypothesized that together with *Masturus* and *Mola* spp., this clade represents the sister group to *Ranzania* (Fig. 7).

The monophyly of the crown group Molidae (the clade formed by four genera with beak-like jaws fused along the midline: *Ranzania*, *Austromola*, *Masturus* and *Mola*), is supported by more than 25 primarily skeletal features (see Santini and Tyler 2002). Within this clade, each genus can be easily separated from the others based on a series of autapomorphies. *Ranzania* is defined by numerous features (Tyler 1980, Santini and Tyler 2002), most notably: a laterally compressed and elongate body, a long and thin opercle, a rod-like interopercle contacting the subopercle, five branchiostegal rays, 18 vertebrae, a supraoccipital crest considerably extended posteriorly behind the head to contact the first dorsal-fin pterygiophore, falcate pectoral fins, rays of the clavus highly branched distally, and a bony carapace formed by relatively large polygonal plates (for a more complete list of autoapomorphic features of *Ranzania* spp. see Santini and Tyler 2002). As reported by Fraser-Brunner (1951) and followed by others (e.g., Tyler 1980), the lips and mouth form a vertically oriented funnel-like structure that can be laterally compressed into a slit; however, this may apply only to some juvenile stages, and the matter of the form of the mouth needs further investigation because others reliably report that the mouth becomes stiffened into an inflexibly fixed opening in adults (e.g., Smith et al. 2010, Nyegaard et al. 2017, Bemis et al. 2020 [Chapter 4]). *Austromola* exhibits unbranched caudal-fin rays

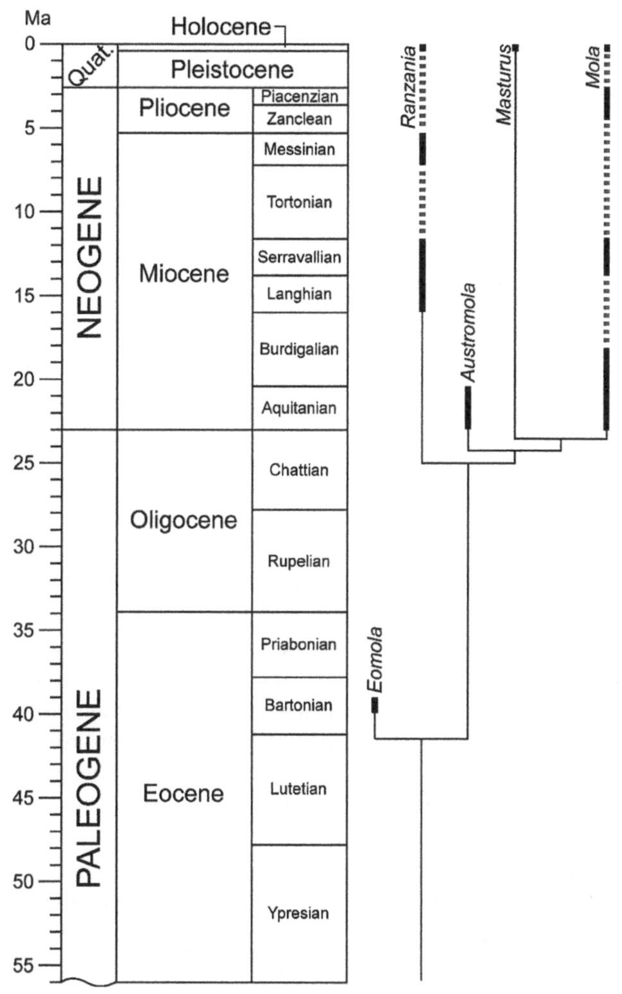

Figure 7. Time calibrated phylogeny of the Molidae illustrating the divergence of molid genera based on fossil evidence.

and hypaxial proximal pterygiophores of the clavus distinctly longer than the epaxial ones (Gregorova et al. 2009). *Masturus* spp. has 16 vertebrae, median-fin rays of the clavus not supported by modified pterygiophores and forming a pronounced lobe, non-ossified interopercle, and posteriormost vertebral centrum bearing a well-developed hemal spine (Tyler 1980, Santini and Tyler 2002). *Mola* spp. have 17 vertebrae, interopercle rod-like and clearly separated from the subopercle, and bony plates present distally in most rays of the clavus (Tyler 1980, Santini and Tyler 2002).

Concerning vertebral numbers in molids, we are well aware of the finely detailed study by Britz and Johnson (2005) of the early life history stages of *Masturus lanceolatus* and *Ranzania laevis* in which they made the phylogenetically important discovery that a primordium of the anteriormost vertebrae in young larvae is incorporated entirely and seamlessly into the basioccipital, with loss of its individuality. Our abdominal vertebral counts for *Masturus* and *Ranzania* spp. do not include this basioccipital-incorporated larval element of a first vertebra. Especially because *Mola* spp. is sister to *Masturus* spp., it is likely or at least entirely possible that one or more of the three species of *Mola* may also have an anlage incorporated into the basioccipital, but the rarity of well-preserved young larval materials of these three species will make it difficult to obtain comparable data in the foreseeable future. In order to make our molid vertebral counts comparable between species, and with those of the other gymnodonts, we follow the lead of Bemis et al. 2020 [Chapter 4] and refer to the molid

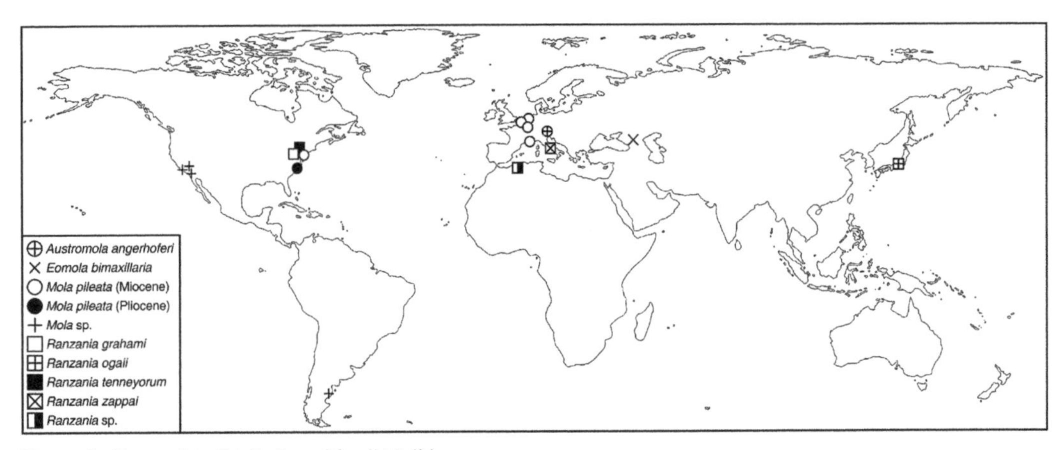

Figure 8. Geographic distribution of fossil Molidae.

counts as being of the functional vertebral units that, from small juveniles to large adults, are separate, articulated, axial, and easily-visible structures. The vast majority of vertebral counts for molids in the literature are of these functional vertebrae, and of course that will probably always be the case for fossils. We note that molids are not alone among the tetraodontiforms in reduced numbers of vertebrae from the 20 found in the two most basal extant families, Triacanthodidae and Triacanthidae, with, for example, 18 in balistids, aracanids, and ostraciids, and in many tetraodontids (e.g., *Canthigaster*); (See Table 2 in Tyler (1980) for vertebral counts in a wide range of tetraodontiforms). It remains to be seen if some of these reductions are by a process of vertebral anlagen absorption into the basioccipital similar to that found in the two species of molids that have been examined for this condition.

The phylogeny of the Molidae clearly evidences the fragmentary and highly incomplete nature of the fossil record of this family (Fig. 8).

The earliest representatives of *Ranzania* spp. date back to the Langhian, *Austromola* is restricted to the Aquitanian as also probably is the oldest putative *Mola* spp. and, finally, *Masturus* spp. is still unknown as a fossil. Moreover, there is a considerable gap in the record extending for more than 16 Ma between *Eomola* and the earliest molid of modern type (with jaws fused along the midline). The topology of the molid tree necessarily indicates that the origin and early divergence of the crown group Molidae likely occurred sometime during the Oligocene and that at least *Ranzania* spp. and *Austromola* were already in existence well before the beginning of the Miocene (Fig. 7). In this context, the *Masturus* and *Mola* spp. split likely took place in the latest Oligocene or in the earliest part of the Miocene, before the divergence estimate (7.2–18.95 Ma) proposed by Bass et al. (2005) based on mitochondrial genes.

Evolutionary Paleoecology of the Miocene Diversification

The integrative analysis of the fragmentary fossil record of the Molidae illustrated above (Fig. 7) clearly indicates that an increased diversity characterized this group during the Miocene. As discussed in the previous section, the phylogenetic relationships between the molid genera and their distribution in time unquestionably suggest that the three extant genera (*Masturus*, *Mola* and *Ranzania*) and the extinct *Austromola* were all in existence at the beginning of the Miocene. In particular, the genus *Ranzania* shows a marked increase of diversity during the Middle Miocene, with four known species described to date. Considering that these Miocene fossils derive from sediments originated in different marine paleobiotopes, including inner and outer shelf, slope, and deep-sea, the increased abundance and diversity of molids during the Miocene seems to represent a genuine pattern.

The identification of the evolutionary driving forces of the crown molid radiation is obviously problematic. As argued by Norris (2000), drivers of marine speciation are difficult to elucidate due to the uncertain distribution of marine organisms through time and because oceans are characterized by few evident geographical barriers and are three-dimensionally continuous. Moreover, many marine species, including those of molids, can disperse widely and quickly, and this results in a constant mixing of gene pools that can inhibit evolutionary changes by preventing the requirements for the isolation necessary for allopatric speciation (the so-called "marine speciation paradox"; Bierne et al. 2003). However, the timing of the crown molid radiation seems to coincide with those of many other groups of fishes and marine mammals, which have been interpreted to be related with global changing patterns of productivity in the oceans with the emergence of high food concentrations, especially during the Miocene (e.g., Berger 2007, Norris et al. 2013, Santini et al. 2013b, 2014, Bisconti et al. 2019). During part of the Oligocene and, most notably, during the Miocene, a global enhancement of trophic resources in the world oceans that resulted from the increased productivity associated with a general rearrangement of oceanic and atmospheric circulation patterns (see Zachos et al. 2001), led to the development of shortened food chains that supported a dramatic proliferation of organisms associated with extreme pelagic production. Therefore, it is reasonable to attribute the early divergence and the diversification of the crown Molidae to this considerable Oligocene and Miocene global remarkable ecosystem shift in the oceans.

Remaining Questions

The fragmentary fossil record of the Molidae described and discussed above provides clear evidence of how little is known about the evolutionary history of this group. As a consequence, there are many open questions derived from the largely incomplete fossil archive of this family. It is clear that additional work is required to elucidate the fundamental question of the origin of the molid body plan and then of the tempo and mode of divergence of this group within the gymnodonts. Nearly all of the very cryptic pre-Miocene history of the Molidae is completely unknown, as well as the anatomy of the relatively common Miocene species *Mola pileata*, and of the species of *Ranzania* that are solely known from isolated beaks and dermal plates. Finally, there is no evidence in the record of the evolutionary history of *Masturus* and, more generally, of the origin of the extant molid species. It is our opinion that future research efforts should address these remaining questions

Acknowledgments

We are particularly obliged to Alexandre F. Bannikov [Borisyak Paleontological Institute of the Russian Academy of Sciences (PIN), Moscow], David Bohaska [National Museum of Natural History, Smithsonian Institution (USNM), Washington], Gäel Clement [Museum National d'Histoire Naturelle (MNHN), Paris], Stephen J. Godfrey [Calvert Marine Museum (CMM), Solomons], Mathias Harzhauser [Naturhistorisches Museum Wien (NHMW), Vienna], Naoki Kohno [Department of Geology and Paleontology, National Museum of Nature and Science (NMNS), Tokyo], Olivier Lambert [Institut Royale des Sciences Naturelles de Belgique (IRSNB), Bruxelles], and Chiara Sorbini [Museo di Storia Naturale, Università di Pisa (MSNUP), Calci] for access to material in their care and for providing the photographs of fossil molids used in this chapter. We thank Alberto L. Cione (Universidad Nacional de la Plata, La Plata) for discussions about fossil molids from Argentina. We also thank Kim A. Smith (Department of Primary Industries and Regional Development, Western Australia) for generously providing information and photographs of the inflexible mouths of numerous specimens of *Ranzania laevis* stranded on the west coast of Australia. The manuscript benefited from a review by Katherine E. Bemis (Virginia Institute of Marine Science, Gloucester Point). The research of GC and LP was supported by grants (ex-60% 2018 and 2019) from the Università degli Studi di Torino.

References

Alfaro, M.E., F. Santini and C.D. Brock. 2007. Do reefs drive diversification in marine teleosts? Evidence from pufferfish and their allies (order Tetraodontiformes). Evolution 61: 2104–2126.

Arcila, D., R.A. Pyron, J.C. Tyler, G. Ortí and R. Betancur-R. 2015. An evaluation of fossil tip-dating versus node-age calibrations in tetraodontiform fishes (Teleostei: Percomorphaceae). Mol. Phylogen. Evol. 82: 13182: 131–145.

Arcila, D. and J.C. Tyler. 2017. Mass extinction in tetraodontiform fishes linked to the Palaeocene-Eocene thermal maximum. Proc. R. Soc. B 284: 20171771.

Bannikov, A.F., J.C. Tyler, D. Arcila and G. Carnevale. 2017. A new family of gymnodont fish (Tetraodontiformes) from the earliest Eocene of the Peri-Tethys (Kabardino-Balkaria, northern Caucasus, Russia). J. Syst. Palaeont. 15: 129–146.

Barnard, K.H. 1927. A monograph of the marine fishes of South Africa, Part 2. Ann. South Afr. Mus. 21: 419–1065.

Bass, A.L., H. Dewar, T. Thys, J.T. Streelman and S.A. Karl. 2005. Evolutionary divergence among lineages of the ocean sunfish family, Molidae (Tetraodontiformes). Mar. Biol. 148: 405–414.

Bemis, K.E., J.C. Tyler, E.J. Hilton and W.E. Bemis. 2020. Overview of the anatomy of ocean sunfishes (Molidae: Tetraodontiformes). pp. 55–71. *In*: T.M. Thys, G.C. Hays and J.D.R. Houghton [eds]. The Ocean Sunfishes: Evolution, Biology and Conservation, CRC Press. Boca Raton, FL, USA.

Beneden, P.-J. van. 1881. Sur un poisson fossile nouveau des environs de Bruxelles et sur certain corps énigmatiques du crag d'Anvers. Bull. Acad. R. Sci. Lett. Beaux-Arts Belgique (3eme série) 1: 116–126.

Beneden, P.-J. van. 1883. Sur quelques forms nouvelles des terrains du pays. Bull. Acad. R. Sci. Lett. Beaux-Arts Belgique (3eme série) 6: 132–134.

Berger, W.H. 2007. Cenozoic cooling, Antarctic nutrient pump, and the evolution of whales. Deep-Sea Res. II 54: 2399–2421.

Berry, C.T. 1941. The dentary of *Syllomus crispatus* Cope. Am. Mus. Novit. 1132: 1–2.

Bierne, N., F. Bonhomme and P. David. 2003. Habitat preference and the marine-speciation paradox. Proc. R. Soc. London. B 270: 1399–1406.

Bisconti, M., D.K. Munsterman and K. Post. 2019. A new balaenopterid whale from the late Miocene of the Southern North Sea Basin and the evolution of balaenopterid diversity (Cetacea, Mysticeti). PeerJ 7: e6915.

Bonaparte, C.L. 1841. Iconografia della fauna italica per le quattro classi degli animali vertebrati. Tomo III. Pesci. Tipografia Salviucci, Roma.

Bosch, M. van den, M.C. Cadée and A.W. Janssen. 1975. Lithostratigraphical and biostratigraphical subdivision of Tertiary deposits (Oligocene-Pliocene) in the Winterswijk-Almelo region (eastern part of the Netherlands). Scripta Geol. 29: 1–167.

Breder, C.M. and E. Clark. 1947. A contribution to the visceral anatomy, development, and relationships of the Plectognathi. Bull. Am. Mus. Nat. Hist. 88: 287–319.

Britski, H.A., R.D. Andreucci, N.A. Menezes and J. Carneiro. 1985. Coalescence of teeth in fishes. Rev. Bras. Zool. 2: 459–482.

Britz, R. and G.D. Johnson. 2005. Occipito-vertebral fusion in ocean sunfishes (Teleostei: Tetraodontiformes: Molidae) and its phylogenetic implications. J. Morphol. 266: 74–79.

Britz, R. and G.D. Johnson. 2012. The caudal skeleton of a 20 mm *Triodon* and homology of the components. Proc. Biol. Soc. Wash. 125: 66–73.

Caldera, E.J., J.L. Whitney, M. Nyegaard, E. Ostalé-Valriberas, L. Kubicek and T.M. Thys. 2020. Genetic insights regarding the taxonomy, phylogeography and evolution of the ocean sunfishes (Molidae: Tetraodontiformes). pp. 37–54. *In*: T.M. Thys, G.C. Hays and J.D.R. Houghton [eds.]. The Ocean Sunfishes: Evolution, Biology and Conservation, CRC Press. Boca Raton, FL, USA.

Carnevale, G. 2007. Fossil fishes from the Serravallian (Middle Miocene) of Torricella Peligna. Palaeont. It. 91: 1–67.

Carnevale, G. and F. Santini. 2007. Record of the slender mola, genus *Ranzania* (Teleostei, Tetraodontiformes), in the Miocene of the Chelif Basin, Algeria. C.R. Palevol. 6: 321–326.

Carnevale, G. and S.J. Godfrey. 2018. Miocene bony fishes from the Calvert, Choptank, St. Marys and Eastover Formations, Chesapeake Group, Maryland and Virginia. Smithson. Contrib. Paleobio. 100: 161–212.

Carroll, R.L. 1988. Vertebrate Paleontology and Evolution. W.H. Freeman and Company, New York.

Cione, A.L. 1978. Aportes paleoictiológicos al conocimiento de la evolución de las paleotemperaturas en el área austral de América del Sur durante el Cenozoico. Aspectos zoogeográficos y ecológicos conexos. Ameghiniana 15: 183–208.

Close, R.A., Z. Johanson, J.C. Tyler, R.C. Harrington and M. Friedman. 2016. Mosaicism in a new Eocene pufferfish highlights rapid morphological innovation near the origin of crown tetraodontiforms. Palaeontology 59: 499–514.

Cuitiño, J.I., M.M. Pimentel, R. Ventura Santos and R.A. Scasso. 2012. High resolution isotopic ages for the "Patagoniense" transgression in southwest Patagonia: stratigraphic implications. J. South Amer. Earth Sci. 38: 110–122.

Cuitiño, J.I., R. Ventura Santos, P.J. Alonso Muruaga and R.A. Scasso. 2015. Sr-stratigraphy and sedimentary evolution of early Miocene marine foreland deposits in the northern Austral (Magallanes) Basin, Argentina. Andean Geol. 42: 364–385.

Cuitiño, J.I., M.T. Dozo, C.J. del Río, M.R. Buono, L. Palazzesi, S. Fuentes and R.A. Scasso. 2017. Miocene marine transgressions: Paleoenvironments and paleobiodiversity. pp. 47–84. *In*: P. Bouza and A. Bilmes [eds.]. Late Cenozoic of Peninsula de Valdés, Patagonia Argentina: an interdisciplinary approach. Springer, Cham.

Cuvier, G. 1817. Le Règne Animal, Vol. 2. Deterville, Paris.

Davenport, J., N.D. Phillips, E. Cotter, L.E. Eagling and D.R. Houghton. 2018. The locomotor system of the ocean sunfish *Mola mola* (L.): role of gelatinous exoskeleton, horizontal septum, muscles and tendons. J. Anat. 233: 347–357.

Deinse, A.B. van. 1953. Fishes in Upper Miocene and Lower Pleistocene deposits in the Netherlands. Meded. Geol. Sticht. N.S. 7: 5–12.

Fierstine, H.L., R.W. Huddleston and G.T. Takeuchi. 2012. Catalog of the Neogene bony fishes of California: A systematic inventory of all published accounts. Occ. Pap. Calif. Acad. Sci. 159: 1–206.

Forsgren, K., R.S. McBride, T. Nakatsubo, T.M. Thys, C.D. Carson, E.K. Tholke et al. 2020 Reproductive biology of the ocean sunfishes. pp. 87–104. *In*: T.M. Thys, G.C. Hays and J.D.R. Houghton [eds.]. The Ocean Sunfishes: Evolution, Biology and Conservation, CRC Press. Boca Raton, FL, USA.

Fraser-Brunner, A. 1951. The ocean sunfishes (family Molidae). Bull. Brit. Mus. (Nat. Hist.) Zool. 1(6): 89–121.

Gill, T.N. 1897. The distinctive characters of the Molinae and Ranzaniinae. Science 156: 966–967.

Gregorova, R., O. Schultz, M. Harzhauser, A. Kroh and S. Ćoric. 2009. A giant early Miocene sunfish from the North Alpine Foreland Basin. J. Vert. Paleont. 29: 359–371.

Gregory, W.K. and H.C. Raven. 1934. Notes on the anatomy and relationships of the ocean sunfish (*Mola mola*). Copeia 1934: 145–151.

Grunert, P., M. Harzhauser, F. Rögl, R. Sachsenhofer, R. Gratzer, A. Soliman et al. 2010. Oceanographic conditions as a trigger for the formation of an Early Miocene (Aquitanian) *Konservat-Lagerstätte* in the Central Paratethys Sea. Palaeo., Palaeo., Palaeo. 292: 425–442.

Gudger, E.W. 1928. Capture of an ocean sunfish. Sci. Monthly 26: 257–261.

Gudger, E.W. 1937a. The natural history and geographical distribution of the pointed-tailed ocean sunfish (*Masturus lanceolatus*), with notes on the shape of the tail. Proc. Zool. Soc. London 1937: 353–396.

Gudger, E.W. 1937b. The structure and development of the pointed tail of the ocean sunfish, *Masturus lanceolatus*. Ann. Mag. Nat. Hist. 19: 1–46.

Hakel, M. and J.D. Stewart. 2002. First fossil Molidae (Actinopterygii: Tetraodontiformes) in western North America. J. Vert. Paleont. 22(Suppl.): 62A.

Hays, G.C., J.D.R. Houghton, T.M. Thys, D.H. Adams, A.E. Ahuir-Baraja, J. Alvarez et al. 2020. Unresolved questions about the ocean sunfishes, Molidae—A family comprising some of the world's largest teleosts. pp. 280–296. *In*: T.M. Thys, G.C. Hays and J.D.R. Houghton [eds.]. The Ocean Sunfishes: Evolution, Biology and Conservation, CRC Press. Boca Raton, FL, USA.

Hooper, S.N., M. Paradis and R.G. Ackman. 1973. Distribution of trans-6-hexadecenoid acid, 7-methyl-7-hexadecenoic acid and common fatty acids in lipids of ocean sunfish *Mola mola*. Lipids 8: 509–516.

Johnson, G.D. and R. Britz. 2005. Leis' Conundrum: Homology of the clavus of the ocean sunfishes. 2. Ontogeny of the median fins and axial skeleton of *Ranzania laevis* (Teleostei, Tetraodontiformes, Molidae). J. Morphol. 266: 11–21.

Katayama, E. and K. Matsuura. 2016. Fine structure of scales of ocean sunfishes (Actinopterygii, Tetraodontiformes, Molidae): Another morphological character supporting phylogenetic relationships of the molid genera. Bull. Natl. Mus. Nat. Sci. A 42: 95–98.

King, C. 2016. A revised correlation of Tertiary rocks in the British Isles and adjacent area of NW Europe. Geol. Soc. London, Spec. Rep. 27: 1–724.

Lambert, O. and S. Louwye. 2006. *Archaeoziphius microglenoideus*, a new primitive beaked whale (Mammalia, Cetacea, Odontoceti) from the Middle Miocene of Belgium. J. Vert. Paleont. 26: 182–191.

Lauder, G.V. and K.F. Liem. 1983. The evolution and interrelationships of the actinopterygian fishes. Bull. Mus. Comp. Zool. 150: 95–197.

Leriche, M. 1907. Révision de la faune ichthyologique des terrains néogènes du bassin du Rhône. pp. 335–352. *In*: Association française pour l'avancement des sciences, fuisonnée avec l'association scientifique de France. Compte Rendu de la 35eme session, Lyon 1906, Notes et Mémoires. Lyon, France.

Leriche, M. 1926. Les Poissons Néogènes de la Belgique. Mem. Mus. R. Hist. Nat. 32: 367–472.

Nakae, M. and K. Sasaki. 2006. Peripheral nervous system of the ocean sunfish *Mola mola* (Tetraodontiformes: Molidae). Ichthyol. Res. 53: 233–246.

Nakamura, I. and K. Sato. 2014. Ontogenetic shift in foraging habit of ocean sunfish *Mola mola* from dietary and behavioral studies. Mar. Biol. 161: 1263–1273.

Norris, R.D. 2000. Pelagic species diversity, biogeography, and evolution. Paleobiology 26: 236–258.

Norris, R.D., S. Kirtland Turner, P.M. Hull and A. Ridgwell. 2013. Marine ecosystem responses to Cenozoic global change. Science 341: 492–498.

Nyegaard, M., N. Loneragan and M.B. Santos. 2017. Squid predation by slender sunfish *Ranzania laevis* (Molidae). J. Fish Biol. 90: 2480–2487.

Nyegaard, M., E. Sawai, N. Gemmell, J. Gillum, N.R. Loneragan, Y. Yamanoue et al. 2018. Hiding in broad daylight: molecular and morphological data reveal a new sunfish species (Tetraodontiformes: Molidae) that has eluded recognition. Zool. J. Linn. Soc. 182: 631–658.

Parenti, P. 2003. Family Molidae Bonaparte 1832—molas or ocean sunfishes. Calif. Acad. Sci. Annotated Checklists of Fishes 18: 1–9.

Patterson, C. 1993. Osteichthyes: Teleostei. pp. 621–656. *In*: M.J. Benton [ed.]. The Fossil Record 2. Chapman & Hall, London.

Phillips, N.D., C. Harrod, A.R. Gates, T.M. Thys and J.D.R. Houghton. 2015. Seeking the sun in deep, dark places: mesopelagic sightings of ocean sunfishes (Molidae). J. Fish Biol. 87: 1118–1126.

Porcasi, J.F. and S.L. Andrews. 2001. Evidence for a prehistoric *Mola mola* fishery on the southern California Coast. J. Calif. Great Basin Anthrop. 23: 51–66.

Purdy, R.W., V.P. Schneider, S.P. Applegate, J.H. McLellan, R.L. Meyer and B.H. Slaughter. 2001. The Neogene sharks, rays, and bony fishes from Lee Creek Mine, Aurora, North Carolina. Smithson. Contrib. Paleobiol. 90: 71–202.

Raven, H.C. 1939a. Notes on the anatomy of *Ranzania truncata*. A plectognath fish. Am. Mus. Novit. 1038: 1–7.

Raven, H.C. 1939b. On the anatomy and evolution of the locomotor apparatus of the nipple-tailed ocean sunfish (*Masturus lanceolatus*). Bull. Am. Mus. Nat. Hist. 76: 143–150.

Regan, C.T. 1903. On the classification of the fishes of the Suborder Plectognathi; with notes and descriptions of new species from specimens in the British Museum collection. Proc. Zool. Soc. London 1902: 284–303.

Romer, A.S. 1966. Vertebrate Paleontology (3rd ed.). University of Chicago Press, Chicago.

Santini, F. and J.C. Tyler. 2002. Phylogeny of the ocean sunfishes (Molidae, Tetraodontiformes), a highly derived group of teleost fishes. Ital. J. Zool. 69: 37–43.

Santini, F. and J.C. Tyler. 2003. A phylogeny of the families of fossil and extant tetraodontiform fishes (Acanthomorpha, Tetraodontiformes), Upper Cretaceous to Recent. Zool. J. Linn. Soc. 139: 565–617.

Santini, F., G. Carnevale and L. Sorenson. 2013a. First molecular scombrid timetree (Scombridae, Percomorpha) shows recent radiation of tunas following invasion of pelagic habitats. Ital. J. Zool. 80: 210–221.

Santini, F., L. Sorenson and M.E. Alfaro. 2013b. A new phylogeny of tetraodontiform fishes (Tetraodontiformes, Acanthomorpha) based on 22 loci. Mol. Phylogen. Evol. 69: 177–187.

Santini, F., G. Carnevale and L. Sorenson. 2014. First multi-locus timetree of seabreams and porgies (Sparidae, Percomorpha). Ital. J. Zool. 81: 55–71.

Sawai, E., M. Nyegaard and Y. Yamanoue. 2020. Phylogeny, taxonomy and size records of ocean sunfishes. pp. 18–36. *In*: T.M. Thys, G.C. Hays and J.D.R. Houghton [eds.]. The Ocean Sunfishes: Evolution, Biology and Conservation, CRC Press. Boca Raton, FL, USA.

Scasso, R.A. and L.N. Castro. 1999. Cenozoic phosphatic deposits in North Patagonia, Argentina: phosphogenesis, sequence-stratigraphy and paleoceanography. J. South Amer. Earth Sci. 12: 471–487.

Schmidt, J. 1921. New studies of sun-fishes made during the "Dana" Expedition, 1920. Nature 107: 76–79.

Smith, K.A., M. Hammon and P.G. Close. 2010. Aggregation and stranding of elongate sunfish (*Ranzania laevis*) (Pisces: Molidae) (Pennant, 1776) on the southern coast of Western Australia. J. R. Soc. West. Aust. 93: 181–188

Sousa, L.L., R. Xavier, V. Costa, N.E. Humphries, C. Trueman, R. Rosa et al. 2016. DNA barcoding identifies a cosmopolitan diet in the ocean sunfish. Sci. Rep. 6: 28762.

Sousa, L.L., I. Nakamura and D.W. Sims. 2020. Movements and foraging behavior of ocean sunfish. pp. 129–145. *In*: T.M. Thys, G.C. Hays and J.D.R. Houghton [eds.]. The Ocean Sunfishes: Evolution, Biology and Conservation, CRC Press. Boca Raton, FL, USA.

Steenstrup, J. and C. Lütken. 1898. Bidrag til kundskab om klumpeller maanefiskene (Molidae). Spolia Atlantica 9: 5–102.

Tyler, J.C. 1980. Osteology, phylogeny, and higher classification of the fishes of the order Plectognathi (Tetraodontiformes). NOAA Tech. Report, NMFS Circ. 434: 1–422.

Tyler, J.C. and A.F. Bannikov. 1992. New genus of primitive ocean sunfish with separate premaxillae from the Eocene of Southwest Russia (Molidae, Tetraodontiformes). Copeia 1992: 1014–1023.

Tyler, J.C. and L. Sorbini. 1996. New superfamily and three new families of tetraodontiform fishes from the Upper Cretaceous: the earliest and most morphologically primitive plectognaths. Smithson. Contrib. Paleobiol. 82: 1–59.

Tyler, J.C. and F. Santini. 2002. Review and reconstructions of the tetraodontiform fishes from the Eocene of Monte Bolca, Italy, with comments on related Tertiary taxa. St. Ric. Giacim. Terz. Bolca 9: 47–119.

Uyeno, T. and K. Sakamoto. 1994. *Ranzania ogaii*, a new Miocene slender mola from Saitama, Japan (Pisces: Tetraodontiformes). Bull. Natn. Sci. Mus., Tokyo, Ser. C. 20: 109–117.

Vogt, P., R.E. Eschleman and S.J. Godfrey. 2018. Calvert Cliffs: Eroding mural escarpment, fossil dispensary, and paleoenvironmental archive in space and time. Smithson. Contrib. Paleobiol. 100: 3–44.

Ward, L.W. and N.L. Gilinsky. 1993. Molluscan assemblages of the Chowan River Formation, Part A, Biostratigraphic analysis of the Chowan River Formation (Upper Pliocene) and adjoining units, the Moore House Member of the Yorktown Formation (Upper Pliocene) and the James City Formation (Lower Pleistocene). Virginia Mus. Nat. Hist. Mem. 3, Part A: 1–32.

Watanabe, Y. and K. Sato. 2008. Functional dorsoventral symmetry in relation to lift-based swimming in the ocean sunfish *Mola mola*. PLoS ONE 3: e3446.

Watanabe, Y.Y. and J. Davenport. 2020. Locomotory systems and biomechanics of ocean sunfish. pp. 72–86. *In*: T.M. Thys, G.C. Hays and J.D.R. Houghton [eds.]. The Ocean Sunfishes: Evolution, Biology and Conservation, CRC Press. Boca Raton, FL, USA.

Weems, R.E. 1974. Middle Miocene sea turtles (*Syllomus*, *Procolpochelys*, *Psephophorus*) from the Calvert Formation. J. Paleont. 48: 278–303.

Weems, R.E. 1985. Miocene and Pliocene Molidae (*Ranzania*, *Mola*) from Maryland, Virginia, and North Carolina (Pisces: Tetraodontiformes). Proc. Biol. Soc. Wash. 98: 422–438.

Winterbottom, R. 1974. The familial phylogeny of the Tetraodontiformes (Acanthopterygii: Pisces) as evidenced by their comparative myology. Smithson. Contrib. Zool. 155: 1–201.

Woodward, A.S. 1901. Catalogue of Fossil Fishes in the British Museum (Natural History). Vol. IV. The Trustees of the British Museum, London.

Yamanoue, Y., M. Miya, K. Matsuura, M. Katoh, H. Sakai and M. Nishida. 2004. Mitochondrial genomes and phylogeny of the ocean sunfishes (Tetraodontiformes: Molidae). Ichthyol. Res. 51: 269–273.

Yoshita, Y., Y. Yamanoue, K. Sagara, M. Nishibori, H. Kuniyoshi, T. Umino et al. 2009. Phylogenetic relationship of two *Mola* sunfishes (Tetraodontiforme: Molidae) occurring around the coast of Japan, with notes on their geographical distribution and morphological characteristics. Ichthyol. Res. 56: 232–244.

Zachos, J., M. Pagani, L. Sloan, E. Thomas and K. Billups. 2001. Trends, rhythms, and aberrations in global climate 65 Ma to present. Science 292: 686–693.

Chapter 2

Phylogeny, Taxonomy and Size Records of Ocean Sunfishes

Etsuro Sawai,[1,*] *Marianne Nyegaard*[2] and *Yusuke Yamanoue*[3]

Introduction

Ocean sunfishes, also called molids, belong to the family Molidae (Order Tetraodontiformes). These peculiar fishes have attracted the interest of people for centuries because of their unique shape (a complete lack of a caudal fin) and large size [> 3 m total length (TL) and > 2,000 kg] (e.g., Sawai 2017, Sawai et al. 2018a). The study of molid morphology and taxonomy dates back to at least the 16th century and encompasses a long legacy of taxonomic confusion (Fraser-Brunner 1951, Sawai et al. 2018a). In recent years, due to the development of genetics, the taxonomy of the genus *Mola* has greatly advanced. In 2017, a new species of the genus *Mola* was described (Nyegaard et al. 2018b), and the scientific name of another *Mola* species was changed (Sawai et al. 2018a). More undescribed species of Molidae may still be discovered in the future. Currently, the family Molidae has five valid species (Nyegaard et al. 2018b, Sawai et al. 2018a): slender sunfish *Ranzania laevis* (Pennant 1776), sharptail sunfish *Masturus lanceolatus* (Liénard 1840), hoodwinker sunfish *Mola tecta* Nyegaard et al. 2017 giant sunfish *Mola alexandrini* (Ranzani 1839), and ocean sunfish *Mola mola* (Linnaeus 1758) (Figs. 1A–E). Here we examine the phylogeny, taxonomy and size records, including summaries of our extensive literature review (up to February 2020) and unpublished data, of these five species. It is our hope that this chapter will serve as a basis for progressing the taxonomic understanding of molids in the future.

[1] Ocean Sunfishes Information Storage Museum (online), C-102 Plaisir Kazui APT, 13-6 Miho, Shimizu-ku, Shizuoka-shi, Shizuoka 424-0901, Japan.

[2] Auckland War Memorial Museum Tamaki Paenga Hira, Natural Sciences, The Domain, Private Bag 92018, Victoria Street West, Auckland 1142, New Zealand.
Email: mnyegaard@aucklandmuseum.com

[3] The University Museum, The University of Tokyo, 7-3-1 Hongo, Bunkyo-ku, Tokyo 113-0033, Japan.
Email: yamanouey@yahoo.co.jp

* Corresponding author: sawaetsu2000@yahoo.co.jp

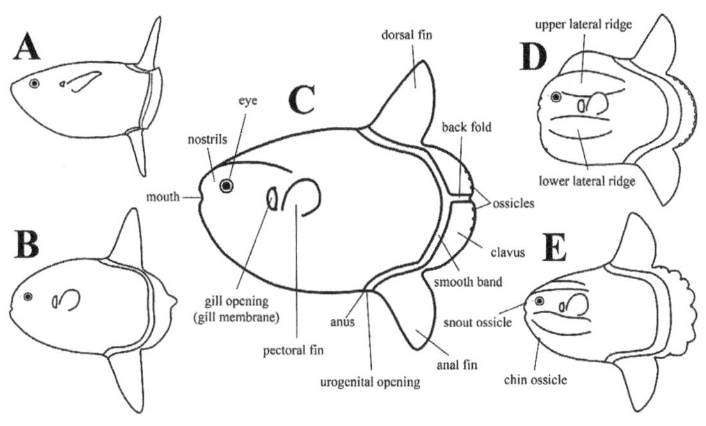

Figure 1. Generalized molid-shape body forms of five species in the Molidae (representative of adult to senescent stages). A: *Ranzania laevis*. B: *Masturus lanceolatus*. C: *Mola tecta*. D: *Mola alexandrini*. E: *Mola mola*.

Morphology

While information on early molid development is lacking, we do know they undergo significant morphological changes, particularly during early life stages, and continue to undergo morphological changes as they grow larger (e.g., Fraser-Brunner 1951, Martin and Drewry 1978, Hellenbrecht et al. 2019). Other important life history traits, such as size at maturity, are also not well understood. Establishing key diagnostic characteristics for each species at each developmental stage is vital for proper fisheries and molid resource management. This chapter focuses on the adult to senescent stages of life. Morphological characteristics are summarized based on the following literature and our unpublished data: Fraser-Brunner (1951), Tyler (1980), Nolf and Tyler (2006), Nakae and Sasaki (2006), Sawai (2016, 2017), Davenport et al. (2018), Nyegaard et al. (2018b), Sawai et al. (2018a, b, 2019).

Morphological Characteristics

Molids have a highly specialized morphology among fishes, with the posterior part of their body apparently absent (Fig. 1). They completely lack a caudal fin (Johnson and Britz 2005, Nakae and Sasaki 2006), and instead have a posterior structure most commonly termed a clavus (Fraser-Brunner 1951) meaning rudder in Latin. This structure has also been labeled a pseudocaudal fin (e.g., Tyler 1980) and gephyrocercal tail (e.g., Iwai 2005). Descriptions of the other body parts of molids are given in Table 1 (also see Fig. 1). This recognizable molid form (hereinafter called molid-shape) is characteristic of the adult to senescent stages and differs from the early developmental stages where the body is rounded, covered in distinct dermal spines and has moderate pigmentation on the eyes and dorsum (Thys et al. 2020b [Chapter 7], Watson 1996, Lyczkowski-Shultz 2006).

Although some researchers have suggested sexual dimorphism in molids (e.g., Fraser-Brunner 1951), it is actually difficult to visually distinguish between sexes by external morphology (Sawai 2017). Internally, the shape of gonads differs in males and females of *Mola* (*M. mola*; Sawai 2017, *M. tecta*; Nyegaard et al. 2018b, *M. alexandrini*; Sawai et al. 2018a) and *Masturus* (Martin and Drewry 1978; E. Sawai, personal observations): the testes of the males are paired long and cylindrical while the ovaries are unpaired and round. In *Ranzania*, however, the male (Jardas and Knežević 1983) and female (Purushottama et al. 2014) gonads are paired long and cylindrical and similar in shape (E. Sawai, personal observations).

Table 1. Basic morphological characteristics common to the family Molidae.

Body part	Position and shape
Body	Compressed laterally
Body scales (denticles)	Cover the body surface
Hypodermis	Covers the entire body except the pectoral, dorsal and anal fins
Mouth	Small and terminal
Teeth	Fused with both jaws, and beak-like
Nostrils	Small, paired holes, flush with surface close to the eyes
Eyes	Round
Gill membrane	Cover the gill opening in front of pectoral fin
Lateral lines	Present, at least on head
Fins (pectoral, dorsal, clavus and anal fins)	Spineless; caudal fin and pelvic fins absent
Urogenital opening	Immediately in front of anal fin base
Anus (anal opening)	Immediately in front of urogenital opening

Key to Molidae Species (Adult to Senescent Stage)

Morphological interspecies differences in molids (from adult stage to senescent stage) and species characteristics are summarized in Table 2 (updated version of Fraser-Brunner 1951). The full complement of fin rays and ossicles (e.g., Fig. 2) are also important meristic characters for species identification in molids (summarized in Table 3). Several morphological characteristics in *Mola* develop with size as they grow into adults and should be observed with caution, such as head bump, chin bump, wavy clavus, ossicles and body scale structure (see Nyegaard et al. 2018b, Sawai et al. 2018a). The following identification key is based on typical adult stage morphology (see Table 2).

1 Body elongate, oblong, moderately deep; body scales on clavus absent; white subdermal layer (hypodermis) very thin; vertical oval mouth; shape of gill openings somewhat triangular and rounded; pectoral fins falcate..***Ranzania laevis*** **(Fig. 1A)**

Body elongate ovoid or orbicular, deep; body scales on clavus present; hypodermis much thicker than for *Ranzania*; horizontal oval mouth; shape of gill openings oval; pectoral fins rounded..2

2 Body elongate ovoid; lower jaw protrudes slightly anteriorly beyond upper jaw; upper part of the central clavus protrudes posteriorly....................................***Masturus lanceolatus*** **(Fig. 1B)**

Body orbicular; no prominent lower jaw; no clavus protrusion 3

3 Lateral ridges absent, front view tapered; a smooth band back-fold on clavus present; clavus margin rounded with an indent at the back-fold; conical body scales without branching tips ....***Mola tecta*** **(Fig. 1C)**

Lateral ridges present, front view gourd-shaped; no smooth band back-fold on clavus, or very faint; body scales not conical..4

4a Head bump and chin bump present; clavus margin round; rectangular body scale; ossicles on paraxial clavus fin rays separate...***Mola alexandrini*** **(Fig. 1D)**

4b Head bump variable and chin bump variable; clavus margin wavy, conical body scales with branching tips; ossicles on paraxial clavus fin rays united.........................***Mola mola*** **(Fig. 1E)**

Table 2. Species level key morphological characters of Molidae (adult stage to senescent stage).

Species	*Ranzania laevis*	*Masturus lanceolatus*	*Mola tecta*	*Mola alexandrini*	*Mola mola*
Shape of main body	Elongate, oblong, moderately deep	Orbicular (ovoid shape more pronounced than *Mola*), deep	Orbicular, deep	Orbicular, deep	Orbicular, deep
Lateral ridges (front view)	Absent (tapered)	Slight (tapered)	Absent (tapered)	Present (gourd-shaped)	Present (gourd-shaped)
Head bump	Absent	Absent	Absent	Present	Variable
Chin bump	Absent	Absent	Absent	Present	Variable
Snout ossicle and chin ossicle	Absent	Absent	Absent	Present	Present
Shape of body scale (viewed from the side)	Pot lid-shaped (some blunt spines in the center of the broad plane)	Conical with branching of tips (smaller than *M. mola* of same size)	Conical without branching of tips	Rectangular	Conical with branching of tips (except region under eye)
Shape of body scale (viewed from above)	Hexagonal shaped	Jagged dot shaped (smaller than *M. mola* of same size)	Simple dot shaped	Line shaped	Jagged dot shaped
Body scales on clavus	Absent	Present	Present	Present	Present
Hypodermis	Just slightly thicker than body scale	Significantly thicker than body scale	Significantly thicker than body scale	Significantly thicker than body scale	Significantly thicker than body scale
Mouth	Vertical oval	Horizontal oval	Horizontal oval	Horizontal oval	Horizontal oval
Snout profile	funnel-like, lips protrude beyond the beak	Rounded without protruding snout	Rounded without protruding snout	Snout protrudes in some specimens	Snout protrudes in some specimens
Protruding lower jaw (beyond upper jaw)	Absent	Present (not pronounced)	Absent	Absent	Absent
Shape of gill openings	Somewhat triangular, roundish	Oval	Oval	Oval	Oval
Otoconia or otolith	Otolith	Otoconia	Unknown	Otoconia	Otoconia
Shape of pectoral fin	Falcate	Rounded	Rounded	Rounded	Rounded
Shape of the clavus margin	Oblique	Median projection (various extending lengths)	Rounded with an indent	Rounded (slightly wavy in some specimens)	Wavy/lobed/scalloped (convex protuberance)
Ossicles on paraxial clavus fin rays	Absent	Absent	Separate	Separate	United
Smooth band back-fold on clavus	Absent	Absent	Present (forms a Y-shaped band on the clavus)	Usually absent (faint when present)	Usually absent (faint when present)
Shape of testis	Paired, elongated, rod-like	Paired, elongated, rod-like	Paired, elongated, rod-like	Paired, elongated, rod-like	Paired, elongated, rod-like
Shape of ovary	Paired, elongated, rod-like	Single, ball shaped	Single, ball shaped	Single, ball shaped	Single, ball shaped

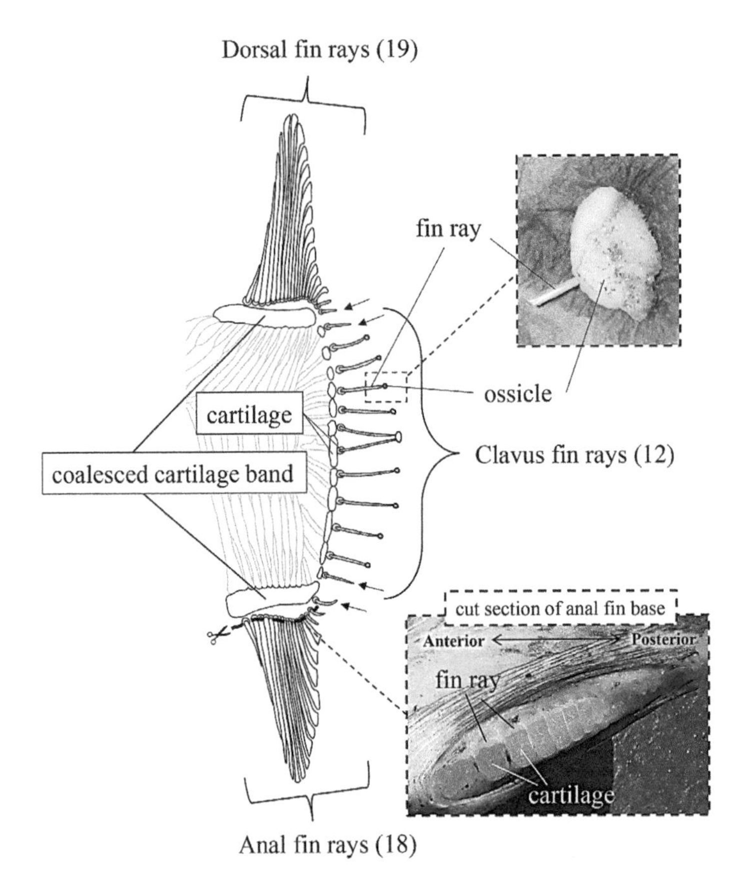

Figure 2. Generalized skeletal illustration of the posterior part of *Mola mola* (> 60 cm total length). Arrows indicate fin rays located at the border of the dorsal fin and clavus, and the anal fin and clavus. Scissors and bold dotted line indicate the cutting surface of the anal fin base.

Table 3. Species level key meristic characters of Molidae (full complement).

Species	*Ranzania laevis*	*Masturus lanceolatus*	*Mola tecta*	*Mola alexandrini*	*Mola mola*
Meristic characters (means)					
Pectoral fin rays	13–14 (13.1)*	10–11 (10.2)*	11–12 (11.9)**	11–12 (11.7)**	10–13 (11.8)**
Dorsal fin rays	18–20 (18.9)*	18–21 (19.6)*	17–19 (18.1)**	16–19 (17.6)**	18–19 (18.4)**
Anal fin rays	18–21 (19.7)*	17–20 (18.2)*	16–18 (17.1)**	15–17 (16.5)**	17–18 (17.4)**
Clavus fin rays	17–23 (20.6)*	19–25 (20.9)*	15–17 (15.9)**	14–24 (17.3)**	11–14 (12.3)**
Dorsal+Clavus+Anal fin rays	53–62 (59.2)*	55–64 (58.8)*	50–52 (51.3)**	48–57 (52.0)**	47–50 (48.5)**
Clavus ossicles	0*	Usually 0; rare, maximum 4 at tip of clavus projection*	5–7 (5.8)**	8–15 (11.8)**	8–9 (8.6)**

* Unpublished data (researched > 40 specimens; E. Sawai).
** Data from Sawai et al. (2018a).

Important Research Points for Studying Taxonomy of Molidae

When conducting Molidae taxonomic research, it is important to pay attention to intraspecific variation between individuals, and consider the potential for deformation in taxidermic specimens. Further, standard morphological measurements and unambiguious methods for fin ray counts are needed to ensure consistency between researchers. Here, we propose standardized methods and comment on important research points for consideration when conducting Molidae taxonomic research.

Individual Variation

In order to classify organisms, it is important to be familiar with intraspecific morphological variation such as ontogenetic changes, sexual dimorphism, and natural variation between individuals. Morphological variations may also include abnormalities caused by malformations, parasites, predators and artificial wounds (e.g., Sawai et al. 2009, 2019, Romanov et al. 2014, Nyegaard et al. 2018b). In one instance, for example, a morphologically abnormal individual was described as a new species of *Mola* (Ayres 1859). In addition to genetics, detailed morphological data are needed to distinguish between intraspecific individual variation and distinct species.

Taxidermic Specimens

Adult molids are valuable for taxonomic studies as their species characteristics are well developed, however, for logistical reasons very large specimens are difficult to store as immersion specimens at museum-related facilities. Therefore, in many cases, large specimens of molids are stored as mounted skins created by taxidermists, as is the case for both the holotype of *Mola alexandrini* and the neotype of *M. mola* (Sawai et al. 2018a). For mounted skins, the reproducibility of the shape of the fresh specimen largely depends on the skill of the taxidermist. Many mounted specimens are slightly shrunken and deformed. Additionally, the morphology of the body scales on the skin, such as the tip shape, may be distorted from chemical coatings during the preservation process (Sawai et al. 2018a).

Counting Clavus Fin Rays

The number of clavus fin rays is an important taxonomic species-level character. However, counting them is not always straight forward as the hypodermis becomes thick in large specimens, which also hampers x-ray transmission. For fin ray counts, it is therefore helpful to cut off the fins and clavus of dead specimens.

Since the clavus is formed by parts of the dorsal and anal fins during the early ontogeny process (Johnson and Britz 2005), there is no clear delineation of the fin rays 'belonging' to each of these areas (arrows in Fig. 2). Consequently, dorsal, anal and clavus fin ray counts from the same specimen may differ slightly between researchers. Here we propose a standardized method of fin ray delineation, which was also used to obtain the fin ray counts in Table 3: fin rays located near the border of the clavus–dorsal fin (or clavus–anal fin) should not be included in the clavus if the base of the fin ray is located adjacent to the coalesced cartilage band (even if the tip of the fin ray faces backwards) (see Fig. 2). More specifically, in the dorsal area, only fin rays below the coalesced cartilage band (in the anal area, only fin rays above the coalesced cartilage band) are counted as the clavus fin rays. This method is also effective when counting fin rays after cutting off the fins and clavus (Fig. 2). Our method differs slightly from that used by Tyler (1980), who counted 18 dorsal, 15 clavus and 16 anal fin rays on figure 306 in Tyler (1980). Using the method described here, we counted 19 dorsal, 12 clavus and 18 anal fin rays on the same figure of Tyler (1980).

Measurement Methods

Two morphometric measurements, TL (total length) and PCBL (pre-clavus band length), are commonly used to specify the body size of molids (Fig. 3; see also Yoshita et al. 2009, Sawai 2016, Sawai and Yamada 2018). Although PCBL is also termed preclaval length (PCL) in Martin and Drewry (1978) and Hellenbrecht et al. (2019) and standard length (SL) in Liu et al. (2009) and Chang et al. (2018), these terms indicate the same measurement. Therefore, here they are integrated into the name 'PCBL'.

Use of the term SL (length from the snout tip to the hypural bone), commonly used for bony fishes, is unsuitable for molids (Martin and Drewry 1978). The anterior margin of the smooth band on the clavus approximately coincides with the posterior end of the last vertebra, but SL can not technically be applied to the Molidae, which completely lacks the hypural bone and caudal fin (Fig. 2). Therefore, we advocate using the term PCBL instead of SL.

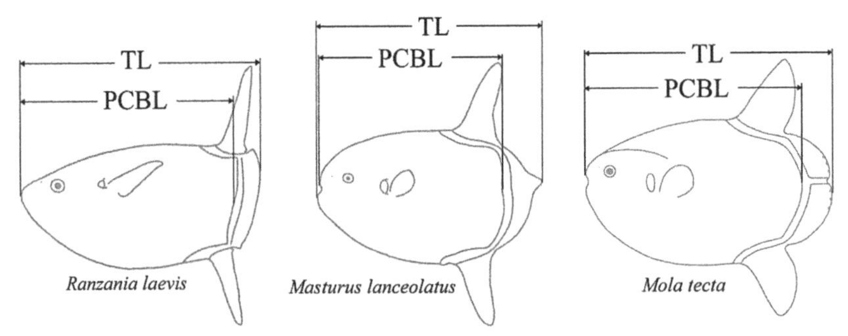

Figure 3. The morphometric measurements of molids. Total length* (TL, length from the most anterior part of the body to clavus tip), pre-clavus band length* (PCBL, length from snout tip to anterior margin of smooth band on clavus). Asterisks indicate a distance along the body axis, not a point to point distance.

Taxonomy and Phylogeny

The Molidae literature is extensive, exceeding 1,000 publications (Sawai et al. 2018a). Most of this literature comprises research related to taxonomy, morphology and distribution records, too extensive to list here. Instead, we summarize the Molidae taxonomic history by listing the major taxonomic milestones in chronological order, focusing on the history, phylogenetic relationship and etymology of the current five accepted species. Details of the wider Molidae taxonomic history, including a total of 56 nominal species described and named through time, is available on the internet site "Eschmeyer's Catalog of Fishes" (Fricke et al. 2020). Molecular methods have greatly aided Molidae taxonomy, and here we divide our review into two periods accordingly (before and after the application of molecular techniques).

Morphologic Taxonomic History

The taxonomic history of Molidae begins with Carolus Linnaeus (Linnaeus 1758), 'the father of modern taxonomy'. He introduced the binomial nomenclature (two-term naming system) for scientific names. All scientific names used prior to Linnaeus (1758) are defined as invalid by the International Code of Zoological Nomenclature (ICZN 1999). The Molidae taxonomic history from the Linnean era until the introduction of molecular studies are outlined in Table 4. During this time the taxonomic studies were based solely on morphology.

Table 4. Key milestones in the morphologic taxonomic history of the currently three valid genera and five valid species in the family Molidae.

Year	Author	Species ID/Taxonomic contribution	Notes
1758	Linnaeus	Described the first Molidae species, *Mola mola* (as *Tetraodon mola*) based on literature	Linnaeus confused *Mola* and *Ranzania* when describing *M. mola* (see Sawai et al. 2018a); No *M. mola* type specimen designated, but later a neotype was designated by Sawai et al. (2018a)
1766	Koelreuter	Established the first Molidae genus, *Mola*	Established based on the Molacanthiform (morphology of early life stages) of *Mola*
1776	Pennant	Described the second Molidae species, *Ranzania laevis* (as *Ostracion laevis*) based on a figure	Pennant considered *R. laevis* to be different at the genus level from *M. mola*
1835	Bonaparte	Proposed family Molidae (as Orthagoriscidae) as an independent family	See Van der Laan et al. (2014) for a family level taxonomic review of the molids
1839	Ranzani [literature review]	Described the third Molidae species, *Mola alexandrini* (as *Orthragoriscus alexandrini*) based on a figure	Ranzani accepted 6 genera and 17 species (including 1 suspected species) in Molidae, and his naming later gave rise to many synonyms; holotype of *M. alexandrini* described by Ranzani was rediscovered by Sawai et al. (2018a)
1840	Liénard	Described the fourth Molidae species, *Masturus lanceolatus* (as *Orthagoriscus lanceolatus*) based on brief description	More detailed description including a figure of the type specimen of *Ma. lanceolatus* was published by Liénard (1841)
1840	Nardo	Established the second Molidae genus, *Ranzania*	See Parenti (2003) and Fricke et al. (2020) for a genus level taxonomic review of the molids
1870	Günther [literature review]	Reduced the number of nominal species in Molidae, accepted three species: *M. mola* (as *Orthagoriscus mola*), *Ma. lanceolatus* (as *Orthagoriscus lanceolatus*), and *R. laevis* (as *Orthagoriscus truncatus*)	Günther treated the tiny, thorny fish (Molacanthiform) considered by many researchers as an 'independent species' (e.g., Pennant 1776), as an early life stage of *Mola*; Also noticed that *M. alexandrini* (as *Orthragoriscus alexandrini*) might be a different species from *M. mola*
1884	Gill	Established the third Molidae genus, *Masturus*	See Parenti (2003) and Fricke et al. (2020) for a genus level taxonomic review of the molids
1951	Fraser-Brunner [literature review]	Reexamined the vast number of nominal species in the literature, and accepted three genera and five species (including two subspecies) in Molidae: *Ranzania laevis laevis*, *Ranzania laevis makua*, *Masturus lanceolatus*, *Masturus oxyuropterus*, *Mola mola*, *Mola ramsayi*	Until the application of DNA analysis, Fraser-Brunner's proposed classification was widely used; the proposed species by Fraser-Brunner have now changed to: *R. laevis laevis* and *R. laevis makua* =>*R. laevis*; *Ma. lanceolatus* and *Ma. oxyuropterus* =>*Ma. lanceolatus*; *M. mola*; *M. ramsayi* = > *M. alexandrini*
2003	Parenti [literature review]	Made a checklist of scientific names of Molidae from species level to genus level	Checklist of Parenti (2003) became the basis of Fricke et al. (2020)

Morphologic and Phylogenetic Taxonomic History

The application of molecular techniques provided important insights into the phylogenetics of Molidae (Table 5). In particular, molecular studies have helped advance and clarify the taxonomy of the genus *Mola*, with the flow of recent taxonomic reexamination of *Mola* shown in Fig. 4.

For genera *Ranzania* and *Masturus*, molecular phylogenetic studies to date suggest that these are monotypic (e.g., Bass et al. 2005, Nyegaard et al. 2018a, b).

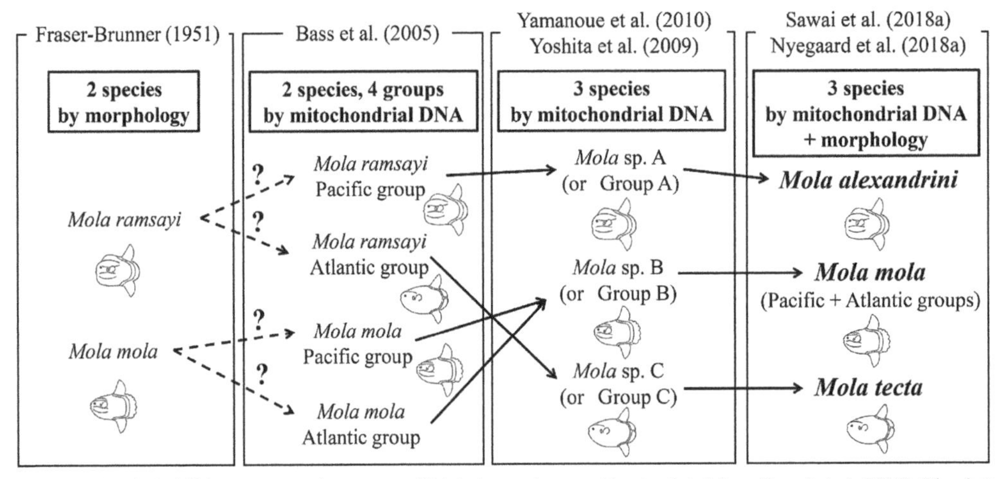

Figure 4. Historical shift between species names of *Mola* in previous studies (updated from Sawai et al. 2017). The dotted arrows indicate unclear relationships. The solid arrows indicate clear relationships.

Interspecific Relationships

Previous studies of phylogenetic relationships among the three genera in Molidae indicated that *Mola* and *Masturus* are morphologically (e.g., Fraser-Brunner 1951, Tyler 1980, Santini and Tyler 2002) and genetically (e.g., Yamanoue et al. 2004, Yoshita et al. 2009, Nyegaard et al. 2018b) more closely related to each other than to *Ranzania* (Fig. 5B). In *Mola*, based on mtDNA, *M. tecta* and *M. alexandrini* are genetically more closely related to each other than to *M. mola* (Bass et al. 2005, Yoshita et al. 2009, Nyegaard et al. 2018a, b; Fig. 5B). Additionally, the two genetic clades within *M. mola*, first suggested by Bass et al. (2005) as Pacific and Atlantic populations (Fig. 4) are currently treated as one species since there is insufficient morphological and genetic data to determine the potential status as separate species or subspecies (Caldera et al. 2020 [Chapter 3], Nyegaard et al. 2018a, b, Sawai et al. 2018a). On the other hand, the genetically divided subtropical and temperate populations within *M. alexandrini* first suggested by Yoshita et al. (2009) were not supported at the putative species level (Nyegaard et al. 2018a).

Phylogenetic Position

The monophyly of the family Molidae within the Tetraodontiform is supported by morphological and molecular studies of various researchers (e.g., Matsuura 2015). The phylogenetic position of Molidae in Tetraodontiformes, examined through analysis of taxonomic characteristics (morphology and molecular), remains uncertain (Matsuura 2015). However, convincing evidence places Molidae as a sister group to the pufferfishes (Tetraodontidae) and porcupine fishes (Diodontidae) (Matsuura 2015, Arcila and Tyler 2017; Fig. 5A). To clarify the phylogenetic position of Molidae in Tetraodontiformes, whole genome sequencing of all molid species and detailed morphological studies including fossils are required.

Table 5. Key milestones in the recent morphologic and genetic taxonomic history of Molidae.

Year	Author	Species ID/Taxonomic contribution	Notes
2003	Streelman et al.	Developed five nuclear DNA microsatellite markers for *Mola*	First molecular study on Molidae; after this study molecular studies of Molidae focused on mitochondrial DNA (mtDNA)
2004	Yamanoue et al.	First determined complete nucleotide sequences of the mitogenomes of three species in Molidae: *Ranzania laevis*, *Masturus lanceolatus*, and *Mola alexandrini* (as *M. mola*)	
2005	Sagara et al.	Identified two genetically separate *Mola* populations in Japanese waters	
2005	Bass et al.	Using molecular tools identified four genetically separate clades (populations) of *Mola*	Prior to this, at a global level, *Mola* was considered to belong to the two species proposed by Fraser-Brunner (1951)
2009	Yoshita et al.	Found species of *Mola* worldwide divided into three groups (A–C); Concluded the nucleotide sequence diversity of the three groups corresponded to species level differences	Conducted molecular phylogenetic analysis combining data (D-loop in mtDNA) from previous data and author's new samples; differences in the morphology of the two groups (A–B) in Japanese waters were found
2009	Sawai et al.	In response to Yoshita et al. (2009), treated three groups as three species (*Mola* spp. A, B, and C)	Morphology of *Mola* sp. C was unknown, so assignment of scientific names to three *Mola* species was put it on hold
2010	Yamanoue et al.	Developed Multiplex PCR-based genotyping of mtDNA for *Mola* spp. A and B in Japanese waters	
2018b	Nyegaard et al.	Revealed morphological characteristics of *Mola* sp. C; *Mola* sp.C was named the fifth Molidae species, *Mola tecta*	First new species of *Mola* in 125 years; Used mtDNA analysis and morphological research of previous data and author's new samples and data in the context of an extensive taxonomic literature review
2018a	Sawai et al.	Established species names of remaining two species (*Mola* spp. A and B): *Mola alexandrini* (sp. A) and *Mola mola* (sp. B)	Used data on species characteristics from Yoshita et al. (2009) and morphological characteristics researched by Fraser-Brunner (1951) along with the author's new data in the context of an extensive taxonomic literature review

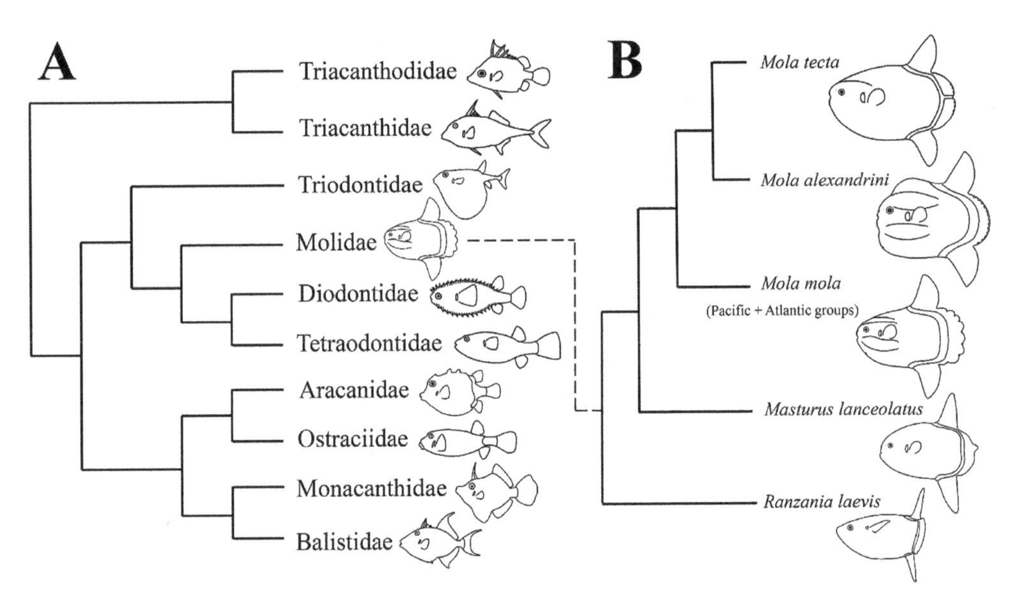

Figure 5. Phylogenetic relationships in A: the Tetraodontiformes (family level) based on Arcila and Tyler (2017), and B: the Molidae (species level) based on Nyegaard et al. (2018b).

Possible Existence of Hybrids

To date, no hybrids of molids have been reported. However, hybridization between closely related molids is not inconceivable, as hybrids have been reported from other families in the Tetraodontiformes, including Tetraodontidae (e.g., Takahashi et al. 2017), Diodontidae (e.g., Doi et al. 2015), Balistidae (e.g., Randall et al. 2002) and Monacanthidae (e.g., Yamamoto 2008). Hybrids generally show intermediate morphological characteristics between the two parent species (Randall et al. 2002, Yamamoto 2008, Takahashi et al. 2017). In order to examine potential hybridization in Molidae, molecular analysis is also required to identify paternal and maternal origins (e.g., Doi et al. 2015, Takahashi et al. 2017). Hybrids may occur due to changes in the distribution of parental species by environmental changes, and overlap of spawning sites (e.g., Takahashi et al. 2017).

Etymology and Names

During the past six centuries [e.g., since Salviani (1554)], molid species have been given a wide variety of names. The meaning and origin of these names are important to understand the taxonomic history of the Molidae, and the relationship between humans and molids. Table 6 summarises the etymology of the scientific and common names of each of the five valid Molidae species (see also Thys et al. 2020a [Chapter 14]). Generally, these names are derived from the morphology.

The English name is most generally used in the world as a common name for organisms. However, an English name is not applied under a similar methodical system to the scientific binomial nomenclature. For example, *M. mola* have multiple English names such as 'ocean sunfish', 'common mola', and 'head fish', with no standard English name. Further, *M. mola* is often simply referred to as 'sunfish', however, 'sunfish' is also used by some to refer to basking sharks *Cetorhinus maximus* (Gunnerus 1765) as well as the family Centrarchidae (e.g., Pennant 1776, Barnard 1947). Such ambiguity can lead to misunderstandings; for example, several researchers (e.g., Liénard 1841) mistaking for a molid, a 25 feet specimen (probably a ca. 7.6 m TL basking shark) from Ireland reported by Smith (1746). Although this is an extreme example, such confusion can be avoided by establishing unique standard English names for all Molidae species.

Table 6. Etymology of scientific name and standard English name in the Molidae.

Name	Meaning [language]	Origin	Reference
Scientific name			
Ranzania	Person's name	Camillo Ranzani (1775–1841)	Nardo (1840)
Masturus	Mastos (breast, nipple) + oura (tail) [Greek]	Clavus shape with a protrusion near the center	Gill (1884), Gudger (1937)
Mola	Millstone [Latin]	Rounded body shape	E.g., Koelreuter (1766), Gudger (1937)
laevis	Smooth [Latin]	Smooth body surface	Pennant (1776)
lanceolatus	Small lance [Latin]	Clavus shape with a protrusion near the center	Gudger (1937)
tecta	Disguised, hidden [Latin]	Evaded discovery for centuries	Nyegaard et al. (2018b)
alexandrini	Person's name	Antonio Alessandrini (1786–1861)	Ranzani (1839)
Standard English name			
Slender sunfish		Elongated body (=slender)	Unknown*
Sharptail sunfish		Clavus shape with a protrusion near the center (=sharptail)	Unknown*
Hoodwinker sunfish		Evaded discovery for centuries (=hoodwinker)	Nyegaard et al. (2018b)
Giant sunfish		The heaviest species in the Molidae (=giant)	This study
Ocean sunfish		Saltwater fish (=ocean) + habit of basking at the sea surface (=sunfish)	E.g., Thys (1994)

* We could not find the specific origin of these names, however, they have been used extensively in the literature.

In Japan, 43 local Japanese names for *M. mola* have been reported (e.g., 'Babarabo', 'Kinabo', 'Manzairaku', 'Shikiri', 'Ukiki') (Sawai 2017), however, one standard Japanese name for each species has been established for use in fish books and scientific papers (e.g., *M. mola* is 'Manbo'). To promote unambiguous nomenclature, we propose the standard English names for each currently valid molid species in Table 6 provided along with the origin of each name. Slender sunfish for *Ranzania laevis* and sharptail sunfish for *Masturus lanceolatus* are common names used in many studies, dating back to at least the 1980s, however, we could not find the origin of these English names in the literature (Table 6). The standard English name for *M. tecta* (Nyegaard et al. 2018b) is a new English name proposed when the species was recently described (Table 6). The English name 'bump-head sunfish' was proposed for *M. alexandrini* when this species was recently re-described (Sawai et al. 2018a), pertaining to the large head bump that develops with size. Although Sawai et al. (2018a) suggested at the time that some individual *M. mola* may possess a small head bump, it has since been pointed out that Atlantic *M. mola* populations might have more developed head bumps than Pacific *M. mola* populations (Wirtz and Biscoito 2019, Caldera et al. 2020 [Chapter 3]). To avoid misidentification from confusion between species based on morphology and name, we here propose changing the English name of *M. alexandrini* from bump-head sunfish to the new standard English name of giant sunfish.

Maximum Size and Distribution

Size Records

Numerous maximum records of molid body sizes exist in the literature. Many are based on estimated weights and/or lengths rather than actual measurements. In addition, due to the long history of taxonomic confusion surrounding the Molidae, some records may have been assigned to the wrong species. Consequently, it is unclear just how large each species can grow. Here, we review actual maximum measurements for each of the five molid species, based on scientifically reliable reports of individual specimens in the literature (Tables 7–8). If there is scientifically credible information that we have overlooked or misidentified here, we hope that it will be updated in future research. The following comments are supplemental to Tables 7–8.

Table 7. Scientifically Reliable Maximum Length Records for Molidae.

Species	Date	Location	Length	Weight (kg)	Method	Sex	Source
Ranzania laevis	ND	Atlantic seas [mari Atlantico]	74 cm length [2+1/2 pedes = 2+1/2 Roman feet]*	ND	ND	ND	Bloch and Schneider (1801)
Masturus lanceolatus	5 January 1912	South Beach, St. Augustine, Florida, USA	304.8 cm long [10 feet]*	ND	Stranding	ND	Gudger and MacDonald (1935), Gudger (1937)
Mola tecta	1961	Otago Harbour, Dunedin, New Zealand	242 cm TL	ND	Stranding	ND	Nyegaard et al. (2018b)
Mola alexandrini	11 August 2004	Off Aji Island, Miyagi Prefecture, Japan	332 cm TL	ND	Set net	Female	Yoshita et al. (2009), Sawai et al. (2018a)
Mola mola	3 September 1919	Off Santa Catalina Island, California, USA	332.7 cm TL [10 feet 11 inches]	ND	Harpoon	ND	Heilner (1920)

Square brackets: original text notation.

ND no data; TL total length.

*possibly TL.

Table 8. Scientifically Reliable Maximum Weight Records for Molidae.

Species	Date	Location	Length	Weight (kg)	Method	Sex	Source
Ranzania laevis	5 March 1998	Near Mandapam in Palk Bay, India	66 cm TL	8.2	Shore seine	Female	Victor et al. (1998)
Masturus lanceolatus	January 2004	Off Eastern Taiwan	195 cm PCBL	409	Set net or drift net or longline	Female	Liu et al. (2009)
Mola tecta	25 December 2015	Off North Taranaki Bight, North Island, New Zealand	101.1 cm TL	52	Trawl	Male	Nyegaard et al. (2018b), this study
Mola alexandrini	16 August 1996	Off Kamogawa, Chiba Prefecture, Japan	272 cm TL	2,300	Set net	Female	Sawai et al. (2018a)
Mola mola	15 November 2007	Off Mihonoseki Lighthouse, Shimane Prefecture, Japan	265 cm TL	1,320	Encircling net	Female	Kawakami et al. (2008), Yamanoue et al. (2010)

PCBL pre-clavus band length; TL total length.

For *Ranzania laevis*, our review revealed a credible maximum length of 74 cm (Table 7), although three larger maximum lengths were found in the literature. The earliest records of these values are, Risso (1827) (= 90 cm), Kamohara (1950) (= 100 cm), and Fraser-Brunner (1951) (= 80 cm). However, all three records are given without descriptions, references and/or photographs of specific individuals, and the scientific credibility is considered uncertain.

For *Masturus lanceolatus*, an approximately 2,000 kg specimen reported by Heemstra (1986) is often cited in the literature as the maximum weight of this species. However, when we enquired with the author of Heemstra (1986), he was unable to confirm the source of the record. It is feasible, therefore, that this *Ma. lanceolatus* record was in fact for *Mola* (P. Heemstra, personal communication). Chang et al. (2018) reported a 382.9 cm PCBL specimen of *Ma. lanceolatus* from Taiwanese waters. However, the reliability of these data is uncertain since the fishermen did not provide photographs, nor any other contextual information confirming the capture of such an enormous individual (C.-T. Chang, personal communication). Tortonese (1990) also reported a 337 cm specimen of *Ma. lanceolatus* without supporting information (e.g., location, date, method) or photographs which calls into question the scientific credibility of the record.

Previously identified as *Mola mola*, the current world weight record holder for extant bony fish species weighed 2,300 kg. This individual was recently re-identified as *Mola alexandrini* (Sawai et al. 2018a, Guinness World Records 2019; Table 8), and the official record for the world's heaviest bony fish was updated accordingly on the Guinness World Records website on 14 September 2018 (Sawai et al. 2019).

The longest *M. mola* recorded here is 332.7 cm TL from Santa Catalina Island, California (Heilner 1920; Table 7). Note, this specimen is likely the same one recounted in Claro (1994), which was recorded as 332.7 cm TL but with no details as to its origin. Several maximum length records of *Mola mola* have been reported in previous studies [e.g., 400 cm length (possibly TL) from Japan (Kamohara 1950); 360 cm TL from New Zealand (McCann 1961)], but the scientific credibility of such claims is unverified due to the lack of accurate measurement and/or photographs with defining features of the actual specimens.

Limited data exist for *Mola tecta*. To date, the formally recorded heaviest specimen is the holotype at 52 kg (101.1 cm TL) (Table 8). However, this species is known to reach at least 242 cm TL (Nyegaard et al. 2018b; Table 7) and as such can most certainly weigh more than 52 kg.

Species Distribution after Recent Taxonomic Revisions

Molids are distributed in temperate and tropical waters around the world excluding the polar regions. Due to the historic taxonomic confusion, detailed species-level distribution (in light of recent taxonomic revisions) is currently not well understood (especially for the three *Mola* species) (Nyegaard et al. 2018b, Sawai et al. 2018a). The fragmentary distributions based on confirmed species-level records are summarized in Table 9. The occurrence of *M. tecta* in the Netherlands (Lidth de Jeude 1890) was initially treated as uncertain (Nyegaard et al. 2018b), but recent observations in California (e.g., Leachman 2019, Monterey Bay Aquarium 2019) adds credibility to occasional northern hemisphere occurrences of this species. Further studies including genetic confirmations are needed to establish species-level distributions for all Molidae species (For all current genetically confirmed global sightings, see Caldera et al. 2020 [Chapter 3]).

Conclusions

Establishing a deeper understanding of molid taxonomy is possible through continued morphological and genetic research. Acquiring further samples and data will help clarify morphological and ontogenetic changes from egg to senescence. While preserving whole-body specimens is difficult

Table 9. Reported distribution of current species in the Molidae.

Species	Distribution	Reference
Ranzania laevis	North and South Pacific, North and South Atlantic, and North and South Indian Ocean; rarely encountered, although large aggregations and stranding events have been reported	E.g., Fraser-Brunner 1951, Parenti 2003, Smith et al. 2010, Froese and Pauly 2019
Masturus lanceolatus	North and South Pacific, North and South Atlantic, and North and South Indian Ocean; predominantly found in tropical and sub-tropical seas	E.g., Gudger 1937, Fraser-Brunner 1951, Martin and Drewry 1978, Jawad et al. 2013, Nyegaard et al. 2018a.
Mola tecta	New Zealand, Australia, South Africa, Chile, Peru, [predominantly occurs in the Southern Hemisphere, but has also been found in the Netherlands and USA (California)]	Lidth de Jeude 1890, Nyegaard et al. 2018a, b, Leachman 2019, Mangel et al. 2019, Monterey Bay Aquarium 2019, Caldera et al. 2020 [Chapter 3]
Mola alexandrini	Japan, New Zealand, Australia, Chile, Ecuador (Galápagos Islands), Indonesia, India, Iran, United Arab Emirates, Oman, Maldive Islands, Spain, South Africa, Côte d'Ivoire, Portugal, Malta, Libya, Adriatic Sea, Argentina–Uruguay; prefer tropical and subtropical seas	E.g., Sawai et al. 2018a, Nyegaard et al. 2018a
Mola mola	Japan, Taiwan, New Zealand, Australia, South Africa, USA, Sweden, Denmark, United Kingdom, Portugal, Italy, Adriatic Sea; rare in the Indian Ocean and the southern hemisphere	E.g., Sawai et al. 2018a, Nyegaard et al. 2018a

due to the large size of adults, digital photographs and preservation of certain body parts (e.g., a skin sample with scales for morphogical identification) can provide crucial taxonomic information and serve as a common resource for additional non-taxonomic research.

Molecular studies (e.g., mitochondrial and nuclear DNA: Caldera et al. 2020 [Chapter 3]) will help identify species-level differences in Molidae, and lend insight into sex determination, potential hybridization, population size estimates, distribution, and evolutionary factors such as fish morphological specialization and growth capacity. Molids have a long history of species confusion which continues to this day. Careful morphological work coupled with genetic techniques and the collaboration of fishermen and researchers to procure samples worldwide will help advance our knowledge of this cosmopolitan group.

Remaining Questions

Two major issues covered in this chapter require resolution. Firstly, it is crucial to clarify how many valid, extant species exist. For example, are *M. mola* populations from the Pacific and Atlantic, showing slight genetic separation, the same species? Or are there clear morphological differences between the two populations that correspond to distinct species/subspecies (such as a head bump in Atlantic populations)? And are *Ranzania* and *Masturus* monotypic genera? Secondly, additional ontogenetic research is required to clarify the developmental stages of molids and create more accurate identification keys particularly for early life stages in the genus *Mola*.

To resolve these issues, collaboration between fishermen, aquariums and research institutions globally can facilitate data and sample collections (e.g., Phillips et al. 2018, Sawai 2019, Howard et al. 2020 [Chapter 13]). Solving these problems will provide a firmer basis for all molid research and lead to a more accurate ecological understanding of these fishes. If we can accurately identify molid species at various developmental stages, we will also be able to deepen our understanding of the evolutionary background of these fishes (e.g., morphological comparison of related species), food web dynamics (e.g., identification of stomach contents of predators), and fisheries management (e.g., estimating population size).

Acknowledgments

We would like to thank the following persons and institutions for assistance: C.-T. Chang, M. Freese, P. Heemstra, Y. Kai, K. Matsuura, and staff of the Tottori Prefectural Museum and the International Coastal Research Center at the Atmosphere and Ocean Research Institute at the University of Tokyo. The research was supported by the Sasakawa Scientific Research Grant from The Japan Science Society (23-512, 24-506 and 25-504). Thank you, the Heisei era, and welcome to the Reiwa era in Japan.

References

Arcila, D. and J.C. Tyler. 2017. Mass extinction in tetraodontiform fishes linked to the Palaeocene-Eocene thermal maximum. Proc. R. Soc. B. 284: 20171771. DOI 10.1098/rspb.2017.1771.

Ayres, W.O. 1859. On new fishes of the Californian coast. Proc. Calif. Acad. Nat. Sci. 1858–1862 (Ser. 1), 2: 25–32.

Barnard, K.H. 1947. A pictorial guide to South African fishes, marine and freshwater. Maskew Miller, Cape Town, South Africa.

Bass, A.L., H. Dewar, T. Thys, J.T. Streelman and S.A. Karl. 2005. Evolutionary divergence among lineages of ocean sunfish family, Molidae (Tetraodontiformes). Mar. Biol. 148: 405–414.

Bloch, M.E. and J.G. Schneider. 1801. Systema ichthyologiae iconibus cx illustratum. Bibliopolio Sanderiano Commissum, Berolini, Germany.

Bonaparte, C.L. 1835. Prodromus systematis ichthyologiae. Nuovi Annali delle Scienze naturali Bologna (Ser. 1), 2(4): 181–196, 272–277.

Caldera, E.J., J.L. Whitney, M. Nyegaard, E. Ostalé-Valriberas, L. Kubicek and T.M. Thys. 2020. Genetic insights regarding the taxonomy, phylogeography and evolution of the ocean sunfishes (Molidae: Tetraodontiformes). pp. 37–54. *In*: T.M. Thys, G.C. Hays and J.D.R. Houghton [eds.]. The Ocean Sunfishes: Evolution, Biology and Conservation, CRC Press. Boca Raton, FL, USA.

Chang, C.T., C.H. Chih, Y.C. Chang, W.C. Chiang, F.Y. Tsai, H.H. Hsu et al. 2018. Seasonal variations of species, abundance and size composition of Molidae in Eastern Taiwan. J. Taiwan Fish. Res. 26: 27–42.

Claro, R. 1994. Characterísticas generales de la ictiofauna. pp. 55–71. *In*: R. Claro [ed.]. Ecología de los preces marinos de Cuba. Instituto de Oceanología (Academia de Ciencias de Cuba), Cuba; and Centro de Investigaciones de Quintana Roo, Quintana Roo, Mexico.

Davenport, J., N.D. Phillips, E. Cotter, L.E. Eagling and J.D.R. Houghton. 2018. The locomotor system of the ocean sunfish *Mola mola* (L.): role of gelatinous exoskeleton, horizontal septum, muscles and tendons. J. Anat. 233: 347–357.

Doi, H., Y. Zenke, H. Takahashi, H. Sakai and T. Ishibashi. 2015. Hybridization of burrfish between *Chilomycterus antillarum* and *Chilomycterus schoepfii* in captivity revealed by AFLP and mtDNA sequence analyses. Ichthyol. Res. 62: 516–518.

Fraser-Brunner, A. 1951. The ocean sunfishes (Family Molidae). Bull. Br. Mus. (Nat. Hist.). Zool. 1: 89–121.

Fricke, R., W.N. Eschmeyer and R. van der Laan [eds.]. 2020. Eschmeyer's Catalog of Fishes: genera, species, references. Online version, updated 3 February 2020. http://researcharchive.calacademy.org/research/ichthyology/catalog/fishcatmain.asp (Accessed 11 February 2020).

Froese, R. and D. Pauly [eds.]. 2019. FishBase, version (08/2019). Online version, updated February 2019. https://www.fishbase.se/search.php (Accessed 13 December 2019).

Gill, T.N. 1884. Synopsis of the plectognath fishes. Proc. U.S. Natl. Mus. 7: 411–427.

Gudger, E.W. and S.M. MacDonald. 1935. The rarest of the ocean sunfishes. Sci. Month. 41: 396–408.

Gudger, E.W. 1937. The natural history and geographical distribution of the pointed-tailed ocean sunfish (*Masturus lanceolatus*), with notes on the shape of the tail. Proc. Zool. Soc. Lond. A107: 353–396, pls. 1–5.

Guinness World Records. 2019. Heaviest bony fish. p. 42. *In*: Guinness World Records 2020. Guinness World Records, London, UK.

Günther, A. 1870. Molina. pp. 317–320. *In*: Catalogne of the Fishes in the British Museum. Vol. 8. Catalogue of the Physostomi, containing the families Gymnotidae, Symbranchidae, Muraenidae, Pegasidae, and of the Lophobranchii, Plectognathi, Dipnoi, Ganoidei, Chondropterygii, Cyclostomata, Leptocardii, in the British Museum. Order of the Trustees, London, UK.

Heemstra, P.C. 1986. Family No.270: Molidae. pp. 907–908. *In*: M.M. Smith and P.C. Heemstra [eds.]. Smiths' sea fishes, 6th edn. Springer-Verlag, New York, USA.

Heilner, V.C. 1920. Notes on the taking of an ocean sunfish (*Mola mola*) off Santa Catalina island, California, September 3, 1919. Bull. New York Zool. Soc. 23: 126–127.

Hellenbrecht, L.M., M. Freese, J.D. Pohlmann, H. Westerberg, T. Blancke and R. Hanel. 2019. Larval distribution of the ocean sunfishes *Ranzania laevis* and *Masturus lanceolatus* (Tetraodontiformes: Molidae) in the Sargasso Sea subtropical convergence zone. J. Plankton Res. 41: 595–608.

Howard, M.J., T. Nakatsubo, J.P. Correia, H. Batista, N. Baylina, C. Taura et al. 2020. Sunfish on display: husbandry of the ocean sunfish *Mola mola.* pp. 243–262. *In*: T.M. Thys, G.C. Hays and J.D.R. Houghton [eds.]. The Ocean Sunfishes: Evolution, Biology and Conservation, CRC Press. Boca Raton, FL, USA.

[ICZN] International Code of Zoological Nomenclature. 1999. International Code of Zoological Nomenclature, 4th edition. The International Trust for Zoological Nomenclature, London, UK.

Iwai, T. 2005. Introduction of Ichthyology [Gyogaku Nyuumon]. Kouseisha-Kouseikaku, Tokyo, Japan.

Jardas, I. and B. Knežević. 1983. A contribution to the knowledge of the Adriatic ichthyofauna—*Ranzania laevis* (Pennant, 1776) (Plectognathi, Molidae). Bilješke-Notes, Institut za oceanografiju i ribarstvo, Split, 51: 1–8.

Jawad, L.A., J.M. Al-Mamry and L.H. Al-Kharusi. 2013. A record of sharp-tail mola, *Masturus lanceolatus* (Liénard, 1840) (Molidae) in the Sea of Oman. J. Appl. Ichthyol. 29: 242–244.

Johnson, G.D. and R. Britz. 2005. Leis's conundrum: homology of the clavus of the ocean sunfishes. 2. Ontogeny of the median fins and axial skeleton of *Ranzania laevis* (Teleostei, Tetraodontiformes, Molidae). J. Morphol. 266: 11–21.

Kamohara, T. 1950. Description of the fishes from the provinces of Tosa and Kishu, Japan. Kochiken Bunkyo Kyokoi, Kochi, Japan.

Kawakami, Y., K. Ichisawa and S. Ando. 2008. Records of marine animals stranded on the coast of Tottori Prefecture, Honshu, Japan from 2006 to 2007, with notes on rare animals in the Sea of Japan. Bull. Tottori Pref. Mus. 45: 17–22.

Koelreuter, I.T. 1766. Piscium rariorum e mus. petrop. exceptorum descriptiones continuatae. Novi. Comment. Acad. Sci. Imp. Petropol. 10: 329–351, pl. 8.

Leachman, S. 2019. Rare hoodwinker sunfish, never before seen in Northern Hemisphere, washes up at Coal Oil Point Reserve. UC Santa Barbara, updated 27 February 2019. https://www.news.ucsb.edu/2019/019361/hoodwinked (Accessed 13 December 2019).

Linnaeus, C. 1758. Systema naturae per regna tria naturae, secundum classes, ordines, genera, species, cum characteribus, differentiis, synonymis, locis. Tomus I. Editio decima, reformata. Laurentii Salvii, Holmiae, Sweden.

Lidth de Jeude, T.W.V. 1890. On a large specimen of *Orthragoriscus* on the Dutch coast. Not. Leyden Mus. 12: 189–195.

Liénard, E. 1840. Description d'une nouvelle espèce du genre mole (*Orthagoriscus*, Schn.) découverte à l'île Maurice. Rev. Zool. 3: 291–292.

Liénard, E. 1841. Description d'une nouvelle espèce du genre mole (*Orthagoriscus*, Schn.) découverte à l'île Maurice et nommée *Orthagoriscus lanceolatus*. Mag. Zool. (Ser. 2). 3: 1–8, poissons pl. 4.

Liu, K.M., M.L. Lee, S.J. Joung and Y.C. Chang. 2009. Age and growth estimates of the sharptail mola, *Masturus lanceolatus*, in waters of eastern Taiwan. Fish. Res. 95: 154–160.

Lyczkowski-Shultz, J. 2006. Molidae: Molas, ocean sunfishes. pp. 2457–2465. *In*: W.J. Richards [ed.]. Early Stages of Atlantic Fishes: An Identification Guide for the Western Central North Atlantic. Volume II. CRC Press, Boca Raton, Florida, USA.

Mangel, J.C., M. Pajuelo, A. Pasara-Polack, G. Vela, E. Segura-Cobeña and J. Alfaro-Shigueto. 2019. The effect of Peruvian small-scale fisheries on sunfishes (Molidae). J. Fish Biol. 94: 77–85.

Martin, F.D. and G.E. Drewry. 1978. Development of fishes of the mid-Atlantic Bight: an atlas of egg, larval and juvenile stages. VI. Stromateidae through Ogcocephalidae. U.S. Dept. Int. Fish. Wildl. Serv., Washington DC, USA.

Matsuura, K. 2015. Taxonomy and systematics of tetraodontiform fishes: A review focusing primarily on progress in the period from 1980 to 2014. Ichthyol. Res. 62: 72–113.

McCann, C. 1961. The sunfish, *Mola mola* (L.) in New Zealand waters. Rec. Domin. Mus. 4: 7–20.

Monterey Bay Aquarium. 2019. Major *Mola* moment: First confirmed hoodwinker sunfish photographed in Monterey Bay. Tumblr. Online, updated 5 September 2019. https://montereybayaquarium.tumblr.com/post/187517538903/major-mola-moment-first-confirmed-hoodwinker (Accessed 13 December 2019).

Nakae, M. and K. Sasaki. 2006. Peripheral nervous system of the ocean sunfish *Mola mola* (Tetraodontiformes: Molidae). Ichthyol. Res. 53: 233–246.

Nardo, G.D. 1840. Considerazioni sulla famiglia dei pesci *Mola*, e sui caratteri che li distinguono. Ann. Sci. R. Lombardo-Veneto. 10: 105–112.

Nolf, D. and J.C. Tyler. 2006. Otolith evidence concerning interrelationships of caproid, zeiform and tetraodontiform fishes. Bull. L'Inst. Royal Sci. Nat. Belgique, Biol. 76: 147–189.

Nyegaard, M., N. Loneragan, S. Hall, J. Andrew, E. Sawai and M. Nyegaard. 2018a. Giant jelly eaters on the line: Species distribution and bycatch of three dominant sunfishes in the Southwest Pacific. Estuar., Coast. Shelf Sci. 207: 1–15.

Nyegaard, M., E. Sawai, N. Gemmell, J. Gillum, N.R. Loneragan, Y. Yamanoue et al. 2018b. Hiding in broad daylight: molecular and morphological data reveal a new ocean sunfish species (Tetraodontiformes: Molidae) that has eluded recognition. Zool. J. Linn. Soc. 182: 631–658.

Parenti, P. 2003. Family Molidae Bonaparte 1832—molas or ocean sunfishes. Calif. Acad. Sci. Annot. Checklists Fishes. 18:1–9.

Pennant, T. 1776. British zoology, 4th edition. Vol. III. Class III. Reptiles. IV. Fish. Benjamin White, London, UK.

Phillips, N.D., L. Kubicek, N.L. Payne, C. Harrod, L.E. Eagling, C.D. Carson et al. 2018. Isometric growth in the world's largest bony fishes (genus *Mola*)? Morphological insights from fisheries bycatch data. J. Morphol. 279: 1312–1320.

Purushottama, G., B. Anulekshmi, C. Ramkumar, S. Thakurdas and B. Mhadgut. 2014. A rare occurrence and biology of the slender sunfish, *Ranzania laevis* (Actinopterygii: Tetraodontiformes: Molidae), in the coastal waters of Mumbai, north-west coast of India. Indian J. Geo-Mar. Sci. 43: 1554–1559.

Randall, J.E., R.F. Myers and R. Winterbottom. 2002. *Melichthys indicus* x. *vidua*, a hybrid triggerfish (Tetraodontiformes: Balistidae) from Indonesia. Aqua J. Ichthyol. Aquat. Biol. 5: 77–80.

Ranzani, C. 1839. Dispositio familiae Molarum in genera et in species. Novi Comment. Acad. Sci. Inst. Bonon. 3: 63–82, pl 6, foldout table.

Risso, A. 1827. Histoire naturelle des principales productions de l'Europe méridionale et particulièrement de celles des environs de Nice et des Alpes maritimes. Tome troisième. F. G. Levrault, Paris, France.

Romanov, E.V., A. Sharp and P. Bach. 2014. Amputation of a major propulsor would not be an insuperable obstacle for survivorship of the slender sunfish *Ranzania laevis* (Molidae:Tetraodontiformes). Mar. Biod. Rec. 7: e20. DOI 10.1017/S1755267214000232.

Salviani, I. 1554. Aquatilium animalium historiae, liber primus, cum eorumdem formis, aere excusis. Hippolytum Salvianum, Romae, Italy.

Sagara, K., Y. Yoshita, M. Nishibori, H. Kuniyoshi, T. Umino, Y. Sakai et al. 2005. Coexistence of two clades of the ocean sunfish *Mola mola* (Molidae) around the Japan coast. Jpn. J. Ichthyol. 52: 35–39.

Santini, F. and J.C. Tyler. 2002. Phylogeny of the ocean sunfishes (Molidae, Tetraodontiformes), a highly derived group of teleost fishes. Ital. J. Zool. 69: 37–43.

Sawai, E., Y. Yamanoue, Y. Sakai and H. Hashimoto. 2009. On the abnormally morphological forms in *Mola* sunfish (*Mola* spp. A and B) taken from Japanese coastal waters. J. Grad. Sch. Biosp. Sci. Hiroshima Univ. 48: 9–17.

Sawai, E. 2016. Morphological identifications of preserved specimens of *Mola* in the Kagoshima University Museum. Nat. Kagoshima 42: 343–347.

Sawai, E. 2017. The mystery of ocean sunfishes [Manbou no Himitsu]. Iwanami Shoten Publishers, Tokyo, Japan.

Sawai, E., Y. Yamanoue, L. Jawad, J. Al-Mamry and Y. Sakai. 2017. Molecular and morphological identification of *Mola* sunfish specimens (Actinopterygii: Tetraodontiformes: Molidae) from the Indian Ocean. Species Divers. 22: 99–104.

Sawai, E. and M. Yamada. 2018. A little knowledge about morphology of young *Ranzania laevis* (Tetraodontiformes: Molidae) from off the Kagoshima mainland, southern Japan. Nat. Kagoshima 44: 5–8.

Sawai, E., Y. Yamanoue, M. Nyegaard and Y. Sakai. 2018a. Redescription of the bump-head sunfish *Mola alexandrini* (Ranzani 1839), senior synonym of *Mola ramsayi* (Giglioli 1883), with designation of a neotype for *Mola mola* (Linnaeus 1758) (Tetraodontiformes: Molidae). Ichthyol. Res. 65: 142–160.

Sawai, E., Y. Yamanoue, T. Sonoyama, K. Ogimoto and M. Nyegaard. 2018b. A new record of the bump-head sunfish *Mola alexandrini* (Tetraodontiformes: Molidae) from Yamaguchi Prefecture, western Honshu, Japan. Biogeography 20: 51–54.

Sawai, E. 2019. Rearing and research of ocean sunfishes in aquariums in Japan [Manbou wa ue wo muite nemuru no ka]. Poplar Publishing, Tokyo, Japan.

Sawai, E., H. Senou and T. Takeshima. 2019. A mounted specimen of *Mola alexandrini* (Ranzani, 1839) exhibited in the Kanagawa Prefectural Museum of Natural History. Bull. Kanagawa Prefect. Mus. (Nat. Sci.). 48: 37–42.

Smith, C. 1746. The ancient and present state of the county and city of Waterford: Being a natural, civil, ecclesiastical, historical and topographical description thereof. A. Reilly, Dublin, Ireland.

Smith, K.A., M. Hammond and P.G. Close. 2010. Aggregation and stranding of elongate sunfish (*Ranzania laevis*) (Pisces: Molidae) (Pennant, 1776) on the southern coast of Western Australia. J. R. Soc. West. Aust. 93: 181–188.

Streelman, J.T., C. Puchulutegui, A.L. Bass, T. Thys, H. Dewar and S.A. Karl. 2003. Microsatellites from the world's heaviest bony fish, the giant *Mola mola*. Mol. Ecol. Notes. 3: 247–249.

Takahashi, H., A. Toyoda, T. Yamazaki, S. Narita, T. Mashiko and Y. Yamazaki. 2017. Asymmetric hybridization and introgression between sibling species of the pufferfish *Takifugu* that have undergone explosive speciation. Mar. Biol. 164: 1–11.

Thys, T. 1994. Swimming heads. Natural History 103: 36–39.

Thys, T.M., M. Nyegaard and L. Kubicek. 2020a. Ocean sunfishes and society. pp. 263–279. *In*: T.M. Thys, G.C. Hays and J.D.R Houghton [eds.]. The Ocean Sunfishes: Evolution, Biology and Conservation, CRC Press. Boca Raton, FL, USA.

Thys, T.M., M. Nyegaard, J.L. Whitney, J.P. Ryan, I. Potter, T. Nakatsubo et al. 2020b. Ocean sunfish larvae: detections, identification and predation. pp. 105–128. *In*: T.M. Thys, G.C. Hays and J.D.R Houghton [eds.]. The Ocean Sunfishes: Evolution, Biology and Conservation, CRC Press. Boca Raton, FL, USA.

Tortonese, E. 1990. Molidae. pp. 1077–1079. *In*: J.C. Quéro, J.C. Hureau, C. Karrer, A. Post and L. Saldanha [eds.]. Check-list of the Fishes of the Eastern Tropical Atlantic (CLOFETA). Vol 2. JNICT, Lisbon, Portugal; SEI, Paris, France; and UNESCO, Paris, France.

Tyler, J.C. 1980. Osteology, phylogeny, and higher classification of the fishes of the order Plectognathi (Tetraodontiformes). NOAA Tech. Rep. NMFS Circ. 434: 1–422.

Van der Laan, R., W.N. Eschmeyer and R. Fricke. 2014. Family-group names of recent fishes. Zootaxa. 3882: 1–230.

Victor, A.C.C., D. Kandasamy and N. Ramamoorthy. 1998. On the large sunfish landed near Mandapam. Mar. Fish. Inf. Serv. Tech. Ext. Ser. 157: 26–27.

Watson, W. 1996. Molidae: Molas. pp. 1439–1441. *In*: H.G. Moser [ed.]. The Early Stages of Fishes in the California Current Region. California Cooperative Oceanic Fisheries Investigations (CalCOFI) Atlas No. 33. South West Fisheries Centre, La Jolla, California, USA.

Wirtz, P. and M. Biscoito. 2019. The distribution of *Mola alexandrini* in the subtropical Eastern Atlantic, with a note on *Mola mola*. Bocagiana. 245: 1–6.

Yamamoto, K. 2008. *Thamnaconus hypargyreus* × *T. modestus* hybrids (Monacanthidae) collected from the East China Sea. Jpn. J. Ichthyol. 55: 17–26.

Yamanoue, Y., M. Miya, K. Matsuura, M. Katoh, H. Sakai and M. Nishida. 2004. Mitochondrial genomes and phylogeny of the ocean sunfishes (Tetraodontiformes: Molidae). Ichthyol. Res. 51: 269–273.

Yamanoue, Y., K. Mabuchi, E. Sawai, Y. Sakai, H. Hashimoto and M. Nishida. 2010. Multiplex PCR-based genotyping of mitochondrial DNA from two species of ocean sunfish from the genus *Mola* (Tetraodontiformes: Molidae) found in Japanese waters. Jpn. J. Ichthyol. 57: 27–34.

Yoshita, Y., Y. Yamanoue, K. Sagara, M. Nishibori, H. Kuniyoshi, T. Umino et al. 2009. Phylogenetic relationships of two *Mola* sunfishes (Tetraodontiformes: Molidae) occurring around the coasts of Japan, with notes on their geographical distribution and morphological characteristics. Ichthyol. Res. 56: 232–244.

Chapter 3

Genetic Insights Regarding the Taxonomy, Phylogeography and Evolution of Ocean Sunfishes (Molidae: Tetraodontiformes)

Eric J. Caldera,[1,*] *Jonathan L. Whitney,*[2] *Marianne Nyegaard,*[3] *Enrique Ostalé-Valriberas,*[4] *Lukas Kubicek*[5] and *Tierney M. Thys*[6]

Introduction

The extraordinary body size, potential for fast growth and high reproductive output of ocean sunfishes (Molidae: Tetraodontiformes) beg important questions about the evolutionary history and development of these charismatic fishes. They have been referred to as "natural mutants" due to their uniquely truncated morphology, and while they may appear ancient to some, they hold an evolutionarily derived position among teleost fishes (the lineage of ray-finned fishes that comprises 96 percent of all extant fish species). The use of mitochondrial DNA (mtDNA), including *d-loop* (control-region) fragments and fully sequenced mitochondrial genomes, has revealed robust species groupings. Significant progress has been made in recent years to morphologically describe these clades; however, establishing taxonomically informative morphologies has proved surprisingly challenging. Combined with a long legacy of taxonomic confusion, the endeavor to tie phylogenetic clades to nomenclature has been a complicated and iterative process. Here, we review the taxonomy of the family Molidae (generically termed molids) in light of recent phylogenetic advances. Concurrently, we investigate whether one of the important morphological traits of the recently re-described *Mola alexandrini* (Ranzani

[1] University of California Los Angeles, Department of Ecology and Evolutionary Biology, Los Angeles 90095 CA, USA.
[2] Joint Institute for Marine and Atmospheric Research, University of Hawai'i at Mānoa, Honolulu, HI 96822 USA; Pacific Islands Fisheries Science Center, Honolulu, HI 96818, USA.
Email; jw2@hawaii.edu
[3] Auckland War Memorial Museum Tamaki Paenga Hira, Natural Sciences, The Domain, Private Bag 92018, Victoria Street West, Auckland 1142, New Zealand.
Email: mnyegaard@aucklandmuseum.com
[4] Laboratorio de Biología Marina/Estación de Biología Marina del Estrecho (Ceuta),Universidad de Sevilla, Avda. Reina Mercedes 6, Sevilla 41012, Spain.
[5] Aussermattweg 22, 3532, Zäziwil, Switzerland.
Email: molamondfisch@gmail.com
[6] California Academy of Sciences, San Francisco, CA 94118, USA.
Email: tierneythys@gmail.com
* Corresponding author: ericcaldera@gmail.com

1839) [*sensu* Sawai et al. 2018, senior synonym for *M. ramsayi* (Giglioli 1883), common name "bump-head sunfish"], strictly comports with phylogenetic tree topologies, or if it can also occur in *M. mola*, as recently reported (Wirtz and Biscoito 2019, Nyegaard et al. 2018b, Sawai et al. 2018 and L. Kubicek, personal observation). We also discuss the variety of molecular genetic tools available for informing the evolutionary ecology of ocean sunfishes. From this vantage point, we examine recent evolutionary and developmental genetic patterns of natural selection within the recently sequenced *M. mola* genome and suggest directions for future research.

Taxonomy: From Morphology to Mitochondrial DNA

Substantial progress has been made in the use of molecular genetic sequences to resolve the fish Tree of Life (fishtreeoflife.org; Betancur-R et al. 2017). While lower-level taxonomy in recent years has typically relied on phylogenetic analysis to support classic morphology-based taxonomy, the two most frequently used sources for classification used by ichthyologists lack a global high-level phylogenetic framework (JS Nelson's volumes of *Fishes of the World* and W. Eschmeyer's *Catalog of Fishes*). Consequently, much of fish classification continues to lean heavily on morphological characters to determine higher level groups. Long before the discoveries of Mendelian inheritance and deoxyribonucleic acid (DNA), both Linnaeus and Darwin acknowledged that grouping organisms based on appearance alone would likely ignore a type of shared history (Linnaeus 1751, Darwin 1859), even though neither could have predicted that DNA would be the biological material to uncover such evolutionary relationships. The incorporation of recent phylogenetic studies into the classification of bony fishes continues to reveal new lineages and clarify taxonomic confusion (Betancur-R et al. 2017). Indeed, the Molidae story is no different, although the taxonomic history is particularly convoluted (Fraser-Brunner 1951, Nyegaard et al. 2018b, Sawai et al. 2018). In the past twenty years (2000–2020), phylogenetic studies have clarified the evolutionary relationships among genera within the Molidae family and identified an undescribed species within the *Mola* genus. As we note in the next section, the vast growth spectrum of the large species of sunfishes, combined with a number of variable morphological characters, continue to make species-level taxonomy challenging.

Molidae Genera

The Molidae family resides within the Tetraodontiformes, an order composed primarily of reef-dwelling puffer and filefishes (Matsuura 2015). The Tetraodontiformes group gets its name from the four (tetra) teeth (odont) that comprise the beak-like mouths of its members, including pufferfish (Diodontidae), triggerfish (Balistidae), boxfish (Ostraciidae) and molids (Molidae). In the first comprehensive review of the molids, Fraser-Brunner (1951) recognized three genera (*Ranzania* Nardo 1840, *Masturus* Gill 1884, and *Mola* Koelreuter 1766). These genera correspond to the common names slender sunfish, sharptail sunfish and ocean sunfish, respectively. Fraser-Brunner (1951), and subsequently Tyler (1980), hypothesized that *Mola* and *Masturus* were more closely related to each other than *Ranzania* based on similarity in morphology. This hypothesis was further supported by a cladistic study utilizing 48 osteological characters (Santini and Tyler 2002). To determine whether these groupings based on morphology were consistent with the molecular genetic evolutionary history of the Molidae, Yamanoue et al. (2004) sequenced the mtDNA genomes of *R. laevis*, *Ma. lanceolatus* and *M. alexandrini*. Note the latter was reported as *M. mola* in Yamanoue et al. (2004), but later found to be a *M. alexandrini* (Yoshita et al. 2009, Sawai et al. 2018), and acquired the mtDNA genomes of three Tetraodontiformes (*Sufflamen fraenataum*, *Stephanolepis cirrhifer* and *Takifugu rubripes*), using a caproid (*Antigonia capros*) as an outgroup for rooting the tree topology. Phylogenetic reconstructions of the mtDNA genomes demonstrated that *Masturus* and *Mola* were in fact sister genera, and that *Ranzania* was basal to the *Masturus*-*Mola* clade with high statistical support from Bayesian posterior probabilities, Maximum Likelihood Bootstrap support and Maximum Parsimony scores (Yamanoue et al. 2004). This study was particularly useful as it demonstrated that the morphology-based taxonomy of the Molidae family, at that time, was consistent with molecular phylogenetics, albeit at the level of genera. While Yamanoue et al. (2004) could not resolve the Tetraodontiformes lineage sister to the Molidae, the recent update to the phylogenetic classification of bony fishes suggests that the Balistoidei sub-order may be sister to the ocean sunfishes (Betancur-R et al. 2017). Yamanoue et al. (2004) also confirmed the placement of the Molidae within the order Tetraodontiformes in addition to providing mtDNA genomic comparisons to the teleosts.

The number (37) and arrangement of genes within the mito-genome of the molids is the same as that of other less specialized teleosts (Yamanoue et al. 2004). Gene lengths are also similar to other teleosts, including the large and small rRNA genes, 13 protein-coding genes, and 22 tRNA genes (Yamanoue et al. 2004). The majority of genes (29 of 37) are coded on the heavy strand, also a feature typical of teleosts (Boore 1999). The mitochondria's circular genome origin of replication is located in the control-region, which is also referred to as the *d-loop* region because it forms a displacement loop structure when two strands of mtDNA are temporarily held open by a separately replicated, single-strand of DNA. The hypervariable nature of this non-coding region is evolutionarily informative (Kocher and Stepien 1997) and is commonly used in phylogenetic studies. The *d-loop* region is 808 base pairs (bp) in *R. laevis*, 810 bp in *Ma. lanceolatus* and 817 bp in *M. alexandrini* and varies similarly with the overall size of the mitochondrial genome (16,478 bp in *R. laevis*, 16,481 bp in *Ma. lanceolatus*, 16,488 bp in *M. alexandrini*), indicating that the molid mitochondrial genome has slowly continued to expand since diverging from its pufferfish ancestors (Yamanoue et al. 2004).

Mola species

A study by Bass et al. (2005) drew upon the phylogenetically informative genetic variation of the control-region and used *d-loop* sequence fragments to conduct an important initial investigation into the phylogenetics of the Molidae family that incorporated broad geographic sampling. This previous study confirmed the basal positioning of *R. laevis* and also showed that *Ma. lanceolatus* samples (n = 4) formed a single species clade, even though the sample locations ranged from Florida to Taiwan and Indonesia. The status of *Ma. lanceolatus* as a monotypic taxon was later confirmed by Yoshita et al. (2009) and Nyegaard et al. (2018a), with an expanded sampling area now also encompassing Australia (Pacific and Indian ocean basins, and Timor Sea), Japan and the tropical North Pacific. While Fraser-Brunner's (1951) retention of the three Molidae genera (*Ranzania, Masturus,* and *Mola*), has held up to phylogenetic analysis, the finding of a single *Masturus* species clade conflicts with his retention of two species within the *Masturus* genus. Phylogenetic analysis has since revealed that at least one of the species-specific characters proposed by Fraser-Brunner (1951) appears to be variable intraspecifically. More precisely, the clavus peduncle of two *Masturus* specimens (Fig. 1), clustering within the same putative species clade in Nyegaard et al. (2018a), superficially resemble those of *Ma. lanceolatus* and *Ma. oxyuropterus*, respectively, as Fraser-Brunner (1951) described and depicted.

For the analysis of *Mola* spp., Bass et al. (2005) inferred the links between phylogenetic clades and nomenclature, as they did not include morphological data in their study. Within *M. mola*, the authors found two robust clades that corresponded to the Pacific and Atlantic ocean basins (n = 5 per basin), suggesting a degree of geographic isolation to gene flow. Similarly, *M. ramsayi* (now *M. alexandrini*) formed two clades that corresponded to the Pacific and Atlantic Ocean basins (n = 2 per basin). Interestingly, the putative *M. ramsayi* Atlantic clade had an increased substitution rate in comparison to the remaining three clades. This clade was subsequently shown to differ at the species level (see below for details).

In later studies, which included both phylogenetics and morphology, it became somewhat problematic to match up the morphology of genetic clades in *Mola* with Fraser-Brunner's (1951) species descriptions (Sagara et al. 2005, Yoshita et al. 2009). Consequently, some researchers adopted the interim labels of *Mola* group or species A, B, and C (for a review see Sawai et al. 2020 [Chapter 2])

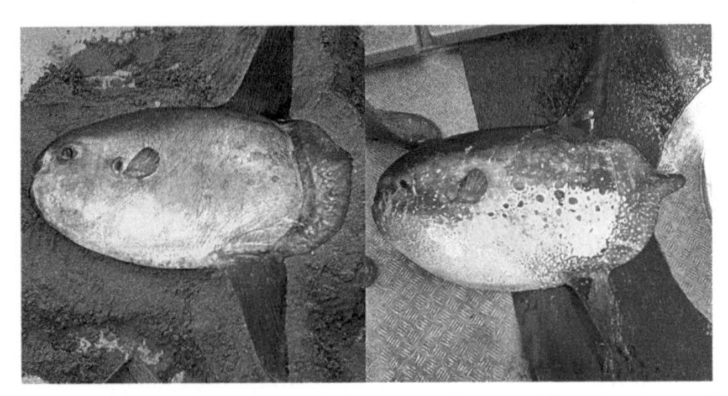

Figure 1. Two specimens with differing clavus peduncles, superficially resembling the short and long clavus peduncles of Fraser-Brunner's (1951) *Masturus oxyuropterus* and *Ma. lanceolatus*, respectively. Left: NTM S.15520-001, 106.5 cm TL, right: E02, 98 cm TL, both clustering in the putative *Ma. lanceolatus* d-loop species clade in Nyegaard et al. 2018a. Photographed by Gavin Dally, Museum and Art Gallery of the Northern Territory (left) and an Australian Fisheries Management Authority fisheries observer (right), reprinted with permission.

which corresponded to statistically robust phylogenetic clades in the *d-loop* region (Sagara et al. 2005, Yoshita et al. 2009, Yamanoue et al. 2010, Sawai et al. 2017). While Yoshita et al. (2009) established that the morphology of *Mola* species B corresponded well with the description of *M. mola* of Fraser-Brunner (1951), *Mola* A was only tentatively tied to *M. ramsayi* because the observed morphology of *Mola* A did not entirely correspond with the description of *M. ramsayi* by Fraser-Brunner (1951) (Sagara et al. 2005, Yoshita et al. 2009, Yamanoue et al. 2010, Sawai et al. 2017). As no morphological information was available at the time for *Mola* C, it was not possible to resolve the nomenclature further.

Subsequently, researchers resumed the task of verifying the proposed link between *Mola* clade B and *M. mola*, and definitively tying *Mola* A and C to nomenclature, a task greatly complicated by the long legacy of taxonomic confusion within Molidae, dating back several centuries (reviewed in Sawai et al. 2020 [Chapter 2]). In 2017, based on phylogenetic and morphologic analysis, combined with an extensive review of 37 nominal *Mola* species from the extensive Molidae taxonomic legacy, Nyegaard et al. (2018b) proposed that *Mola* C was an undescribed species, and named it *M. tecta*—the first new addition to the *Mola* genus in over 125 years. This new species' Latin name (from 'tectus' meaning hidden or disguised) and its common name (hoodwinker) derive from the fact that it evaded taxonomic identification for so long, despite "hiding in broad daylight." In fact, the putative Atlantic basin *M. ramsayi* clade samples (South Africa, accession numbers: AY940816, AY940826) in Bass et al. (2005) were shown to cluster with *Mola* C/*M. tecta*. One possible reason for this deception is *M. tecta's* lack of bump-head and bump-chin characters (more prominent in larger/older *M. mola* and *M. alexandrini*) in combination with the possibly smaller average size of *M. tecta* compared to *M. mola* and *M. alexandrini* may contribute to observational misidentification of mature *M. tecta* individuals as either *M. mola* and/or *M. alexandrini* in earlier developmental stages (e.g., see Fig. 1 in Hays et al. 2020 [Chapter 15]).

A major challenge in describing the morphology of *Mola* species has been, and still is, the vast growth spectrum of the sunfishes, over which they change their body morphology drastically (Sawai et al. 2020 [Chapter 2], Thys et al. 2020 [Chapter 7]). Much of the modern taxonomic work builds on Fraser-Brunner (1951), who discussed a variety of morphological characters that distinguish *M. alexandrini* and *M. mola*. While some characters have largely stood the test of time, such as fin ray and ossicle counts, other characters range in subtlety and overlap between the species and become more prominent with size. One particularly dichotomous character is the clavus, the structure that replaced their ancestral caudal fin over evolutionary history and gives the ocean sunfish its truncated shape. However, the general body morphology is more challenging to anchor.

Fraser-Brunner (1951), Yoshita et al. (2009) and Sawai et al. (2018) proposed that *M. alexandrini* does not have a wavy clavus, whereas in *M. mola* the clavus becomes wavy with growth (Fig. 2, A

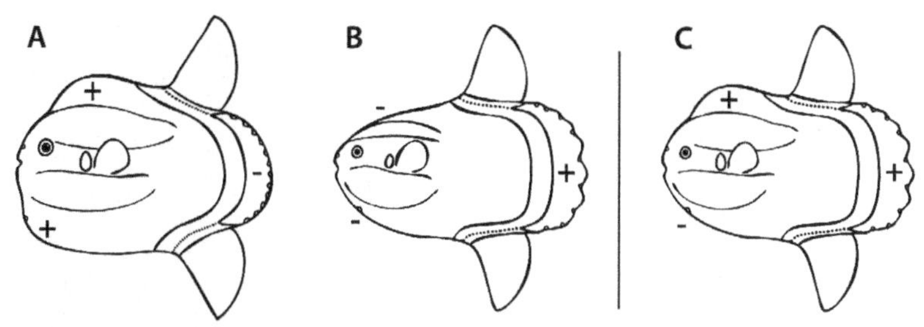

Figure 2. Proposed morphological features of *Mola alexandrini* (A) and *Mola mola* (B) (adapted from Sawai et al. 2018). (C): morphological features of Ceuta, Spain molids. +/– indicates presence/absence of a bump head, bump chin, and wavy clavus.

and B). However, it is worth noting that this character on its own is unreliable, as smaller specimens mostly have a smooth clavus (Sawai et al. 2020 [Chapter 2]). In addition, the clavus can take on abnormal morphologies due to congenital malformation or damage from predation, parasites, and other anthropogenic factors (Sawai et al. 2009, Nyegaard et al. 2018b, 2019).

Sawai et al. (2018) further proposed that the presence of a bump on the head of large individuals is a distinguishing character for *M. alexandrini*. The bump-head length is measured by first envisioning a line connecting the snout tip to the base of the dorsal fin, then measuring the greatest perpendicular distance from said line to the outermost edge of the head (Yoshita et al. 2009). Capitalizing on this unique "bump-head" character Sawai et al. (2018) proposed that *M. alexandrini* should be assigned the new English name (i.e., common name) "the bump-head sunfish", despite some *M. mola*, as mentioned earlier, may also exhibit this trait (see also Nyegaard et al. 2018b).

A problematic issue with emphasizing the bump-head feature of *M. alexandrini* as its common name, however, is that common names can bias species identification, and potentially impact monitoring efforts (Sarasa et al. 2012, Duckworth and Pine 2003). This name has since been reconsidered in Sawai et al. (2020 [Chapter 2]).

In an effort to elucidate the bump-head problem, here we review a population of molids off the coast of Ceuta, Spain, which robustly presents both the *M. alexandrini*-associated bump-head and the *M. mola*-associated wavy clavus (Fig. 2 A–C). Hereafter, such bump-head, wavy clavus individuals are referred to as "BH-WC". To determine whether these BH-WC individuals cluster with phylogenetic clade A or B (*M. alexandrini* or *M. mola*), we sequenced the *d-loop* region and made a phylogenetic reconstruction following both amplification and Bayesian/Maximum Likelihood tree construction methods similar to Bass et al. (2005), but with slightly modified PCR master mix reagents and a hot-start thermocycler profile. BH-WC individuals (ranging from 36–195 cm TL, 2.2–324.6 kg) were sampled from Ceuta, Spain in June of 2019. In addition, we included individuals putatively identified as *Mola* A from the Galápagos Islands, and off the coasts of mainland Ecuador and Peru to determine species clade placement. We also sequenced an individual from Esperance, Western Australia, that washed up on the beach in 2008. To facilitate the identification of clades A, B and C, we included previously published representative sequences of clade A (*M. ramsayi/M. alexandrini)*, clade B (*M. mola*) and clade C (*M. tecta*) from GENBANK, the NIH database repository and annotated collections of all publicly available DNA sequences.

Our phylogenetic reconstruction placed the BH-WC individuals within *M. mola* and not within the anticipated *M. alexandrini* (Fig. 3). Importantly, this highlights that the bump-head trait is not appropriate to use in isolation for the identification of *M. alexandrini* and must always be used in combination with other traits. The presence of a bump-head in our specimens here may be due to a number of factors; perhaps only older *M. mola* individuals develop it, or it might be a variable trait within species (but persists at a lower frequency in *M. mola*), or perhaps it is a trait associated with sexual dimorphism. Our results suggest that the bump-head trait is synapomorphic to *M. alexandrini* and *M. mola* and is therefore inappropriate to use in isolation for species-level differentiation within the genus. Further, this highlights the importance of continuing phylogenetic and morphological

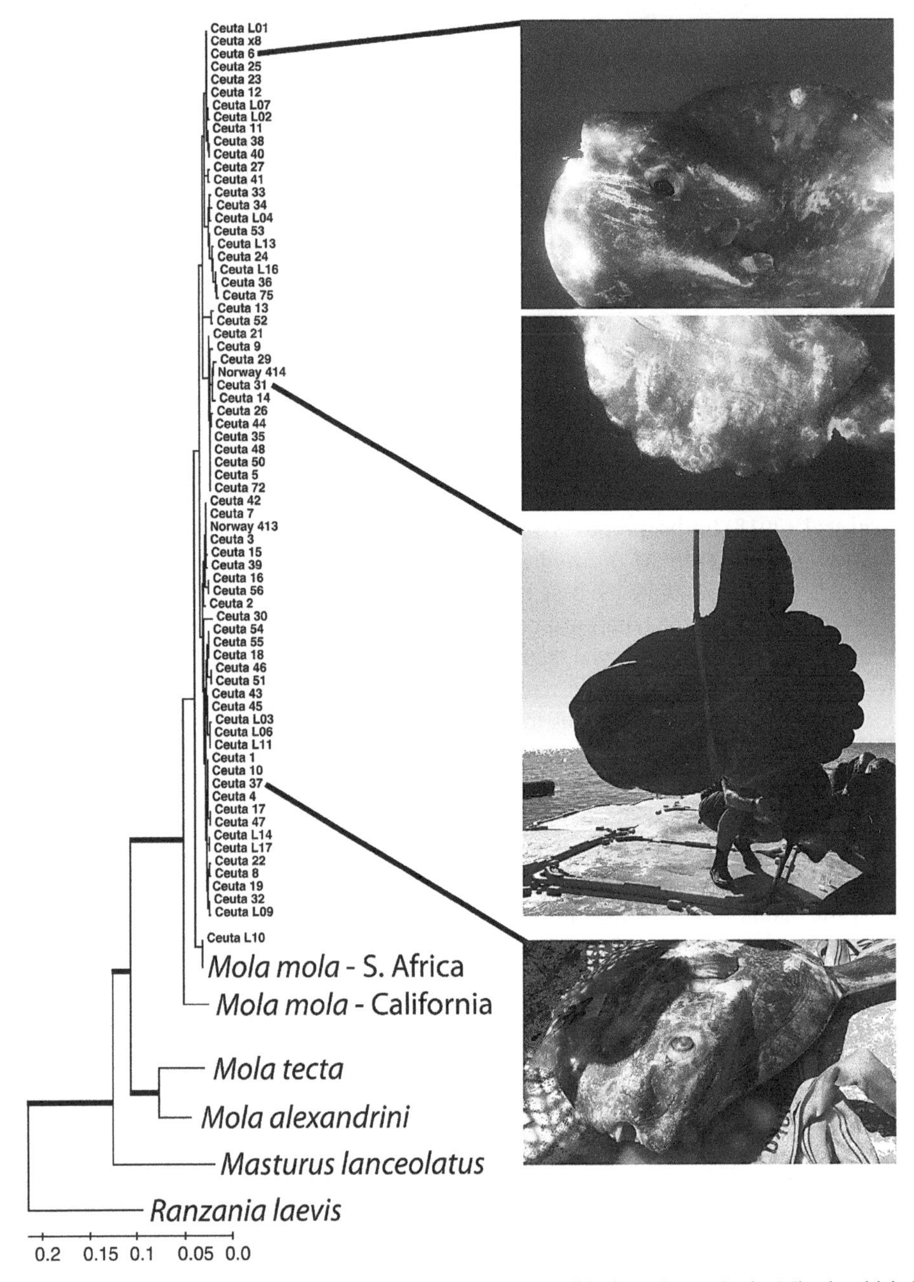

Figure 3. Phylogenetic identification of the bump-head, wavy clavus molids from Ceuta, Spain. Mitochondrial *d-loop*, maximum likelihood phylogeny (scale bar indicates substitutions per site), including photographs of three individuals weighing 23 kg, 325 kg, and 16 kg (top to bottom, respectively). Thick lines (on cladogram) indicate clades with Bayesian posterior probabilities > 0.8 and ML bootstrap support > 70%. Photos taken and used with permission by Lukas Kubicek.

research across the entire size spectrum of all species across their distribution ranges, to improve species keys that may be applied reliably to all (or most) size ranges. These observations suggest that associating the simpler English name of 'giant ocean sunfish' with *M. alexandrini* might serve as a less-confusing common name (Sawai et al. 2020 [Chapter 2]).

In our analysis, individuals from Peru clustered almost exclusively with *M. tecta* (clade C), with only one individual being placed with *M. alexandrini* (clade A; Fig. 4). Thus, these data suggest the

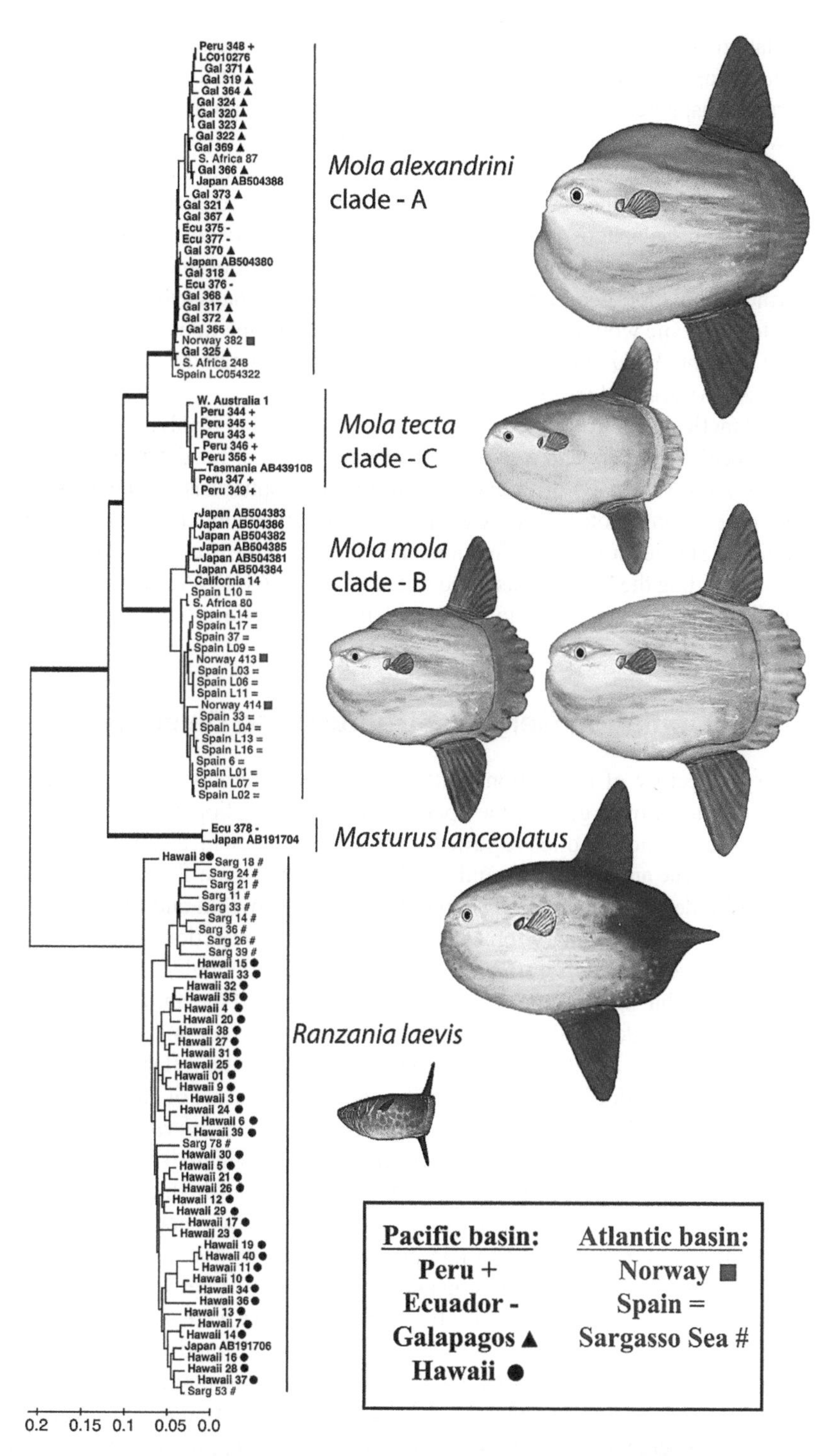

Figure 4. Maximum likelihood phylogenetic placement of molids spanning the Pacific (black) and Atlantic (gray) ocean basins using mitochondrial *d-loop* sequences (scale bar indicates substitutions per site, symbols emphasize population origin). Mature individuals from Peru, mainland Ecuador and the Galápagos Islands were putatively identified as *Mola alexandrini*. Larval individuals from Hawai'i and the Sargasso Sea were putatively identified as *Ranzania laevis*. Morphological identifications of Norwegian samples were confirmed. Spain samples (near Ceuta) are a subset of Fig. 3. Representative *Ranzania*, *Masturus*, and *Mola* (species clades A, B, C) GENBANK accession numbers included. Thick lines indicate clades with Bayesian posterior probabilities > 0.8 and ML bootstrap support > 70%. Drawing of molids by Jamie Watt.

Peruvian population sampled indeed belongs to the newly described *M. tecta*, confirming the core distribution range proposed by Nyegaard et al. (2018b) as widely distributed in the temperate waters of the Southern Hemisphere. The *M. tecta* (clade C) distinction is phylogenetically quite robust, in part because we observed that all *M. tecta* share a 27 bp insertion that is missing in the A and B clades. Interestingly, our single specimen from Esperance turned out to be *M. tecta*, making this a first confirmed record of this species in Western Australia (Nyegaard et al. 2018a). The majority of mainland Ecuador and Galápagos Islands individuals were *M. alexandrini* (clade A). We did not observe any phylogeographic structuring between mainland Ecuador and Galápagos in the *d-loop* region, but further population-geographic structuring analysis, perhaps using variable markers from the diploid nuclear genome would be more appropriate for capturing population-scale variation. We also discovered one non-*Mola* individual in our dataset, from mainland Ecuador, that turned out to be sharptail sunfish, *Masturus lanceolatus* (Fig. 4, sample 378).

Finally, one practical genetic tool to emerge from the problem of resolving *Mola* species was the development of specific primers and a PCR multiplexing protocol for determining *M. alexandrini* (clade A) and *M. mola* (clade B) differences from gel electrophoresis fragment patterns. This is both a cost and time-saving protocol as it does not require sequencing amplicons (Yamanoue et al. 2010). Furthermore, species-level identifications among *R. laevis*, *Ma. lanceolatus*, and *Mola* spp. can be made by barcoding the highly conserved mitochondrial 16s rRNA gene, which is particularly useful for small and/or damaged larvae, where identification via morphology is not an option (Hellenbrecht et al. 2019).

Population Genetic Structure and Diversity

The evolutionary trajectory of populations within a species is largely determined by the genetic structuring of individuals over geographic space. For example, allopatric speciation (divergence caused by geographic isolation) is often regarded as the most common mechanism of speciation (Mayr 1963, Lande 1980, Coyne and Orr 2004), and varying amounts of gene-flow among populations can either facilitate or hinder local adaptation (Lenormand 2002, Kawecki and Ebert 2004). Moreover, the relative impact of the evolutionary forces of genetic drift and selection are greatly influenced by the standing genetic diversity and N_e (effective population size) of populations. Thus, understanding the genetic and geographic structuring of species populations, and the amount of genetic diversity they harbor, is critical to understanding both their evolutionary history and future trajectory.

The study by Yoshita et al. (2009) used the variation held in the mtDNA *d-loop* for assessing sequence divergence, nucleotide/haplotype diversity, and F_{ST} (a measure of population differentiation due to genetic structure) in *Mola* B (now *M. mola*) and *Mola* A (now *M. alexandrini*) occurring around the coast of Japan. The F_{ST} between *M. mola* and *M. alexandrini* was high (0.835). Among populations within a species, F_{ST} typically ranges from 0 to 0.3, with 0.3 representing high population differentiation, so the high F_{ST} (0.835) indicates clear genetic isolation between the two species. Sequence divergence was substantially higher between this pair (8.35 percent) than within *M. alexandrini* (1.00 percent) or *M. mola* (1.63 percent). Sequence divergence was higher still between the *Mola* genus and *Ma. lanceolatus* (12.3 percent) and *R. laevis* (16.1 percent). There were similar levels of haplotype diversity in *M. alexandrini* and *M. mola* (1.00 ± 0.016, 0.991 ± 0.003) and nucleotide diversity (0.016 ± 0.008, 0.019 ± 0.009), respectively, indicating that each species has maintained a reasonably stable effective population size and standing genetic variation. In contrast, *M. alexandrini* and *M. mola* had substantially different degrees of geographic structure. All *M. mola* F_{ST} values among the Pacific coast of Japan, Sea of Japan, and North Pacific ocean basin were not significant, and the lack of geographic structure in North Pacific *M. mola* is consistent with their supposed distribution being linked to the Kuroshio Current. Conversely, *M. alexandrini* F_{ST} between the Pacific coast of northeastern Japan and subtropical waters was substantial (0.209, $P < 0.001$). However, further research by Nyegaard et al. (2018a) did not find apparent spatial or thermal separation between these two *M. alexandrini* groups off west coast Australia versus east coast Australia and northern New Zealand, and further research is needed to resolve the driver of the apparent genetic structuring within this species.

Phylogeographic Patterns in *Mola, Ranzania,* and *Masturus*

In light of several phylogenetic analyses (including new results presented here), the phylogeographic patterns of the three *Mola* species are becoming clearer (Yamanoue et al. 2004, Bass et al. 2005, Sagara et al. 2005, Yoshita et al. 2009, Thys et al. 2013, Ahuir-Baraja et al. 2017, Sawai et al. 2017, Sawai et al. 2018, Nyegaard et al. 2018a,b). Based on the records of individuals identified genetically, *M. alexandrini* appears to have a wide distribution in the northern and southern hemispheres in both temperate and tropical areas of the Pacific, Atlantic, and Indian ocean basins (Fig. 5A). In the Atlantic, genetically identified individuals are recorded in the North Sea (as far north as Norway,

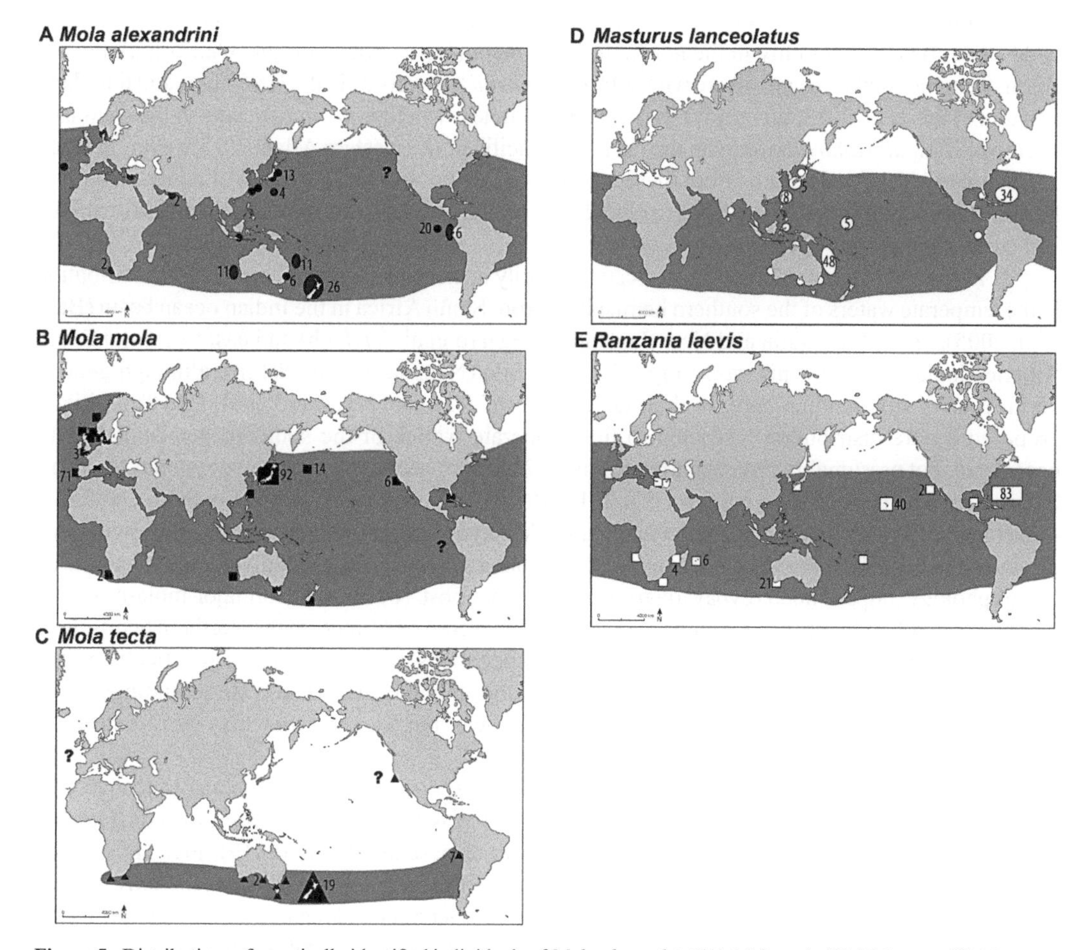

Figure 5. Distributions of genetically identified individuals of *Mola alexandrini* (A) *Mola mola* (B) *Mola tecta* (C) *Masturus lanceolatus* (D), and *Ranzania laevis* (E). Black and white circles, squares and triangles indicate point records of genetically identified specimens for each species, as confirmed by the literature and this study. The dark gray shaded areas represent preliminary distributional ranges drawn manually based on geographic extent of genetically and morphologically identified specimens. Question marks highlight areas of uncertainty with regards to the distribution, as mentioned in the main text (Yamanoue et al. 2004, Bass et al. 2005, Sagara et al. 2005, Yoshita et al. 2009, Yamanoue et al. 2010, Zhang and Hanner 2011, Landi et al. 2014, Prakash et al. 2016, Steinke et al. 2016, Ahuir-Baraja et al. 2017, Sawai et al. 2017, Nyegaard et al. 2018ab, Hellenbrecht et al. 2019, Knorrn 2019). Unpublished data from: Australian Museum, Sydney (AMS), J. Bergsten, E.J. Caldera, S.A. James, L. Johnson, E. Keskin, J.C. Mangel, M. Nyegaard, L.K. Phan, B. Rinkevich, South African Institute for Aquatic Biodiversity (SAIAB), T.M. Thys, T.T.V. Tran, M. Valdez-Moreno, and J.L. Whitney).

this study), Mediterranean and South Africa and into the Indian ocean basin including Oman. In the Pacific, genetically confirmed occurrences are widespread in the Indo-West Pacific (from Australia to Japan), but in the Eastern Pacific are currently limited to the Galápagos and coastal Ecuador (Fig. 5).

Mola mola (clade B, Fig. 5B) is globally distributed in both temperate and tropical areas of the Pacific, Atlantic, and Indian ocean basins (Yamanoue et al. 2004, Bass et al. 2005, Sagara et al. 2005, Yoshita et al. 2009, Thys et al. 2013, Ahuir-Baraja et al. 2017, Sawai et al. 2017, Sawai et al. 2018, Nyegaard et al. 2018a,b), although they appear to be relatively rare off Australia and New Zealand (Nyegaard et al. 2018a). In the Atlantic, genetically identified individuals are recorded as far north as Sweden in the North Sea, into the Mediterranean, south to South Africa and west to Florida (Fig. 5B). Phylogenetic relationships based on *d-loop* sequences (Bass et al. 2005, Sawai et al. 2017, Nyegaard et al. 2018a, this study) suggest there are two genetically distinct populations in the North Pacific (Japan to California) and Atlantic (from the North Sea to South Africa) and possibly across the Indian ocean basin to Western Australia. Overall *M. mola* and *M. alexandrini* have similar widespread distributions, with a couple of notable exceptions. In the central and east Pacific, genetic confirmations of *M. mola* to date occur only in the northern hemisphere (Japan, Aleutians and California), but are absent from the southern hemisphere where *M. alexandrini* has records in the Galápagos and mainland Ecuador (Thys et al. 2013, this study). It is currently unclear whether *M. mola* occurs in the southeast Pacific or if *M. alexandrini* occurs in the northeast Pacific or the western Atlantic. However, the lack of genetically identified records from these locations could possibly be the result of under-sampling. Further genetic surveys will be needed to elucidate the precise distribution of these broadly distributed and co-occurring *Mola* species.

M. tecta (clade C, Fig. 5C) appears to occur primarily in the Indo-Pacific. There is a core distribution in the temperate waters of the southern hemisphere from South Africa in the Indian ocean basin (Bass et al. 2005), around Australia and New Zealand (Nyegaard et al. 2018a,b) and east to coastal South America including Peru (this work; Fig. 5C) and Chile (Nyegaard et al. 2018b), although genetic verification is still pending here. In their description of *M. tecta*, Nyegaard et al. (2018b) initially proposed a core distribution restricted to the temperate waters of the southern hemisphere with occasional, but presumably rare, occurrences in the Northern hemisphere. Such occasional Northern hemisphere occurrences were confirmed in 2019 where a *M. tecta* washed up in Southern California (Leachman 2019) and was confirmed genetically (M. Nyegaard, unpublished data). Following this discovery, several *M. tecta* have been reported in Monterey Bay by SCUBA divers, identified based on morphology (https://montereybayaquarium.tumblr.com/post/187517538903/major-mola-moment-first-confirmed-hoodwinker). Currently, there is only one record of *M. tecta* in the Northern hemisphere Atlantic from 1889 (Nyegaard et al. 2018b), but two specimens have been verified genetically off the tip of South Africa at the confluence of the Atlantic and Indian ocean basins. These findings prove that *M. tecta* are not restricted to the southern hemisphere, but suggests they may be rare north of the equator.

Considering the phylogeographic patterns of *Mola* species, it seems reasonable to question the monotypic status of the circumglobal *Ranzania* and *Masturus* genera. *R. laevis* is considered a single cosmopolitan species (i.e., temperate to tropical, Froese and Pauly (2019), Fig. 5D and 5E) however, no phylogeographic study has ever been completed on the genus. To date, the only genetic comparisons of any *Ranzania* are limited to preliminary accounts. Nyegaard et al. (2018b) found little variation in COI (cytochrome oxidase I) among 12 individuals sampled across a wide range including North Atlantic, South Africa, and Hawai'i. In contrast, Yoshita et al. (2009), reported a 5 percent sequence divergence in *d-loop* (n = 2) between an individual from Japan and another from an unknown location (Yamanoue et al. 2004). Could there be a cryptic species of *Ranzania* in the North Pacific or elsewhere? Interestingly, Fraser-Brunner (1951) proposed two subspecies of *Ranzania laevis* based on several morphological differences between ocean basins (*R. laevis laevis* in the Atlantic and *R. laevis makua* in the North Pacific), with intermediate forms in the Indian Ocean. Such patterns of a morphological continuum across a geographic gradient are suggestive of potential reductions in gene flow across ocean basins, potentially similar to that found within *Mola* clades. Recently,

Gaither et al. (2016) reviewed how molecular evidence has revealed cryptic species complexes in several circumtropical taxa. A thorough phylogeographic survey of *Ranzania* across all ocean basins and regions is certainly warranted and could reveal yet undetected cryptic diversity in this lineage.

Similarly, *Ma. lanceolatus* is considered a single circumtropical species (Froese and Pauly 2019) (Fig. 5D). However, no comprehensive phylogeographic survey has been completed. The only published phylogeographic analyses on *Masturus* to date were those of Bass et al. (2005), Yoshita et al. (2009), and Nyegaard et al. (2018b), where *Masturus* were essentially used as the outgroup. Bass et al. (2005) using *d-loop*, and Nyegaard et al. (2018b) using both *d-loop* and COI similarly observed that all individuals (n = 5 in each study) formed a single species clade and found no obvious structure between the Western Pacific (Japan, Taiwan, Indonesia) and Atlantic (Florida). In another study, Nyegaard et al. (2018a) compared 51 new *d-loop* sequences from *Ma. lanceolatus* sampled off Australia (east coast n = 49, west coast n = 1, and north coast n = 1) to 19 sequences from Japan, Taiwan and the tropical North Pacific (Yoshita et al. 2009, Yamanoue et al. 2010), and similarly found a single species clade with no obvious geographical structure. As mentioned earlier, this conflicts with the revision by Fraser-Brunner (1951) who was convinced that there were two morphological forms (*Ma. lanceolatus* and *Ma. oxyuropterus*). Although these preliminary analyses support a single species with no evidence of population structuring among ocean basins, there is an obvious need to explore the cohesiveness of this cosmopolitan species with more comprehensive sampling and informative genetic markers.

As discussed above, the mtDNA *d-loop* region has been a useful marker for resolving species-level differences, and also captures a degree of geographic distinction within species. But additional genetic markers more appropriate for examining population genetic structure are available. The haploid nature of the mitochondrial genome renders it less informative for population genetic structure analysis than diploid markers in the nuclear genome. Diploid loci can harness the power of assessing Hardy-Weinberg and linkage equilibrium to determine population genetic boundaries over geographic space without *a priori* information about sampling location using programs such as STRUCTURE (Pritchard et al. 2000, Falush et al. 2003). DNA microsatellites, which are composed of short tandem repeats of small base pair sequences (e.g., TCG, TCG, TCG, TCG, or TA, TA, TA..., etc.), are particularly useful markers for population genetics, given their propensity to vary due to slippage during replication. Visualizing microsatellite repeats is also cost-effective, as it requires only the determination of fragment-length differences, as opposed to sequencing base pairs. PCR reactions of microsatellites with non-overlapping lengths can be multiplexed, further reducing reagent/sequencing costs. Streelman et al. (2003) developed a suite of five microsatellite markers with high heterozygosity for *M. mola*, meaning they should be informative for population-level structuring within *Mola* species. These markers are particularly useful as they all share a similar di-repeat (CA) motif that simplifies fragment length scoring of chromatograms, which can become complicated when di, tri and/or other microsatellite motifs are intermingled. Future studies should take advantage of these informative, cost-effective markers. An additional genetic tool with great potential to inform molid species distributions and seasonality is that of environmental DNA, often referred to as eDNA (Taberlet et al. 2012). eDNA work involves amplifying and sequencing fish DNA sequences from seawater in combination with bioinformatic processing (e.g., Curd et al. 2019) to determine species presence/absence. eDNA also has an important utility of reducing sampling biases, as the technique negates the need for expertise in morphological identification or a mechanism to capture fish.

One of the most popular metabarcoding PCR primer sets for marine fishes (MiFish) captures a 163–185 bp region of the 12s rRNA gene (Miya et al. 2015). The small fragment length of MiFish allows for high-throughput, next generation sequencing technology. Here, we barcoded *M. alexandrini*, *M. mola*, *M. tecta*, *R. laevis* and *Ma. lanceolatus* using MiFish to determine whether the marker holds the appropriate genetic variation to separate out the species. Indeed, each species possessed SNP level differences (ranging from one to six), indicating that this single barcode is adequate for capturing molid species diversity using eDNA. It is worth noting that the least genetic variability occurred between *M. alexandrini and M. tecta*—only one fixed SNP level difference was observed.

Moving forward, these MiFish barcodes will help resolve database underrepresentation, which can lead to biases in species identification. This is particularly important in light of the 2018 *M. tecta* and *M. alexandrini* descriptions (Nyegaard et al. 2018b, Sawai et al. 2018). For example, mtDNA *COI* (cytochrome *c* oxidase 1) barcodes in the commonly used BOLD database (Barcode of Life Database; www.boldsystems.org) contain no representative *M. alexandrini* or *M. tecta* sequences as of this writing (i.e., a Molidae eDNA study using this database would be biased against detecting *M. alexandrini* or *M. tecta*).

The Genome and Development of the *Mola mola*

The first fully sequenced molid genome was recently published by Pan et al. (2016), providing some important initial insights into the unique morphological development of the "natural mutant," *M. mola*. Extraordinary features of the *M. mola* include its truncated tail, fast captive growth rate, large body size, and spongy, largely un-ossified endoskeleton. To begin understanding the genomic determinants of these characteristics, putatively associated genes and pathways were compared against seven fish genomes spanning teleost fish diversity, including two other Tetraodontiformes (*Takifugu rubripes* aka Fugu, and *Tetraodon nigroviridis* aka Tetraodon).

Genome Size & Content

The *M. mola* genome was originally estimated to be 730 Mb using the k-mer method (Binghang et al. 2013) but was determined to be 642 Mb upon assembly. In short, the k-mer method estimates genome size independent of genome assembly by assessing how the frequency of multiple-length "mers," vary with genome length. Mers represent nucleotide base pair combination variants of a particular length (e.g., a 3-mer represents ATG, AGT, TGA, TAG, GTA, GAT). The Tetraodontiformes and Stickleback (*Gasterosteus aculeatus*) formed a monophyletic clade, and within this group, *M. mola* had the largest genome (Stickleback 447 Mb, Tetraodon 359 Mb, and Fugu 393 Mb), which suggested that the molid genome had expanded since diverging from the pufferfishes (Tetraodontidae) 68 million years ago (Near et al. 2013, Arcila et al. 2015). By contrast, all sister taxa to this clade had genomes over 800 Mb, including Spotted Gar (*Lepisosteus oculatus*) 869 Mb, Zebrafish (*Danio rerio*) 1,412 Mb, Tilapia (*Oreochromis niloticus)* 817 Mb, Medaka (*Oryzias latipes*) 869 Mb. The *M. mola* genome was composed of 11 percent repetitive sequences (transposable elements, tandem repeats, and simple-sequence repeats), which was similar to the genome content of Fugu.

With the fully sequenced *M. mola* genome, researchers can probe into the genetic underpinnings of unique molid morphological traits. One such trait is the large body weight that *Mola* and *Masturus* can achieve. *M. mola* was once touted as the world's heaviest bony fish, weighing 2.3 tons (Carwardine 1985)—this claim to fame is now held by *M. alexandrini* (Sawai et al. 2020 [Chapter 2]). This switch however is not without controversy (Thys et al. 2020 [Chapter 7]). Regardless of the record holder's species identity, the endoskeleton of *Mola* spp. are composed of spongy un-ossified material labeled as cartilage by past researchers (e.g., Cleland 1862). Bemis et al. 2020 [Chapter 4], however, questions this identification and mentions the need for further research. To better understand the genetic basis of this strange skeletal material, Pan et al. (2016) searched for secretory calcium-binding phosphoprotein (SCPP) genes. SCPP genes are reasonable candidates since they have a crucial role in vertebrate bone formation. Bony fishes through mammals contain multiple categories of SCPP genes (Kawasaki 2009) but they are absent in cartilaginous fishes (Kawasaki 2011). As such the absence or modification of these genes (compared with other bony fishes) could potentially explain the unossified skeleton in *M. mola*. For example, most bony vertebrates contain multiple categories of SCPP genes (Kawasaki 2009), but no SCPP genes have been identified in any of the cartilaginous chondrichthyans (sharks, skates and rays) to date (Kawasaki 2011).

Interestingly, *M. mola* does contain multiple SCPP genes, but two genes that are found in zebrafish and Fugu, are missing in *M. mola* (*fa93e10* and *scpp7*). In addition, the gene *scpp4* (present in Fugu and Medaka) is a pseudogene in *M. mola* due to a single base pair insertion, providing evidence that this gene lost its function after splitting from the pufferfish lineage. The functional consequence of losing *scpp4* is still unclear in teleosts. For example, *scpp4* has been lost in zebrafish (which have a bony skeleton), but their highly duplicated genome may compensate for this loss with lineage specific SCPP genes (Kawasaki 2009). Thus, the role of SCPP genes in forming the cartilaginous skeleton of *M. mola* remains unclear. But regulatory changes to the extracellular matrix could also play a role (see Patterns of Selection section in this chapter).

Perhaps the most uniquely defining characteristic of *M. mola* is its truncated posterior region called the clavus (Latin for rudder). HOX gene analyses were conducted by Pan et al. (2016) to characterize genomic determinants of this intriguing morphology. They hypothesized that loss of HOX genes would play a role in the truncated tail shape since these genes specify segmental identities along the anterior-posterior axis of developing vertebrate embryos and shape limb structure (Zakany and Duboule 2007). It does not appear, however, that HOX gene loss plays a role in the *M. mola* truncated shape. Instead, *M. mola* has more HOX clusters than Fugu in addition to possessing functional copies of HOX genes that are pseudogenes in Fugu. It remains possible that altered expression patterns of HOX genes and/or changes in *cis*-regulatory elements (non-coding DNA that regulates the transcription of neighboring genes) are responsible for the *M. mola* truncated phenotype.

Genetic Diversity

Among the *M. mola* genome, 489,800 heterozygous single nucleotide polymorphisms (SNPs) were identified, estimating heterozygosity at 0.78×10^{-3}, which is lower than other marine fishes (e.g., Whale Shark 0.9×10^{-3} and Atlantic cod 2.09×10^{-3}; Star et al. 2011, Marra et al. 2019), but greater than higher mammals (e.g., humans 0.69×10^{-3}, Wang et al. 2008), which have notoriously low effective population sizes, N_e. Briefly, heterozygosity and N_e are measures of the genetic diversity/variability held within populations. Pan et al. (2016) additionally inferred historical changes in *M. mola* N_e using a coalescent approach (Li and Durbin 2011). PSMC analysis suggested that there was an increase in N_e from ~ 3 to ~ 0.9 million years ago. This putative increase in N_e may have been aided by an asteroid strike approximately 2.15 mya that caused a massive marine extinction event (Barash 2011), possibly opening up habitats that facilitated *M. mola* population expansion. Similar divergences have occurred in triggerfish (Tetraodontiformes: Balistidae) genera (*Melichthys*, *Abalistes*, *Acanthaluteres*, *Thamnaconus*) within the past two million years (McCord and Westneat 2015). N_e is thought to have stopped expanding (decreasing slightly) around 0.9 mya, which could be related to the mid-Pleistocene climate transition (Hayward et al. 2007).

Patterns of Selection

Using a sub-set of five teleost (*M. mola*, Fugu, Tetraodon, Medaka, and Zebrafish) and a dataset of 10,660 homologous teleost genes, Pan et al. (2016) conducted tests to identify genes under positive selection. Briefly, positive selection is determined by examining the ratio of genetic non-synonymous substitutions (dN) to synonymous substitutions (dS) in protein coding sequences—an elevated dN/dS ratio implies positive selection. Using this approach, 1,117 genes (10.5 percent) were found to be under positive selection, and 1,067 of these were found to be evolving fastest in the *Mola* lineage compared to pairwise comparisons of all five other lineages. In particular, important hormones for regulating growth, growth hormone and insulin-like growth factor 1(GH/IGF-1) axis, were found to be under positive selection and fast evolving. The GH/IGF family of genes is of particular interest as it may contain genomic links to fast growth and massive body size in *Mola*. The GH/IGF-1 axis had several components: insulin-like growth factors (IGFs), IGF receptors (IGFRs), IGF-binding proteins (IGFBPs), growth hormone (GH), growth hormone receptor (GHR) and the insulin receptor (INRS). GH is released by the pituitary and subsequently IGF-1 is released by the liver as a result of GH stimulation. GH and IGF-1 impacts are realized via their downstream effects on GHR, IGFR and INSR, which results in an overall increase in growth and a decrease in differentiation or suppression of apoptosis (programmed cell death). Within the GH/IGF-1 axis, the following genes were found to have dN/dS ratios higher than background: *ghr1, igflra, ifglrb, grb2, akt3, irs2a.* Interestingly, positively selected sites within *igflra* and *ifglrb* occurred within functional domains of the receptors, suggesting that positive selection in these sites may have enhanced the function of these genes, thus contributing to the massive body size of the *M. mola* and their rapid captive growth rate capacity (e.g., gaining nearly 400 kg in 15 months as reported by Powell, 2001). The extracellular matrix (ECM) may play a role in the development of the strange *M. mola* skeleton, as this microenvironment provides bulk, shape and strength to bone and cartilage, as well as contains components required for the conversion of cartilage to bone and its homeostasis (Han 2011). Numerous ECM genes were observed with varying degrees of elevated dN/dS—the most interesting of which were two copies of COL2A1 encoding type II collagen (*col2a1a, col2a1b*), which normally represent approximately 80–90 percent of the collagen in the cartilage matrix. COL2A1, along with aforementioned genes represent promising candidates for comparative genomics among the other sunfish genera, *Masturus* and *Ranzania.*

Future Directions

Pan et al. (2016) provided the first genomic insights into a molid species by comparing the *M. mola* genome to that of other teleost fishes. However, additional population-genetic scale insights (such as geographic structure, gene-flow, hybridization, effective population size, local selection) can be gleaned from sequencing additional molid genomes. Indeed, the assembly of future genomes will be facilitated by scaffolding with the Pan et al. (2016) genome, thus allowing for less costly low-pass genome sequencing using next-generation sequencing technology. To date, no studies have examined the genetic expression patterns of any of the ocean sunfishes. It is worth mentioning, however, that obtaining a fully sequenced genome is the first step to conducting such studies. Moving forward, the sequenced *M. mola* genome could be used to identify candidate genes, whose function can be subsequently verified using transcriptomic approaches such as RNA-seq (Wang et al. 2009).

Unraveling the evolutionary history of the unique morphology of ocean sunfishes would be further benefited by comparative genomics within the family. Researchers should consider sequencing the nuclear genomes of *Ranzania*, *Masturus*, *M. tecta* and *M. alexandrini*. Finally, while next-generation sequencing technology has substantially reduced sequencing costs, assuring that several genomes and transcriptomes will continue to emerge (e.g., Bian et al. 2019), older, less expensive technologies such as Sanger/ABI sequencing used for obtaining mtDNA *d-loop* sequences and/or microsatellites are still useful tools for rapid species identification. Population tracking will remain useful for foundational research, fisheries management and conservation alike.

In summary, genetics can be used as a tool to not only plumb the evolutionary history of molids and unveil mechanisms underlying their strangely unique features, but also help researchers, aquarists and fisheries managers identify wild molids in the field by helping distinguish those features specific to each group. Phylogenetic studies have clarified the evolutionary relationships within the Molidae family and have been critical in the identification of new species within the *Mola* genus. Genetic evidence (combined with morphometrics) has been critical in helping to clear up the muddied taxonomic history, most notably in distinguishing the three *Mola* species and elucidating their geographic ranges. As more individuals are genotyped, we continue to gain a sharper picture of the delineations between species and populations. In the future, phylogeographic surveys with comprehensive sampling and genome-wide markers hold promise for revealing finer-scale patterns of gene flow and potentially cryptic diversity in the ocean sunfishes (summarized in Table 1).

Table 1. Genetic tools for molid discoveries.

	Genetic Tools	**Utility & Discoveries**
Past	• Mitochondrial genome(s) • *d-loop* mtDNA sequences • PCR/Gel electrophoresis	• Phylogenetic placement of Molidae family & evolutionary history of Molidae genera • Species-level resolution of taxa • Determine Mola A & B species clades (no sequencing required)
Present	• Nuclear genome (*Mola mola*) • Microsatellite DNA	• Developmental insights, candidate genes, comparative genomics, heterozygosity, effective population size • Population genetic structure & diversity (inexpensive management/ conservation monitoring)
Future	• Additional molid nuclear genome(s) • Low pass nuclear genomes • Transcriptome(s) • eDNA (environmental DNA)	• Evolutionary emergence of large body size, fecundity, cartilaginous endoskeleton, clavus, etc. • See above (cost-prohibitive deep sequencing not required) • Gene expression & confirmation of candidate genes • Species distributions and seasonality (no morphological identification bias, ease of sampling)

Acknowledgements

Special thanks to Steve Karl for initiating this genetic research and the Paul Barber lab at UCLA for continuing it. A global team of collaborators provided molid samples and contributed studies to Genbank. Marko Freese and Lea Hellenbrecht collected *Ranzania* larva in the Sargasso Sea, Anuschka Faucci, Bruce Mundy, and Jonathan Whitney collected Hawai'i *Ranzania* larvae, Pro Delphinus (Francisco Bernedo, Joanna Alfaro-Shigueto and Jeff Mangel) collected Peruvian *Mola* spp., Mike de Maine and staff from the Two Oceans Aquarium collected South African *Mola* sp., A. Hearn, J. Alvarez, K. Weng, J. Ryan, A. Dove and T. Thys collected Galápagos *Mola* sp., Lukas Kubicek and Enrique Ostalé collected Spanish and Mediterranean *Mola* sp., and Torkild Bakken collected Norwegian *Mola* sp. sample 382. Norway samples 413 and 414 were provided by Tromsø Museum, Helgeland museum. N. M. Skaggs and R. Orduna assisted with DNA extraction and amplification of mtDNA *d-loop* sequences. Z.J. Gold assisted with MiFish barcoding. We would also like to thank additional collectors who have generously contributed specimens for genetic analyses over the years including: D. Adams, Abderrahman Ali Ahmed, Joanna Alfaro-Shigueto, April Balwin, Clive Brown, Robin Horn Capfish, Steve Cook, Chuck Farwell, Ju Heerdero, Eddie Higgins, John Hyde, M. Irey, Laith Jawad, Kamagawa Sea World, Jess Lee, Álvaro Sabino Lorenzo, Richard Lord, Jeff Mangel, Annamaria Mariot, Tom Mason, Dana Morton, Mibu Murray, Franga Potter, Evgeny Romanov, I.P. Smith, Kim Smith, Arnold Suzimoto Tringali, and Adam Zandie. Finally, we express our gratitude to José Carlos García Gómez and José Manuel Ávila Ribera for supporting the Estación de Biología Marina del Estrecho (Ceuta) investigation. Galápagos genetic samples were obtained under the framework genetics contract MAE-DNB-CM-2016-0041. "The role of oceanic islets in the conservation of migratory marine species" and under research permit PC-69-16. Mainland Ecuador samples were obtained under the framework genetics contract MAE-DNB-CM-2016-0046-M-003.

References

Ahuir-Baraja, A.E., Y. Yamanoue and L. Kubicek. 2017. First confirmed record of *Mola* sp. A in the western Mediterranean Sea: morphological, molecular and parasitological findings. J. Fish Biol. 90(3): 1133–1141. doi: 10.1111/jfb.13247.

Arcila, D., R.A. Pyron, J.C. Tyler, G. Orti and R. Betancur-B. 2015. An evaluation of fossil tip-dating versus node-age calibrations in tetraodontiform fishes (Teleostei: Percomorphaceae). Mol. Phylogenet. Evol. 82: 131–145. doi: 10.1016/j.ympev.2014.10.011.

Barash, M.S. 2011. Environmental changes in the neogene and the biotic response. Oceanology 51(2): 306–314. doi: 10.1134/s0001437011020019.

Bass, A.L., H. Dewar, T. Thys, J.T. Streelman and S.A. Karl. 2005. Evolutionary divergence among lineages of the ocean sunfish family, Molidae (Tetraodontiformes). Mar. Biol. 148(2): 405–414. doi: 10.1007/s00227-005-0089-z.

Bemis, K.E., J.C. Tyler, E.J. Hilton and W.E. Bemis. 2020. Overview of the anatomy of ocean sunfishes (Molidae: Tetraodontiformes). pp. 55–71. *In*: T.M. Thys, G.C. Hays and J.D.R. Houghton [eds.]. The Ocean Sunfishes: Evolution, Biology and Conservation, CRC Press. Boca Raton, FL, USA.

Betancur-R, R., E.O. Wiley, G. Arratia, A. Acero, N. Bailly, M. Miya et al. 2017. Phylogenetic classification of bony fishes. BMC Evol. Biol. 17: 40. doi: 10.1186/s12862-017-0958-3.

Bian, C., Y. Huang, J. Li, X.X. You, Y.H. Yi, W. Ge et al. 2019. Divergence, evolution and adaptation in ray-finned fish genomes. Sci. China Life Sci. 62(8): 1003–1018. doi: 10.1007/s11427-018-9499-5.

Binghang, L., Y. Shi, J. Yuan, Y. Galaxy, H. Zhang, N. Li et al. 2013. Estimation of genomic characteristics by analyzing k-mer frequency in de novo genome projects.

Boore, J.L. 1999. Animal mitochondrial genomes. Nucleic Acids Res. 27(8): 1767–1780. doi: 10.1093/nar/27.8.1767.

Cawardine, M. 1985. Guinness Book of Animal Records. Guinness Publisher. Middlesex, UK.

Cleland, J. 1862. On the anatomy of the short sunfish (*Orthragoriscus mola*). Nat. Hist. Rev. 2: 170–185.

Coyne, J.A. and A.H. Orr. 2004. Speciation. Sunderland, Massachusetts: Sinauer Associates.

Curd, E.E., Z. Gold, G.S. Kandlikar, J. Gomer, M. Ogden, T. O'Connell et al. 2019. Anacapa Toolkit: An environmental DNA toolkit for processing multilocus metabarcode datasets. Methods Ecol. Evol. 10(9): 1469–1475. doi: 10.1111/2041-210x.13214.

Darwin, C. 1859. On the Origin of Species by Means of Natural Selection, Or the Preservation of Favoured Races in the Struggle for Life. London: John Murray, Albemarle Street.

Duckworth, J.W. and R.H. Pine. 2003. English names for a world list of mammals, exemplified by species of Indochina. Mamm. Rev. 33(2): 151–173. doi: 10.1046/j.1365-2907.2003.00012.x.

Falush, D., M. Stephens and J.K. Pritchard. 2003. Inference of population structure using multilocus genotype data: Linked loci and correlated allele frequencies. Genetics 164(4): 1567–1587.

Fraser-Brunner, A. 1951. The ocean sunfishes (family Molidae). Bull. Br. Mus. Nat. Hist. Zool. 1: 87–121.

Froese, R. and D. Pauly. 2019. FishBase. World Wide Web electronic publication.www.fishbase.org.

Gaither, M.R., B.W. Bowen, L.A. Rocha and J.C. Briggs. 2016. Fishes that rule the world: circumtropical distributions revisited. Fish Fish (Oxf)17(3): 664–679. doi: 10.1111/faf.12136.

Giglioli, H.H. 1883. Zoology at the fisheries exhibition. II—Notes on the vertebrata. Nature 28: 313–316.

Gill, H. 1884. Synopsis of the plectognath fishes. Proceedings of the United States National Museum. 7: 411–427.

Han, L., A.J. Grodzinsky and C. Ortiz. 2011. Nanomechanics of the Cartilage Extracellular Matrix. Annu. Rev. Mater. Sci. 41: 133–168. doi: 10.1146/annurev-matsci-062910-100431.

Hays, G.C., J.D.R. Houghton, T.M. Thys, D.H. Adams, A.E. Ahuir-Baraja, J. Alvarez et al. 2020. Unresolved questions about the ocean sunfishes, Molidae—a family comprising some of the world's largest teleosts. pp. 280–296. *In*: T.M. Thys, G.C. Hays and J.D.R. Houghton [eds.]. The Ocean Sunfishes: Evolution, Biology and Conservation, CRC Press. Boca Raton, FL, USA.

Hayward, B.W., S. Kawagata, H.R. Grenfell, A.T. Sabaa and T. O'Neill. 2007. Last global extinction in the deep sea during the mid-Pleistocene climate transition. Paleoceanography 22(3): 14. doi: 10.1029/2007pa001424.

Hellenbrecht, L.M., M. Freese, J.-D. Pohlmann, H. Westererg, T. Blancke and R. Hanel. 2019. Larval distribution of the ocean sunfishes Ranzania laevis and Masturus lanceolatus (Tetraodontiformes: Molidae) in the Sargasso Sea subtropical convergence zone. J. Plankton Res. 41(5): 595–608.

Kawasaki, K. 2009. The SCPP gene repertoire in bony vertebrates and graded differences in mineralized tissues. Dev. Genes Evol. 219(3): 147–157. doi: 10.1007/s00427-009-0276-x.

Kawasaki, K. 2011. The SCPP Gene family and the complexity of hard tissues in vertebrates. Cells Tissues Organs 194(2-4): 108–112. doi: 10.1159/000324225.

Kawecki, T.J. and D. Ebert. 2004. Conceptual issues in local adaptation. Ecol. Lett. 7(12): 1225–1241. doi: 10.1111/j.1461-0248.2004.00684.x.

Knorrn, A.H. 2019. Larval development and distribution of Ranzania laevis (Pennant, 1776) and Masturus lanceolatus (Liénard, 1840) across thermal frontal zones in the Sargasso Sea. Unpublished bachelor thesis. Johann Wolfgang Goethe University of Frankfurt, Germany.

Kocher, T.D. and C.A. Stepien. 1997. Molecular Systematics of Fishes. New York: Academic Press.

Koelreuter, I.T. 1766. Piscium rariorum e mus. Petrop. Exceptorum descriptiones continuatae. Novi Commmentarii Academiae Scientiarum Imperialis Petropolitanae 10 pro Anno MCDDLXIV 329-351.

Lande, R. 1980. Genetic variation and phenotypic evolution during allopatric speciation. Am. Nat. 116(4): 463–479.

Landi, M., M. Dimech, M. Arculeo, G. Biondo, R. Martins, M. Carneiro et al. 2014. DNA Barcoding for Species Assignment: The Case of Mediterranean Marine Fishes. PloS One 9: e106135. doi.org/10.1371/journal.pone.0106135.

Leachman, S. 2019. Hoodwinked: rare hoodwinker sunfish, never before seen in Northern Hemisphere, washes up at Coal Oil Point Reserve. The Current.

Lenormand, T. 2002. Gene flow and the limits to natural selection. Trends Ecol. Evol. 17(4): 183–189.

Li, H. and R. Durbin. 2011. Inference of human population history from individual whole-genome sequences. Nature 475(7357): 493–U84. doi: 10.1038/nature10231.

Linnaeus, C. 1751. Philosophia Botanica. Stockholm & Amsterdam.

Marra, N.J., M.J. Stanhope, N.K. Jue, M.H. Wang, Q. Sun, P.P. Bitar et al. 2019. White shark genome reveals ancient elasmobranch adaptations associated with wound healing and the maintenance of genome stability. Proc. Natl. Acad. Sci. U.S.A. 116(10): 4446–4455. doi: 10.1073/pnas.1819778116.

Matsuura, K. 2015. Taxonomy and systematics of tetraodontiform fishes: a review focusing primarily on progress in the period from 1980 to 2014. Ichthyol. Res. 62(1): 72–113. doi: 10.1007/s10228-014-0444-5.

Mayr, Ernst. 1963. Animal species and evolution. Cambridge, MA: Harvard University Press.

McCord, C.L. and M.W. Westneat. 2016. Phylogenetic relationships and the evolution of BMP4 in triggerfishes and filefishes (Balistoidea). Mol. Phylogenet. Evol. 94: 397–409. doi: 10.1016/j.ympev.2015.09.014.

Miya, M., Y. Sato, T. Fukunaga, T. Sado, J.Y. Poulsen, K. Sato et al. 2015. MiFish, a set of universal PCR primers for metabarcoding environmental DNA from fishes: detection of more than 230 subtropical marine species. R Soc Open Sci. 2(7): 150088. doi: 10.1098/rsos.150088.

Nardo, G.D. 1840. Considerazioni sulla famiglia dei pesci Mola, e sui caratteri chi li distinguono. Annali delle Scienze del Regno Lombardo-Vento. 10: 105–112.

Near, T.J., A. Dornburg, R.I. Eytan, B.P. Keck, W.L. Smith, K.L. Kuhn et al. 2013. Phylogeny and tempo of diversification in the superradiation of spiny-rayed fishes. Proc. Natl. Acad. Sci. U.S.A. 110(31): 12738–12743. doi: 10.1073/pnas.1304661110.

Nyegaard, M., N. Loneragan, S. Hall, J. Andrew, E. Sawai and M. Nyegaard. 2018a. Giant jelly eaters on the line: Species distribution and bycatch of three dominant sunfishes in the Southwest Pacific. Estuar. Coast. Shelf Sci. 207: 1–15.

Nyegaard, M., E. Sawai, N. Gemmell, J. Gillum, N.R. Loneragan, Y. Yamanoue et al. 2018b. Hiding in broad daylight: molecular and morphological data reveal a new ocean sunfish species (Tetraodontiformes: Molidae) that has eluded recognition. Zool. J. Linn. Soc. 182(3): 631–658. doi: 10.1093/zoolinnean/zlx040.

Nyegaard, M., S. Andrzejaczek, C.S. Jenner and M.N.M. Jenner. 2019. Tiger shark predation on large ocean sunfishes (Family Molidae)—two Australian observations. Environ. Biol. Fishes 102(12): 1559–1567.

Pan, H.L., H. Yu, V. Ravi, C. Li, A.P. Lee, M.M. Lian et al. 2016. The genome of the largest bony fish, ocean sunfish (*Mola mola*), provides insights into its fast growth rate. GigaScience 5. doi: 10.1186/s13742-016-0144-3.

Powell, D.C. 2001. A fascination for fish: adventures of an underwater pioneer. 2nd ed. Berkeley: University of California Press, USA.

Prakash, S., T.T. Ajithkumar and M. Thangaraj. 2016. DNA barcoding of sharp tail sunfish *Masturus lanceolatus* Lienard, 1840 (Tetraodontiformes: Molidae). Proc. Zool. Soc. 69(1): 153–156.

Pritchard, J.K., M. Stephens and P. Donnelly. 2000. Inference of population structure using multilocus genotype data. Genetics 155(2): 945–959.

Sagara, K., Y. Yoshito, M. Nishibori, H. Kuniyoshi, T. Umino, Y. Sakai et al. 2005. Coexistence of two clades of the ocean sunfish *Mola mola* (Molidae) around the Japan coast. Jpn. J. Ichthyol. 52(1): 35–39.

Santini, F. and J.C. Tyler. 2002. Phylogeny of the ocean sunfishes (Molidae, Tetraodontiformes), a highly derived group of teleost fishes. Ital. J. Zool. 69(1): 37–43. doi: 10.1080/11250000209356436.

Sarasa, M., S. Alasaad and J.M. Perez. 2012. Common names of species, the curious case of Capra pyrenaica and the concomitant steps towards the 'wild-to-domestic' transformation of a flagship species and its vernacular names. Biodivers. Conserv. 21(1): 1–12. doi: 10.1007/s10531-011-0172-3.

Sawai, E., Y. Yamanoue, Y. Sakai and H. Hashimoto. 2009. On the abnormally morphological forms in Mola sunfish (*Mola* spp. A and B) taken from Japanese coastal waters. J. Grad. Sch. Biosp. Sci., Hiroshima Univ. 48: 9–17.

Sawai, Etsuro, Y. Yamanoue, L. Jawad, J. Al-Mamry and S. Yoichi. 2017. Molecular and morphological identification of Mola sunfish specimens (Actinopterygii: Tetraodontiformes: Molidae) from the Indian Ocean. Species Divers. 22.

Sawai, E., Y. Yamanoue, M. Nyegaard and Y. Sakai. 2018. Redescription of the bump-head sunfish Mola alexandrini (Ranzani 1839), senior synonym of Mola ramsayi (Giglioli 1883), with designation of a neotype for *Mola mola* (Linnaeus 1758) (Tetraodontiformes: Molidae). Ichthyol. Res. 65(1): 142–160. doi: 10.1007/s10228-017-0603-6.

Sawai, E., M. Nyegaard and Y. Yamanoue. 2020. Phylogeny, taxonomy and size records of ocean sunfishes. pp. 18–36. *In*: T.M. Thys, G.C. Hays and J.D.R. Houghton [eds.]. The Ocean Sunfishes: Evolution, Biology and Conservation, CRC Press. Boca Raton, FL, USA.

Star, B., A.J. Nederbragt, S. Jentoft, U. Grimholt, M. Malmstrom, T.F. Gregers et al. 2011. The genome sequence of Atlantic cod reveals a unique immune system. Nature 477(7363): 207–210. doi: 10.1038/nature10342.

Steinke, D., A.D. Connell and P.D.N. Hebert. 2016. Linking adults and immatures of South African marine fishes. Genome 59(11): 959–967.

Streelman, J.T., C. Puchulutegui, A.L. Bass, T. Thys, H. Dewar and S.A. Karl. 2003. Microsatellites from the world's heaviest bony fish, the giant *Mola mola*. Mol. Ecol. Notes 3(2): 247–249. doi: 10.1046/j.1471-8286.2003.00413.x.

Taberlet, P., E. Coissac, M. Hajibabaei and L.H. Rieseberg. 2012. Environmental DNA. Mol. Ecol. 21(8): 1789–1793. doi: 10.1111/j.1365-294X.2012.05542.x.

Tyler, JC. 1980. Osteology, phylogeny, and higher classification of the fishes of the order Plectognathi (Tetraodontiformes). National Oceanic and Atmospheric Administration.

Thys, T. and R. Williams. 2013. Ocean Sunfish in Canadian Pacific Waters: Summer Hotspot for a Jelly-eating Giant? 2013 Oceans—San Diego. doi: 10.23919/OCEANS.2013.6740966.

Thys, T.M., J.P. Ryan, H. Dewar, C.R. Perle, K. Lyons, J. O'Sullivan et al. 2015. Ecology of the Ocean Sunfish, *Mola mola*, in the southern California Current System. J. Exp. Mar. Biol. Ecol. 471: 64–76. doi: 10.1016/j.jembe.2015.05.005.

Thys, T.M., M. Nyegaard, J. Whitney, J.P. Ryan, I. Potter, T. Nakatsubo et al. 2020. Ocean sunfish larvae: detections, identification and predation. pp. 105–128. *In*: T.M. Thys, G.C. Hays and J.D.R. Houghton [eds.]. The Ocean Sunfishes: Evolution, Biology and Conservation, CRC Press. Boca Raton, FL, USA.

Wang, J., W. Wang, R.Q. Li, Y.R. Li, G. Tian, L. Goodman et al. 2008. The diploid genome sequence of an Asian individual. Nature 456(7218): 60-U1. doi: 10.1038/nature07484.

Wang, Z., M. Gerstein and M. Snyder. 2009. RNA-Seq: a revolutionary tool for transcriptomics. Nat. Rev. Genet. 10(1): 57–63. doi: 10.1038/nrg2484.

Writz, P. and M. Biscoito. 2019. The distribution of *Mola alexandrini* in the Subtropical Eastern Atlantic, with a note on *Mola mola*. Bocagiana 245: 1–6.

Yamanoue, Y., M. Miya, K. Matsuura, M. Katoh, H. Sakai and M. Nishida. 2004. Mitochondrial genomes and phylogeny of the ocean sunfishes (Tetraodontiformes: Molidae). Ichthyol. Res. 51(3): 269–273. doi: 10.1007/s10228-004-0218-6.

Yamanoue, Yusuke, Kohji Mabuchi, Etsuro Sawai, Yoichi Sakai, Hiroaki Hashimoto and Mutsumi Nishida. 2010. Multiplex PCR-based genotyping of mitochondrial DNA from two species of ocean sunfish from the genus *Mola* (Tetradontiformes: Molidae) found in Japanese waters. Jpn. J. Ichthyol. 57(1): 27–34.

Yoshita, Y., Y. Yamanoue, K. Sagara, M. Nishibori, H. Kuniyoshi, T. Umino et al. 2009. Phylogenetic relationship of two Mola sunfishes (Tetraodontiformes: Molidae) occurring around the coast of Japan, with notes on their geographical distribution and morphological characteristics. Ichthyol. Res. 56(3): 232–244. doi: 10.1007/s10228-008-0089-3.

Zakany, J. and D. Duboule. 2007. The role of Hox genes during vertebrate limb development. Curr. Opin. Genet. Dev. 17(4): 359–366. doi: 10.1016/j.gde.2007.05.011.

Zhang, J.-B. and R. Hanner. 2010. DNA barcoding is a useful tool for the identification of marine fishes from Japan. Biochem. Syst. Ecol. 39: 31–42.

Chapter 4

Overview of the Anatomy of Ocean Sunfishes (Molidae: Tetraodontiformes)

Katherine E. Bemis,[1,2,*] *James C. Tyler,*[3] *Eric J. Hilton*[1] and *William E. Bemis*[4,5]

Introduction

The anatomy of Molidae has been the focus of research and commentary since the mid-1500s (e.g., Fig. 1; Rondelet 1554, Gesner 1558). Since that time, many authors have contributed to the understanding of the anatomy of these fishes generally (e.g., Harting 1865, Steenstrup and Lütken 1898, Rosén 1912–1916, Gregory and Raven 1934, Raven 1939a, b, Fraser-Brunner 1951, Tyler 1980). Several studies focused on specific anatomical systems have also incorporated comparative data from molids (cited throughout this review). Progress in understanding evolutionary relationships within tetraodontiforms based on muscles (Winterbottom 1974), larvae (Leis 1984, Thys et al. 2020 [Chapter 7]), fossils (Santini and Tyler 2003, Arcila and Tyler 2017, Carnevale et al. 2020 [Chapter 1]), genetics (e.g., Yamanoue et al. 2004, 2008, Bass 2005, Yoshita et al. 2009, Caldera et al. 2020 [Chapter 3]), and ontogeny (e.g., Britz and Johnson 2005a, Johnson and Britz 2005, Konstantinidis and Johnson 2012) continue to inform our understanding of molid anatomy, as do advances in behavior and ecology (e.g., Watanabe and Sato 2008, Houghton et al. 2009, Dewar et al. 2010, Nakamura et al. 2015, Sousa et al. 2020 [Chapter 8]) and taxonomy (Nyegaard et al. 2018, Sawai et al. 2018 and Sawai et al. 2020 [Chapter 2]).

This chapter reviews published work on the anatomy of Molidae beginning with external anatomy and general body form followed by sections on anatomical regions and organ systems. We do not treat the muscular or reproductive systems because these are subjects of other chapters of this book (see Watanabe and Davenport 2020 [Chapter 5] and Forsgren et al. 2020 [Chapter 6], respectively). We also direct the reader to a more detailed study of the anatomy of Molidae from which we abstracted this chapter and which supplements the information available in the literature with new observations on each of the three extant genera (Bemis et al. in prep.).

[1] Department of Fisheries Science, Virginia Institute of Marine Science, William & Mary, Gloucester Point, Virginia, USA.
[2] Department of Vertebrate Zoology, National Museum of Natural History, Smithsonian Institution, Washington, District of Columbia, USA.
[3] Department of Paleobiology, National Museum of Natural History, Smithsonian Institution, Washington, District of Columbia, USA.
[4] Department of Ecology and Evolutionary Biology, Cornell University, Ithaca, NY, USA.
[5] Cornell University Museum of Vertebrates, Ithaca, NY, USA.
* Corresponding author: bemisk@si.edu

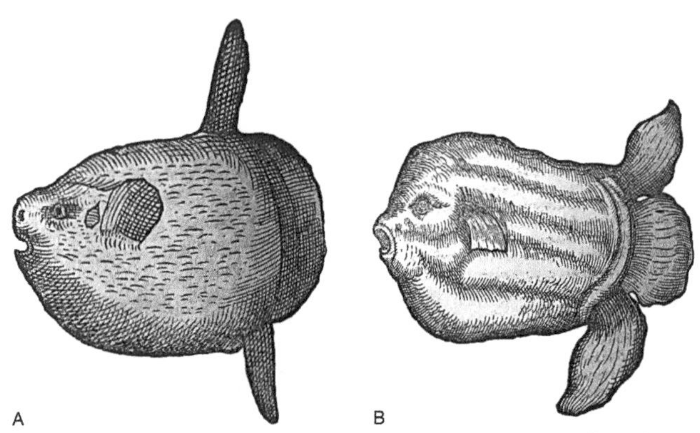

Figure 1. Early wood cuts depicting Molidae (which appear to represent *Mola* spp.) from Gesner (1558). **A.** This image was copied by Gesner from Rondelet (1554; as cited by Johnson and Britz 2005) and shows a fish with a nostril that is large and dorsal and anal fins that are small for molids; the general body form and restricted gill opening of molids, however, are shown well. **B.** *Mola* cf. *alexandrini*. Although not identified, this image depicts a fish with a distinct separation between the dorsal and anal fins and the clavus, which is illustrated as being almost a distinct caudal fin. The profile of the head suggests that this represents *M. alexandrini*. The pectoral fin is unusual, although it may be damaged. In both illustrations, the region of flexion between the body and the dorsal and anal fins and the clavus is shown well.

External Anatomy and Larval Forms

The unusual body form of Molidae makes these fishes among the most distinctive and immediately recognizable of any bony fish family. Adult *Masturus lanceolatus* and *Mola* spp. are among the largest and heaviest of all bony fishes, with *Mola alexandrini* reaching a size of 2,300 kg and 272 cm TL (Sawai et al. 2018); *Ranzania laevis* is smaller, reaching a maximum of 90 cm TL (Schmidt 1921: 1). Extant molids are laterally compressed, with *R. laevis* being more so than the other two genera. The bodies of *M. lanceolatus* and *Mola* spp. are rounded both anteriorly and posteriorly, whereas that of *R. laevis* is more angular anteriorly and truncate posteriorly (Tyler 1970). The pectoral fin of *R. laevis* is elongate, whereas the pectoral fins of *M. lanceolatus* and *Mola* spp. are short and round. The dorsal and anal fins of all genera are elongate and stiff (Winterbottom 1974, Tyler 1980, Nyegaard et al. 2018). The posterior portion of the body (i.e., posterior to the dorsal and anal fins) is the clavus, which is homologous to posterior portions of the dorsal and anal fins because molids lack a true caudal fin (see below and Johnson and Britz 2005). *Masturus lanceolatus* has a distinct central lobe of the clavus (Fraser-Brunner 1951, Tyler 1970, 1980). There are differences in the scales that cover much of the body and those at the bases of the median fins and the clavus, which form bands to allow the fins to flex (see Sawai et al. 2020 [Chapter 2] and Nyegaard et al. 2018 for discussion of these bands as a species-diagnostic feature). Unlike *R. laevis* or *M. lanceolatus*, species of *Mola* have large dermal ossicles along the edge of the clavus, and their structure is diagnostic at the species level (Sawai et al. 2018, Nyegaard et al. 2018). Head and claval shape have been suggested to be sexually dimorphic in *M. mola* (e.g., the head is suggested to be deeper in females; Fraser-Brunner 1951); however this awaits further study.

Zoologists have studied ontogenetic variation and metamorphosis of larval Molidae since the 18th century (e.g., Fig. 2; Koelreuter 1766: plate VIII, figs. 2, 3). The morphology of larval molids has also been the source of taxonomic confusion, with some larvae given the generic name *Molacanthus* and even placed in their own family for a period (Putnam 1871, Gill 1884; subsequently recognized as larval *Mola* spp.). The confusion stems from the very different shapes of small molid larvae, which have round to ovoid bodies with variably shaped and sized spines (Lyczkowski-Shultz 2005). Collection of larvae of different sizes eventually made it possible to associate larval molids with adults (Steenstrup and Lütken 1863, Schmidt 1921). We understand larval development best for *Ranzania laevis* because specimens are readily available (Leis 1977, Hellenbrecht et al. 2019), whereas larval

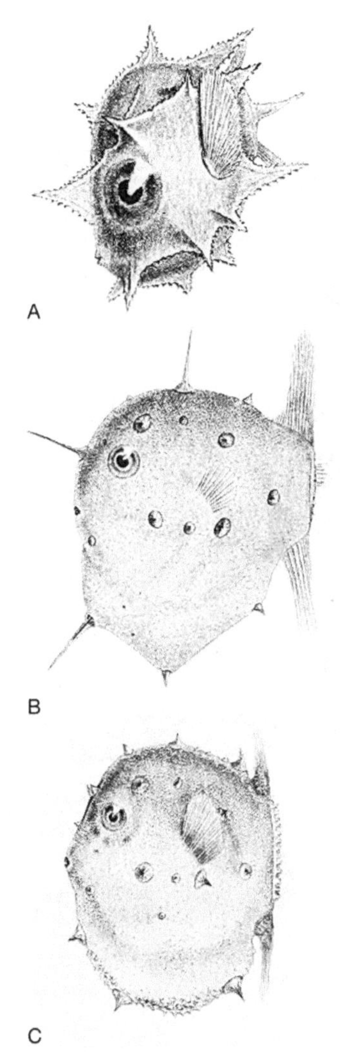

Figure 2. Illustrations of larval Molidae modified from Steenstrup and Lütken (1898). Note variation in spines and the tuft of fin rays in the center of the clavus of *Masturus*. **A.** *Ranzania laevis*, 3.33 mm TL. **B.** *Masturus lanceolatus*, 23.3 mm TL. **C.** *Mola* sp., 20.0 mm TL.

specimens of *Masturus lanceolatus* and particularly for the species of *Mola* are rare in collections (see Thys et al. 2020 [Chapter 7] for a review of molid larval collections and distribution). Molid larvae do not show notochordal flexion, but rather the posterior portion of the body forms the clavus (Leis 1977, Johnson and Britz 2005). Once the fin rays begin to ossify, the prominent body spines begin resorption (Leis 1977).

Skeletal System

The skeleton is the best studied organ system of Molidae, and aspects of it have been described by Leydig (1857), Kölliker (1859, 1860), Harting (1865), Goette (1879), Trois (1883–1884), Steenstrup and Lütken (1898), Stephan (1900), Supino (1904), Nowikoff (1910), Kaschkaroff (1914a, b, 1916) and Studnicka (1916). Tyler (1980) provided a detailed description of the skeleton of all three genera (Figs. 3, 4 and 5). All accounts note the peculiar consistency of the bone at a tissue level (e.g., Harting 1865); the detailed anatomy of molid bone, however, requires further study. We interpret the lightly

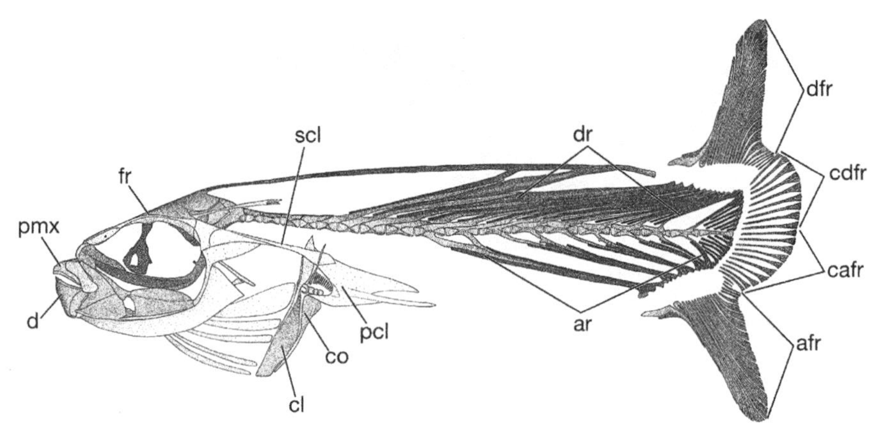

Figure 3. Skeleton of *Ranzania laevis*, ANSP 109435, 65.1 mm SL, anterior to left. Modified from Tyler (1980: fig. 316). Abbreviations: afr, anal fin rays; ar, anal fin radials; cafr, claval anal fin rays; cdfr, claval dorsal fin rays; cl, cleithrum; co, coracoid; d, dentary; dfr, dorsal fin rays; dr, dorsal fin radials; fr, frontal; pcl, postcleithrum; pmx, premaxilla; scl, supracleithrum.

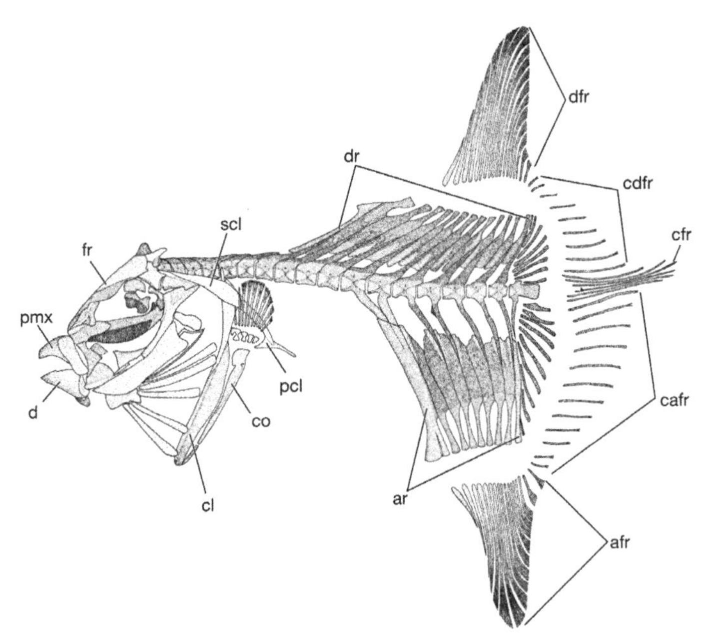

Figure 4. Skeleton of *Masturus lanceolatus*, USNM 117330, 127 mm SL, anterior to left. Modified from Tyler (1980: fig. 315). Abbreviations: afr, anal fin rays; ar, anal fin radials; cafr, claval anal fin rays; cdfr, claval dorsal fin rays; cfr, caudal fin rays; cl, cleithrum; co, coracoid; d, dentary; dfr, dorsal fin rays; dr, dorsal fin radials; fr, frontal; pcl, postcleithrum; pmx, premaxilla; scl, supracleithrum.

ossified bone as a reduction, consistent with Molidae having a reduced number and size of skeletal elements (e.g., loss of coronomekelian; relatively small opercular elements), in addition to elements generally lost in tetraodontiforms (e.g., infraorbital bones, ribs). The skeleton of *Ranzania laevis* is more heavily ossified than that of *Masturus lanceolatus* or *Mola mola* and we consider the condition in *R. laevis* to be plesiomorphic to Molidae (Raven 1939a, Carnevale and Santini 2007).

Our brief overview and discussion of the skeleton of Molidae below is based largely on Tyler (1980) and three recent studies (Britz and Johnson 2005a, Johnson and Britz 2005, Konstantinidis and Johnson 2012). The reader is referred to these papers and to Bemis et al. (in prep.) for fuller accounts of the skeletal system of Molidae.

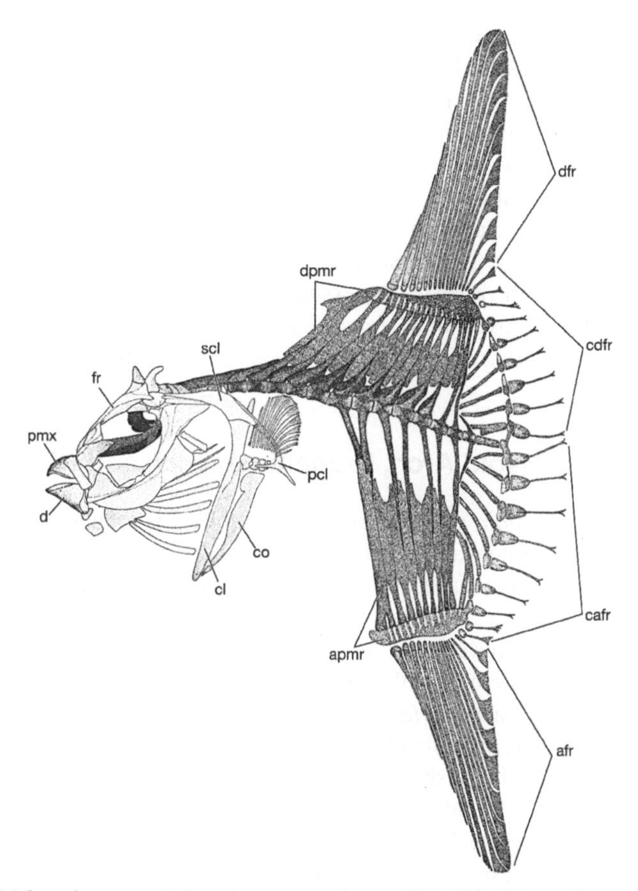

Figure 5. Skeleton of *Mola mola*, composite based on two specimens, SU 16438, 306 mm SL and SU 16441, 310 mm SL, anterior to left. Modified from Tyler (1980: Fig. 306). Abbreviations: afr, anal fin rays; apmr, anal proximal-middle radials; cafr, claval anal fin rays; cdfr, claval dorsal fin rays; cl, cleithrum; co, coracoid; d, dentary; dfr, dorsal fin rays; dpmr, dorsal proximal-middle radials; fr, frontal; pcl, postcleithrum; pmx, premaxilla; scl, supracleithrum.

Skull Roof and Braincase

The largest bones of the cranium are the frontals dorsally, the parasphenoid ventrally, and the pterotics laterally. The supraoccipital of *Ranzania laevis* is relatively larger than that of *Masturus lanceolatus* and *Mola mola*; the supraoccipital, pterotic, and epioccipitals of *R. laevis* bear elongate posterior extensions. In *M. mola*, the epioccipital crests are horn-like. The basioccipitals of all Molidae are large and extended dorsally to exclude the exoccipitals from the foramen magnum. In the midline of the orbit is a median element that has variously been termed the basisphenoid (Tyler 1980) or the pterosphenoid (Britz and Johnson 2012). Tyler (1980) reported the vomer as absent in *R. laevis*; Konstantinidis and Johnson (2012) described and illustrated the vomer in a 2.5 mm NL specimen and, in the next stage that they examined, the vomer was expanded anterodorsally.

Oral Jaws, Suspensorium, and Opercular Series

The oral jaws of Molidae have premaxillary and dentary beaks. The left and right premaxillae are fused, as are the left and right dentaries; no sutures between the right and left sides are visible (see Carnevale et al. 2020 [Chapter 1] and Tyler and Bannikov 1992 for discussion of the beak in fossil molids). Posterior to the premaxillary and dentary beaks are specialized dental structures used for crushing known as triturating teeth (Tyler 1980).

The suspensorium is firmly bound to the neurocranium anteriorly via the palatine, rendering it immobile (Breder and Clark 1947, Konstantinidis and Johnson 2012). Raven (1939a) incorrectly reported that the symplectic is absent. The hyomandibula is large in all molids and has a broad dorsal articulatory head; the shaft of the hyomandibula tapers strongly (Tyler 1980). Raven (1939a: 4) speculated that the elongate posterior extension of the hyomandibula in *Ranzania laevis* "prevents the collapse of the branchial chamber."

In all Molidae, the preopercular bone is the largest of the opercular series. The opercle itself is a small, rectangular (*Ranzania laevis*, *Mola mola*) or oval (*Masturus lanceolatus*) element. The greatly reduced subopercle of all genera varies in shape (straight in *R. laevis*; sigmoid in *M. lanceolatus* and *M. mola*). The interopercle is absent in *M. lanceolatus*, is a short needlelike bone in *M. mola*, and is most fully developed in *R. laevis*, in which it contacts the subopercle and extends to the anterior tip of the preopercle.

Ventral Portion of the Hyoid Arch, Branchiostegals, and Gill Arches

In *Masturus lanceolatus* and *Mola mola* the anterior ceratohyal is rounded anteriorly, and broadly separated from the posterior ceratohyal (Tyler 1980), whereas the anterior ceratohyal tapers posteriorly in *Ranzania laevis*. There are six branchiostegals in *M. lanceolatus* and *M. mola* and only five in *R. laevis*, the posteriormost of which is enlarged (Figs. 3, 4 and 5); Tyler (1980) suggested that the sixth branchiostegal of *R. laevis* may have fused to the fifth to form the enlarged element.

The ceratobranchials are the only gill arch elements that support gill tissue (Tyler 1980). All ceratobranchials are of similar size in *Masturus lanceolatus* and *Mola mola*, whereas in *Ranzania laevis* the fifth ceratobranchial is more slender than the preceding ones. Conversely, the epibranchials of *R. laevis* are all slender, similarly shaped elements, whereas those of *M. lanceolatus* and *Mola* spp. vary in shape and size. All Molidae have pharyngobranchials 2, 3, and 4, which bear elongate, recurved teeth (Harting 1865: plate 3, fig. 7, Steenstrup and Lütken 1898: 94, Suyehiro 1942: 193, fig. 140, Berkovitz and Shellis 2017: fig. 4.114) perhaps related to feeding on jellyfishes and other soft-bodied prey.

Postcranial Axial Skeleton

The vertebral column is short in Molidae, even as larvae (i.e., the notochord is relatively short; Leis 1977, 1984). As noted above, the caudal portion of the notochord fails to go through flexion during metamorphosis (Leis 1977, Johnson and Britz 2005), and thus the vertebral column is truncated. Relative to other fishes, the simplified vertebrae lack ribs and parapophyses (Gregory and Raven 1934, Tyler 1980). The neural and hemal spines, which are more sharply inclined in *Ranzania laevis* than in *Masturus lanceolatus* and *Mola mola* (Raven 1939a), strongly interdigitate with the proximal radials of the anterior portions of the dorsal and anal pterygiophores, forming a tightly bound inflexible sheet of bone. The terminal centrum of all molids has a cartilaginous cap that articulates with the claval pterygiophores (Johnson and Britz 2005).

Britz and Johnson (2005a) demonstrated a fascinating ontogenetic fusion in early larvae of the first vertebra with the basioccipital. Here, we distinguish between the fused first vertebrae and those that we term "functional vertebrae" (i.e., ones not fused into the skull) in our vertebral counts. This is because the fusion of the first vertebra with the basioccipital is so complete in juvenile and adult molids. There are reports of eight functional abdominal and eight functional caudal vertebrae in *Mola mola*, by Cleland (1862), Steenstrup and Lütken (1898), and Kaschkaroff (1914a), but Tyler (1980) reported eight functional abdominal vertebrae and nine caudal vertebrae for this species. *Masturus lanceolatus*, however, has eight caudal vertebrae, so perhaps the earlier authors misidentified their specimens at the generic level, or the number of vertebrae varies in *M. mola* (Britz and Johnson 2005a). Fraser-Brunner (1951) reported that *M. lanceolatus* and *M. mola* have nine abdominal and eight caudal vertebrae; Tyler (1980) suspected that this was a miscount based on radiographs.

Dorsal and Anal Fins, Skeletal Supports, and Clavus

In Molidae, both the anterior "fin-like" portions of the dorsal and anal fins, as well as the claval portion of the fins, are supported by a series of proximal, middle, and distal radials. In the anterior portion, the middle radials are fused to each other to form a block-like cartilage or a weakly ossified bone that is marked laterally by grooves or canals that surround the tendons of the median fin musculature (Tyler 1980). The proximal and middle radials were interpreted as being continuous (i.e., proximal-middle radials) by Johnson and Britz (2005). The distal radials of both the anterior portions and claval portion of the fins are large, ovoid, and clearly separate elements deeply embraced by the fin rays (Tyler 1980).

The two hypotheses for the homology of the clavus are that it represents a modified caudal fin (Goodsir 1841, Cleland 1862, Schmidt 1921, Gudger 1937a, b, Winterbottom 1974a) or that it is homologous to the posterior portions of the dorsal and anal fins (Ryder 1886, Boulenger 1904, Regan 1903, Raven 1939a, Fraser-Brunner 1951, Tyler 1970, 1980, Santini and Tyler 2002). In an ontogenetic study, Britz and Johnson (2005b) and Johnson and Britz (2005) clearly showed that the clavus is formed by elements of the dorsal and anal fins; the innervation of claval fin rays also resembles that of the dorsal and anal fins of other teleosts (Nakae and Sasaki 2006).

The greatly expanded fin rays of the anterior portions of the dorsal and anal fins are extensively branched, with up to 50 tips per fin ray; the more posterior fin rays bend at a sharp angle near their distal ends (Tyler 1980). The number of fin rays in the anterior portions of the fins varies within Molidae, but ranges within individual taxa are not well known. A variable number of fin rays support the clavus (19 in *Ranzania laevis*, 21 in *Masturus lanceolatus*, and 16 in *Mola mola*; Tyler 1980). Fraser-Brunner (1951) interpreted that *M. lanceolatus* is the only molid to retain fin rays homologous to the caudal fin rays of other fishes. Several fin rays (four to eight; see Tyler 1980: 391) intercalate between the upper and lower lobes of the clavus in *M. lanceolatus*; unlike the other claval rays, these are not supported by radial elements.

Appendicular Skeleton

The posttemporal is absent in all Molidae, and the supracleithrum articulates directly with the pterotic (Tyler 1980). In *Ranzania laevis*, the supracleithrum is slender compared to *Masturus lanceolatus* and *Mola mola*, and the postcleithrum is posteriorly expanded; in all genera the postcleithra have an anteroventral extension that lies lateral to the pectoral radials. *Ranzania laevis* has the most pectoral fin rays of any molid (14, compared to 10 in *M. lanceolatus* and 12 in *M. mola*; Tyler 1980), although variation in fin ray counts is poorly known. As in most tetraodontoids, molids lack all elements of the pelvic girdle and fins (Tyler 1962).

Integument and Hypodermis

Masturus lanceolatus and *Mola mola* have rounded or elliptical scales that interdigitate with scattered small scales; all scales in these genera have a central spine (Katayama and Matsuura 2016). In contrast, the interdigitated scales of *Ranzania laevis* are larger, approximately hexagonal, and bear a row of one to four short, blunt spines (Tyler 1980, Katayama and Matsuura 2016). Scale morphology is diagnostic at the species level (Sawai et al. 2020 [Chapter 2]).

The body wall of molids is formed by a thickened hypodermal layer of tissue immediately below the skin. The hypodermis is much thicker than the skin, and, together with the skin, the body wall can be four or more centimeters thick; Goodsir (1841) reported the body wall of a 1888 mm specimen of *Mola mola* to be 152 mm. The body wall of *Masturus lanceolatus* and *M. mola* is much thicker than that of *Ranzania laevis*, which is thinner over much of the body (*R. laevis* does have a thicker body wall supporting its midventral keel, although even this is thinner than that of other molids). The identity of the hypodermis has caused confusion in the literature, and several names have been applied to this structure (e.g., "Tissue of a very peculiar kind," Goodsir 1841: 189; "Subdermal connective tissue," Green 1899: 321; "Collagenous material... rubber like armor," Gregory and Raven 1934: 150; "Layer of subcutaneous tissue," Roon and Pelkwijk 1939: 68; "Greatly thickened collagenous dermis," incorrectly interpreted as the stratum compactum of the dermis by Logan and Odense 1974: 1040; "Thick layer of collagenous connective tissue," Tyler 1980: 379; "Subcutaneous gelatinous tissue," Watanabe and Sato 2008; and "Capsule," Davenport et al. 2018: 1). The thickened hypodermis of *M. mola* and presumably *M. lanceolatus* likely serves to decrease the density of the fish over ontogeny (Watanabe and Davenport 2020 [Chapter 5]), and may be related to its diving behavior (Watanabe and Sato 2008, Davenport et al. 2018, Watanabe and Davenport 2020 [Chapter 5]); Nakamura et al. (2015) suggested that the thickened hypodermis provided insulation during deep dives.

Brain, Nerves, and Sense Organs

Several historical (e.g., Vignal 1881, Burr 1928: figs. 40–41) and recent works (Nakae and Sasaki 2006, Uehara et al. 2015: fig. 2) describe the brain of molids. Perhaps most notable is the illustration in Harting (1865: plate 3, fig. 1) that depicts the brain within the cranial cavity (Fig. 6A). In other figures, Harting illustrated aspects of the brain and cranial nerves, including the large optic nerves and optic tecta, and the hypophysis. Unlike most other vertebrates, the optic decussation is anterior to the diencephalon (Kappers et al. 1936). Although the lateral-line system of molids is relatively poorly developed, the so-called "lateral line lobe" of the brain identified by Burr (1928) is surprisingly large. Nakae and Sasaki (2006: 244, fig. 8a) state: "The reduced number of foramina for emergence of the nerves from the cranium is a distinct feature of *M. mola*."

The spinal cord of molids is relatively short (Owen 1866, Chanet et al. 2013), as in other tetraodontiforms (Uehara and Ueshima 1986). In Molidae, the spinal cord lies entirely within the cranium (Uehara et al. 2015). The spinal canal, which is restricted to only the first 13 vertebrae, contains only the filum terminale and cauda equina (Uehara et al. 2015: fig. 2d). As in other tetraodontiforms, molids lack Mauthner cells (Uehara et al. 2015: 299). These large neurons, which extend from the hindbrain into the spinal cord, mediate escape responses such as C-starts in other teleosts.

The reduced lateral-line system of Molidae consists of only a few superficial neuromasts associated with specialized scales (Nakae and Sasaki 2006: fig. 1). The trunk line extends in an abbreviated arch dorsal to the pectoral fin and terminates just anterior to the level of the insertions of the dorsal and anal fins (Nakae and Sasaki 2006). The trunk line is continuous with the cranial lateral lines, although some lines of the head remain separate (e.g., the supraorbital and infraorbital lines) or shifted in position from that of a typical teleost (e.g., the mandibular ramus of the anteroventral lateral line nerve innervates the anterior portion of the preopercular line, making it the likely equivalent to the mandibular lateral line of other fishes; see Nakae and Sasaki 2006).

Few details on the anatomy and variation of the reduced olfactory system of molids are available (Tyler 1980, Jinxiang and Wanduan 1982). Thus, the capacity of molids to detect and process water borne cues is largely unknown, warranting further investigation.

The eyes of Molidae are large in juveniles and adults (Phillips et al. 2015), set in a deep conjunctival recess, and in *Mola mola* are reported to be asymmetrical in their position (Harting 1865: plate 2, fig. 1, Meek 1904). *Mola mola* also has a nictitating membrane (Cuvier 1805b, Brimley 1939). The highly mobile eyes of molids are used to search for prey (Kino et al. 2009: fig. 4, Nakamura et al. 2015, Phillips et al. 2015).

Several authors (Cuvier 1805b, Harting 1865, Thompson 1889, Meek 1904) described and illustrated the inner ear and three semicircular ducts of *Mola mola* (Cleland 1862 incorrectly described that *M. mola* only has two semicircular ducts). A pillar passes through the arc of the posterior vertical duct (Fig. 6B; Harting 1865, Meek 1904). Thompson (1889) incorrectly described this pillar as passing through the arc of the anterior vertical duct. Otoliths of Molidae are small (not

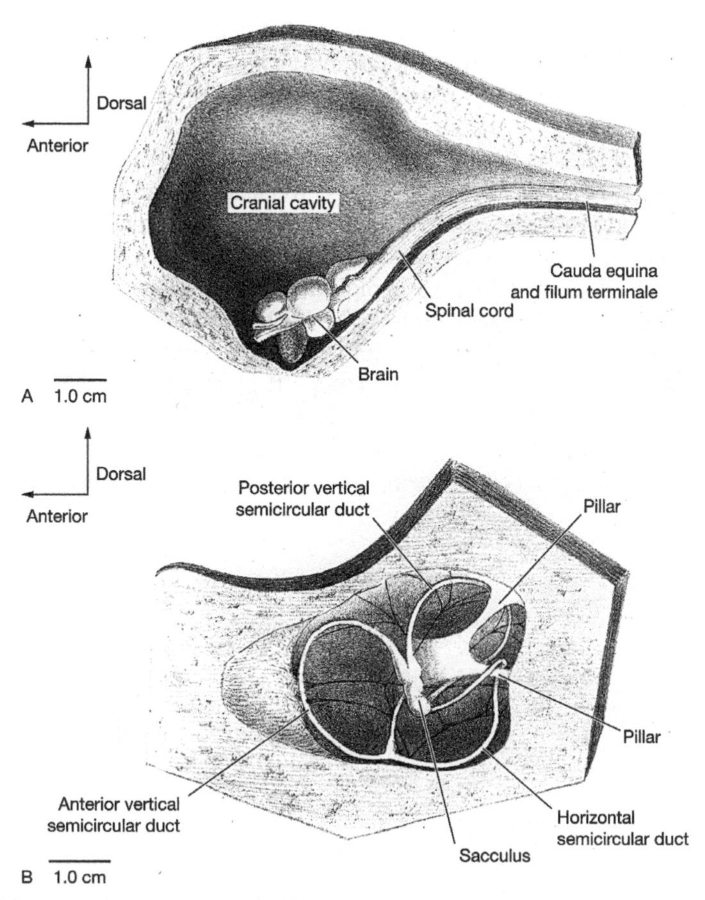

Figure 6. Brain and inner ear of *Mola mola* modified from Harting (1865: plate 3, fig. 1; Harting referred to this specimen as *Orthragoriscus ozodura*). Figures have been flipped so that anterior faces to left. **A.** Dissection of head in medial view showing brain and spinal cord contained within the cranial cavity. **B.** Semicircular ducts, sacculus, and pillars of connective tissue. Authors who have studied the inner ear of molids have figured the arrangement of the ducts and pillars differently (e.g., Harting 1865, Thompson 1889, Meek 1904).

absent, as claimed by Cleland 1862), weakly formed, irregular in shape, and described as gelatinous (Cuvier 1805b) or as "chalk dust" (Nolf and Tyler 2006; see also Nolf 1985; disc-shaped otoconia in the sagitta of *Mola* crumble easily Gauldie 1990, which may explain the description of "chalk dust"). There is apparent taxonomic variation in the structure of otoliths within Molidae because *Ranzania laevis* has spine-like crystal extrusions on its otoliths, unlike those of *M. mola* (see Nolf and Tyler 2006: 155, plate 18, fig. 10). The mineral in the otoliths of Molidae is vaterite (Gauldie 1990), a form of calcium carbonate known to occur in taxa with reduced sound perception (Oxman et al. 2007, Reimer et al. 2016).

Viscera

Several works describe portions of the digestive system of Molidae (e.g., Agassiz 1857, Gregory and Raven 1934, Raven 1939a, b, Suyehiro 1942, Chanet et al. 2012). The "stomach" of molids is continuous with the intestine (Fig. 7; straight, or I-shaped; Wilson and Castro 2011), and it lacks esophageal and pyloric valves in *Mola mola* (Cuvier 1805c); an esophageal valve is present in *Ranzania laevis*, although a pyloric sphincter appears to be absent (Raven 1939a). The intestine is longer in *Masturus lanceolatus* and *M. mola* than in *R. laevis*, although in all molids the coiled intestine occupies much of the abdominal cavity (Fig. 7, Chanet et al. 2012). Two lobes are present in the liver of *R. laevis* and *M. mola*, whereas that of *M. lanceolatus* is a single lobe. In *M. mola*, the gallbladder is on the right side of the abdominal cavity between the two liver lobes (Chanet et al. 2012: figs. 3, 5, and 6).

The family Molidae has long been reported as lacking a swim bladder (e.g., Fraser-Brunner 1943, Breder and Clark 1947, McCune and Carlson 2004, Watanabe and Sato 2008, Chanet et al. 2012); Suyehiro's (1942) report of a swim bladder in adult specimens is in error. Fraser-Brunner (1951: 151, fig. 9b) illustrated and described a swim bladder in a 21 mm specimen of *Masturus lanceolatus* and suggested that the loss of the swim bladder in molids occurs during early ontogeny. The absence of the swim bladder in adults is a synapomorphy of Molidae (Santini and Tyler 2002).

The paired kidneys of molids are large retroperitoneal structures. Most teleosts have fused kidneys, and Chanet et al. (2012: table 1, 2013) suggested that a paired kidney is a synapomorphy of tetraodontiforms and lophiiforms. The urinary bladder of Molidae lies along the posterior wall of the abdominal cavity (Fig. 7B; Owen 1866: 536, Gregory and Raven 1934). It empties together with the gonads into a common urogenital opening immediately posterior to the anus (Fig. 7; Gregory and Raven 1934).

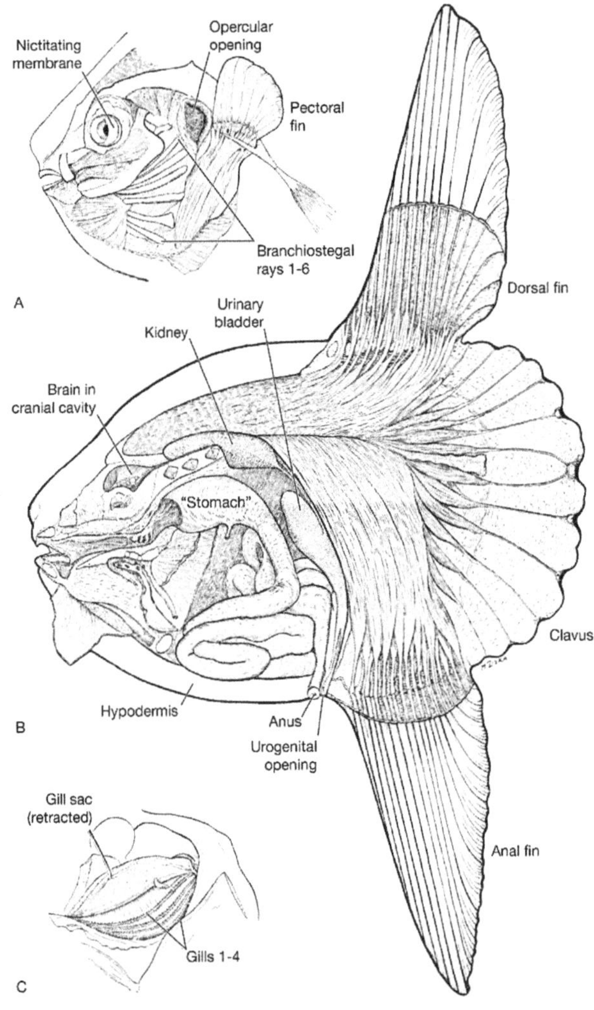

Figure 7. Dissection of *Mola mola* from Gregory and Raven (1934). The fish was reported to be 47 inches (= 1194 mm) TL. **A.** Head showing nictitating membrane and features of opercular region. **B.** Aspects of internal anatomy. **C.** Detail of left gill region with gill sac retracted to show enlarged gills.

Circulatory System and Respiration

The circulatory system of Molidae has not been studied in detail for more than 100 years. The large heart has thick walled chambers. Rosén (1912: 7) reported that the number of atrioventricular leaflets ranged from three to five based on the literature. Not having a specimen to dissect, Rosén concluded that there are four because this was the most commonly reported number, and also noted that no other living tetraodontiform has as many atrioventricular leaflets. Harting (1865: plate 3 fig. 6) noted that the ventriculo-bulbar valve has four leaflets in *Mola mola*; the number of leaflets in the ventriculo-bulbar valve of *M. mola* is also greater than in other tetraodontiforms (Rosén 1912).

The stout ventral aorta gives rise to four large afferent branchial arteries on each side of the body to serve gills 1–4 (Fig. 8A). An unusual feature concerns the pattern of branching of afferent branchial arteries serving gills 3 and 4. The three larger gills receive correspondingly large afferent branchial arteries; the afferent branchial artery for the fourth gill is about half the diameter of the other three (Milne-Edwards 1858, Parker 1900). There are three coronary arteries (Milne-Edwards 1858, Parker 1900).

Adeney and Hughes (1977: fig. 4) described the double afferent system of vessels serving the gills in *Mola mola*. In this system, each afferent branchial artery divides into two vessels before entering the gill filaments and eventually the secondary lamellae as dorsal and ventral afferent filament arteries. These give rise to efferent filament arteries, which eventually join to form the efferent branchial artery proper. The efferent branchial arteries exit the gills and join to form the major arterial trunks in a circulus cephalicus (Ridewood 1899).

The unusual anatomy and circulation of molid gills received attention more than 150 years ago in a paper by Alessandrini (1839) with beautiful color illustrations (Fig. 8C). In the intervening years,

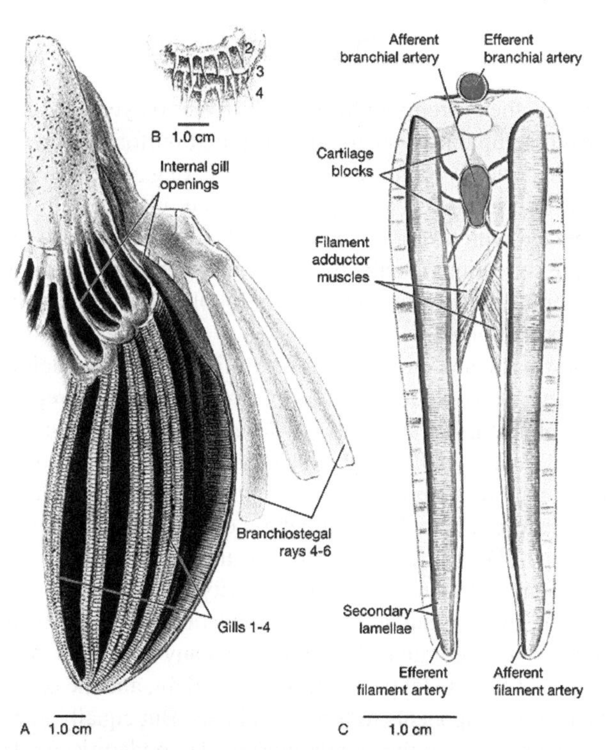

Figure 8. Gill arches of molids. **A.** Gill arches showing internal gill openings and gills 1–4, modified from Harting (1865). **B.** Pharyngeal teeth with pharyngobranchials 2, 3, and 4 labeled, modified from Harting (1865). **C.** Two hemibranchs attached at their bases supported by cartilaginous blocks and circulatory pattern through the secondary lamellae, modified from Alessandrini (1839). Red indicates oxygenated blood; blue indicates deoxygenated blood. Terminology of cartilaginous blocks based on Adeney and Hughes (1977).

an account by Adeney and Hughes (1977) stands out as contributing new information such as the first estimate of gill surface area (see Hughes 1984 which provides a review of gill surface area). Adeney and Hughes' (1977) estimates of gill surface area suggested that *Mola mola* falls within the intermediate activity group of fishes; however, this was based only on two individuals of very different sizes. It would be interesting to collect a larger dataset of gill surface area to reevaluate the conclusion of intermediate activity based on new data on swimming performance. Although Adeney and Hughes (1977) did not cite Alessandrini's (1839) account, it supports similar findings.

Adeney and Hughes (1977) divided the gills of molids into ventral, middle, and dorsal regions. Ceratobranchial cartilages support only the middle region, whereas the ventral region extends anteriorly, and the dorsal region extends posterodorsally. The ventral and dorsal regions are supported by cartilages that Adeney and Hughes (1977) termed upper and lower cartilaginous blocks, which are distinct from the ceratobranchial proper. Mineralized primary gill filaments provide additional structural support for the array of filaments (Adeney and Hughes 1977).

Endocrine Organs

There is little published information about the endocrine organs of molids. Harting (1865: plate 3, fig. 1) and Burr (1928: fig. 13) illustrated aspects of the hypophysis of *Mola mola*; Burr (1928: fig. 40) also illustrated the epiphysis and paraphysis. The compact thyroid gland is dorsal to the ventral aorta (Chanet et al. 2012, 2013: fig. 8). The urophysis (i.e., caudal neurosecretory system) is absent in Molidae based on light microscopy (Hamana 1962; Fridberg and Bern 1968 suggested further study using electron microscopy). The absence of the urophysis is likely due to the reduction of the caudal region associated with the evolution of the clavus.

Discussion

There has been research on the anatomy of Molidae for over 500 years, with substantial study since the mid-1800s. However, comparative anatomical information for the three extant genera remains incomplete. Specimens historically identified as *Mola mola* (e.g., since Cuvier 1805a–e), and recent advances in taxonomic discrimination among species, may ultimately reveal that morphological variation has been conflated with taxonomic variation. The specimen base for anatomical studies is also problematic because collections usually retain smaller specimens due to logistical constraints of preserving and storing large adults. As a result, the morphology of large adult *Masturus lanceolatus* and *Mola* spp. is very poorly known.

We should view new basic anatomical data for Molidae (Britz and Johnson 2005a, b, Johnson and Britz 2005, Nakae and Sasaki 2006, 2010, Nyegaard et al. 2018, Nyegaard and Sawai 2018) in context of advances in molid behavior and biology. *Mola mola* was, for much of its history, interpreted as "slow" and "sluggish", with "inefficient jaws" (Binney 1842: 93; also see Burr 1928). Recent ecological studies reveal that *M. mola* is, in fact, a rapid and active swimmer that makes deep dives for foraging (e.g., Watanabe and Sato 2008, Houghton et al. 2009, Dewar et al. 2010, Nakamura et al. 2015, Thys et al. 2015). We may be able to better understand many aspects of molid anatomy (e.g., circulatory system) in the framework of high activity metabolism.

Many aspects of molid anatomy suggest loss and reduction as primary themes in molid evolution, and indeed molids have reduced, eliminated, or modified many structures. Among these, perhaps the most important is the loss of the caudal peduncle and caudal fin, and the corresponding evolution of modified dorsal and anal fin components to form the clavus. But equally impressive are anatomical characters unique to Molidae, such as the greatly thickened hypodermis. As we learn more about the comparative anatomy of the organ systems of Molidae and view them in the context of new ecological and behavioral discoveries, it is certain that there will be many fascinating surprises.

Remaining Questions and Future Goals

Although our understanding of molid anatomy has deepened considerably in recent years, a number of key questions and challenges still remain: (1) Are there phylogenetically informative characters based on the viscera and other soft tissue structures, which have been little studied in comparison to the skeletal system? (2) How do we characterize the unusual molid bone histologically and what might be the functional significance of such unusual bone? (3) Extremely large specimens are rarely studied. Do they have significant anatomical differences, and what can we learn from studying complete ontogenetic series? (4) What elements support the claval extension in *Masturus lanceolatus*? (5) How does the high activity level of molids relate to their unique morphology, particularly the heart, gills, and circulation? In parallel with advances in behavioral and trophic ecology, such information will open up new avenues of discovery.

Acknowledgments

Mann Library at Cornell University, VIMS Hargis Library, William & Mary Swem Library, Breck Bartholomew, and Biodiversity Heritage Library provided many references and images. Support for this project provided by a National Science Foundation GRFP, John E. Olney Sr. Ichthyology Award, Virginia Institute of Marine Science Office of Academic Studies, Clyde D. and Lois W. Marlatt, Jr. Fellowship, Susan and Philip Bartels, and the Tontogany Creek Fund. We thank Tierney Thys, Jonathan Houghton, and Graeme Hays for inviting us to contribute this chapter, which is contribution No. 3853 of the Virginia Institute of Marine Science, William & Mary.

References

Adeney, R.J. and G.M. Hughes. 1977. Some observations on gills of oceanic sunfish, *Mola mola*. J. Mar. Biol. Assoc. U.K. 57: 825–837.

Agassiz, A. 1857. *Orthragoriscus mola*, on grounds of internal anatomy, does not belong in the same family with *Diodon* and *Tetraodon*, but is the type of a distinct family. Untitled paper, Proc. Am. Acad. Arts Sci. 3: 319.

Alessandrini, A. 1839. De piscium apparatu respirationis tum speciatim Orthragorisci = *Orthragoriscus alexandrini* Ranzani. Nuovi Coram. Acad. Sci. Inst. Bonon. (Accad. Sci. Istit. Bologna) 3: 359–382.

Arcila, D. and J.C. Tyler. 2017. Mass extinction in tetraodontiform fishes linked to the Palaeocene–Eocene thermal maximum. Proc. R. Soc. B. 284: 1–8.

Bass, A.L., H. Dewar, T. Thys, J.T. Streelman and S.A. Karl. 2005. Evolutionary divergence among lineages of the ocean sunfish family, Molidae (Tetraodontiformes). Mar. Biol. 148: 405–414.

Bemis, K.E., J.C. Tyler, E.J. Hilton and W.E. Bemis. In prep. Comparative anatomy and ontogeny of Ocean Sunfishes (Tetraodontiformes: Molidae). Submission planned for NOAA Professional Papers.

Berkovitz, B.K.B. and R.P. Shellis. 2017. The Teeth of Non-mammalian Vertebrates. Amsterdam, the Netherlands, Elsevier.

Binney, A. 1842. Observations made during two successive summers at Nahant, on the habits of the *Orthogariscus mola* or short sunfish. Proc. Boston Soc. Nat. Hist. 1: 93.

Boulenger, G.A. 1904. Teleostei. pp. 539–727. *In*: S.F. Harmer and A.E. Shipley [eds.]. The Cambridge Natural History, vol. VII. Hemichordata, Ascidians and Amphioxus, Fishes. London: Macmillan.

Breder, C.M. and E. Clark. 1947. A contribution to the visceral anatomy, development and relationships of the Plectognathi. Bull. Amer. Mus. Nat. Hist. 88: 287–319.

Brimley, H.H. 1939. The ocean sun-fishes on the North Carolina coast: The Pointed-Tailed *Masturus lanceolatus* and the Round-Tailed *Mola mola*. J. Elisha Mitchell Soc. 55: 295–303.

Britz, R. and G.D. Johnson. 2005a. Occipito-vertebral fusion in ocean sunfishes (Teleostei: Tetraodontiformes: Molidae) and its phylogenetic implications. J. Morph. 266: 74–79.

Britz, R. and G.D. Johnson. 2005b. Leis' conundrum: Homology of the clavus of the ocean sunfishes. 1. Ontogeny of the median fins and axial skeleton of *Monotrete leiurus* (Teleostei, Tetraodontiformes, Tetraodontidae). J. Morph. 266: 1–10.

Burr, H.S. 1928. The central nervous system of *Orthagoriscus mola*. J. Comp. Neuro. 45: 33–128.

Caldera, E.J., J.L. Whitney, M. Nyegaard, L. Kubicek and T.M. Thys. 2020. Genetic insights regarding the taxonomy, phylogeography and evolution of ocean sunfishes (Molidae: Tetraodontiformes). pp. 37–54. *In*: T.M. Thys, G.C. Hays and J.D.R. Houghton [eds.]. The Ocean Sunfishes: Evolution, Biology and Conservation, CRC Press. Boca Raton, FL, USA.

Carnevale, G. and F. Santini. 2007. Record of the slender mola, genus *Ranzania* (Teleostei, Tetraodontiformes), in the Miocene of the Chelif Basin, Algeria. C.R. Palevol 6(2007): 321–326.

Carnevale, G., L. Pellegrino, J.C. Tyler. 2020. Evolution and fossil record of the ocean sunfishes. pp. 1–17. *In*: T.M. Thys, G.C. Hays and J.D.R. Houghton [eds.]. The Ocean Sunfishes: Evolution, Biology and Conservation, CRC Press. Boca Raton, FL, USA.

Chanet, B., C. Guintard, T. Boisgard, M. Fusellier, C. Tavernier, E. Betti et al. 2012. Visceral anatomy of Ocean Sunfish (*Mola mola* (L., 1758), Molidae, Tetraodontiformes) and Angler (*Lophius piscatorius* (L., 1758), Lophiidae, Lophiiformes) investigated by non-invasive imaging techniques. Comptes Rendus Biologies 335: 744–752.

Chanet, B., C. Guintard, E. Betti, C. Gallut, A. Dettai and G. Lecointre. 2013. Evidence for a close phylogenetic relationship between the teleost orders Tetraodontiformes and Lophiiformes based on an analysis of soft anatomy. Cybium. 37: 179–198.

Cleland, J. 1862. On the anatomy of the short sunfish (*Orthagoriscus mola*). Nat. Hist. Rev. 170–185.

Cuvier, G. 1805a–e. Leçons d'anatomie comparée. 5 vol. Paris. [Vol I, 518 p.; Vol. II, 697 p.; Vol. III, 558 p.; Vol. IV, 539 p.; Vol. V, 368 p.]

Davenport, J., N.D. Phillips, E. Cotter, L.E. Eagling and J.D.R. Houghton. 2018. The locomotor system of the ocean sunfish *Mola mola* (L.): Role of gelatinous exoskeleton, horizontal septum, muscles and tendons. J. Anat. 233: 347–357.

Dewar, H., T. Thys, S.L.H. Teo, C. Farwell, J. O'Sullivan, T. Tobayama et al. 2010. Satellite tracking the world's largest jelly predator, the Ocean Sunfish, *Mola mola*, in the Western Pacific. J. Exp. Mar. Biol. Ecol. 393: 32–42.

Forsgren, K., R.S. McBride, T. Nakatsubo, T.M. Thys, C.D. Carson, E.K. Tholke et al. 2020 Reproductive biology of the ocean sunfishes. pp. 87–104. *In*: T.M. Thys, G.C. Hays and J.D.R. Houghton [eds.]. The Ocean Sunfishes: Evolution, Biology and Conservation, CRC Press. Boca Raton, FL, USA.

Fraser-Brunner, A. 1943. Notes on the plectognath fishes. VIII. The classification of the suborder Tetraodontoidea, with a synopsis of the genera. Ann. Mag. Nat. Hist. Ser. 11. 10: 1–18.

Fraser-Brunner, A. 1951. The ocean sunfishes (family Molidae). Bull. Br. Mus. (Nat. Hist.) Zool. 1: 89–121.

Fridberg, G. and H.A. Bern. 1968. The urophysis and the caudal neurosecretory system of fishes. Biol. Rev. 43: 175–199.

Gauldie, R.W. 1990. Vaterite otoliths from the Opah, *Lampris immaculatus* and two species of Sunfish, *Mola mola* and *M. ramsayi*. Acta Zoologica (Stockholm) 71: 193–199.

Gesner, K. 1558. Historiae animalium, Liber IV. Qui est de piscium and aquatilium animantium natura. Tigurum (Zurich).

Gill, T.N. 1884. Synopsis of the plectognath fishes. Proc. U.S. Nat. Mus. 7(26-27)(art. 448): 411–427.

Goette, A. 1879. Beiträge zur vergleichenden Morphologie des Skeletsystems der Wirbelthiere. II. Die Wirbelsaule und ihre Anhänge. 5. Die Teleostier. Arch. Mikrosk. Anat. (Bonn) 16: 117–142.

Goodsir, J. 1841. On certain peculiarities in the structure of the Short Sun-fish (*Orthagoriscus mola*). Edinburgh New Philos. J. 30: 188–194.

Green, E.H. 1899. The chemical composition of the sub-dermal connective tissue of the Ocean Sunfish. Bull. U.S. Fish Comm. 19: 321–324.

Gregory, W.K. and H.C. Raven. 1934. Notes on the anatomy and relationships of the Ocean Sunfish (*Mola mola*). Copeia 1934(4): 145–151.

Gudger, E.W. 1937a. The natural history and geographical distribution of the Pointed-tailed Ocean Sunfish (*Masturus lanceolatus*), with notes on the shape of the tail. Proc. Zool. Soc. London, Ser. A, p. 363–396.

Gudger, E.W. 1937b. The structure and development of the pointed tail of the Ocean Sunfish, *Masturus lanceolatus*. Ann. Mag. Nat. Hist. Ser. 10, 19: 1–46.

Hamana, A.K. 1962. Über die Neurophysis spinalis caudalis bei Fischen. 3. Kyoto Pref. Med. Univ. 71: 478–490.

Harting, P. 1865. Notices zoologiques, anatomiques et histiologiques, sur l'*Orthragoriscus ozodura*; suivies de considérations sur l'ostéogénèse des téléostiens en général. Verb. K. Akad. Wet. (Amsterdam) ll(2): 1–48.

Hellenbrecht, L.M., M. Freese, J.-D. Pohlmann, H. Westerberg, T. Blancke and R. Hanel. 2019. Larval distribution of the ocean sunfishes *Ranzania laevis* and *Masturus lanceolatus* (Tetraodontiformes: Molidae) in the Sargasso Sea subtropical convergence zone. J. Plankton Res. 41: 595–608.

Houghton, D.R., N. Liebsch, T.K. Doyle, A.C. Gleiss, M.K.S. Lilley, R.P. Wilson et al. 2009. Harnessing the sun: Testing a novel attachment method to record fine scale movements in Ocean Sunfish (*Mola mola*). Tagging and

Tracking of Marine Animals with Electronic Devices, Reviews: Methods and Technologies in Fish Biology and Fisheries 9, DOI 10.1007/978-1-4020-9640-2 14.

Hughes, G.M. 1984. General anatomy of the gills. Fish Physiology 10: 1–72.

Jinxiang, S. and L. Wanduan. 1982. A study of patterns of the olfactory organ of tetraodontiform fishes in China. Acta Zoologica Sinica 28: 389–98.

Johnson, G.D. and R. Britz. 2005. Leis' conundrum: Homology of the clavus of the ocean sunfishes. 2. Ontogeny of the median fins and axial skeleton of *Ranzania laevis* (Teleostei, Tetraodontiformes, Molidae). J. Morph. 266: 11–21.

Kappers, C.U.A.-, G.C. Huber and E.C. Crosby. 1936. The Comparative Anatomy of the Nervous System of Vertebrates, Including Man, Vol. 1. Hafner, New York.

Kaschkaroff, D.N. 1914a. Materialien zur vergleichenden Morphologic der Fische. Vergleichendes Studium der Organisation von Plectognathi. Bull. Soc. Imp. Nat. (Moscow), New Ser. 27: 263–370.

Kaschkaroff, D.N. 1914b. Zur Kenntnis des feineren Baues und der Entwickelung des Knochens bei Teleostiern. I. Die Knochenentwickelung bei *Orthagoriscus mola*. Anat. Anz. 47: 113–138.

Kaschkaroff, D.N. 1916. La structure et le developement de l'os d' *Orthagoriscus mola*. Zool. Zh. (Rev. Zool. Russe, Moscow) 1: 136–138.

Katayama, E. and K. Matsuura. 2016. Fine structures of scales of ocean sunfishes (Actinopterygii, Tetraodontiformes, Molidae): Another morphological character supporting phylogenetic relationships of the molid genera. Bull. Natl. Mus. Nat. Sci., Ser. A 42: 95–98.

Kino, M., T. Miayzaki, T. Iwami and J. Kohbara. 2009. Retinal topography of ganglion cells in immature Ocean Sunfish, *Mola mola*. Environmental Biology of Fishes 85: 33–38.

Koelreuter, I.T. 1766. Piscium rariorum e museo Petropolitano exceptorum descriptiones continuatae. Novi Commentarii Academiae Scientiarum Imperialis Petropolitanae v. 10 (for 1764): 329–351, Pl. 8.

Kölliker, R.A. 1859. On the different types in the microscopic structure of the skeleton of osseous fishes. Ann. Mag. Nat. Hist., Ser. 3, 4: 67–77.

Kölliker, R.A. 1860. Ueber die Knochen von *Orthagoriscus*. Sitzungber. Phys.-Med. Ges. (Wtirtzburg), 1859: 38.

Konstantinidis, P. and G.D. Johnson. 2012. Ontogeny of the jaw apparatus and suspensorium of the Tetraodontiformes. Acta Zool. 93: 351–366.

Leis, J.M. 1977. Development of eggs and larvae of Slender Mola, *Ranzania laevis* (Pisces, Molidae). Bull. Mar. Sci. 27: 448–466.

Leis, J.M. 1984. Tetraodontiformes: Relationships. pp. 459–463. *In*: H.G. Moser, W.J. Richards, D.M. Cohen, M.P. Fahay, A.W. Kendall and S.L. Richardson [eds.]. Ontogeny and Systematics of Fishes. Allen Press, Lawrence, KS, USA.

Leydig, F. 1857. Lehrbuch der Histologie des Menschen und der Thiere. Frankfurt.

Logan, V.H. and P.H. Odense. 1974. Integument of Ocean Sunfish (*Mola mola* L.) (Plectognathi) with observations on lesions from 2 ectoparasites, *Capsala martinieri* (Trematoda) and *Philorthagoriscus serrates* (Copepoda). Can. J. Zool. 52(8): 1039–1045.

Lyczkowski-Shultz, J. 2005. Molidae: Molas, ocean sunfishes. *In*: W.J. Richards [ed.]. Early Stages of Atlantic Fishes: An Identification Guide for the Western Central North Atlantic. CRC Press, Boca Raton, FL, USA.

McCune, A.R. and R.L. Carlson. 2004. Twenty ways to lose your bladder: Common natural mutants in zebrafish and widespread convergence of swim bladder loss among teleost fishes. Evol. Dev. 6: 246–259.

Meek, A. 1904. Notes on the auditory organ and the orbit of *Orthagoriscus mola*. Anat. Anz. 25: 217–219.

Milne-Edwards, H. 1858. Leçons sur la physiologie et l'anatomie compare de l'homme et des animaux. Tome 3. Paris.

Nakae, M. and K. Sasaki. 2006. Peripheral nervous system of the Ocean Sunfish *Mola mola* (Tetraodontiformes: Molidae). Ichthyol. Res. 53: 233–246.

Nakae, M. and K. Sasaki. 2010. Lateral line system and its innervation in Tetraodontiformes with outgroup comparisons: Descriptions and phylogenetic implications. J. Morph. 271: 559–579.

Nakamura, I., Y. Goto and K. Sato 2015. Ocean Sunfish rewarm at the surface after deep excursions to forage for siphonophores. J. Ani. Ecol. 84: 590–603.

Nolf, D. 1985. Otolithi Piscium. *In*: H.P. Schultze [ed.]. Handbook of Paleoichthyology. 10. Fischer, Stuttgart and New York. 1–145.

Nolf, D. and J.C. Tyler. 2006. Otolith evidence concerning interrelationships of caproid, zeiform and tetraodontiform fishes. Bulletin de L'Institut Royal des Sciences Naturelles de Belgique, Biologie 76: 147–189.

Nowikoff, M. 1910. Ueber den Bau des Knochens von *Orthagoriscus mola*. Anat. Anz. 37: 97–106.

Nyegaard, M. and E. Sawai. 2018. Species identification of sunfish specimens (Genera *Mola* and *Masturus*, Family Molidae) from Australian and New Zealand natural history museum collections and other local sources. Data Brief. 19: 2404–2415.

Nyegaard, M., E. Sawai, N. Gemmell, J. Gillum, N.R. Loneragan, Y. Yamanoue et al. 2018. Hiding in broad daylight: molecular and morphological data reveal a new ocean sunfish species (Tetraodontiformes: Molidae) that has eluded recognition. Zool. J. Linnean Soc. 182: 631–658 [First published online on 19 July 2017].

Owen, R. 1866. On the Anatomy of Vertebrates. Vol. 1. Fishes and reptiles. Longmans, Green and Company, London.

Oxman, D.S., R. Barnett-Johnson, M.E. Smith, A. Coffin, D.L. Miller, R. Josephson et al. 2007. The effect of vaterite deposition on sound reception, otolith morphology, and inner ear sensory epithelia in hatchery-reared Chinook salmon (*Oncorhynchus tshawytscha*). Can. J. Fish. Aquat. Sci. 64: 1469–1478.

Parker, G.H. 1900. Note on the blood vessels of the heart in the Sun-fish (*Orthagoriscus mola* Linn.). Anatomischer Anzeiger 17: 313–316.

Phillips, N.D., C. Harrod, A.R. Gates, T.M. Thys and J.D.R. Houghton. 2015. Seeking the sun in deep, dark places: mesopelagic sightings of ocean sunfishes (Molidae). J. Fish. Bio. 87: 1118–1126.

Putnam, F.W. 1871. On the young of *Orthagoriscus mola*. Proc. Amer. Soc. Adv. Sci. for 1870, 19 meet : 255–260.

Raven, H.C. 1939a. Notes on the anatomy of *Ranzania truncata*. Amer. Mus. Novitates 1038: 1–7.

Raven, H.C. 1939b. On the anatomy and evolution of the locomotor apparatus of the Nipple-tailed Ocean Sunfish (*Masturus lanceolatus*). Bull. Amer. Mus. Nat. Hist. 76: 143–150.

Regan, C.T. 1903. On the classification of the fishes of the suborder Plectognathi; with notes and descriptions of new species from specimens in the British Museum collection. Proc. Zool. Soc. Lond. 1902: 284–303.

Reimer, T., T. Dempster, F. Warren-Myers, A.J. Jensen and S.E. Swearer. 2016. High prevalence of vaterite in sagittal otoliths causes hearing impairment in farmed fish. Sci. Rep. 6: 25249.

Ridewood, W.G. 1899. On the relations of the efferent branchial blood vessels to the circulus cephalicus in teleostean fishes. Proc. Zool. Soc. London. 1899: 939–956, pl. LXIII–LXV.

Rondelet, G. 1554. Libri de piscibus marinis, in quibus verae piscium effigiens expressae sunt. Quae in tota piscium historia contineantur, indicat elenchus pagina nona et decima. Postremo accesserunt indices necessarii. Lugdunum (Lyon): Matthias Bonhomme.

Roon, J.M. van and J.J. ter Pelkwijk. 1939. Mechanism of the jaw and body muscles of *Orthagoriscus mola* L. Zool. Meded. (Leiden) 22: 66–75.

Rosén, N. 1912. Studies on the plectognaths. 1. The blood-vascular system. Ark. Zool. 7: 1–24.

Rosén, N. 1913a. Studies on the plectognaths. 2. The air-sac, with notes on other parts of the intestines. Ark. Zool. 7: 1–23.

Rosén, N. 1913b. Studies on the plectognaths. 3. The integument. Ark. Zool. 8: 1–29.

Rosén, N. 1913c. Studies on the plectognaths. 4. The body-muscles. Ark. Zool. 8: 1–14.

Rosén, N. 1916a. Studies on the plectognaths. 5. The skeleton. Ark. Zool. 10: 1–28.

Rosén, N. 1916b. Ueber die Homologie der Fischschuppen. Ark. Zool. 10: 1–36.

Ryder, J.A. 1886. On the origin of heterocercy and the evolution of the fins and fin-rays of fishes. Rep. U.S. Fish Comm. 1884: 981–1085.

Santini, F. and J.C. Tyler. 2002. Phylogeny of the ocean sunfishes (Molidae, Tetraodontiformes), a highly derived group of teleost fishes. Ital. J. Zool. 69: 37–43.

Santini, F. and J.C. Tyler. 2003. A phylogeny of the families of fossil and extant tetraodontiform fishes (Acanthomorpha, Tetraodontiformes), Upper Cretaceous to Recent. Zool. J. Linnean Soc. 139: 565–617.

Sawai, E., Y. Yamanoue, M. Nyegaard and Y. Sakai. 2018. Redescription of the Bump-head Sunfish *Mola alexandrini* (Ranzani 1839), senior synonym of *Mola ramsayi* (Giglioli 1883), with designation of a neotype for *Mola mola* (Linnaeus 1758) (Tetraodontiformes: Molidae). Ichthyol. Res. 65: 142–160.

Sawai, E., M. Nyegaard and Y. Yamanoue. 2020. Phylogeny, taxonomy and size records of ocean sunfishes. pp. 18–36. *In*: T.M. Thys, G.C. Hays and J.D.R. Houghton [eds.]. The Ocean Sunfishes: Evolution, Biology and Conservation, CRC Press. Boca Raton, FL, USA.

Schmidt, J. 1921. Contributions to the knowledge of the young of the sunfishes (*Mola* and *Ranzania*). Meddel. Komm. Havunders. Ser. Fiskeri 6: 1–16.

Sousa, L.L., I. Nakamura and D.W. Sims. 2020. Movements and foraging behavior of the ocean sunfishes. pp. 129–145. *In*: T.M. Thys, G.C. Hays and J.D.R. Houghton [eds.]. The Ocean Sunfishes: Evolution, Biology and Conservation, CRC Press. Boca Raton, FL, USA.

Steenstrup, J. and C. Lütken. 1863. Oversigt over det kgl. Danske Vidensk Selsk Forhandl. Copenhagen.

Steenstrup. J. and C. Lütken. 1898. Spolia Atlantica. Bidrag til kundskab om klump-eller maanefiskene (Molidae). Dan. Vidensk. Selsk. Skr. 6te Raek. Naturvidensk. Math. Afd. Copenhagen. 9: 1–102.

Stephan, P. 1900. Recherches histologiques sur la structure du tissus osseux des poissons. Bull. Biol. Sci. Fr. Belg. Paris. Ser. 5, 2: 281–429.

Studnicka, F.K. 1916. Ueber den knochen von *Orthagoriscus*. Anat. Anz. 49: 151–169, 177–194.

Supino, F. 1904. Contributo allo studio del tessuto osseo dell'*Orthagoriscus*. Atti R. Accad. Lincei, Rend., Cl. Sci. Fis., Mat. Nat., Ser. 5, 13: 118–121.

Suyehiro, Y. 1942. A study on the digestive system and feeding habits of fish. Jap. J. Zool. 10: 1–303.

Thompson, D'A.W. 1889. On the auditory labyrinth of *Orthagoriscus mola* L. Anat. Anz. 3: 93–96.

Thys, T.M., J.P. Ryan, H. Dewar, C.R. Perle, K. Lyons, J. O'Sullivan et al. 2015. Ecology of the Ocean Sunfish, *Mola mola*, in the southern California current system. J. Exp. Mar. Biol. Ecol. 471: 64–76.

Thys, T.M., M. Nyegaard, J.L. Whitney, I. Potter, J. Ryan, T. Nakatsubo et al. 2020. Ocean sunfish larvae: detections, identification and predation. pp. 105–128. *In*: T.M. Thys, G.C. Hays and J.D.R. Houghton [eds.]. The Ocean Sunfishes: Evolution, Biology and Conservation, CRC Press. Boca Raton, FL, USA.

Trois, E.F. 1883–1884. Ricerche sulla struttura della *Ranzania truncata*. Parts I, II. Atti del R. Istituto Veneto, Tom. II, (6), Part I, pp. 1269–1306; Part II, pp. 1543–1559.

Tyler, J.C. 1962. The pelvis and pelvic fin of plectognath fishes; a study in reduction. Proc. Acad. Nat. Sci. Phila. 114: 207–250.

Tyler, J.C. 1970. The progressive reduction in number of elements supporting the caudal fin of fishes of the order Plectognathi. Proc. Acad. Nat. Sci. Philadelphia. 122: 1–85.

Tyler, J.C. 1980. Osteology, phylogeny, and higher classification of the fishes of the order Plectognathi (Tetraodontiformes). NOAA Technical Report NMFS Circular 434: 1–422.

Tyler, J.C. and A.F. Bannikov. 1992. New genus of primitive ocean sunfish with separate premaxillae from the Eocene of southwest Russia (Molidae, Tetraodontiformes). Copeia, 1992: 1014–1023.

Uehara, M. and T. Ueshima. 1986. Morphological studies of the spinal cord in Tetraodontiformes fishes. J. Morph. 190: 325–333.

Uehara, M., Y.Z. Hosaka, H. Doi and H. Sakai. 2015. The shortened spinal cord in tetraodontiform fishes. J. Morph. 276: 290–300.

Vignal, W. 1881. Note sur l'anatomie des centres nerveux du mole, *Orthagoriscus mola*. Arch. Zool. Exp. Gen. 9: 369–386.

Watanabe, Y. and K. Sato. 2008. Functional dorsoventral symmetry in relation to lift-based swimming in the Ocean Sunfish *Mola mola*. Plos One 3(10): e3446.

Watanabe, Y.Y. and J. Davenport. 2020. Locomotory systems and biomechanics of ocean sunfish. pp. 72–86. *In*: T.M. Thys, G.C. Hays and J.D.R. Houghton [eds.]. The Ocean Sunfishes: Evolution, Biology and Conservation, CRC Press. Boca Raton, FL, USA.

Winterbottom, R. 1974. The familial phylogeny of the Tetraodontiformes (Acanthopterygii: Pisces) as evidenced by their comparative myology. Smithson. Contrib. Zool. 155: 1–201.

Wilson, J.M. and L.F.C. Castro. 2011. Morphological diversity of the gastrointestinal tract in fishes. pp. 1–55. *In*: M. Grosell, A.P. Farrell and C.J. Brauner [eds.]. Fish Physiology 30: The Multifunctional Gut of Fish. Elsevier Inc., London.

Yamanoue, Y., M. Miya, K. Matsuura, M. Katoh, H. Sakai and M. Nishida. 2004. Mitochondrial genomes and phylogeny of the ocean sunfishes (Tetraodontiformes: Molidae). Ichthyol. Res. 51(3): 269–273.

Yamanoue, Y., M. Miya, K. Matsuura, M. Katoh, H. Sakai and M. Nishida. 2008. A new perspective on phylogeny and evolution of tetraodontiform fishes (Pisces: Acanthopterygii) based on whole mitochondrial genome sequences: Basal ecological diversification? BMC Evolutionary Biology 8: 212–226.

Yoshita, Y., Y. Yamanoue, K. Sagara, M. Nishibori, H. Kuniyoshi, T. Umino et al. 2009. Phylogenetic relationship of two *Mola* sunfishes (Tetraodontiformes: Molidae) occurring around the coast of Japan, with notes on their geographical distribution and morphological characteristics. Ichthyol. Res. 56: 232–244.

Chapter 5

Locomotory Systems and Biomechanics of Ocean Sunfish

Yuuki Y. Watanabe[1,2,*] and *John Davenport*[3]

Introduction

The ocean sunfish *Mola mola* (Linnaeus 1758) has an unusual appearance for a fish, with a massive, laterally-compressed body, exaggerated dorsal and anal fins, and an absent caudal fin. The first question someone might pose upon seeing this species is 'how does this strange-looking fish swim?' The first published scientific observations on sunfish swimming were made in the 1800s by American zoologist John Ryder (Ryder 1885). Ryder reported that ocean sunfish almost exclusively use the dorsal and anal fins as locomotive appendages, and their tail as a rudder, while their pectoral fins function as balancers (Ryder 1885). Since then, few studies on the locomotor systems and biomechanics of ocean sunfish have been conducted, primarily due to their cumbersome size (typically 30–300 kg) and scant commercial value (at least in the Western Hemisphere). However, detailed anatomical studies (Bemis et al. 2020 [Chapter 4]), and behavioral studies using electronic tags to record the species' natural swimming behavior, have recently been published (e.g., Watanabe and Sato 2008, Houghton et al. 2009, Davenport et al. 2018) and our understanding of their locomotory systems and biomechanics has significantly improved.

Here, we review how ocean sunfish, specifically *M. mola*, swim from the kinematic, morphological and anatomical points of view. We then place our knowledge into an ecological context to hypothesize how this strange-looking fish with its unusual swimming style may have evolved. It is noteworthy however, that most of our knowledge comes from the single species, *M. mola*, so care should be taken when extrapolating to other Molidae species. All dissections and biomechanical study conducted so far have been attributed to *M. mola* and it is likely, but cannot be assumed, that other externally similar members of the genus, e.g., *Mola alexandrini* (Ranzani 1839), senior synonym of *Mola ramsayi* (Giglioli 1882), and *Mola tecta* Nyegaard et al. 2017, are internally similar in structure to *M. mola*. For the purposes of this investigation, our use of the term ocean sunfish refers only to *M. mola* unless otherwise stated.

[1] National Institute of Polar Research, Tachikawa, Tokyo 190-8518, Japan.
[2] Department of Polar Science, The Graduate University for Advanced Studies, SOKENDAI , Tachikawa, Tokyo 190-8518, Japan.
[3] School of Biological, Earth and Environmental Sciences and Environmental Research Institute, University College Cork, Ireland; Email: J.Davenport@ucc.ie
* Corresponding author: watanabe.yuuki@nipr.ac.jp

Fin Shape and Movement

As described in late 19th century (Ryder 1885), ocean sunfish swim by simultaneously flapping their exaggerated dorsal and anal fins laterally (Fig. 1). The fins' longitudinal axes are set at angles to the direction of water flow over them ('angle of attack') and produce lift-based thrust forces. In support of this observation, acceleration signatures recorded from free-swimming ocean sunfish are similar to those of free-swimming penguins (Watanabe and Sato 2008), except that flapping is lateral in sunfish and dorsoventral in penguins. In cross section, the shape of dorsal and anal fins of ocean sunfish is streamlined, with a rounded leading edge and tapered trailing edge (Fig. 1A). The elongate dorsal and anal fins are similar in shape to each other within individuals (Fig. 1B). Interestingly, the shape of the two fins change ontogenetically, with larger individuals having wider (i.e., lower aspect ratio) fins; nevertheless, two fins remain symmetrical to each other within individuals during growth (Fig. 1C).

As such, the dorsal and anal fins of ocean sunfish can be considered as paired, vertical wings. This is interesting, because the two fins have different developmental origins in fishes in general and are asymmetrical in most other teleost fish species (Lauder and Drucker 2004). The propulsive arrangement of ocean sunfish is unique even among aquatic vertebrates. All other underwater fliers (e.g., manta rays, sea turtles, sealions, and penguins) use their symmetrical appendages (e.g., flippers) located at both sides of the body as paired horizontal as opposed to vertical wings. Symmetry of dorsal and anal fins in ocean sunfish is considered important for balanced swimming, because lift

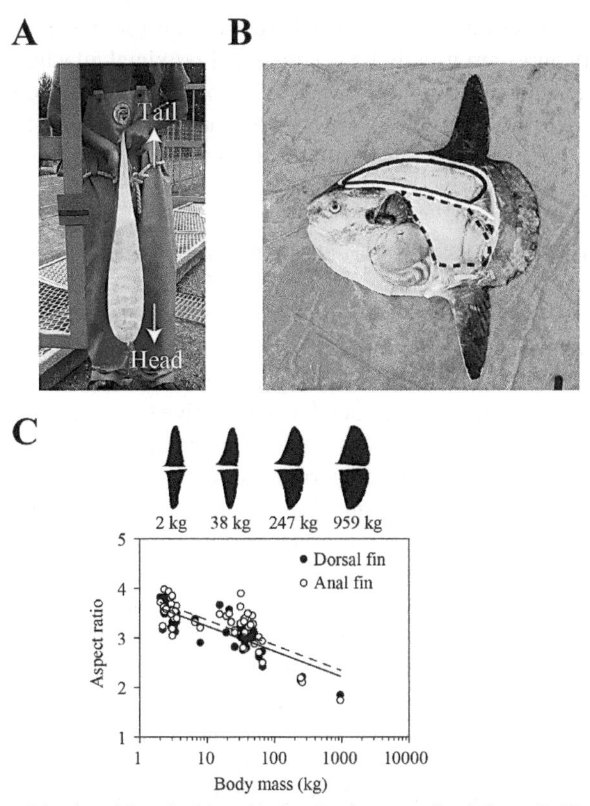

Figure 1. A. Cross section of the dorsal fin of a *M. mola*, showing its streamlined shape. B. The arrangement of dorsal fin muscle (solid line) and anal fin muscles (dashed line). The two muscles are separated by the horizontal septum (white line). C. Relationship between body mass (BM) (on log scale) and aspect ratio (AR) (i.e., the ratio of the length to the mean width, calculated as length squared divided by projected area) for dorsal fins (filled circles, with solid regression line) and anal fins (open circles, with dashed regression line). Best-fit regression lines are: AR = 3.75–0.51 × $\log_{10}$ (BM) (N = 49, R^2 = 0.61) for dorsal fins, and AR = 3.88–0.51 × $\log_{10}$ (BM) (N = 49, R^2 = 0.54) for anal fins (BM range 2–959 kg). Examples of the fins (outlined from photos) from four mass groups are also shown. B and C are redrawn from Watanabe and Sato (2008).

forces produced by a wing are proportional to wing area, and mechanical efficiency depends on wing shape, especially aspect ratio (i.e., the ratio of the span (span = length of fin) to the mean chord (= mean width of fin); Alexander 2003). The two fins are driven by different sets of muscles (Fig. 1B), as discussed in detail below. The two sets of muscles have similar masses within individuals despite markedly different shapes, suggesting that the two fins are driven by nearly equal power (Watanabe and Sato 2008).

Ocean sunfish oscillate their pectoral fins to balance their bodies in the water column (Fig. 1B). Sunfish also have a unique and evolutionarily novel tail-like structure, the clavus which has been said to function as a simple rudder (Fraser-Brunner 1951). Although its embryological origin was much debated in the 20th century, the clavus has now been shown, in another molid species (viz. *Ranzania laevis*, Pennant 1776), to be derived from dorsal and anal fin elements (Johnson and Britz 2005).

White Muscles

The musculature of *M. mola* related to swimming has long been known to be markedly different from the general teleost pattern (e.g., Gregory and Raven 1934), because it is dominated by the muscles driving the dorsal and anal fins. In contrast, more typical teleosts (e.g., salmon and cod) have a dominant myotomal axial musculature that drives the caudal fin. Davenport et al. (2018) recently investigated the musculoskeletal anatomy of *M. mola* in greater detail and their findings are outlined here.

In general, vertebrate muscles are attached to structures at their two ends, and their contraction tends to bring the structures closer to one another. By convention (derived primarily from mammalian studies, where the structures are usually bony), the fixed proximal attachment is called the origin, while the mobile distal attachment (known as the insertion) moves with contraction. Such muscles are attached to origins and insertions by collagenous tendons. These nomenclatures have weaknesses when applied to fish, in which muscles can be attached to elastic, mobile structures at each end, as well as to the elastic envelope of the skin, and may or may not exhibit typical tendons (Westneat and Wainwright, 2001).

When an ocean sunfish has its skin and underlying hypodermis removed, the lateral surfaces of the dorsal and anal fin musculature are revealed (Figs. 1B, 2). They are cream/white in color with little vascularization and presumably anaerobic white muscles. Almost all anal fin white muscles have broad origins on the ventral surfaces of a thick, tough, multi-layered elastic fibrous sheet called the horizontal septum (Fig. 2C; see also schematic Fig. 3. B,C). This septum runs dorsal and lateral to the vertebral column to which it is firmly bound by connective tissue and is also firmly bound to the inner surfaces of the hypodermis. In this position, the horizontal septum acts as an elastic diaphragm between the dorsal and anal fin muscle compartments. It is non-gelatinous and much more elastic than the hypodermis. The horizontal septum of conventional teleosts is known to be a collagenous tensile element that acts as a major force transmitter from the axial musculature to the axial skeleton (Westneat and Wainwright 2001); in *M. mola,* it has been co-opted into transmitting forces from the dorsal and anal fin musculature to the dorsal and anal fins. The importance of the horizontal septum in *M. mola* was not recognized by Gregory and Raven (1934), whose drawings incorrectly implied that dorsal and anal fin muscles were attached to the vertebral column.

Almost all anal white muscle origins occupy the full length and width of the ventral surface of the horizontal septum from the rear of the visceral cavity to the end of the vertebral column (Figs. 3, 4); a small number of anal fin white muscles have origins on the interior surface of the hypodermis. Mainly, the muscle and tendons are directed dorsoventrally, though the anterior muscles are rather longer and directed caudally as well as dorsoventrally, so that their contraction elevates the fin as well as flaps it to the side.

The anal fin white muscles are inserted (via long tendons that pass through the hemal radial cartilages) onto processes at the proximal ends of the bony rays (lepidotrichia) of the anal fin (Fig. 4). Manipulation of the muscles indicates that they are primarily inclinators that serve to move

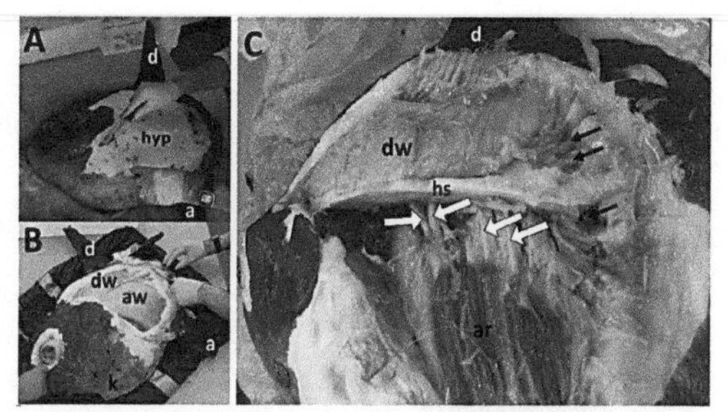

Figure 2. Dissection of *Mola mola* (redrawn from Davenport et al. 2018). A. Oblique view of fish from left-hand side and from ventral aspect. Key: dorsal fin (d), anal fin (a) hypodermis (also known as capsule in Davenport et al. 2018) (hyp). B. Oblique view of fish from anterior and ventral aspects, with hypodermis removed to reveal white muscles of dorsal (dw) and anal (aw) fins. The keel (k) is also labelled. C. Lateral view of fish. Note that the image exhibits barrel distortion, with head, medial fins and clavus curving away from the central part of the image. White anal fin muscles have been removed. Key: dorsal fin white muscles (dw), anal fin red muscles (ar), fibrous horizontal septum (hs). Black arrows indicate claval muscles; white arrows indicate hemal spines.

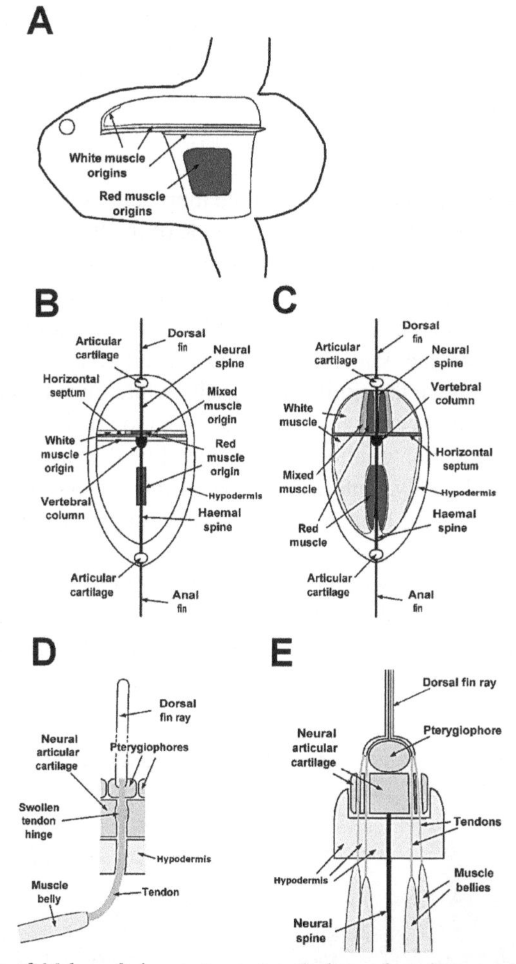

Figure 3. Schematic diagrams of *Mola mola* locomotor system (redrawn from Davenport et al. 2018). A. Lateral view to indicate location of origins of white and red muscles. B. Transverse section through muscle compartments to indicate location of origins of white muscles, red muscles and mixed red and white muscles. C. Transverse section through muscle compartments to indicate location of white, red and mixture of red and white muscle blocks. D. Simplified diagram of relationship between muscle, tendon, hypodermis, articular cartilage and dorsal fin ray from lateral aspect. E. Simplified transverse section diagram of relationship between muscle bellies, tendons, hypodermis, articular cartilage and dorsal fin ray.

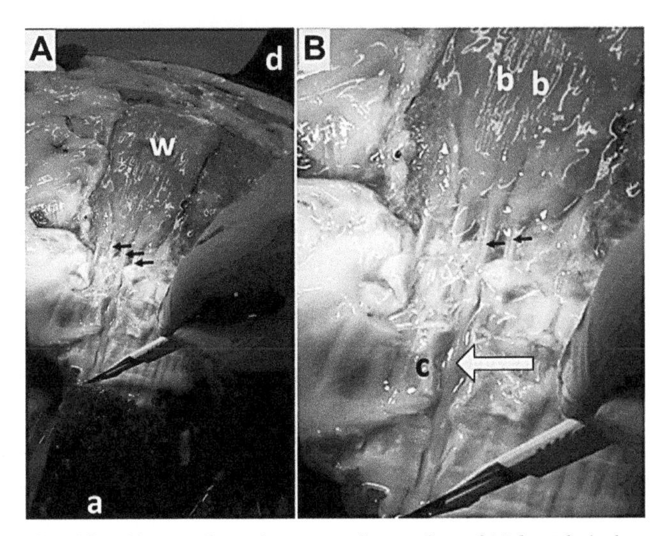

Figure 4. Arrangement of anal fin white muscles and corresponding tendons of *Mola mola* (redrawn from Davenport et al. 2018). A. Lateral view, hypodermis mostly removed. Key: anal fin (a), dorsal fin (d), anal fin white muscle (w). Black arrows indicate tendons. B. Close-up of basal area of anal fin. Key: bellies of white muscles (b), hemal radial cartilage (c). Black arrows indicate tendons; large light grey arrow indicates swollen portion of tendon sheath within radial cartilage; point of scalpel indicates distal part of tendon.

the rays from a central position to the side, though the more anterior muscles also served to elevate (and thereby spread) the anal fin. The muscles on one side of the fish are antagonistic to the muscles on the other side, so that alternate contraction causes the anal fin to flap from side-to-side.

Many of the dorsal fin white muscles have origins on the dorsal surface of the horizontal septum, the surface of which runs anteriorly above the skull and acts as the floor to a chamber (semi-circular section) in the hypodermis above the skull. Some of the white muscles have origins in the hypodermis, laterally and in the chamber above the skull (Fig. 5). The white muscles are connected via tendons (that run through the neural radial cartilages; see Fig. 3D, E) to the fin rays of the dorsal fin, but the length of the muscles and their orientation varies considerably. Posteriorly, the muscles and tendons are short and directed ventrodorsally. Anteriorly, many of the muscles are long and directed almost parallel with the vertebral column; their tendons curve through hypodermis channels and radial cartilages to meet the fin rays.

Figure 6 illustrates the complexity of the dorsal fin white muscles. In some cases, at the anterior end of the dorsal chamber, multiple short white muscle bellies (bipennate muscles) are attached to shared tendons (Fig. 6A); these bellies have origins on both hypodermis and horizontal septum. When pennate muscles contract and shorten, their pennate angle increases, transferring force to the

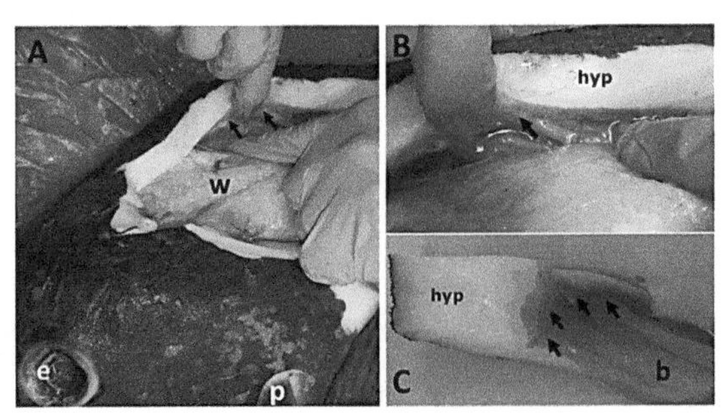

Figure 5. Muscle origins on hypodermis of *Mola mola* (redrawn from Davenport et al. 2018). A. View of muscle chamber above skull. Key: white muscle (w), black arrows indicate position of origins. B., C. Close-ups of white muscle origins (arrowed). Key: hypodermis (hyp), muscle belly (b).

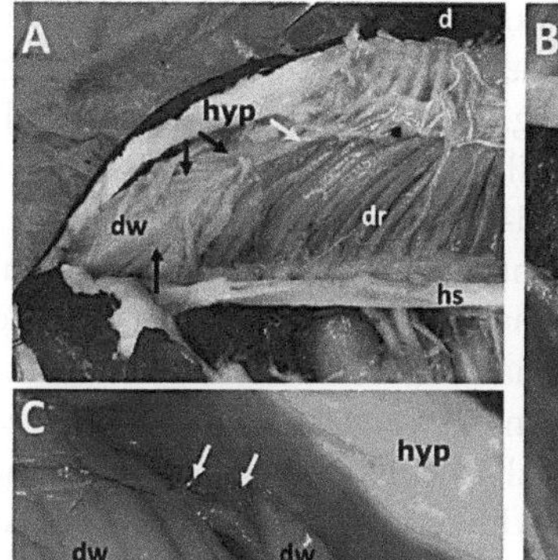

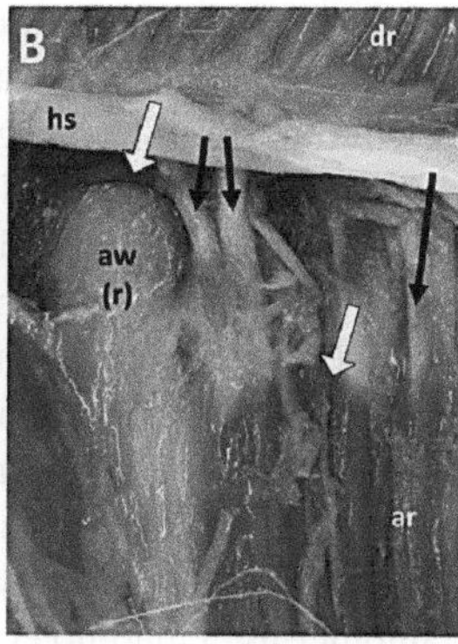

Figure 6. Detail of arrangements of locomotory muscles of dorsal and anal fins of *M. mola* (redrawn from Davenport et al. 2018). A. Muscle chamber above skull (most dorsal fin white muscles removed). Key: dorsal fin (d), hypodermis (hyp), dorsal fin white muscles (dw), dorsal fin red muscles (dr), horizontal septum (hs). Black arrows indicate separate white muscle bellies connected to a single tendon (indicated by white arrow), forming a bipennate muscle. B. Close-up of midsection of horizontal septum (hs), all white muscles removed from left side of fish. Key: dorsal fin red muscles (dr), anal fin red muscles (ar). Medial surface of anterior anal fin white muscles of right side of fish (aw(r)). White arrows indicate blood vessels, black arrows indicate hemal spines. C. Close-up of dorsal fin muscle origins at anterior of muscle chamber. Red muscle origins (indicated by white arrows) are medial to those of dorsal fin white muscles (dw).

tendon. Pennate muscles are known from terrestrial vertebrates (particularly mammals) and are also found in the chelipeds of crabs (Alexander 1979). These types of muscles generally allow higher force production, but a smaller range of movement (Martini and Ober 2006). Alexander (1979) demonstrated (for crab claws) that the bipennate arrangement allowed more powerful muscles to be packed into smaller spaces than is the case for conventional muscles in which the muscle fibers and tendons are parallel. Bipennate muscles have not been reported in any other fish species. In *M. mola*, the bipennate arrangement may help to facilitate a similar muscle length throughout the dorsal white muscle mass.

Hidden Red Muscles

Medial to the anal fin white muscles, Davenport et al. (2018) found red aerobic muscles (not previously described) that are entirely separate from the white musculature and brown/red in color (Figs. 2 and 6); they are well-vascularized with numerous arteries and veins visible (Fig. 6). They have origins on the lateral surfaces of ventral bony projections (hemal spines) that link the vertebral column with the anal fin radial cartilages. These are the only muscles driving the dorsal and anal fins that have origins on skeletal elements; all other origins are on the upper or lower surfaces of the horizontal septum or the inner surfaces of the hypodermis (see Fig. 3B, C). The anal fin red muscles are much shorter than the overlying white muscles. Their insertions (via long tendons) are on the anal fin rays. The muscles are not connected either to the vertebral column, or to the horizontal fibrous septum. As with the anal fin white muscles, they operate primarily as inclinators. All are directed dorsoventrally and are similar in length.

The dorsal fin white muscles also overlay more medial, dark-colored red muscles (Figs. 3 and 6). However, the dorsal fin red muscles have a different arrangement from that of the anal fin red muscles. They are more medially-distributed than the white muscles that hide them, but their origins (on the horizontal septum and collagenous hypodermis) are similar in location to those of the overlying white muscles. Hence the red muscles vary greatly in length, being long and axially-orientated anteriorly, short and ventrodorsally orientated at the posterior end of the fin; curving of muscles and tendons to connect with the fin rays is like that of the white muscles. The short red muscles that drive the posterior part of the dorsal fin are separate from the more lateral white muscles. The longer anterior red muscles that drive the anterior part of the dorsal fin exhibit a slightly different pattern. There is not a clear separation between them and the more lateral white muscles. Instead, there is an intervening zone of mixed red and white muscles (see Fig. 3). Although neural vertebral spines run from the vertebral column towards the neural radial cartilages of the dorsal fin bases, no muscles have origins on them. Red muscles of the dorsal fin are well-vascularized.

The medial positioning of red propulsive muscles in *M. mola* has some similarities to that found in highly active, endothermic fishes (e.g., tunas and some lamnid sharks) (Carey et al. 1971). It implies that they will exert less force on the fin rays than the more lateral white muscles, as they are more closely aligned with the axes of the fin rays than the white muscles (c.f. tuna red muscles: Syme and Shadwick 2011). However, there may be thermal benefits, as in endothermic fishes, especially in large individuals of *M. mola*. By being buried deep within the fish, they may vary less in temperature than the white muscles (gigantothermy: Paladino et al. 1990). Muscle temperature has been measured in 1–2 m long sunfish and showed thermal buffering occurred during diving cycles (Nakamura et al. 2015), but it is not known whether the temperature probes recorded from red or white muscle. Red muscle endothermy however is unlikely in *M. mola* as its cruising speed (0.2–0.7 m/s; Nakamura and Sato 2014) is comparable with that of marlin (Genus *Makaira*; which are not endothermic) and lower than endothermic tunas and lamnid sharks that cruise typically at 1–2 m/s (Watanabe et al. 2015).

The finding that white muscles mostly have origins on the dorsal and ventral surfaces of the fibrous/elastic horizontal septum (Davenport et al. 2018), means that fast swimming must be accompanied by synchronized contractions of those muscles, and hence synchronized vertical fin flapping. Otherwise, there would be inefficient interference/competition between dorsal fin and anal fin white muscle contractions. While much attention has been given to the synchronized fin-flapping in *M. mola*, slow-moving sunfish often show unsynchronized dorsal and anal fin action (e.g., a swimming *M. mola* in captivity). In contrast, slow (and sustained) swimming propulsion will be accomplished predominantly by contraction of red muscles. Since the anal fin red muscles have their origins on the hemal spines, while the dorsal fin red muscles originate on the horizontal septum, no interference/competition occurs when they contract asynchronously.

Tendons

Figures 4 and 6 show details of the tendons of the white muscles of the anal and dorsal fins respectively. From Fig. 4, it is evident that the anal fin white muscle tendons are very long, of similar individual length, and are held distally within sheaths that traverse the radial cartilages. The portions of the sheaths within the cartilages are swollen. Manipulation by Davenport et al. (2018) showed that the swollen sections can be bent easily, effectively acting as tendon hinges, facilitating flapping of the anal fin. Histologically, the swollen sections are characterized by thicker and well-vascularized epitenons (outer connective tissue surrounding tendon bundles).

Figure 6 and the schematic shown in Fig. 3D demonstrate the complexity of the tendon arrangements of the dorsal fin white muscles. The tendons vary greatly in length and most curve dorsally in the hypodermis before entering the tendon sheaths and traversing the neural radial cartilages (swollen tendon hinges are also present within these tendon sheaths). Manipulation of the muscles and tendons of the dorsal fin demonstrates that they can produce substantial lateral movements of the fins (i.e., acting as inclinators), as well as produce changes in fin shape by acting as elevators.

Figure 7 illustrates some of the muscles and tendons of the clavus. The muscles, buried in hypodermis material, are all short, red, and have origins close to one another on the rearmost part of the horizontal septum and/or the posterior end of the vertebral column. The matching tendons pass through cartilaginous material and cross a long, narrow 'hinge' of flexible connective tissue into the clavus itself where they attach to fin rays. Manipulation shows that bending the clavus can act as a simple rudder, presumably if contraction of the claval muscles is synchronized. It is also feasible that the shape of the clavus could be changed by differential contraction of those muscles, with

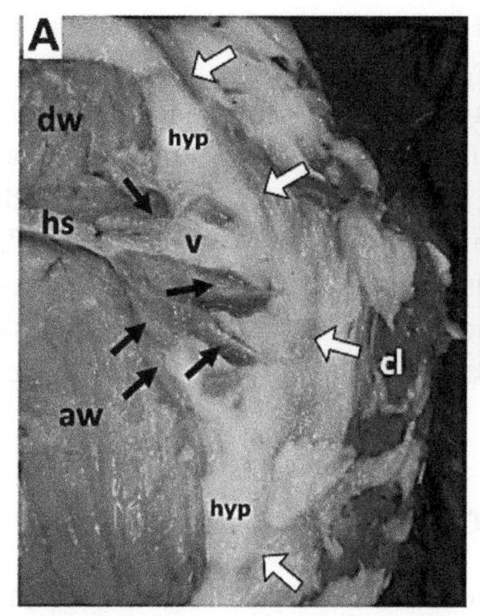

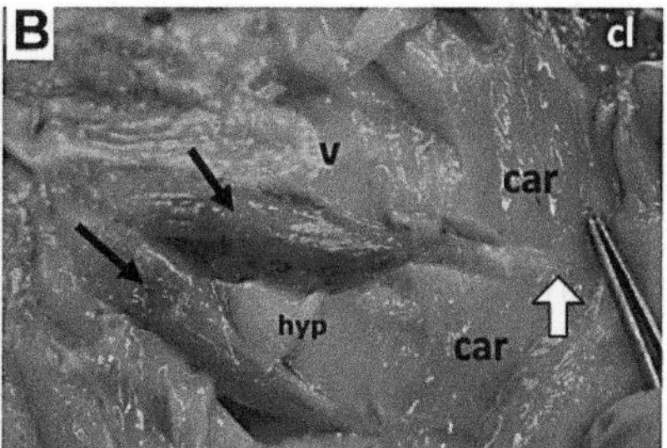

Figure 7. Detail of arrangements of locomotory muscles of clavus of *Mola mola* (redrawn from Davenport et al. 2018). A. View of rear of left-hand side of fish, hypodermal material mostly removed. Key: dorsal fin white muscle (dw), anal fin white muscle (aw), horizontal septum (hs), hypodermis (hyp), clavus (cl), posterior end of vertebral column (v). Black arrows indicate claval muscles; white arrows indicate position of soft 'hinge' of clavus. B. Close-up of two claval muscles (indicated by black arrows) and associated structures. Key: hypodermis (hyp), clavus (cl), caudal end of vertebral column (v), cartilage (car). White arrow indicates position of tendon; tip of forceps indicates position of hinge.

active flapping observed in publicly available film. The function of this behavior is unclear, however since it likely delivers very little propulsion, we suggest it is more important to stability (Davenport unpublished data). Specifically, from observations at slow swim speeds, the clavus appears to function as a rudder, but moves in synchrony with the vertical fins during fast swimming to oppose yaw. This suggests a complex stabilizing role for the clavus with a skeletal arrangement that may facilitate stability in pitch and roll (Davenport, unpublished data).

Muscle Mass

Table 1 shows the mass of the propulsive and claval muscles of a 17 kg specimen of *Mola mola* that had live-stranded on the banks of Lough Foyle, N. Ireland in 2014. Total propulsive muscle mass (one side) was 1.5 kg, implying that the overall propulsive muscle mass was 3 kg, 17.6 percent of total body mass. Dorsal and anal fin white muscles (one side) were of similar mass (0.55 kg and

Table 1. Masses and proportions of unpaired muscles of specimen (wet mass 17 kg, total length 0.67 m) of *Mola mola* (right-hand side only). (Davenport, Phillips and Eagling, unpublished data).

Fin Type	Muscle Type	Mass (kg)	% Total muscle mass	Fin Type	Muscle Type	Mass (kg)	% Total Dorsal+Anal fin muscle mass
Dorsal	White	0.55	36.7	Dorsal	both	0.80	55.2
Dorsal	Red	0.25	16.7	Anal	both	0.65	44.8
Anal	White	0.50	33.3				
Anal	Red	0.15	10.0				
						Red muscle as % propulsive muscle mass	
SUBTOTAL		1.45	96.7				
				Dorsal		31.2	
Claval	Red	0.05	3.3	Anal		23.1	
ALL		1.50	100.0	BOTH		27.5	

0.50 kg respectively), but dorsal fin red muscle mass was greater (0.25 kg) than anal fin red muscle mass (0.15 kg).

Overall propulsive muscle mass as a proportion of total body mass (17.6 percent) is low for a teleost (Johnston 1981). Partly this ratio reflects the large quantity of hypodermis, discussed below, that plays a role in buoyancy regulation but leaves less space for muscle. On the other hand, the quantity of propulsive red muscle mass in relation to total propulsive muscle mass (27.5 percent) is very high, comparable with the highest value previously recorded in any known teleost (29.8 percent in mackerel: Greer-Walker and Pull 1975). The presence of plentiful red aerobic muscles is compatible with the long-distance cruising now known to be characteristic of young sunfish (Pope et al. 2010, Nakamura and Sato 2014).

Hypodermis

The hypodermis has been variously described as inflexible, rubbery, collagenous or (most recently) as gelatinous (Watanabe and Sato 2008). Its composition has a high-water content (89.8 percent water, Tables 2 and 3), and possesses a complex collagen and elastin meshwork that confers structural elasticity and stiffness that varies over different regions over the body (Davenport et al. 2018).

The hypodermis is greasy in texture. Calculations based upon data shown in Table 3 suggest that the material's positive buoyancy might be partly provided by lipid, but no lipid droplets have been observed, and there is no visible matrix structure other than collagen and elastin fibre (Davenport et al. 2018). Thus, a more complete biochemical investigation is needed. An interesting feature of the hypodermal histology is that it is almost devoid of vascularization, suggesting that the tissue has a low turnover rate and minimal energy demand. Hypodermal thickness increases rapidly with body size, from 4 cm for a 2 kg individual to 21 cm for a 959 kg individual, when measured at the belly just anterior to the anus (Watanabe and Sato 2008). Accordingly, the proportion of hypodermis (by mass) also increases with body size, from 26 percent for a 2 kg individual to 44 percent for a 247 kg individual (Fig. 8).

The hypodermis functions as an exoskeleton. Exoskeletons (external protective and supportive structures) are common amongst invertebrates, but relatively uncommon in extant fish. Boxfishes

Table 2. Water content of tissues of *Mola mola* (Davenport et al. 2018) and *Cyclopterus lumpus* (Davenport and Kjørsvik 1986).

Tissue type	**Water content (mean % by mass, n = 3, SD in parentheses)**
Mola mola	
Dorsal fin white muscle	83.5 (3.6)
Dorsal fin red muscle	80.3 (0.2)
Anal fin white muscle	82.2 (1.1)
Anal fin red muscle	79.4 (1.0)
Hypodermis	89.8 (1.1)
Cyclopterus lumpus	
Female	
Axial white muscle	86
Subcutaneous gelatinous tissue	93
Male	
Axial white muscle	64
Hypodermis	89

Table 3. Composition of hypodermis of *Mola mola* (Davenport et al. 2018).

	Mean (n = 3)	SD
Water content as % wet mass	89.8	1.1
Salt content as % dry mass	23.4	4.5
Organic content as % dry mass	76.6	4.5
Salt content as % wet mass	2.4	0.7
Organic content as % wet mass	7.8	0.6

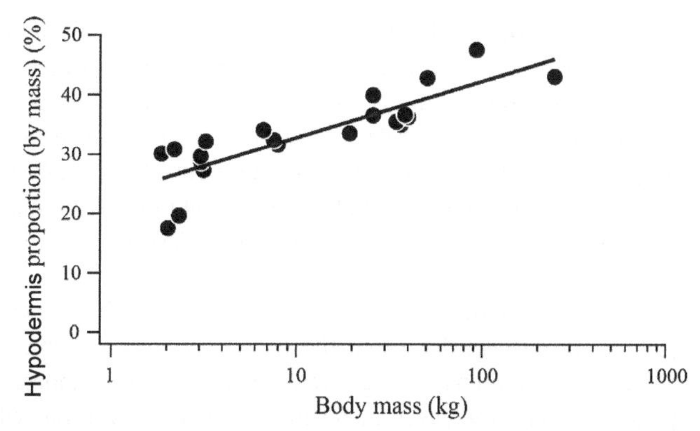

Figure 8. Relationship between body mass (BM) (on log scale) and the proportion (by mass) of the hypodermis in *Mola mola*. The best-fit regression line is: hypodermis proportion (as %) = 23.3 + 9.5 × $\log_{10}$ (BM) (N = 21, R^2 = 0.72, BM range 2–247 kg).

(Ostraciidae, Tetraodontiformes) have calcareous exoskeletons, but the only previous record of a hypodermal exoskeleton amongst teleosts has been that of the lumpfish *Cyclopterus lumpus* (Linnaeus 1758, Davenport and Kjørsvik 1986). The hypodermis of *Mola mola* covers the whole body, except for openings for the mouth, urogenital apertures, anus, opercula, eyes and fins and has a shaping function. In consequence, the sunfish body is essentially rigid, with a blunt, streamlined shape that presumably has a relatively low drag coefficient (c_d) (Lighthill 1971). Streamlining is also evident in the presence of shallow pockets medial to the pectoral fins, which allow the fins to be folded flush with the body surface, similar to arrangements seen in fast-swimming scombrid fish (Bemis et al. 2020 [Chapter 4] for further anatomical details). Future research exploring the drag coefficient of the sunfish body (e.g., modelling their 3D body shape, surface structure and Reynolds numbers) will help provide additional insight into movement mechanics.

The hypodermis's exoskeletal role is evident at the roots of the dorsal and anal fins. The neural and hemal radial cartilages are embedded in especially thick and rigid portions of the hypodermis which provide stable anchorages for the propulsive action of the fins. This region also features long, curving channels (e.g., Fig. 6A) that contain the tendon sheaths which connect muscles with fin rays. The convex and rigid posterior edge of the body hypodermis is also important structurally, since the claval muscles and tendons are embedded within it (Fig. 7) and their action on the clavus (via a hinge) allows an efficient rudder action. Finally, the hypodermis has an exoskeletal role of providing origins for muscles (via short tendons). This is clearly seen for the dorsal fin muscles that have origins on the inner surfaces of the hypodermis above the skull (Figs. 5 and 6). This is not the case, however, for the anal fin white muscles which have only a few such origins.

The hypodermis also functions as a buoyancy aid (Watanabe and Sato 2008) as originally suggested back in 1961 (Arakawa and Masuda 1961). While larval sunfish (7 mm) have swim bladders (see Fraser-Brunner 1951, Chanet et al. 2014), adult ocean sunfish lack them. They are however close to neutral buoyancy (body density, about 1.027 g/ml) in seawater, of which density varies around 1.025 g/ml depending on temperature and salinity. The tissues of *Mola mola* are relatively homogenous in density whereas the hypodermis has a density of 1.015 g/ml and is positively buoyant in sea water. The musculature is relatively watery (ca. 79–84 percent, Table 2), but is negatively buoyant (mean density, 1.041 g/ml, n = 4, Watanabe, unpublished data) in seawater because of a high protein content (protein has a density of about 1.35 g/ml, Fischer et al. 2004). Much of the skeleton of ocean sunfish is composed of cartilage (Clelland 1862). Measurements of cartilage density have rarely been carried out in any vertebrate, but Davenport and Kjørsvik (1986) recorded values of 1.033–1.052 g/ml for parts (skull, gill arch, vertebral column) of the entirely cartilaginous skeleton of the neutrally-buoyant, pelagic lumpfish, *Cyclopterus lumpus*. These values are higher than the density of sea water, so cartilaginous skeletons are presumably negatively buoyant, though much less so than those constructed of calcified bone. The bilobed liver of ocean sunfish is positively buoyant, except for in small individuals (e.g., 1.041 g/ml for a 2 kg individual and 0.992 g/ml for a 247 kg individual). However, the liver is relatively small (2.6 percent of body mass) compared to some sharks (17–30 percent) that are known to maintain neutral buoyancy by their livers (e.g., Corner et al. 1969). Overall, it was estimated that 69–100 percent of the weight in water is supported by the hypodermis, whereas the rest is supported by the liver depending on body size (Watanabe and Sato 2008).

The exoskeletal role of the hypodermis and its role in providing buoyancy both merit further exploration. Although Clelland (1862) described the ocean sunfish's internal skeleton as largely cartilaginous, public museum specimens of large *Mola mola* skeletons appear lightly calcified in places. Although hypodermal mass increases in positive allometric fashion with increasing body mass, Davenport et al. (2018) reported differences in thickness and stiffness of the hypodermis over different parts of the body. Together, these findings suggest that there are ontogenetic changes in the relationship between densities and makeup of the hypodermis and skeleton that could be fruitfully investigated. Likewise, it would be useful to explore how ontogenetic shifts in the hypodermis reflects a move towards a more predominantly gelatinous diet as sunfish grow (e.g., Phillips et al. 2020, Nakamura and Sato 2014, Sousa et al. 2016), paying attention to their nutritional value beyond mere calorific content (Hays et al. 2018).

Ecological Significance of Mola Swimming Style

The order Tetraodontiformes, comprised of approximately 430 extant species, represents one of the most morphologically and ecologically diverse radiations among teleosts (Yamanoue et al. 2008). Most species are found on coral reefs, sea grasses, and other tropical, shallow-water environments, although some species are found in deep sea (e.g., spikefish; family Triacanthodidae). Ocean sunfish (and possibly other species in the same genus) are unusual amongst Tetraodontiformes in that they inhabit coastal waters and the open ocean and can explore the water column actively, sometimes to great depths (e.g., Cartamil and Lowe 2004, Hays et al. 2009, Dewar et al. 2010, Potter and Howell 2011, Nakamura et al. 2015, Phillips et al. 2015, Thys et al. 2017). In our view, their swimming style and locomotory systems, reviewed in this article, are well suited for such a pelagic lifestyle.

Elongated (i.e., high aspect ratio) dorsal and anal fins enable cost-efficient locomotion, because a large amount of water can be moved by the fins at a relatively slow speed for a given thrust force, which decreases induced drag (Alexander 2003). In addition, the rigid body of ocean sunfish protected by the exoskeleton hypodermis has minimal lateral movements during swimming, which further decreases drag (Lighthill 1971). A combination of high aspect ratio fins and minimal body movements are also found in tunas and lamnid sharks, which are cruising specialists (Webb 1984) that maintain elevated swimming speed and migrate greater distances than most other fishes (Watanabe et al. 2015). In contrast to tunas and lamnid sharks, ocean sunfish are apparently ectothermic (i.e., body temperature is not elevated) and thus, less athletic. Nevertheless, functional similarities of the swimming styles between tunas and ocean sunfish suggest that they are cruising specialists. Cost-efficient cruising is particularly important for adult ocean sunfish, because their main prey in oceanic systems (i.e., gelatinous zooplankton) are often distributed patchily, meaning individuals may undergo protracted periods without feeding during well documented, long-distance movements (Sims et al. 2009a,b, Dewar et al. 2010, Thys et al. 2015, Sousa et al. 2016). However, this behavioral generalization of 'cruisers' must be caveated given that ocean sunfish are capable of rapid acceleration during swimming or during breaching events (noted burst speeds of; 2.2–2.4 m/s, Watanabe and Sato 2008, Nakamura and Sato 2014; 6.6 m/s, Thys et al. 2015); although the drivers of such behaviors remain unclear.

Lastly, returning to buoyancy regulation, the hypodermis may be of benefit during forays throughout the water column. Specifically, although gas in swim bladders is much less dense than the tissues of the hypodermis, it is more compressible at depth. Consequently, fishes with swim bladders risk losing buoyancy during descent or have excessive buoyancy (and lose control) during ascents (Alexander 1966, Watanabe et al. 2008). In contrast, the hypodermis is presumed to be incompressible, and ocean sunfish could maintain buoyancy close to neutral over a wide depth range, allowing them to conduct frequent vertical movements with a constant flapping frequency (Watanabe and Sato 2008). Therefore, the unusual adaptation of having a thick hypodermis may further help ocean sunfish search the open ocean three dimensionally for patchily distributed prey.

Remaining Questions

Further investigation of the internal skeleto-muscular anatomy of species of the Molidae other than *Mola mola* is desirable to identify similarities and differences in the structure, function and evolution of the locomotory apparatus. As far as *Mola mola* is concerned, detailed study of the microstructure and mechanical properties of the hypodermis and horizontal septum is needed to better understand the transmission of muscular forces (c.f. Westneat and Wainwright 2001). In particular, the layers and microsepta of the horizontal septum merit investigation, as does the structure of its interface with the hypodermis on either side of the body.

Investigating the biomechanics of fast locomotion in *Mola mola* is difficult, principally because they are large animals that can grow extremely quickly—at least in captivity (Pan et al. 2016, Powell 2001). Existing video in the public domain is mainly for very large animals swimming slowly, often in aquaria, or for smaller specimens filmed in situations where moving cameras, parallax and barrel distortion pose serious analytical problems. One molid species that might be useful in elucidating synchronized fin flapping is the slender sunfish *R. laevis*, which is distributed pantropically, is much smaller, highly streamlined, and a schooling fish as an adult (Robison 1975). Images from video of a swimming school are shown in Fig. 9. Inspection of the video confirms that dorsal and anal fin flapping is synchronized (as indicated by Robison 1975), with mean fin flapping (n = 10) estimated at 2.28 Hz (range 1.58–3.33 Hz). Consequently, the species appears suitable for placing in water tunnels and simultaneously filming its swimming from different angles.

In addition, more biologging studies using accelerometers and video cameras (e.g., Nakamura et al. 2015) are needed to better understand the natural swimming behavior of molids and how it relates to their foraging ecology. Ocean sunfish exhibit dramatic ontogenetic changes in morphology, including the shape of their dorsal and anal fins (Fig. 1C) and the proportion of hypodermal mass (Fig. 8). It is expected, therefore, that ocean sunfish change swimming and foraging behavior with their increasing body size. Exploring the biomechanical, ontogenetic changes of ocean sunfish not only deepens our understanding of their life history but also helps reveal the hidden forces that may have shaped the evolution of one of the ocean's strangest shaped fish.

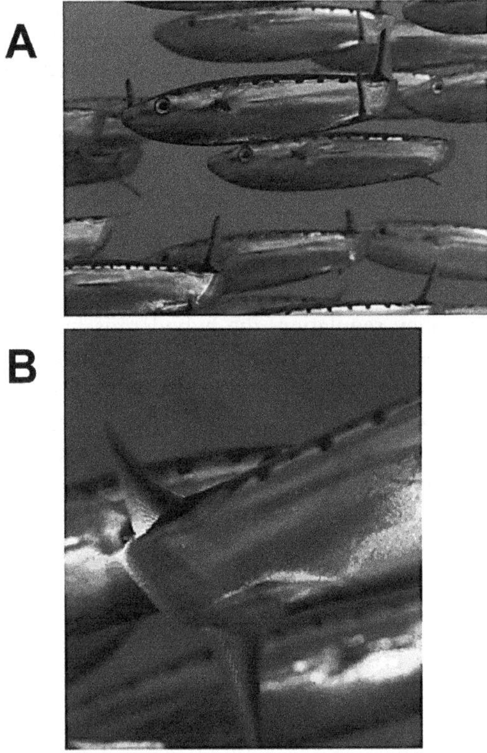

Figure 9. Images of slender sunfish *Ranzania laevis*. A. Partial school of slender sunfish. Note varied positions of dorsal and anal fins. B. Close-up of fully elevated dorsal fin of individual fish.

Acknowledgements

Yuuki Watanabe thanks Drs. Etsuro Sawai and Katsufumi Sato for their help with the collection and dissection of specimens. John Davenport acknowledges his debt to Drs. Natasha Phillips, Poul Larsen, Jon Houghton, Lawrence Eagling and Liz Cotter who have worked with him on investigations of the morphology and biomechanics of ocean sunfish. He also thanks The Fisheries Society of the British Isles for the award of an Alwyn Wheeler travel grant that enabled him to work with those scientists. Additional funding was provided by Queens University Belfast (G. and M. Williams Fund).

References

Alexander, R.M. 1966. Physical aspects of swim bladder function. Biol. Rev. 41: 141–176.

Alexander, R.M. 1979. The Invertebrates. Cambridge University Press, Cambridge and New York.

Alexander, R.M. 2003. Principles of animal locomotion. Princeton University Press, Princeton and Oxford.

Arakawa, K.Y. and S. Masuda. 1961. Some observations on ocean sunfish *Mola mola* (Linnaeus) with reference to its swimming behaviour and feeding habits. JAZA 3(4): 95–96.

Bemis, K.E., J.C. Tyler, E.J. Hilton and W.E. Bemis. 2020. Overview of the anatomy of ocean sunfishes (Molidae: Tetraodontiformes). pp. 55–71. *In*: T.M. Thys, G.C. Hays and J.D.R Houghton [eds.]. The Ocean Sunfishes: Evolution, Biology and Conservation, CRC Press. Boca Raton, FL, USA.

Block, B.A., S.L.H. Teo, A. Walli, A. Boustany, M.J.W. Stokesbury, C.J. Farwell et al. 2005. Electronic tagging and population structure of Atlantic bluefin tuna. Nature 434: 1121–1127.

Carey, F.G., J.M. Teal, J.W. Kanwisher and K.D. Lawson. 1971. Warm-bodied fish. Am. Zool. 11: 137–145.

Cartamil, D.P. and C.G. Lowe. 2004. Diel movement patterns of ocean sunfish *Mola mola* off southern California. Mar. Ecol. Prog. Ser. 266: 245–253.

Chanet, B., C. Guintard, T. Boisgard, M. Fusellier, C. Tavernier, E. Betti et al. 2012. Visceral anatomy of ocean sunfish (*Mola mola* (L., 1758), Molidae, Tetraodontiformes) and angler (*Lophius piscatorius* (L., 1758), Lophiidae, Lophiiformes) investigated by non-invasive imaging techniques. C. R. Biol. 335: 744–752.

Chanet, B., C. Guintard and G. Lecointre. 2014. The gas bladder of puffers and porcupinefishes (Acanthomorpha: Tetraodontiformes): phylogenetic interpretations. J. Morphol. 275: 894–901.

Clelland, J. 1862. On the anatomy of the short sunfish (*Orthragoriscus mola*). Nat. Hist. (Lond.). 2: 170–185.

Corner, E.D.S., E.J. Denton and G. Forster. 1969. On the buoyancy of some deep-sea sharks. Proc. Roy. Soc. Lond. Ser. B 171: 415–429.

Davenport, J. and E. Kjørsvik. 1986. Buoyancy in the lumpsucker *Cyclopterus lumpus* (L.). J. Mar. Biol. Ass. UK 66: 159–174.

Davenport, J., N.D. Phillips, E. Cotter, L.E. Eagling and J.D.R. Houghton. 2018. The locomotor system of the ocean sunfish *Mola mola*: role of gelatinous exoskeleton, horizontal septum, muscles and tendons. J. Anat. 233: 47–357.

Dewar, H., T.M. Thys, S.L.H. Teo, C. Farwell, J. O'Sullivan, T. Tobayama et al. 2010. Satellite tracking the world's largest jelly predator, the ocean sunfish, *Mola mola*, in the Western Pacific. J. Exp. Mar. Biol. Ecol. 393: 32–42.

Fergusson, I.K., L.J. Compagno and M.A. Marks. 2000. Predation by white sharks *Carcharodon carcharias* (Chondrichthyes: Lamnidae) upon chelonians, with new records from the Mediterranean Sea and a first record of the ocean sunfish *Mola mola* (Osteichthyes: Molidae) as stomach contents. Environ. Biol. Fish. 58: 447–453.

Fischer, H., I. Polikarpov and A.F. Craievich. 2004. Average protein density is a molecular-weight-dependent function. Protein Sci. 13: 2825–2828.

Fraser-Brunner, A. 1951. The ocean sunfishes (Family Molidae). Bull. Brit. Mus. (Nat. Hist.). Zool. 1: 87–121.

Gladstone, W. 1988. Killer whale feeding observed underwater. J. Mammal. 69: 629–630.

Greer-Walker, M. and G.A. Pull. 1975. A survey of red and white muscles in marine fish. J. Fish Biol. 7: 295–300.

Gregory, W.K. and H.C. Raven. 1934. Notes on the anatomy and relationships of the ocean sunfish (*Mola mola*). Copeia 1934: 145–151.

Halstead, B.W. 1978. Poisonous and venomous marine animals of the world. Darwin Press, Princeton, New Jersey.

Hays, G.C., M.R. Farquhar, P. Luschi, S.L.H. Teo and T.M. Thys. 2009. Vertical niche overlap by two ocean giants with similar diets: Ocean sunfish and leatherback turtles. J. Exp. Mar. Biol. Ecol. 370: 134–143.

Hays, G.C., T.K. Doyle and J.D.R. Houghton. 2018. A Paradigm Shift in the Trophic Importance of Jellyfish? Trends Ecol. Evol. 33: 874–884.

Houghton, J.D.R., N. Liebsch, T.K. Doyle, A.C. Gleiss, M.K.S. Lilley, R.P. Wilson et al. 2009. Harnessing the sun: testing a novel attachment method to record fine scale movements in ocean sunfish (*Mola mola*). *In*: J.L. Nielsen, H. Arrizabalaga, N. Fragoso, A. Hobday, M. Lutcavage, J. Sibert [eds.]. Tagging and Tracking of Marine Animals with Electronic Devices. Reviews: Methods and Technologies in Fish Biology and Fisheries, vol 9. Springer, Dordrecht.

Johnson, G.D. and R. Britz. 2005. Leis' Conundrum: homology of the clavus of the ocean sunfishes. 2. Ontogeny of the median fins and axial skeleton of *Ranzania laevis* (Teleostei, Tetraodontiformes, Molidae). J. Morphol. 266: 11–21.

Johnston, I.A. 1981. Structure and function of fish muscles. Symp. Zool. Soc. Lond. 48: 71–113.

Lauder, G.V. and E.G. Drucker. 2004. Morphology and experimental hydrodynamics of fish fin control surfaces. IEEE J. Ocean. Eng. 29: 556–571.

Lighthill, M.J. 1971. Large-amplitude elongated body theory of fish locomotion. Proc. Roy. Soc. Lond. Ser. B 179: 125–138.

Martini, F. and W.C. Ober. 2006. Fundamentals of Anatomy and Physiology. Pearson Educational, London.

Nakamura, I. and K. Sato. 2014. Ontogenetic shift in foraging habit of ocean sunfish *Mola mola* from dietary and behavioral studies. Mar. Biol. 161: 1263–1273.

Nakamura, I., Y. Goto and K. Sato. 2015. Ocean sunfish rewarm at the surface after deep excursions to forage for siphonophores. J. Anim. Ecol. 84: 590–603.

Paladino, F.V., M.P. O'Connor and J.R. Spotila. 1990. Metabolism of leatherback turtles, gigantothermy, and thermoregulation of dinosaurs. Nature 344: 858–860.

Pan, H., H. Yu, V. Ravi, C. Li, A.P. Lee, M.M. Lian et al. 2016. The genome of the largest bony fish, ocean sunfish (*Mola mola*), provides insights into its fast growth rate. GigaScience 5: s13742–016–0144–3

Phillips, N.D., C. Harrod, A.R. Gates, T.M. Thys and J.D.R. Houghton. 2015. Seeking the sun in deep, dark places: mesopelagic sightings of ocean sunfishes (Molidae). J. Fish Biol. 87: 1118–1126.

Phillips, N.D., E.A.E. Smith, S.D. Newsome, J.D.R. Houghton, C.D. Carson, J. Alfaro-Shigueto et al. 2020. Bulk tissue and amino acid stable isotope analysis reveal global ontogenetic patterns in ocean sunfish trophic ecology and habitat-use. Mar. Ecol. Prog. Ser. 633: 127–140.

Phleger, C.F. 1998. Buoyancy in marine fishes: direct and indirect role of lipids. Am. Zool. 38: 321–330.

Pope, E.C., G.C. Hays, T.M. Thys, T.K. Doyle, D.W. Sims, N. Queiroz et al. 2010. The biology and ecology of the ocean sunfish *Mola mola*: a review of current knowledge and future research perspectives. Rev. Fish Biol. Fish. 20: 471–487.

Potter, I.F. and W.H. Howell. 2011. Vertical movement and behavior of the ocean sunfish, *Mola mola,* in the northwest Atlantic. J. Exp. Mar. Biol. Ecol. 396(2): 138–146.

Powell, D.C. 2001. A Fascination for Fish: Adventures of an Underwater Pioneer, 2nd ed. University of California Press, Berkeley.

Robison, B.H. 1975. Observations on living juvenile specimens of the slender mola, *Ranzania laevis* (Pisces, Molidae). Pacific Sci. 29: 27–29.

Ryder, J.A. 1885. The swimming-habits of the sunfish. Science 6: 103–104.

Santini, F. and J.C. Tyler. 2002. Phylogeny of the ocean sunfishes (Molidae, Tetraodontiformes), a highly derived group of teleost fishes. Ital. J. Zool. 69: 37–43.

Sims, D.W., N. Queiroz, T.K. Doyle, J.D.R. Houghton and G. Hayes. 2009a. Satellite tracking of the world's largest bony fish, the ocean sunfish (*Mola mola*) in the North East Atlantic. J. Exp. Mar. Biol. Ecol. 370: 127–133.

Sims, D.W., N. Queiroz, N. Humphries, F. Lima and G. Hays. 2009b. Long-term GPS tracking of ocean sunfish, *Mola mola*, offers a new direction in fish monitoring. PLoS One 4(10): e7351.

Sousa, L.L., N. Queiroz, G. Mucientes, N.E. Humphries and D.W. Sims. 2016. Environmental influence on the seasonal movements of satellite-tracked ocean sunfish Mola mola in the north-east Atlantic. Anim. Biotel. 4: 1–19.

Syme, D.A. and R.E. Shadwick. 2011. Red muscle function in stiff-bodied swimmers: there and almost back again. Phil. Trans. Roy. Soc. Ser. B 366: 1507–1515.

Thys, T.M., J.P. Ryan, H. Dewar, C.R. Perle, K. Lyons, J. O'Sullivan et al. 2015. Ecology of the ocean sunfish, *Mola mola*, in the southern California current system. J. Exp. Mar. Biol. Ecol. 471: 64–76.

Thys, T.M., A. Hearn, K.C. Weng, J. Ryan and C.R. Penaherrera. 2017. Satellite tracking and site fidelity of short ocean sunfish, *Mola ramsayi*, in the Galapagos Islands. J. Mar. Biol. 1–10.

Walker, J.A. and M.W. Westneat. 2000. Mechanical performance of aquatic rowing and flying. Proc. Roy. Soc. Lond. B. 267: 1875–1881.

Watanabe, Y. and K. Sato. 2008. Functional dorsoventral symmetry in relation to lift-based swimming in the ocean sunfish *Mola mola*. PLoS ONE 3: e3446.

Watanabe, Y., Q. Wei, D. Yang, X. Chen, H. Du, J. Yang et al. 2008. Swimming behavior in relation to buoyancy in an open swimbladder fish, the Chinese sturgeon. J. Zool. 275: 381–390.

Watanabe,Y., K.J. Goldman, J.E. Caselle, D.D. Chapman and Y.P. Papastamatiou. 2015. Comparative analyses of animal-tracking data reveal ecological significance of endothermy in fishes. Proc. Nat. Acad. Sci. USA 112: 6104–6109.

Webb, P.W. 1984. Body form, locomotion and foraging in aquatic vertebrates. Am. Zool. 24: 107–120.

Weng, K.C., P.C. Castilho, J.M. Morrissette, A.M. Landeira-Fernandez, D.B. Holts, R.J. Schallert et al. 2005. Satellite tagging and cardiac physiology reveal niche expansion in salmon sharks. Science 310: 104–106.

Westneat, M.W. and S.A. Wainwright. 2001. Mechanical design for swimming: muscle, tendon, and bone. Chapter 7. pp. 271–311. *In*: Fish Physiology: Tuna: Physiology, Ecology, and Evolution, Volume 19 (1st Edition). Academic Press, Cambridge, Massachusetts.

Yamanoue, Y., M. Miya, K. Matsuura, M. Katoh, H. Sakai and M. Nishida. 2008. A new perspective on phylogeny and evolution of Tetraodontiform fishes (Pisces: Acanthopterygii) based on whole mitochondrial genome sequences: basal ecological diversification? BMC Evol. Biol. 8: 212.

Chapter 6

Reproductive Biology of the Ocean Sunfishes

Kristy Forsgren,[1,*] *Richard S. McBride*,[2] *Toshiyuki Nakatsubo*,[3] *Tierney M. Thys*,[4] *Carol D. Carson*,[5] *Emilee K. Tholke*,[6] *Lukas Kubicek*[7] and *Inga Potter*[8]

Introduction

A Record in Question

Among their many claims to fame, members of the ocean sunfish family, Molidae, hold the world record for vertebrate fecundity (Carwardine 1995, Collette et al. 2010). Little empirical research, however, has been conducted on molid reproductive biology. Consequently, many basic questions remain, from molid size at sexual maturity to the location and timing of spawning, larval development and metamorphosis and success of larval recruitment. Furthermore, the current fecundity record is based on a single 1.5 m *M. mola* (Linnaeus 1758) female which was estimated to be carrying "no fewer than 300 million small unripe ova" within a single ovary (Schmidt 1921). Unfortunately, no photos were taken of this record-setting ovary and the developmental stage of its small oocytes was not investigated.

In order to properly address this record and sunfish fecundity in general, it is necessary first to define what is meant by fecundity. Fecundity here is defined as the number of yolked oocytes produced within a set time period, such as a year in the case of annual fecundity. This definition provides a measure of productivity instead of just noting the number of ova at one point of time, which is a measure of standing stock. Measuring fecundity as a unit of productivity is more relevant to the overall reproductive potential of an individual fish. In contrast to the term fertility, which typically refers to the ability to reproduce/produce viable offspring, fecundity defines the potential for reproduction [i.e., the ability to produce viable gametes (egg and sperm)]. The term reproduction here simply refers to the biological process of producing new organisms (i.e., offspring).

[1] College of Natural Sciences and Mathematics, California State University, Fullerton CA.
[2] Northeast Fisheries Science Center, National Marine Fisheries Service, Woods Hole, MA USA. Email: richard.mcbride@noaa.gov
[3] ORIX Aquarium Corporation, Nippon Life Hamamatsucho Crea Tower, 2-3-1 Hamamtsu-cho, Minato-ku, Tokyo 105-0013, Japan; Email: t.nakatsubo@aqua.email.ne.jp
[4] California Academy of Sciences, San Francisco, CA, USA; Email: tierneythys@gmail.com
[5] The New England Coastal Wildlife Alliance, Middleboro, MA, USA; Email: krillcarson@mac.com
[6] Integrated Statistics, US National Marine Fisheries Service, Northeast Fisheries Science Center, Woods Hole, MA, USA. Email: Emilee.Tholke@noaa.gov
[7] Aussermattweg 22, 3532, Zäziwil, Switzerland; Email: molamondfisch@gmail.com
[8] Affiliate Faculty, University of New Hampshire, Durham NH USA; Email: ingapotter@gmail.com
* Corresponding author: kforsgren@fullerton.edu

Past Work

Nakatsubo et al. (2008) suggested that *M. mola* under 75 cm are not sexually mature. This proposed threshold aside, it is still unclear if the current record-holding 1.5 m female was fully sexually mature. The total number of oocytes tallied at early stages of ovarian development (e.g., primary or secondary growth stage) can be misleading because more oogonia may become oocytes, and all of the ovarian follicles will not transform into mature ova. Many ova undergo atresia (i.e., degeneration of ovarian follicles) reducing the number of ovarian follicles. Some fail to ovulate (Saidapur 1978, Wallace and Selman 1981, Murua et al. 2003). As such, when quantifying total fecundity, atrestia must be considered.

Previous investigations of sunfish reproduction (Nakatsubo et al. 2007a,b, Kang et al. 2015) suggest that sexually mature *M. mola* have asynchronous ovarian development. In other words, not all of their ova ripen and release at the same time but rather their ovarian follicles develop at different rates within the same ovary. At any one time, a sexually mature molid ovary may in fact contain all developmental stages of ovarian follicles-primary, secondary and maturation stages. Each female presumably releases only a subset of ripe ova (mature yolked oocytes) at a time over the course of her reproductive season.

Additional questions surround the actual spawning events. While molids are broadly distributed (see Sawai et al. 2020 [Chapter 2] and Caldera et al. 2020 [Chapter 3]), only a few studies have documented ocean sunfish spawning seasons or examined spawning frequency in any of the recognized five species of molid [*Mola mola* (Linnaeus 1758), *Mola tecta* Nyegaard et al. 2018, *Mola alexandrini* (Ranzani 1839), *Masturus lanceolatus* (Liénard 1840), *Ranzania laevis* (Pennant 1776)]. Notably off the coast of Japan, Nakatsubo et al. (2007b) reported that *M. mola* spawn multiple times during the year specifically from August to October. Likewise, Kang et al. (2015) identified that *M. mola* spawn between July and October off the island of Jeju, South Korea. Populations of slender mola (*Ranzania laevis*) appear to spawn continuously in the Sargasso Sea during the months of March and April (Hellenbrecht et al. 2019). To date, however, no one has witnessed an actual spawning event in the wild for any molid species (although see later section about a rare egg release in a captive individual).

Questions Addressed

This chapter reviews our current knowledge of molid reproductive biology. Additionally, it offers methods and new results from data collected on the fecundity and ovarian development of *M. mola* from three studies spanning multiple geographic regions [the East Coast of the USA, the Mediterranean (Italy and Portugal) and Japan]. It also offers the first data from a *M. tecta* ovary collected opportunistically from a stranded female on the coast of California.

The five themes of inquiry include: (1) At what size do female mola reach sexual maturity, i.e., when are mature ova produced and thus capable of being fertilized?, (2) What is the time, duration, seasonality, frequency, and geographic location of spawning events for any molids?, (3) What is the most biologically meaningful way of defining molid fecundity?, (4) Under this definition, are sunfishes still the most fecund vertebrates? What is the fertility and reproductive output of mola, i.e., what is the number of actual offspring produced? and lastly, (5) What is molid spawning stock biomass? Beyond these themes, the chapter also highlights remaining gaps in our knowledge and suggests avenues for future research.

Methods

Study 1 presents new previously unpublished *M. mola* data from the east coast of the USA and was compiled by coauthors R. McBride, E. Tholke, and C. Carson; Study 2 includes new *M. mola* data from Italy and Portugal and data from one *M. tecta* from California. It was compiled by K. Forsgren,

T. Thys, and L. Kubicek; Study 3 presents both new and previously published data from *M. mola* individuals collected in Japan and was compiled by T. Nakatsubo.

Study 1: U.S.A. Eastern Seaboard (Western Atlantic)

In the field, data were collected from *M. mola* strandings on New England beaches, especially those on the northern shores of Cape Cod, USA (approx. 41.8°N, 70.3°W). Since 2008, the New England Coastal Wildlife Alliance has responded to ocean sunfish strandings, which begin typically in early September and continue through December. The majority of strandings occur in October and November annually and most stranded fish are dead. Among individuals studied for this chapter, several were measured for total length (TL cm), total weight (TW kg), and ovary weight (OW g). Data were collected for 27 female *M. mola* in total between 2015 and 2018.

In the laboratory, ovarian tissue initially fixed in 10 percent buffered formalin, was processed for histology using standard, wax-mounted methods and stained with Schiff–Mallory trichrome (Press et al. 2014). Histology preparations were used to select and evaluate individuals for oocyte size distributions and to estimate total yolked oocyte fecundity. To identify stage-specific oocyte sizes, about 200 oocyte diameters per fish were measured from whole cells teased apart from a subsample of ovarian tissue, photographed with a digital camera, and measured with ImageJ software (Press et al. 2014). The smaller mode of translucent cells was subsampled randomly to be representative of the size distribution (100–110 cells), whereas the larger mode of opaque cells, evident only in fish MM2015-05, was measured for all cells except those damaged. Fecundity was estimated for one female with an advanced batch of yolked oocytes (fish MM2015-05) where a subsample of ovarian tissue was weighed to the nearest 0.0001 g and teased apart. The larger opaque oocytes were counted, and the fecundity of the individual was expanded gravimetrically. Oocyte number was determined by counting the oocytes in a given weight or volume, and estimating the total oocytes present by the proportion.

Study 2: Europe (Italy and Portugal) and California (Eastern Pacific)

In the field, ovaries were collected from freshly dead (bycatch) females from two regions: Tonnarella nets in Camogli, Italy (n = 6; C1, C14, C19, C24, C49, C59), and the Tunipex Madraque in Olhao, Portugal (n = 3; P5, P9, P13) in 2011–2012. All but one sample from Italy (C59) was analyzed using mtDNA and identified as *M. mola* (Y. Yamanoue unpublished data). Ovarian tissue from a dead stranded *M. tecta* was collected in Santa Barbara, CA [n = 1; CA1, total length (TL): 215 cm, ovary weight (OW): 1.43 kg]. The fish was not weighed but was genetically confirmed as *M. tecta* by M. Nyegaard (unpublished data).

In the laboratory, ovarian tissue was fixed in 10 percent neutral buffered formalin and processed for paraffin wax histology with tissues dehydrated in a graded series of ethanol and xylenes, infiltrated with paraffin wax, and embedded in paraffin blocks. Five-micrometer tissue sections of each sample were cut using a rotary microtome. Serial sections were mounted on slides and stained using hematoxylin and eosin. Tissue sections were analyzed using bright-field microscopy (Olympus BX60). A digital camera (QICAM QImaging Fast 1394, QImaging) and imaging software (Q-Capture Pro 7, QImaging, 2010) were used to take micrographs. The diameter (μm) of oocytes was measured using ImageJ software (version 1.50i, National Institutes of Health). Only oocytes that appeared to be sectioned through the approximate center of the nucleus were measured. The oocyte diameters were measured to the nearest 0.001 mm with three measurements for each oocyte examined in order to obtain an estimate of variation. A total of 20 oocytes/stage/individual were measured. To maintain consistent classification of oocytes, a single observer measured oocyte diameters for each individual, and a second observer confirmed oocyte stages and measurements. Oocytes were staged based on morphological characteristics previously established for teleost fishes (Nagahama 1983, Guraya 1986, Wallace and Selman 1990).

Study 3: Japan

In the field, specimens were collected mainly by set-nets in the Kanto region (e.g., Chiba and Kanagawa prefecture) of Japan. Many of the individuals were captured for display at Kamogawa Sea World (Chiba, Japan). Individuals off the coast of Kanagawa prefecture were collected by Mr. Kawachi (formerly of Kamogawa Sea World) for graduate research at Nihon University. For comparative studies, analysis was conducted separately for wild and reared individuals. Individuals reared in captivity for less than seven days were considered wild.

In the lab, a selection of gonad tissue was fixed with 10 percent seawater formalin or Buan solution, and a paraffin section was prepared according to a conventional method (http://nria.fra.affrc.go.jp/RCFD/histological-methods_e.html). After hematoxylin-eosin double staining, observations were made using an optical Olympus CH-2 microscope. The stage of the cells was classified into three major phases: primary growth (containing chromatin-nucleolus, perinucleolus, and other previtellogenic stages), secondary growth (vitellogenic stages), and maturation (migratory nucleolus stage—ovulation). The most mature oocyte observed was used to determine the maturation stage of the individual (Murayama et al. 1995, Grier et al. 2018).

Results and Discussion

Study 1: U.S.A. Eastern Seaboard of the USA (Western Atlantic)

Size at maturity was evident among stranded females based on the dramatic increase in ovary weight for females from 105 to 215 cm TL (Fig. 1). The inflection of ovary weight values between 150 and 160 cm TL implies that *M. mola* > 1.5 m are maturing or are mature. Closer inspection of the gonad histology of these larger fish confirmed evidence of sexual maturity, either as a larger batch of oocytes distributed among pre-yolked oocytes or in small numbers of larger degrading germ cells that were left-over from a previous spawning period (McBride and Tholke, unpublished data). Previous studies on Japanese *M. mola* conducted by Nakatsubo et al. (2007b), did not consider the mature/immature boundary but as a result of preliminary investigations into seasonality, determined that *M. mola* males (> 165 cm) and females (> 185 cm) were reliably mature specimens. Both these datasets further support the suggestion that the record-holding 1.5 m fish examined by Schmidt (1921) was likely a sexually immature individual without an advanced clutch of oocytes.

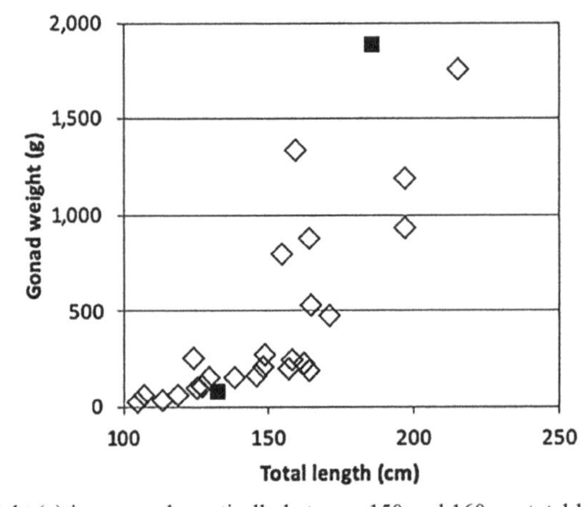

Figure 1. Gonad (ovary) weight (g) increases dramatically between 150 and 160 cm total length in 27 female *Mola mola*. Specimens were found stranded along the shores of Cape Cod, Massachusetts, USA during October-December, 2015–2018. Individuals are marked with a diamond symbol, except two fish marked with a solid square, which are given a closer look in Fig. 2.

A batch of developing, yolked oocytes was evident in at least one female which has the heaviest ovary (nearly 2 kg). In this fish (MM2015-05, TL:186 cm) two batches of oocytes evident in the ovary: a smaller batch of pre-yolked oocytes and a larger batch of yolked oocytes (Fig. 2). By comparison, the smaller fish (MM2017-37, 133 cm TL) had an ovary weighing less than 100 g with only pre-yolked (perinucleolar) oocyte stages which is consistent with it being an immature fish.

Spawning season and frequency were not possible to ascertain from these limited collections. Two hypotheses however, emerged based on the assumption that spawning occurred in late summer-early fall, as documented in Japan (Nakatsubo et al. 2007b, Kang et al. 2015) during the period of maximal sea surface temperature. The first hypothesis is that Fish MM2015-05 was ready to, but had not yet released its eggs. This hypothesis seems unlikely because this fish was collected in October at the end of the spawning season when sea surface temperatures drop (Dewar et al. 2010). The advanced cohort of oocytes had a maximum size < 500 microns, which does not indicate imminent spawning. [Note:

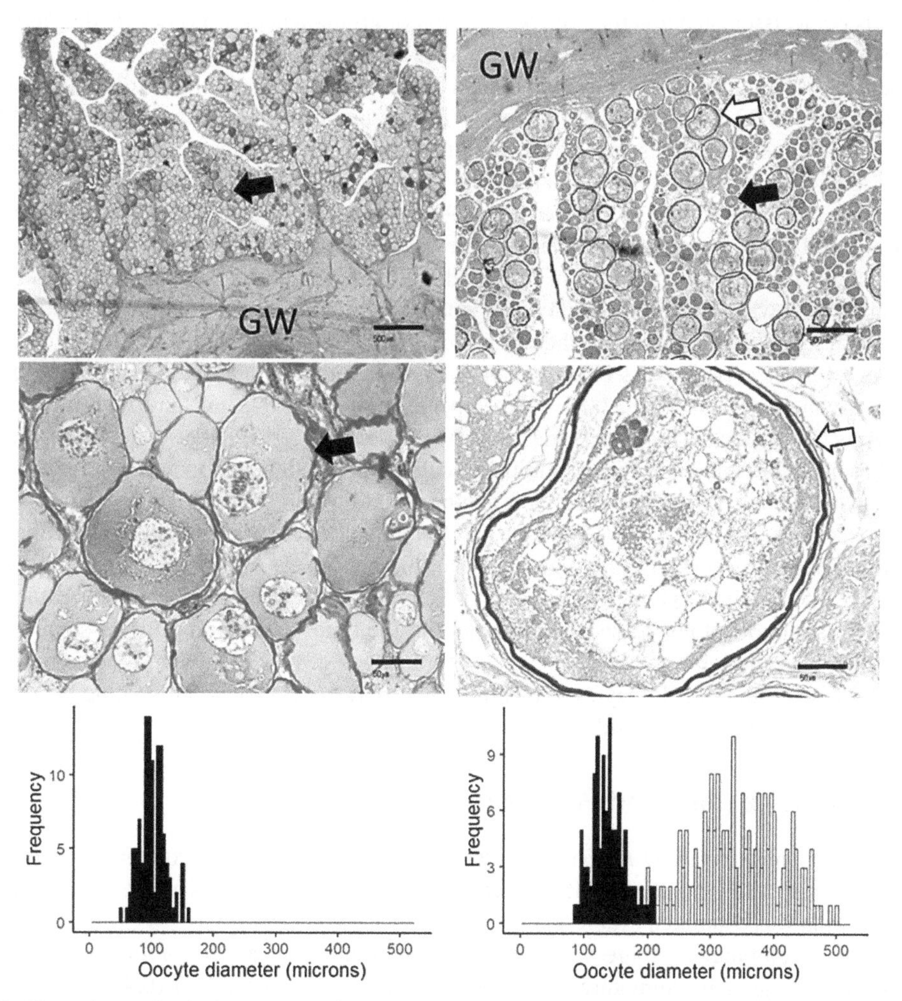

Figure 2. The top images in Fig. 2 are a broad view of gonad histology (scale bar = 500 mm). The middle images are closer views of gonad histology (scale bar = 50 μm). Both views have marks to identify the gonad wall (GW), pre-yolked oocyte stages (filled arrows) and yolked oocyte stages (open arrows). Ovarian tissue was fixed in 10 percent buffered formalin and stained with Schiff–Mallory trichrome (Press et al. 2014). The bottom images depict oocyte diameter distributions for these same two females, where subsamples of ovarian tissue (~ 0.02 g) were teased apart to separate germ cells, photographed with a digital camera, and diameters were measured with ImageJ software (Press et al. 2014). The smaller (filled) mode of translucent cells was subsampled randomly to be representative of the size distribution (100–110 cells), whereas the larger (open) mode of opaque cells, evident in Fish MM2015-05 only, was measured for all cells except those damaged.

M. mola egg size varies between 500–1000 microns. See a rare case of egg release detailed later in this chapter (Fig. 10). Schmidt (1921) estimated *Masturus lanceolatus* eggs to be approximately 1.8 mm in diameter and *Ranzania* sp. eggs to be 1.3–1.4 mm in diameter.]

Hypothesis 2 suggests that, even though we observed a batch of advanced stage oocytes in Fish MM2015-05, these oocytes may not reach maturation until the following summer. In this scenario, the time from initiation of vitellogenesis to ovulation may be up to one year (Press et al. 2014) or even two, as evidenced by some other large fishes (e.g., Kennedy et al. 2011). We determined that the advanced stage oocytes were yolked (or vitellogenic) and were a separate size cohort from the reserve of smaller oocytes in fish MM2015-05. Given these data, this individual was likely group-synchronous (with respect to vitellogenesis). While we cannot extrapolate as to whether *M. mola* ovulates all at once (i.e., a total spawner) or in discrete events within the spawning season (i.e., a batch spawner; Murua and Saborido-Rey 2003, McBride et al. 2015), Nakatsubo (2008) stated that ocean sunfish are multiple (batch) spawners based on the appearance of asynchronous oocyte development. With these traits being associated together, the observation of group-synchronous oocyte development in our samples opens up the possibility of either total or batch spawning patterns.

The size distribution of larger oocytes in Fish MM2015-05 was sufficiently discrete to estimate a total fecundity of 22.5 million yolked oocytes. This estimate accounted for the number of larger, opaque cells counted within a sample of ovarian tissue ($n = 202$), as observed with image analysis software. This estimate was then extrapolated gravimetrically from the sample weight of ovarian tissue (0.0169 g) to the entire weight of the gonad (1879.5 g) [(#oocytes count/subsample weight)*whole ovary weight]. This likely provided an overestimate, as the expansion included the weight of the gonad wall, which is a tissue type without germ cells. Although not weighed in our samples, the gonad wall weight in *M. tecta* was ~ 80 percent of the total gonad weight; discussed subsequently). This estimate is an order of magnitude greater than the batch size of other large pelagic fishes such as dolphinfish (*Coryphaena hippurus*) (Linnaeus 1758), wahoo (*Acanthocybium solandri*) (Cuvier 1832), or albacore (*Thunnus alalunga*) (Bonnaterre 1788)—with upper estimates of 0.62, 1.67, and 2.3 million eggs per batch, respectively (Jenkins and McBride 2009, McBride et al. 2012, Saber et al. 2016). However, these other fishes are known to recruit additional cohorts of vitellogenic oocytes within a spawning season (asynchronous development) so their annual fecundity can be much, much higher.

Intra-specifically, our first approximation of *M. mola* yolked oocyte fecundity (based on a 1.86 m female) is at least an order of magnitude less than the previously published estimate by Schmidt (1921) for a 1.5 m female. This is not to say that Schmidt (1921) was incorrect in his estimate, but rather that his report did not provide information on the type of fecundity. It is assumed that his method estimated the total number of oocytes at all stages of development, whereas our method was an estimate of all yolked oocytes; the latter being more accessible when seeking to convey the reproductive potential of an individual fish.

Study 2: Europe (Italy and Portugal) and California

The ovaries from nine female *M. mola* from the Mediterranean Sea (Italy and Portugal) were examined. Sizes ranged from 35.5–88.3 cm TL [45.63 ± 16.83 cm TL (mean ± stdev)]. One *M. tecta* (TL 215 cm) from California was examined (Table 1).

Oocyte developmental stage varied among individuals but all *M. mola* individuals were determined to be sexually immature. The ovarian tissue from sample C19 (0.434 m TL) did not have any discernible developing ovarian follicles (i.e., only primordial stage oocytes were present). Tissue samples C14, C49, C59, P5, P9, and P13 (0.383 + 0.044 m TL) displayed only primary ovarian follicles at the early perinucleolar stage present (i.e., pre-yolked oocytes; Fig. 3A).

Ovarian tissue from C1 (0.49 m TL) and the largest female *M. mola* sample C24 (0.88 m TL) had ovaries with primary ovarian follicles at the early and late perinucleolar stages present (i.e., pre-yolked oocytes; Fig. 3B). The diameter of early perinucleolar staged primary ovarian follicles

Table 1. Histological comparison of oocyte developmental stages observed in *M. mola* females across three studies involving global collection and analyses presented in the current study.

Study	***n***	**TL (cm)**	**Season**	**Developmental stages of oocytes**
USA Eastern Seaboard	27	105–215	Sep–Dec	One female with batch of yolked oocytes and a batch of pre-yolked oocytes (perinucleolar; Oct)
Italy and Portugal	9	35.5–88.3	April–June	Six females with oocytes in primary stages (early perinucleolar stage); two females with oocytes in primary stages (both early and late perinucleolar stages)
Japan	9	185–272	May–Dec	No final matured oocytes found; two females with oocytes in nuclear migration phase (August); One female w/ degenerated oocyte (Nov)

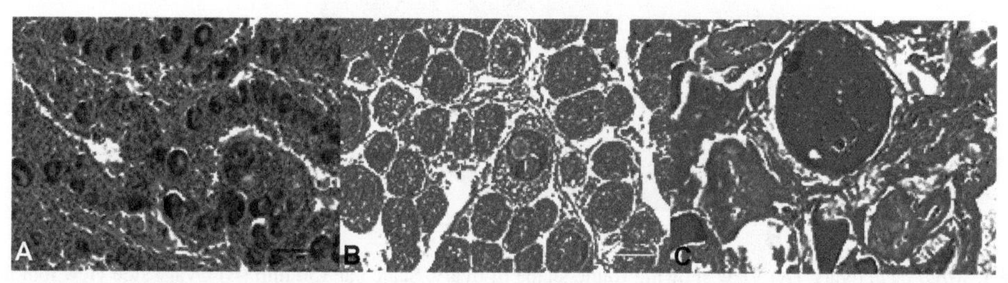

Figure 3. *Mola* ovarian tissue. (A) Ovarian tissue from P13, representative of *M. mola* with only primary oocytes at the early perinucleolar stage; (B) Ovarian tissue from C24, the largest female collected, representative of *M. mola* with primary oocytes at both the early and late perinucleolar stages; (C) Ovarian tissue with early cortical alveolus stage secondary ovarian follicle from CA1, *M. tecta*. Hematoxylin and eosin staining. Scale bar is 5 µm.

was 2.94 ± 0.09 µm (mean ± SEM) with a minimum diameter of 1.62 µm and a maximum diameter of 5.22 (n = 67 primary ovarian follicles measured). Late perinucleolar stage ovarian follicles had a diameter of 4.38 ± 0.13 µm (mean ± SEM) with a minimum diameter of 2.65 µm and a maximum diameter of 6.55 µm (n = 67 ovarian follicles measured).

The staging of ovarian follicles was based on morphological characteristics previously described for fish oogenesis (e.g., Wallace and Selman 1981, Nagahama 1983, Guraya 1986, Wallace and Selman 1990) and not ovarian follicle diameter alone. Thus, as may be expected, the size of early and late perinucleolar primary ovarian follicles overlapped. The overall diameter of *M. mola* primary ovarian follicles (i.e., both early and late perinucleolar stages) was 3.87 ± 0.09 µm. Since vitellogenic ovarian follicles (i.e., yolked) were not observed in these tissue samples, all *M. mola* collected from Italy and Portugal, were determined to be sexually immature.

One *M. tecta* ovary was examined (CA1; 215 cm TL). Perinucleolar (early and late) primary ovarian follicles were 4.08 ± 0.16 µm in diameter. The diameter of primary ovarian follicles of *M. mola* and *M. tecta* were not significantly different ($p < 0.05$). The *M. tecta* ovary also had secondary ovarian follicles at the early cortical alveolus stage present (Fig. 3C). Secondary ovarian follicles were 10.65 ± 0.27 µm in diameter (diameter ranged from 7.29–15.37 µm; n = 44 ovarian follicles). Secondary ovarian follicles do not contain vitellogenin, thus are also considered to be pre-yolked oocytes. Therefore, the *M. tecta* examined was also determined to be sexually immature at the time of sampling. However, it was evident that this female had actively developing ovarian follicles and, had she not stranded, may have continued to develop oocytes that could have been ovulated. There was no evidence that this female had recently spawned (i.e., there were no post-ovulatory follicles present within the ovarian tissue examined).

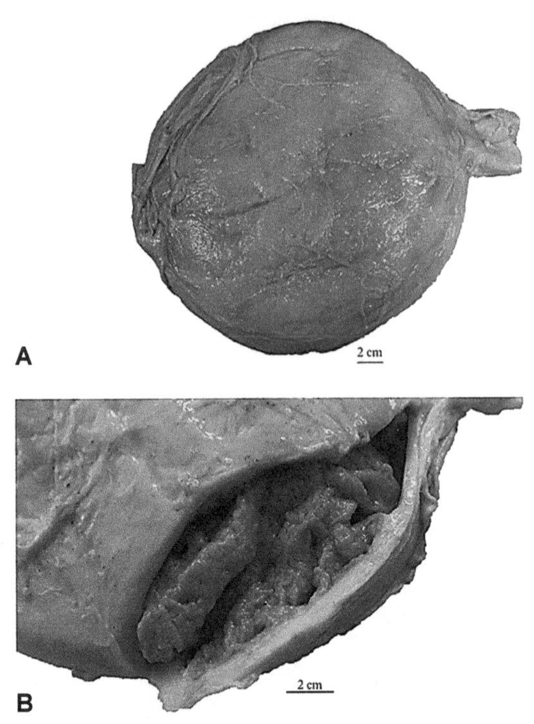

Figure 4. *Mola tecta* ovary. (A) View showing round shape of ovary. Scale bar 150 mm (B) Ovary side view looking into the oviduct. Scale bar 150 mm. Note the thickness of the ovarian wall and central lumen of the ovary. Ovary has also been cut to display the ovarian cortex. Scale bar is 2 cm.

The *M. tecta* ovary was received as a whole structure to examine and a few gross morphological features were noted. The ovary was spherical (Fig. 4A) with a central lumen that appeared to be modified into an oviduct (i.e., entovarian duct) into which mature ova would presumably be shed during ovulation. The ovary was 214.9 mm in diameter, 43.5 mm in thickness, and weighed 1.43 kg. The ovarian wall was relatively thick (6.48 mm) compared to several other marine teleost ovaries studied by Forsgren (personal observation; Fig. 4B) and was dissected from the cortex of the ovary. The ovary wall weighed 1.18 kg and comprised 82.5 percent of the total weight of the ovary (the ovarian cortex weighed 0.54 kg). Given the mass of the ovarian wall, calculation of fecundity based on the number of ovarian follicles for a given weight of tissue should be determined with caution, at least for *M. tecta*. For future studies, we recommend that the ovarian walls of additional *M. tecta* be weighed, and their thickness be measured. Data are also required for other sunfish species to enable intra-specific comparisons of fecundity.

Study 3: Japan

Estimation of Maturity

Maturation of *M. mola* individuals was estimated from a total of 328 (captive 180 and wild 148) individuals (196 females, 132 males; 25–272 cm TL). The gonad index (GI) was calculated using gonad weight (GW) and TL, ($GI = GW(g)/TL(cm)^3 \cdot 10000$) and used as an estimate of maturity (Fig. 5). GI is considered to be more advantageous in analyzing maturity of sunfish than gonadal somatic index (GSI) calculated from body weight ($GSI = GW(g)/BW(kg) \cdot 100$; Nakatsubo et al. 2007a).

Table 2. Data for nine female *Mola mola* and one *Mola tecta* from which ovarian tissue was analyzed. Females were opportunistically collected as bycatch in Camogli, Italy (labeled C) and Olhão Portugal (labeled P) and for *M. tecta* as a stranded individual in Santa Barbara, CA (labeled CA). All individuals were dead upon collection and none were sacrificed for the sake of this study.

Female	*Date*	*TL(cm)*	*Ovarian Mass (g)*
C1	24 May 2011	48.7	4.9
C14	27 May 2011	37.5	3.3
C19	28 May 2011	43.4	4.7
C24	30 May 2011	88.3	26.8
C49	27 June 2011	37.6	2.64
C59	28 April 2012	48.0	3.0
P5	15 May 2011	36.2	2.36
P9	15 May 2011	35.5	2.6
P13	15 May 2011	35.5	2.36
CA1	27 Feb 2019	215	1430

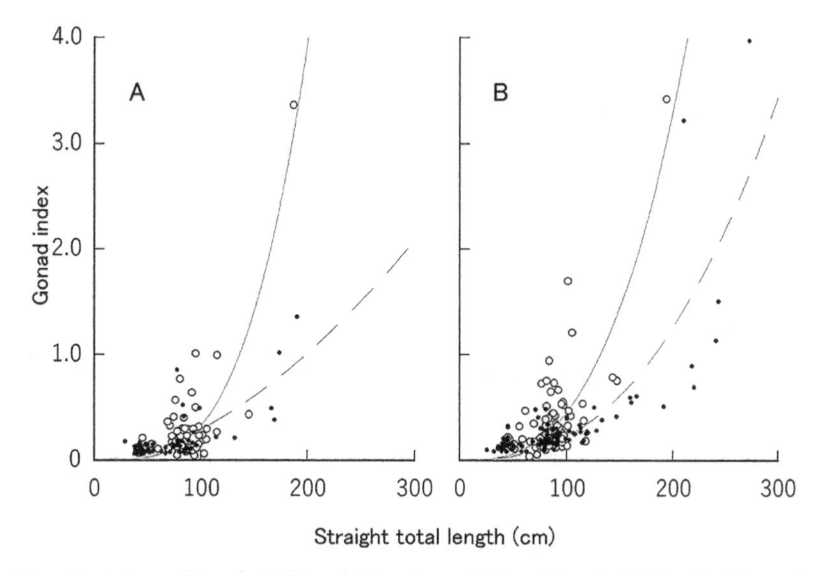

Figure 5. Relationships between *M. mola* total length TL and gonad index. (A) male (B) female. (○): captive individuals. (●): wild individuals. Solid and dashed lines represent the regression for the captive and wild individuals, respectively. The equations are as follows: captive male $GI = 2.40 \cdot 10^{-8} \cdot TL^{3.57}$, $R^2 = 0.74$; wild male, $GI = 8.55 \cdot 10^{-5} \cdot TL^{1.77}$, $R^2 = 0.70$; captive female $GI = 9.49 \cdot 10^{-7} \cdot TL^{2.84}$, $R^2 = 0.62$; wild female $GI = 2.64 \cdot 10^{-6} \cdot TL^{2.47}$, $R^2 = 0.66$. Redrawn with permission and modification from Nakatsubo et al. (2007a).

Seasonal Changes in Gonad Index and Estimation of Spawning Period in Japan

Seasonal changes in TL and maturity were analyzed for 183 wild individuals: 111 females and 72 males (25–272 cm TL, 1–2300 kg) and 151 captive individuals: 87 females and 64 males (42–194 cm TL, 4–297 kg). The total length of wild females tended to increase in summer (Fig. 6A), while their gonadal index levels increased sharply in August (Fig. 6C). Increases in the total length of wild males were less dramatic, although their gonadal index levels tended to increase between May and August (Fig. 6B). In contrast, some captive individuals showed high GI values in both sexes

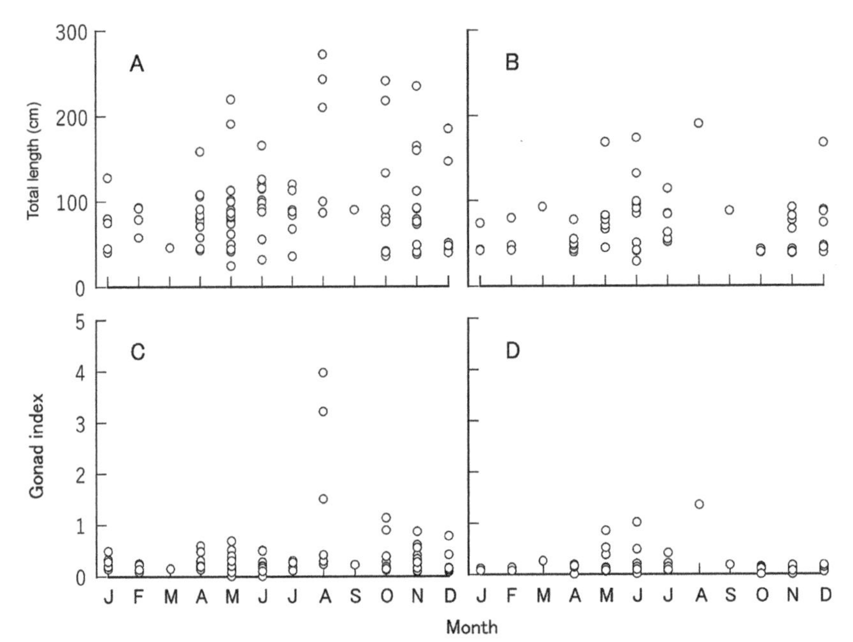

Figure 6. Seasonal changes in the total length and gonad index of wild *Mola mola* A and C females, B and D males. Redrawn with permission and modification from Nakatsubo et al. (2007b).

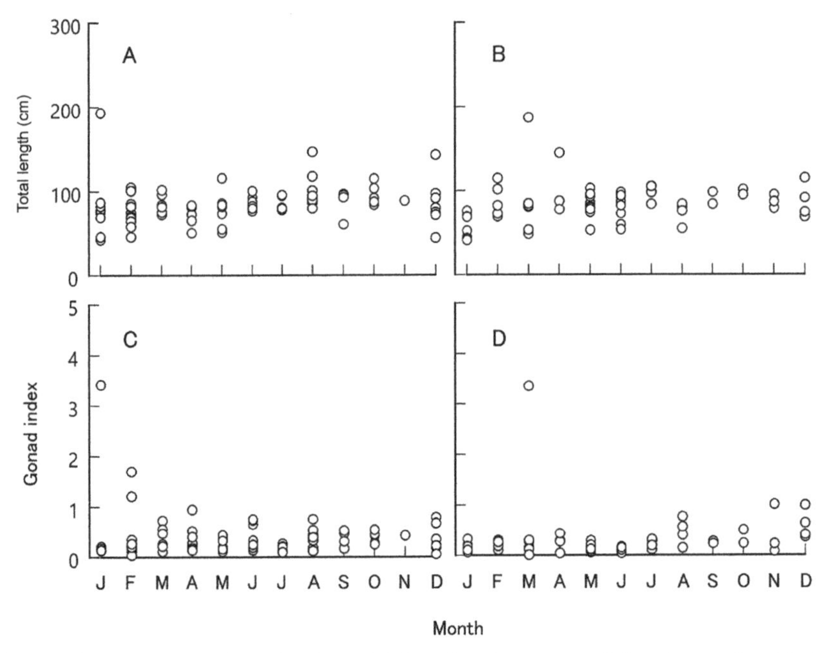

Figure 7. Seasonal changes in the total length and gonad index of captive ocean sunfish *Mola mola* A and C females, B and D males. Redrawn with permission and modification from Nakatsubo et al. (2007b).

but did not seem to show seasonality (Fig. 7A,B,C,D). Captive conditions, which differed from wild environmental conditions in terms of temperature, sunshine and prey, likely influence the timeline of maturity.

Changes in Histological Maturity

Gonad maturity was examined histologically in 13 large individuals (9 females, > 185 cm TL; 4 males, > 165 cm TL) collected from May to December 2006. Two of the females, collected in August 1996 were examined and found to have oocytes in the maturation (i.e., nuclear migration) stage (See Fig. 8A–F). Degenerated oocytes were observed from the ovary of an individual collected at the end of October.

In the four males, histological sections of testes and spermatozoa in testicular tissue fluid were observed in individuals collected in May (191 cm TL, GI: 0.386), June (174 cm TL, GI: 1.019), August (190 cm TL, GI:1.359), and December (168 cm TL, GI: 0.147). Individuals in May, June and August had lobular lumens filled with spermatocytes or spermatids. Only the individual collected in August had motile spermatozoa in the testes. In contrast, the individual collected in December showed many gaps in the lobular lumens, and some spermatocyte masses were observed. There were no spermatozoa in the tissue fluid (Fig. 9A–E).

In the males, motile spermatozoa were observed, but in the females, no mature stage oocytes were observed. The absence of large individuals available for analysis in certain months (e.g., no large females were collected in September; no large males were collected September-November) meant that there was a lack of data for those months. The results suggested that both males and females mature by August, with the gonads regressing in females at the end of October and males in December. Given these findings, the spawning season of ocean sunfish migrating to the Kanto coast of Japan appears to be from the end of August to September.

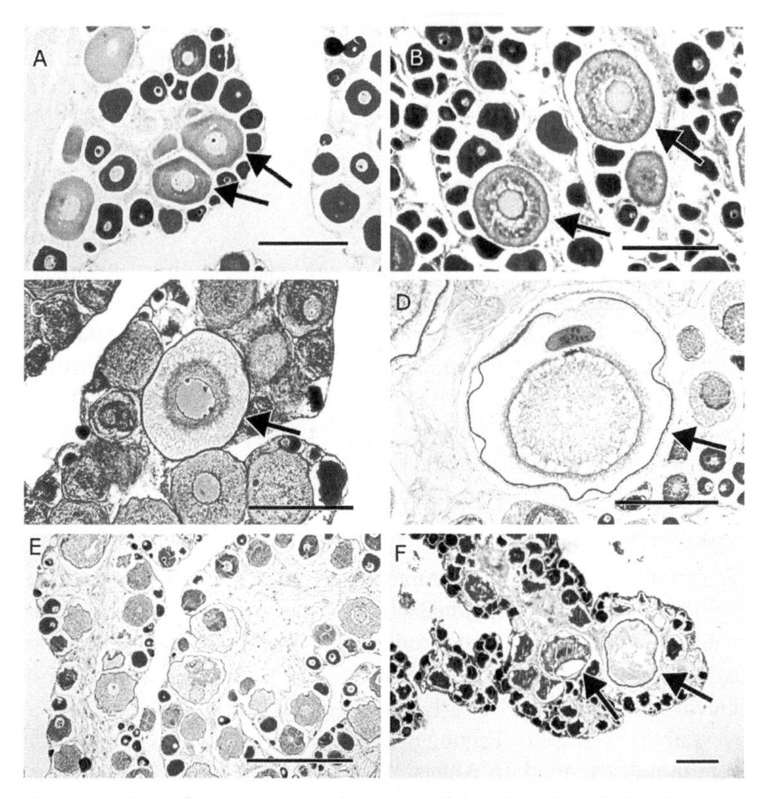

Figure 8. Histological observations of oocyte and ovary in ocean sunfish *Mola mola*. A: Perinucleolus stage (Nov. 2006, 235 cm TL 800 kg BW), B: Yolk vesicle stage (Oct. 2005, 241 cm TL 960 kg BW), C: Yolk globule stage (May 2006, 191 TL 305 BW), D: Migratory nucleus stage (Aug. 1996, 220 cm TL - BW), E: Ovary of migratory nucleus phase (Aug. 1996, 245 cm TL - BW), F: degenerated oocyte (Oct., 2005, 218 cm TL 600 kg BW). Bar scales = 200 μm in A–D, F, and 500 μm in E. HE stained. Redrawn with permission and modification from Nakatsubo et al. (2007b) and Nakatsubo (2008).

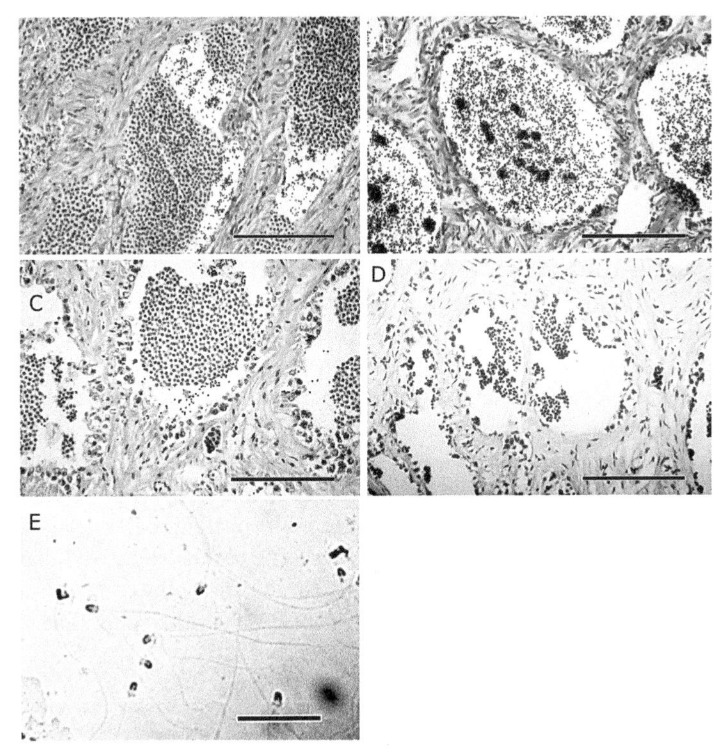

Figure 9. Histological observations on the testes of ocean sunfish *Mola mola*. A. May (169 cm TL), B: June (174 cm TL), C: August (190 cm TL), D: December (168 cm TL), E: Sperm collected from testis of the same specimen as C. Scale bars = 100 μm in A–D, and 20 μm in E. HE (Hematoxylin and Eosin) stained (A–D). Redrawn with permission and modification from Nakatsubo et al. (2007b).

A Rare Case of Mola Egg Release

Kamogawa SeaWorld, Japan recorded a rare case of egg release from a 100 cm TL individual that was on exhibit (T. Nakatsubo, unpublished observation). The individual released a clear jelly-like egg mass over the course of five days (Fig. 10) from 17 to 21 February 1997. A total of 19 releases were observed. Individual eggs were 1.00 ± 0.04 mm (mean ± SD) in diameter. The density was roughly one egg/ml egg mass. A total of 9.4 L of egg mass was collected in 11 releases resulting in an estimated 9,400 eggs. The total number of eggs however was not counted. All eggs were at the same stage of maturity (T. Nakatsubo, unpublished data). The initial moment of egg release was not observed.

Rare Encounters with Large Molids

Three large ocean sunfish were caught 16 August 1996 in Kamogawa, Japan measuring 210 cm, 243 cm and 272 cm TL. The 272 cm individual was reported as the heaviest bony fish in the world and formally weighed at 2,300 kg. Recently, this individual reclassified from *M. mola* to *M. alexandrini* (Sawai et al. 2017, Sawai et al. 2020 [Chapter 2]) based on visual inspection of a photograph (Fig. 11). The clavus, however, is not clearly visible in the photo and as such, this reclassification remains controversial (T. Nakatsubo, Kamogawa SeaWorld, personal communication).

Another large molid, captured 16 August 1996 (2.2 m TL, weight unknown) held an ovary weighing 2.720 kg. This individual had an ovary tunica that was approximately 5 mm in thickness and considered negligible in terms of overall weight. The analysis of egg size composition and fecundity was performed by dissecting out three small sections of tissue (4.0, 4.3, and 7.1 mg each) from a formalin-fixed ovary. Each piece was transferred separately to a petri dish with a small amount of distilled water.

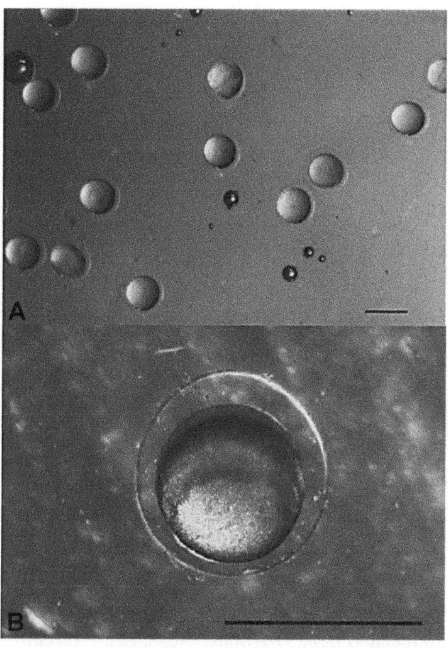

Figure 10. Photos of eggs from a captive ocean sunfish, *Mola mola*, 100 cm TL who released eggs in an exhibit tank of Kamogawa SeaWorld February 1997. (A) Eggs in a jelly-like egg mass (B) Single egg. Scale bars indicate 1000 microns.

Figure 11. World record-holding *Mola* sp. captured Aug 16 1996 Kamogawa Chiba Japan. Photo reprinted with permission from Kamogawa SeaWorld.

Oocytes (ranging in size from 20.5–500.3 um) were separated apart using two dissecting needles while observing under a stereomicroscope (Leica EZ4W). The separated oocytes were photographed at 8X or 35X magnification. The counting and measurements of oocyte was achieved using ImageJ software (version 1.51 k, National Institutes of Health) taking care not to measure the same oocyte twice. Statistical analysis was performed with statistical software EZR (Kanda 2013) (Saitama Medical Center, Jichi Medical University, Saitama, Japan; www.jichi.ac.jp/saitama-sct/SaitamaHP.files/statmedEN.html), a graphical user interface for R (The R Foundation for Statistical Computing). The total number of ova was estimated statistically from the number of oocytes in each piece. 874.3 +/– 127.2 S.D. million ova in different stages of reproduction were recorded (See Fig. 12A–E and Fig. 13).

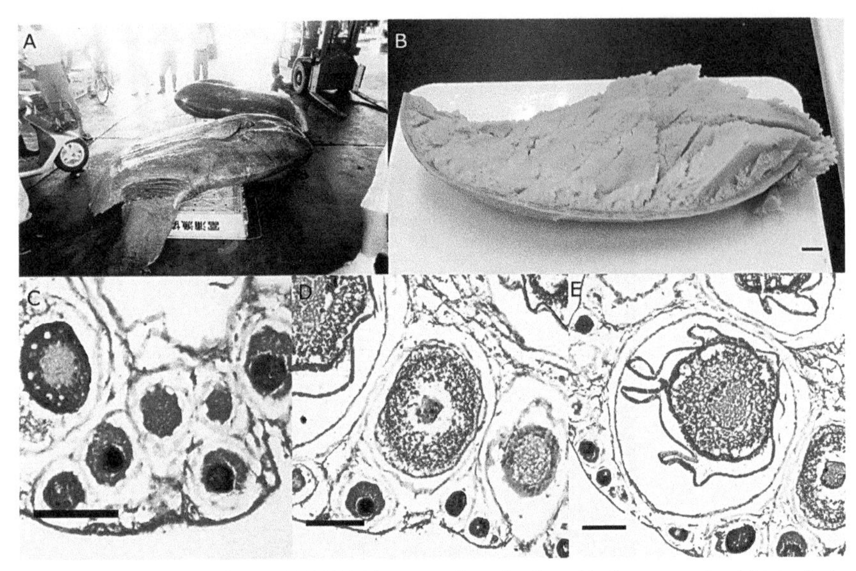

Figure 12. *Mola* sp. 220 cm TL caught Aug 16 1996 (above specimen in photo A). Representative histological examples of examined oocytes from small B, medium C and large D groups. (Scale bar = 100 μm). E. Piece of ovary. Scale bar 10 mm. Photos printed with permission from T. Nakatsubo and Kamogawa Seaworld.

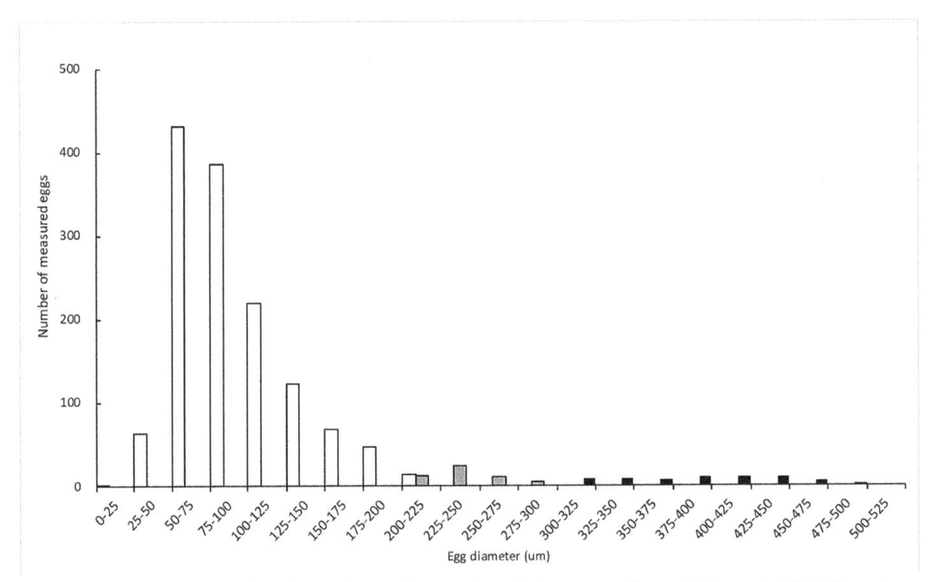

Figure 13. Ova size distribution of *Mola* sp. 2.20 m TL caught off Kamogawa Japan 16 August 1996. Ova size grouping with the small-medium border set at 211.6 um and the medium-large border set at 296.6 um. Small size are white bars. Medium size is light grey and larger size is black.

A third individual (2.10 m TL, no weight taken) was caught on 16 August 1996 and recorded as *M. mola* due to its normal shape and lack of head or chin bumps. Its ovary was photographed, examined and weighed (2.98 kg). From a piece of 1 g ovarian tissue, 12,910 ova were separated out but only large ova (those that seemed to be more transparent were counted). This count was then multiplied by ovary weight. Statistics were not performed on the counts. The phase of maturity was classified as migratory stage and the estimated ova count was 38.49 million. The ova were estimated to be 0.5 mm diameter (Fig. 14A–D).

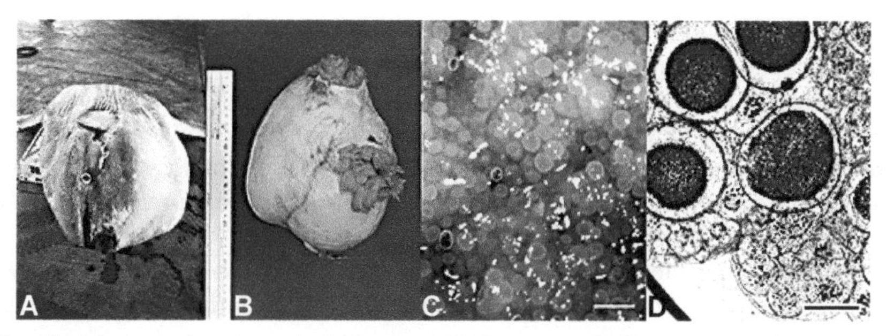

Figure 14. Gravid *Mola* sp. caught and examined 16 August 1996 Kamogawa, Chiba, Japan. A. Individual *Mola* sp. 2.10 m, B. Single ovary weighed 2.98 kg, C. Oocytes (scale bar 250 um), D. Egg mass. Scale bar 500 um.

Key Findings

Results from these studies, while preliminary, represent the most detailed information to date on the spawning season and fecundity of *M. mola*. Collectively, our findings suggested that gonad weight/GI increases with size/body length in female *M. mola* between 150–160 cm (Study 1 and Study 3). Females > 1.5 m were estimated to be sexually mature (Study 1) with a seasonal change in GI in females during the summer (Study 3). No *M. mola* females larger than 90 cm were examined in Study 2 and as such no individuals displayed mature oocytes (i.e., vitellogenic ovarian and/or hydrated follicles). Study 3 supports the suggestion that the *M. mola* spawning season in Japanese waters runs from the end of August through October (Nakatsubo et al. 2007b). This is similar to findings from Kang et al. (2015) who suggested that in South Korea, the spawning period for *M. mola* runs from July to October.

In Studies 1–3, the majority of *M. mola* females displayed immature oocytes. This is likely due, in part, to both the small body size of females (most < 1 m, Study 2 (mean TL 45.6 cm) and the season of sample collection. The examined *M. tecta* was large (2.15 m) but the absence of any post-ovulatory follicles indicated that she had not recently spawned. She likely could have produced mature oocytes/ova had she not been stranded in the winter before her presumed summer/fall spawning season.

The results from Study 3 suggest the gonadal index of *M. mola* females in Japan peaks in August and is relatively low earlier in the summer and later in the fall. The two females with the most mature oocyte development were collected in August, whereas in Study 1, all females were collected in the fall through early winter. Therefore, the timing of oocyte maturation is likely dependent upon local environmental conditions. This first approximation of *M. mola* yolked oocyte (i.e., vitellogenic stage ovarian follicle) fecundity was at least an order of magnitude less than the previously published estimate by Schmidt (1921; Study 1). It should be noted that Schmidt (1921) did not define fecundity and made his estimate based on small, and likely previtellogenic oocytes. Such information is of great value in future studies as we move towards the inclusion of sunfishes in fisheries or ecosystem models.

The examination of three large sunfish (2.10–2.72 m) captured on 16 August 1996 allowed a glimpse into overall egg capacity and comparison to the estimated large number of ova (300 million) reported by Schmidt (1921). While debate exists over the world's most fecund bony fish, the *Mola* genera are certainly still in contention for possessing the largest number of ova in a single individual at one time. The estimated ova count from a 2.2 m TL female sets a new record (874 million ova) for a single individual ocean sunfish. To the best of our knowledge, no other fishes have been recorded carrying more than 300 million ova (previtellogenic or otherwise) at any one time.

The number of yolked ova however provides a better prediction of what will go on to ovulate and be released (spawned) in the next batch (if a total spawner) or in subsequent batches (if a batch spawner). As mentioned earlier, understanding the level of oocyte maturity is integral to establishing functional (realized) fecundity. In that regard, more research is needed.

In summary, the molids studied in this chapter are gonochoric (i.e., separate sexes for life), broadcast, pelagic spawners and oviparous (i.e., releasing their eggs into the water column where they are fertilized externally). The diameter of their developing oocytes can be grouped into three classes: immature (< 500 microns), developing (500–1000 microns), mature/spawning (< 1 mm). Whether they are group synchronous (having two or more distinct populations of oocytes simultaneously and ovulating once in a season) or asynchronous (capable of ovulating on a regular basis, sometimes every day, over a prolonged period) remains unclear. Whether or not they are determinate spawners (with a fixed potential annual fecundity prior to the spawning period) or indeterminate spawners (with unfixed potential annual fecundity and eggs developing at any time during the spawning season) also remains to be determined.

While much work regarding the reproductive biology of ocean sunfishes remains, the new data presented in this chapter provide deeper insights into *Mola* spp. reproductive biology. Looking to the future, assessments of ocean sunfish population resilience need to be underpinned by empirical data on size at maturity, the duration of the reproductive season, daily spawning behavior, and the release of eggs over the span of days and weeks. Attaining such information will require coordinated efforts between researchers from many regions, across different times of the year.

With the help of stranding networks, bycaught samples, individuals caught for display and in fisheries with set nets, this study was able to recover biological samples of numerous specimens to address poorly known aspects of molid reproductive biology. Such opportunistic sampling mechanisms are critical to data acquisition when dealing with elusive species—similar to approaches used in studying sea turtles (Vélez-Rubio et al. 2013, Domènech et al. 2019). As a priority, future research endeavors should attempt to gather reproductive data from a broader size range of fish, in particular individuals > 2 m TL which are poorly represented. Access to a more complete size range will allow for more accurate estimates of: (1) sexual maturity based on body size, (2) the time course of oocyte development, (3) maturation, (4) ovulation and (5) spawning frequency throughout the breeding/spawning season. This information, in turn, will allow for more accurate estimates of batch, annual and lifetime fecundity as well as spawning stock biomass (Lasker 1985), essential for improving the accuracy of data used in the IUCN listing for ocean sunfishes (Liu et al. 2015). Progress on all these fronts would represent a step change in our capacity to manage and conserve this remarkable family of fishes.

Acknowledgements

Thank you to the staff and interns of the New England Coastal Wildlife Alliance who were part of NECWA's Ocean Sunfish Stranding Team. Thanks also to Team Mola participants, trained community-members who assisted NECWA during stranding events. Thank you to Bob Prescott, Director of Mass Audubon as well as Audubon staff and volunteers who also provided stranding response support. A special thank you to NECWA staff members Michael Rizzo and Dr. Tammy Silva, who helped document ocean sunfish carcasses in the field as well as input and analyze stranding data. Thank you to Dana Morton for providing the *Mola tecta* ovary. Thanks to Beverly Macewicz (Southwest Fisheries Center) for initial work on the *M. mola* ovaries. Thanks to Katherine Grier for histological processing and analysis (Mola C1, C14, C19, C24, C49, C59, P5, P9, P13, CA1).

For the reproduction studies from Japan, T. Nakatsubo would like to thank Kamogawa Sea World, Dr. Teruo Tobayama [Former General Manager and Director (Deceased)], Dr. Kazutoshi Arai (General Manager of head office and Former Aquarium Director), Mr. Hiroshi Katsumata (Aquarium Director), Mr. Akihisa Osawa (Curator of Fishes), All other aquarium staff who assisted in this work, College of Bioresource Sciences (Nihon University), Dr. Hitomi Hirose, Dr. Kiyoshi Asahina, Dr. Haruo Sugita, Dr. Nobuhiro Mano, ORIX Aquarium Corporation, Mr. Takaaki Nitanai (President), Mr. Takashi Iwamaru (Executive officer), Ms. Yumiko Tabata (Senior Manager).

References

Caldera, E., J.L. Whitney, T.M. Thys, E. Ostalé-Valriberas and L. Kubicek. 2020. Genetic insights regarding the taxonomy, phylogeography and evolution of the ocean sunfishes (Molidae: Tetraodontiformes). pp. 37–54. *In*: T.M. Thys, G.C. Hays and J.D.R. Houghton [eds.]. The Ocean Sunfishes: Evolution, Biology and Conservation, CRC Press. Boca Raton, FL, USA.

Carwardine, M. 1995. The Guinness book of animal records. Guinness Publishing, Middlesex, UK.

Collette, B.B. 2010. Reproduction and development in epipelagic fishes. pp. 21–63. *In*: K. Cole [ed.]. Reproduction and Sexuality in Marine Fishes: Patterns and Processes. Berkeley: University of California Press.

Domènech, F., F.J. Aznar, J.A. Raga and J. Tomás. 2019. Two decades of monitoring in marine debris ingestion in loggerhead sea turtle, Caretta caretta, from the western Mediterranean. Environ. Pollut. 244: 367–378. doi:https://doi.org/10.1016/j.envpol.2018.10.047.

Grier, H.J., W.F. Porak, J. Carroll and L.R. Parenti. 2018. Oocyte Development and Staging in the Florida Bass, Micropterus floridanus (LeSueur, 1822), with Comments on the Evolution of Pelagic and Demersal Eggs in Bony Fishes. Copeia 106(2): 329–345.

Guraya, SS. 1986. The cell and molecular biology of fish oogenesis. Monogr. Dev. Biol. 18: 1–223.

Hellenbrecht, L.M., M. Freese, J.-D. Pohlmann, H. Westerberg, T. Blancke and R. Hanel. 2019. Larval distribution of the ocean sunfishes *Ranzania laevis* and *Masturus lanceolatus* (Tetraodontiformes: Molidae) in the Sargasso Sea subtropical convergence zone. J. Plankton Res. 41(5): 595–608. https://doi.org/10.1093/plankt/fbz057.

Jenkins, K.L.M. and R.S. McBride. 2009. Reproductive biology of wahoo, *Acanthocybium solandri*, from the Atlantic coast of Florida and the Bahamas. Mar. Freshwater Res. 60(9): 893–897. https://doi.org/10.1071/MF08211.

Kanda, Y. 2013. Investigation of the freely-available easy-to-use software "EZR" (Easy R) for medical statistics. Bone Marrow Transplant. 48: 452–458. Advance online publication 3 December 2012; doi: 10.1038/bmt.2012.244.

Kang, M.J., H.J. Baek, D.W. Lee and J.H. Choi. 2015. Sexual maturity and spawning of ocean sunfish *Mola mola* in Korean Waters. Korean J. Fish. Aquat. Sci. 48(5): 739–744. doi:10.5657/KFAS.2015.0739.

Kennedy, J., A.C. Gundersen, A.S. Høines and O.S. Kjesbu. 2011. Greenland halibut (*Reinhardtius hippoglossoides*) spawn annually but successive cohorts of oocytes develop over 2 years, complicating correct assessment of maturity. Can. J. Fish. Aquat. Sci. 68(2): 201–209. doi:10.1139/F10-149.

Lasker, R. 1985. An egg production method for estimating spawning biomass of pelagic fish: application to the northern anchovy, *Engraulis mordax*. NOAA Technical Report, National Marine Fisheries Service, Special Scientific Report - Fisheries Series 36: 99.

Liu, J., G. Zapfe, K.-T. Shao, J.L. Leis, K. Matsuura, G. Hardy et al. 2015. *Mola mola* (errata version published in 2016). The IUCN Red List of Threatened Species 2015: e.T190422A97667070. Downloaded on 24 December 2019.

McBride, R.S., D.J.G. Snodgrass, D.H. Adams, S.J. Rider and J.A. Colvocoresses. 2012. An indeterminate model to estimate egg production of the highly iteroparous and fecund fish, dolphinfish (*Coryphaena hippurus*). Bull. Mar. Sci. 88: 283–303. doi:10.5343/bms.2011.1096.

McBride, R.S., S. Somarakis, G.R. Fitzhugh, A. Albert, N.A. Yaragina, M.J. Wuenschel et al. 2015. Energy acquisition and allocation to egg production in relation to fish reproductive strategies. Fish Fish. 16(1): 23–57. doi:10.1111/faf.12043.

Murayama, T., I. Mitani and I. Aoki. 1995. Estimation of the spawning period of the Pacific mackerel *Scomber japonicus* based on the changes in gonad index and ovarian histology. Bull. Japan. Soc. Fish. Ocean. 59(1): 11–17.

Murua, H. and F. Saborido-Rey. 2003. Female reproductive strategies of marine fish species of the North Atlantic. J. Northwest Atl. Fish. Sci. 33: 23–31. Retrieved from http://journal.nafo.int/Portals/0/2003-Vol33/murua.pdf.

Murua, H., G. Kraus, F. Saborido-Rey, P.R. Witthames, A. Thorsen and S. Junquera. 2003. Procedures to estimate fecundity of marine fish species in relation to their reproductive strategy. J. Northwest Atl. Fish. Sci. 33: 33–54. Retrieved from https://journal.nafo.int/Portals/0/2003-Vol33/murua2.pdf.

Nagahama, Y. 1983. The functional morphology of teleost gonads. pp. 223–269. *In*: W.S. Hoar, D.J. Randall and E.M. Donaldson [eds.]. Fish Physiology Vol IXA.

Nakatsubo, T., M. Kawachi, N. Mano and H. Hirose. 2007a. Estimation of Maturation in Wild and Captive Ocean Sunfish *Mola Mola*. Aquacul. Sci. 55: 259–264.

Nakatsubo, T., M. Kawachi, N. Mano and H. Hirose. 2007b. Spawning period of ocean sunfish *Mola Mola* in waters of the Eastern Kanto Region, Japan. Aquacul. Sci. 55: 613–618.

Nakatsubo, T. 2008. A Study on the Reproductive Biology of Ocean Sunfish *Mola mola*." Doctoral thesis, Nihon University.

Press, Y.K., M.J. Wuenschel and R.S. McBride. 2014. Time course of oocyte development in winter flounder (*Pseudopleuronectes americanus*) and spawning seasonality for the Gulf of Maine, Georges Bank, and Southern New England stocks. J. Fish. Biol. 85(2): 421–445. doi:10.1111/jfb.12431.

Saber, S., D. Macías, J.O. de Urbina and O.S. Kjesbu. 2016. Contrasting batch fecundity estimates of albacore (*Thunnus alalunga*), an indeterminate spawner, by different laboratory techniques. Fish. Res. 176: 76–85. doi:10.1016/j.fishres.2015.12.013.

Saidapur, S.K. 1978. Follicular atresia in the ovaries of nonmammalian vertebrates. Int. Rev. Cytol. 54: 225–244.

Sawai, E., Y. Yamanoue, M. Nyegaard and Y. Saki. 2017. Redescription of the bumphead sunfish Mola alexandrini (Ranzania 1839), senior synonym of *Mola ramsayi* (Giglioli 1883) with designation of a neotype for Mola mola (Linnaeus 1758) Tetraodontiformes: Molidae. Ichthyol. Res. 1–19. doi: 10.1007/s10228-017-0603-6.

Sawaii, E., M. Nyegaard and Y. Yaminonue. 2020. Phylogeny, taxonomy and size records of the ocean sunfishes. pp. 18–36. *In*: T.M. Thys and J.D.R. Houghton [eds.]. The Ocean Sunfishes: Evolution, Biology and Conservation, CRC Press.

Schmidt, J. 1921. New studies of sun-fishes made during the "Dana" Expedition, 1920. Nature. 107: 76–79. Retrieved from https://www.nature.com/articles/107076a0.

Vélez-Rubio, G.M., A. Estrades, A. Fallabrino and J. Tomás. 2013. Marine turtle threats in Uruguayan waters: insights from 12 years of stranding data. Mar. Biol. 160: 2797–811.

Wallace, R.A. and K. Selman. 1981. Cellular and dynamic aspects of oocyte growth in teleosts. Am. Zool. 21: 325–343.

Wallace, R.A. and K. Selman. 1990. Ultrastructural aspects of oogenesis and oocyte growth in fish and amphibians. J. Electron Microsc. Tech. 16: 175–201.

CHAPTER 7

Ocean Sunfish Larvae

Detections, Identification and Predation

Tierney M. Thys,[1,*] *Marianne Nyegaard*,[2] *Jonathan L. Whitney*,[3] *John P. Ryan*,[4] *Inga Potter*,[5] *Toshiyuki Nakatsubo*,[6] *Marko Freese*,[7] *Lea M. Hellenbrecht*,[8] *Rachel Kelly*,[9] *Katsumi Tsukamoto*,[10] *Gento Shinohara*,[11] *Tor Mowatt-Larssen*[12] and *Lukas Kubicek*[13]

Introduction

The natural history of molid larvae is poorly known despite the molids holding the vertebrate record for having the most ova in a given individual: 300 million in a 1.5 m TL female (Schmidt 1921a) and more recently, 847 million in a 2.2 m TL female (Forsgren et al. 2020 [Chapter 6]). Spawning is equally mysterious and has never been witnessed in the wild for any of the five currently recognized species: *Mola mola* (Linnaeus 1758), *M. alexandrini* (Ranzani 1839), *Mola tecta* Nyegaard et al. 2017, *Masturus lanceolatus* (Liénard 1840) and *Ranzania laevis* (Pennant 1776). Since molid larvae appear rarely in global plankton surveys, scant data exist on the size and timing of egg releases, larval development, dispersal patterns and survival rates through metamorphosis and into adulthood. While steady progress is being made (Forsgren et al. 2020 [Chapter 6]), many aspects of molid reproductive strategies and larval life histories remain unclear. This chapter gathers our current knowledge of these early life stages of the molids.

[1] California Academy of Sciences, San Francisco, CA 94118, USA.
[2] Auckland Memorial Museum Tamaki Paenga Hira, Natural Sciences, The Domain, Private Bag 92018, Victoria Street West, Auckland 1142, New Zealand; Email: mnyegaard@auklandmuseum.com
[3] Joint Institute for Marine and Atmospheric Research, University of Hawai'i at Mānoa,1000 Pope Road, Marine Sciences Building 312, Honolulu, HI USA; Email: Jw2@hawaii.edu
[4] MBARI, 7700 Sandholdt Road, Moss Landing, CA 95039, USA; Email: ryjo@mbari.org
[5] Affiliate faculty, University of New Hampshire, Durham, NH, USA; Email: ingapotter@gmail.com
[6] ORIX Aquarium Corporation, Nippon Life Hamamatsucho Crea Tower, 2-3-1 Hamamtsu-cho, Minato-ku, Tokyo 105-0013, Japan; Email: t.nakatsubo@aqua.email.ne.jp
[7] Thünen Institute of Fisheries Ecology, Herwigstrasse 31, 27572 Bremerhaven, Germany; Email: Marko.Freese@thuenen.de
[8] Department of Biological Sciences, University of Bergen, Thormøhlensgate 53 a/b, Bergen. Email: Lea.Hellenbrecht@gmail.com
[9] Centre for Marine Socioecology, University of Tasmania, Hobart 7005, Tasmania, Australia; Email: rachel19191@gmail.com
[10] The University of Tokyo, 1-1-1 Yayoi, Bunkyo-ku, Tokyo 113-8657, Japan; Email: tsukamoto@marine.fs.a.u-tokyo.ac.jp
[11] Department of Zoology, National Museum of Nature and Science, 4-1-1 Amakubo, Tsukuba-shi, Ibaraki 305-0005, Japan. Email: s-gento@kahaku.go.jp
[12] Scripps Institution of Oceanography, UC San Diego, La Jolla CA, USA; Email: tor.mowatt.larssen@gmail.com
[13] Aussermattweg 22, 3532, Zäziwil, Switzerland; Email: molamondfisch@gmail.com
* Corresponding author: tierneythys@gmail.com

Post-hatching, it is well known that Molidae larvae quickly develop into conspicuous, rotund and distinctly spiky, larvae recognizable at the genus level (Lyczkowski-Shultz 2003). For the *Mola* genus however, the early larval stages are difficult to identify to the species level using diagnostic features. As such, historical and modern records of *Mola* spp. larvae identified mophologically must be viewed with caution. Notably, larvae < 2 mm notochord length (NL) require identification using molecular techniques since they are so visibly similar (Fig. 1; Lyczkowski-Shultz 2003, Hellenbrecht et al. 2019). Larger larvae > 2 mm (NL), on the other hand, are sufficiently distinct between the three genera to allow visual discrimination (Sawai et al. 2020 [Chapter 2], Bemis et al. 2020 [Chapter 4]) at early stages.

The identification of larval fish records can help track underlying changes in fish communities and community health over time (e.g., Marshall et al. 2019, Lasker 1981). Ideally, such data can be considered alongside oceanography (e.g., as a proxy for spawning stock) to explore the drivers of ecosystem change. This information is essential for effective ecosystem management and forecasting of stock biomass (Smith et al. 2018). For ocean sunfishes, large gaps exist in our knowledge base with regards to early life history however, in recent years, new records from research cruises have energized interest in mapping of *R. laevis* and *Ma. lanceolatus* larvae in the Sargasso Sea (Hellenbrecht et al. 2019). Meanwhile, research conducted off Japan and South Korea is shedding light on the spawning season of *Mola* spp. (Nakatsubo 2007a,b, Kang et al. 2015, respectively). Lastly, in terms of trophic ecology, molid larvae are being found to serve as episodic food sources for larger piscivorous predators (e.g., blue marlin, *Makaira nigricans*, Shimose et al. 2013). Drawing together these converging lines of evidence, this chapter summarizes our current knowledge of molid larvae, places this information in a geographical context, provides guidance on species identification and highlights some key areas for future research.

Methods

An extensive review of online and published resources was conducted (up to February 2020) to unearth records of molid larvae and their associated predators. Larval data were solicited from curators of fish collections, natural history museums and plankton surveys worldwide (see Appendix 1 for the full list of organizations that provided larval information).

The original species identification was reviewed for all records, and where possible, categorized to either family, genus or species-level based on Martin and Drewry (1978), Lyczkowski-Shultz (2003) and Hellenbrecht et al. (2019). All material originally identified as *M. mola* was considered as *Mola* spp., as it is currently impossible to visibly distinguish morphological differences between the very early life stages of *M. mola, M. alexandrini* and *M. tecta* (Bemis et al. 2020 [Chapter 4]).

Results and Discussion

Collectively, 452 records were gathered comprising 9,770 larvae in total: 285 *Masturus*, 84 *Mola* spp., 61 unspecified Molidae and 9,340 *Ranzania* (340 of which were eggs).

Early Molid Life Stages (Larval and *Molacanthus* Stages)

Like most marine fishes, molids have complex life histories with morphologically distinct stages, presumably driven by ontogenetic shifts in habitat, diet and predation risk. Sunfish ontogeny, however, includes particularly remarkable metamorphoses wherein the early life stages are quite different morphologically from the adult form (Fraser-Brunner 1951, Martin and Drewry 1978, Leis 1984). The molid life stages caused much confusion in early ichthyology, where several small specimens were initially described as different species to the adults, e.g., reviewed in Nyegaard et al. 2018. Such nominal species include *Molacanthus pallasii* (Swainson 1839) and *Ostracion boops* (Richardson 1844), now synonymized with *M. mola* and *R. laevis*, respectively (Fraser-Brunner 1951, Fricke et al. 2020).

Of the five currently recognized molid species, our understanding of egg and larval development is skewed heavily towards *R. laevis*. This is the only species to have been reared successfully from eggs (Leis 1977, Baensch 2013), and is sampled the most extensively and recently (Hellenbrecht et al. 2019). In comparison, our current knowledge of early *Masturus* and *Mola* ontogeny is based on publications describing individual specimens collected from around the world and collated into growth series with descriptions of morphological development (e.g., Martin and Drewry 1978, Lyczkowski-Shultz 2003). Based on these resources, we briefly describe the early life stages of the molids beginning with *Ranzania*.

At hatching, *R. laevis* larvae are in a relatively advanced stage of development, including a functional mouth, pigmented eyes, and ossification of several pectoral fin rays (Leis 1977, 1984).[1] Notably (given the lack of a caudal fin in adults), the hatchlings possess a well-developed notochord, finfold and post-anal myomeres (Fig. 1). This morphology resembles other Tetraodontiform larvae (Fraser-Brunner 1951, Leis 1977, Martin and Drewry 1978). However, the finfold, post-anal myomeres and distal notochord soon start to atrophy (timeframe uncertain), and the larvae become rotund with a distinctly spiky appearance (Fig. 1). A small, 'spiky' specimen of *R. laevis* was previously described as *Ostracion boops* (op. cit.) giving rise to the taxa-specific term '*Ostracion boops* stage', or 'ostracioniform stage' (e.g., Fraser-Brunner 1951, Leis 1977, Martin and Drewry 1978).

In 'typical' teleosts, the distal notochord does not atrophy but bends dorsally with growth (termed flexion), forming the hypural elements that support the caudal fin rays (e.g., Kendall et al. 1984, Rodríguez et al. 2017). However, in all molid larvae, the distal notochord atrophies completely and neither bends dorsally nor forms the hypural bone, and a caudal fin (or remnants thereof) never develops (e.g., Leis 1977, Johnson and Britz 2005). The common pre-flexion, flexion and post-flexion larval stages for typical bony fishes (e.g., Kendall et al. 1984) are therefore meaningless when applied to larval Molidae. Another marked difference between Molidae and other teleosts is the formation of the taxa-specific clavus in place of a caudal fin. Specifically, as the distal notochord atrophies, the clavus fin rays are formed from elements of the dorsal and anal fin rays, developing inwards and closing the gap between them (Leis 1977, Tyler 1980, Johnson and Britz 2005).

The final complement of fin rays is established relatively early in Molidae, at only a few mm in larvae length (Leis 1977, Lyczkowski-Shultz 2003). In more typical teleosts, this point marks the transformation from larva to juvenile (e.g., Martin and Drewry 1978, Kendall et al. 1984). In Molidae, however, the transformation to juvenile is not easily delineated as it involves extensive, taxa-specific morphological changes during a prolonged stage of metamorphosis before the adult molid shape is reached. Specifically, in *Mola* and *Masturus,* the body is elongated (i.e., deeper than it is long) with a prominent ventral keel and numerous dermal spines (Fig. 1). This stage is neither typical larval nor typical juvenile in its morphology, and is referred to as 'pre-juvenile' or the taxa-specific *'Molacanthus*' stage (Fraser-Brunner 1951, Martin and Drewry 1978, Leis 1984, Lyczkowski-Shultz 2003), or also 'molacanthiforms'. This *Molacanthus* stage occurs from *ca.* 4 and 5 mm TL for *Mola* and *Masturus*, respectively, followed by transformation to juveniles around 55–60 mm pre-claval length (PCL) (Lyczkowski Shultz 2003) where the dermal spines are lost and the typical adult molid shape emerges (Fig. 1). By contrast, development in *R. laevis* is direct with no *Molacanthus* stage (Leis 1977, Lyczkowski-Shultz 2003). The transformation to juvenile appears to occur around 12 mm SAL (snout to anus length) or 15 PCL (Leis 1977, Lyczkowski-Shultz 2003) when the dermal spines are disappearing and adult-like morphology emerges (Fig. 1).

[1] Despite the presence of remnant yolk post-hatching, the term yolk-sac larva is not particularly informative in the context of early molid development, as structures such as a functional mouth and eyes are generally taken as indicators of the end of the yolk sac stage in teleosts (Leis 1977).

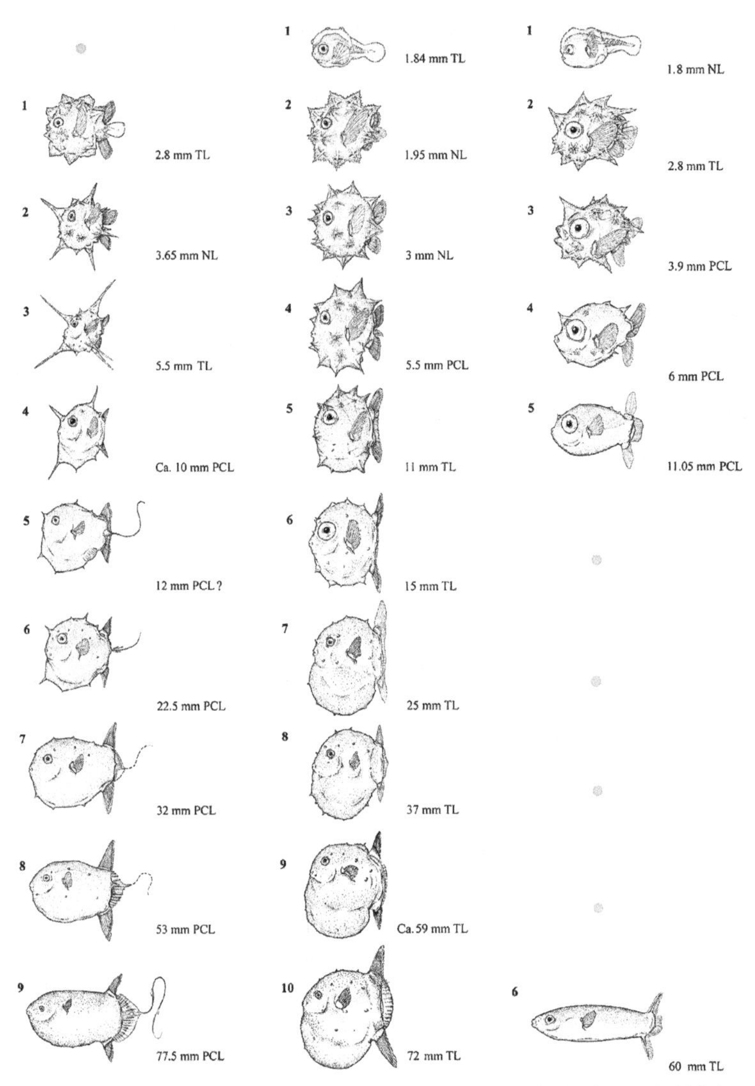

Figure 1. Illustrations of early development stages of *Masturus lanceolatus* (left), *Mola* spp. (middle) and *Ranzania laevis* (right), drawn by artist Nicholas Bezio from literature specimens[2] Notochord length (NL), Pre-clavus length (PCL), Total length (TL).

[2] *Masturus lanceolatus.* **(1)** Larva, Sargasso Sea, Schmidt (1921b) Fig. 5c (lower specimen, reversed). **(2)** Larva, W Atlantic Ocean, Lyczkowski-Shultz (2003) p6(B). **(3)** *Molacanthus*-stage, Sargasso Sea, Schmidt (1921b) Fig. 5a (upper specimen). **(4)** *Molacanthus*-stage, Sargasso Sea, Hellenbrecht et al. (2019), Fig. 2D (right panel). **(5)** *Molacanthus*-stage, NW Pacific, Sokolovskaya and Sokolovskiy (1975) Fig. 2f. **(6)** *Molacanthus*-stage, Japan, Yabe (1950) Fig. 1, Fig. 2(1). **(7)** *Molacanthus*-stage, NW of Tuamotus, Pacific, Günther (1880) Fig. 94 (left specimen, reversed), location from Gudger (1937). **(8)** *Molacanthus*-stage, W Sargasso Sea, Gudger (1935) Fig. 1. **(9)** Juvenile, Japan, Yabe (1950) Fig. 2(3). *Mola* spp. **(1)** Larva, Mediterranean Sea, Sanzo (1939) Tav 7 Fig. 16 (reversed). **(2)** Larva, Gulf of Mexico, Lyczkowski-Shultz (2003) p8(B). **(3)** Larva, W Atlantic Ocean, Lyczkowski-Shultz (2003) p8(C). **(4)** *Molacanthus*-stage, W Atlantic Ocean, Lyczkowski-Shultz (2003) p8(D). **(5)** *Molacanthus*-stage, NW Pacific, Sokolovskaya and Sokolovskiy (1975) Fig. 2g. **(6)** *Molacanthus*-stage, presumably European waters, Steenstrup and Lütken (1898) Tab IV Fig A (reversed). **(7)** *Molacanthus-stage*, South Africa, Smith (1949) Fig S(3) p38. **(8)** *Molacanthus*-stage, presumably European waters, Reuvens (1895) Pl 5 (right), size from Martin and Drewry (1978). **(9)** *Molacanthus*-stage, Japan, Kuronuma (1940) Fig. 1, size from Martin and Drewry (1978). **(10)** juvenile, New South Wales, Australia, museum specimen AMS I.27082-00. *Ranzania laevis.* **(1)** Larva, Hawai'i, Leis (1977) Fig. 2 (top). **(2)** Larva, Mediterranean Sea, Sanzo (1919) Fig. 1. **(3)** Larva, Hawai'i Leis (1977) Fig. 6. **(4)** Larva, Hawai'i Leis (1977) Fig. 7. **(5)** Larva, Hawai'i Leis (1977) Fig. 8. **(6)** juvenile (?), 'E Indies' [=S & SE Asia], Harting (1865) Pl II Fig. 2 (reversed).

The remarkable and unusual morphology of molid larvae and molacanthiforms presumably plays an important role in larval survival as they pass through early life stages, transition to the juvenile stage and lose these features. For example, the primary function of the prominent dermal spines is likely a defense against predators that may no longer pose a threat when the molid attains juvenile size, increased mobility and a potential change in habitat (Leis 1977). Furthermore, *Ma. lanceolatus* has a swim bladder during the *Molacanthus* stage [as reported in a 21 mm TL specimen by Fraser-Brunner (1951)], which may be linked to larval survival and dispersal, and is lost upon maturation to the adult form (Chanet et al. 2012). Other larval features include heavy pigmentation (melanophores) in the dorsum and dorsal part of the gut (Leis 1977) which may play a role in protecting larvae from UV radiation whilst occupying surface waters (Paris et al. 2007).

Collection and examination of more specimens will improve our knowledge of the early life stages of the molids, in particular *Mola* spp. smaller than 30 cm TL (Lyczkowski-Shultz 2003, Nyegaard et al. 2018, Sawai et al. 2018, Hellenbrecht et al. 2019). At present, larvae and *Molacanthus* stage individuals can be identified to genus level based on the shape and arrangement of dermal spines and body shape (e.g., Fraser-Brunner 1951, Martin and Drewry 1978, Lyczkowski-Shultz 2003, Hellenbrecht et al. 2019). For the purposes of this chapter, we use the term *Molacanthus* stage according to Lyczkowski-Shultz (2003) and follow Hellenbrecht et al. (2019) treating it as part of the larval stage in the assessment of larval distribution.

Multiple Spawners

Based on gonadal maturation, Nakatsubo et al. (2007b) suggested that *M. mola* are multiple spawners, breeding in the waters off Japan from August to October. Likewise, Kang et al. (2015) reported spawning events off the island of Jeju, South Korea, from July to October. In the Sargasso, *R. laevis* also appears to be a multiple spawner (Hellenbrecht et al. 2019). For additional details see Forsgren et al. 2020 [Chapter 6].

Spawning Areas and Seasonality

Although spawning has not been observed in Molidae, the location and timing of where larvae have been detected can provide insights to spawning events. Unfortunately, most planktonic records of ocean sunfish larvae lack accompanying environmental metadata to parametrize habitat suitability. As such only general statements can be made with regard to potential spawning locations.

To date, Molidae larvae have been sampled only in the tropics and subtropics (Fig. 2; Appendix 1). *Ranzania laevis* comprises the majority of collected larvae (96 percent) and is the only species found in samples from the Atlantic, Pacific and Indian Ocean basins. Larvae of all three molid genera overlap spatially in the North Atlantic (Lyczkowski-Shulz 2003), particularly the western Sargasso Sea, and the Gulf of Mexico and western Pacific. 110 of the 452 total records (25%) have been reported from the Sargasso Sea (Schmidt 1921a,b, Parin 1968, Ayala et al. 2016, Hellenbrecht et al. 2019, Knorrn 2019), with fewer individuals reported from the Straits of Florida (Richardson et al. 2010), Mediterranean Sea (Sanzo 1919) and off the coast of Brazil (Nogueira et al. 2012). In the Pacific, sunfish larvae are regularly collected around Hawai'i (Sherman 1961, Leis and Miller 1976, Leis 1977), the western North Pacific near the Marianas Archipelago (Sokolovskaya and Sokolovskiy 1975, Kawakami et al. 2010), the western South Pacific [e.g., Vanuatu (Wan and Zhang 2005)], Fiji and French Polynesia (See Fig. 2, Appendix 1).

Collections from the North Pacific Ocean, including the Hawai'ian Islands, reveal that spawning in *Ranzania* occurs from January to May, with most larvae being captured between March and April (Sherman 1961, Sokolovskaya and Sokolovskiy 1975, Leis and Miller 1976, Leis 1977). However, in the western tropical Pacific, the collection of ocean sunfish larvae is reported as late as August (Kawakami et al. 2010) and October (Wan and Zhang 2005). In the Atlantic, larvae are reported

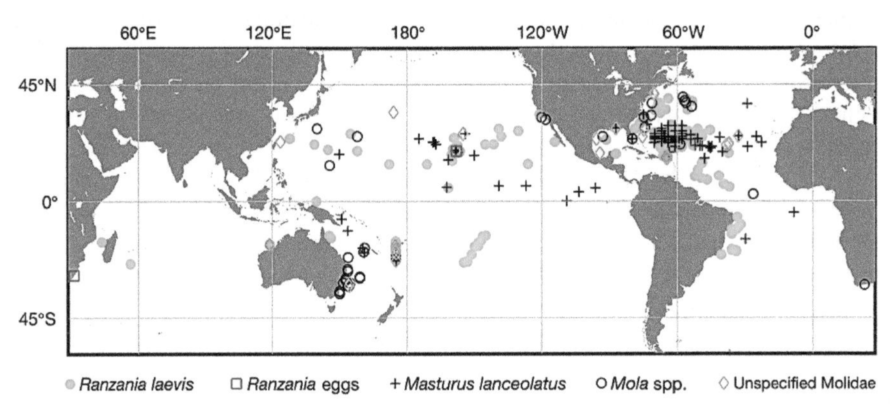

Figure 2. Global detections of ocean sunfish larvae and early juveniles up to 60 mm TL gathered from historical and modern sources catalogued with metadata in Appendix 1.

primarily in March and April in the Sargasso Sea (Ayala et al. 2016, Hellenbrecht et al. 2019, Knorrn 2019) and from January through October in the Straits of Florida (Richardson et al. 2010).

Ma. lanceolatus is the second most sampled species. Larvae have been collected throughout the North Atlantic from the Gulf of Mexico along the U.S. east coast (Lyczkowski-Shulz 2003) and extensively throughout the Sargasso Sea where they overlap with *Ranzania* (Fig. 2; Appendix 1; Schmidt 1921a,b, Ayala et al. 2016, Hellenbrecht et al. 2019). They also occur east towards Africa, as far north as the Azores and have been sampled in the South Atlantic as far south as Brazil (Fig. 2; Appendix 1). In the Pacific, larval *Ma. lanceolatus* are recorded near the Marianas Archipelago (Sokolovskaya and Sokolovskiy 1975), around the Hawai'ian Ridge (Fig. 2; Appendix 1), along the equator and in the western South Pacific (Fourmanoir 1976; Fig. 2; Appendix 1). Like *Ranzania*, most records come from the Sargasso Sea between March and April (Schmidt 1921b, Ayala et al. 2016, Hellenbrecht et al. 2019), and during August in the Pacific hemisphere (Sokolovskaya and Sokolovskiy 1975).

Larvae of the genus *Mola* spp. have been collected in the western Atlantic from the Gulf of Mexico (Lyczkowski-Shulz 2006), to the Sargasso Sea and eastward towards the mid-North Atlantic (Fig. 2, Appendix 1). Only one record was found for the South Atlantic (off the southern coast of South Africa. In the northern Pacific, larval *Mola* spp. have been reported sporadically off the Hawai'ian Ridge (Borets and Sokolovskij 1978), and the United States' northwestern coast. Records from the South Pacific are more extensive, particularly from Australia's east coast (Fig. 2, Appendix 1). With regards to seasonality, *Mola* spp. spawning is estimated to occur between August and October off Japan based on examination of ovaries and testes of adult individuals (Nakatsubo et al. 2007a,b, Forsgren et al. 2020 [Chapter 6]). Areas of spawning activity have also been proposed in the North and South Pacific (Sokolovskaya and Sokolovskiy 1975, Kawakami et al. 2010, Kawakami 2016), and Florida (Richardson et al. 2010).

Environmental and Oceanographic Factors Influencing Detection Sites

While Molidae larvae lack the capacity to actively maintain position in the water column horizontally, they can be distributed broadly throughout the water column, with individuals being collected in plankton tows from the surface to 300 m depth (Schmidt 1921a [100 m], Sherman 1961, Sokolovskaya and Sokolovskiy 1975, Robinson 1975, Leis and Miller 1976, Hellenbrecht et al. 2019 [300 m]). Gathering hydrographic data at the same time as larval collection is essential to understanding larval presence in relation to environmental parameters. Recently, Hellenbrecht et al. (2019) collected *R. laevis* and *Ma. lanceolatus* larvae in the Sargasso Sea concurrently with temperature and salinity measurements and found that *Ranzania* larvae were detected between 22°C and 28°C with the highest abundance found around 25°C. These observations are broadly consistent with Sokolovskaya and Sokolovskiy (1974), who found *Ranzania* larvae were collected when surface waters were between 24°C and 28°C in the upper water column (i.e, 0–25m). With regards to salinity, Knorrn (2019) found high abundances of freshly hatched *Ranzania* larvae in areas of the central Sargasso Sea with surface salinities between 36.9 and 37.2 PSU. These findings support Nogueira et al. (2012) who reported *Ranzania* larvae abundance was associated with higher salinities (~ 35–37 PSU) off eastern Brazil. Whether, these anecdotal records constitute habitat requirements, or are simply environmental covariates, remains unclear. Thus, care should be taken not to report these temperature and salinity measurements as established tolerance ranges. But given the dearth of information of spawning grounds, these works can allow us to move towards habitat suitability models. These efforts, in turn, can help facilitate much needed targeted sampling programs to develop over time.

Predators of Early Molid Life Stages (Larvae, Molacanthiforms and Juveniles)

Our literature review and query of archived samples comprised predation events on early molid life stages (larvae, molacanthiforms and juveniles) between 1862 and 2014. They fell into three main categories: (1) one-off observational events (n = 31), here termed 'discrete observations', where one or a small number of molids were found in the stomach of a single predator, but no information was available on the contribution to the overall stomach contents, (2) descriptive, anecdotal, narrative accounts (n = 2) and; (3) targeted dietary studies (*n* = 13) where several specimens of a predatory species were examined, and small molids were identified among the gut contents. These latter records included information on the frequency of occurrence and contribution relative to other prey. In total 26 different predator species were identified, comprising 17 large pelagic piscivorous fishes from six families, and nine seabirds species from five families (Appendix 2). Unfortunately, few dietary records included size data for both the predator and prey, which combined with a modest sample size, allowed only a descriptive overview at present.

Predation by Teleost Fishes

Combined, the discrete observations represent predation on 14 *Mola* sp. (11 mm TL – 64 mm PCBL), 83 *Ma. lanceolatus* (4.6 mm TL – 130 mm PCBL), 14 *R. laevis* (24 mm PCBL – 100 mm TL) and five unidentified Molidae. These prey were found in quantities of 1 – 7 specimens in each stomach (average of two) with one exception of 45 specimens found in the stomach of a single fish. Fish predation on early molid life stages were found in ten diet studies (Matthews et al. 1977, Brock 1984, Allain et al. 2003, Satoh et al. 2004, Shimose et al. 2010, Young et al. 2010, Choy et al. 2013, Shimose et al. 2013, Oyafuso et al. 2016, Perelman et al. 2017), each providing quantitative data on molid stomach content contribution for several predatory fish (Appendix 2).

A notable difference between the targeted dietary studies and the discrete observations is that the former report almost exclusively *Ma. lanceolatus* and *R. laevis*, with only one exception where *M. mola* is mentioned (Satoh et al. 2004, see further discussion below). Another difference is the size of the molid prey reported; although the size information in the dietary studies is scant, it nevertheless indicates that the size of *R. laevis* were markedly larger than those of the incidental reports. It is even possible that some ingested prey might constitute small adults, as sexual maturity in *R. laevis* likely occurs around 300 mm TL and *ca.* 500 g (Smith et al. 2010).

The two types of records (discrete observations and dietary studies) revealed predation by the same or similar species of large pelagic piscivorous fishes from the families of lancetfishes (Alepisauridae), dolphinfishes (Coryphaenidae), gemfishes (Gemphylidae), billfishes (Istiophoridae), mackerels and tunas (Scombridae) and dragonfishes (sub-family Stomiinae). For full species details refer to Appendix 2).

Overall, the focused dietary studies revealed that early life stage molids generally comprised a minor or negligible proportion of predator gut contents, by both mass and number (Appendix 2). There are however, a small number of notable exceptions where *R. laevis* reportedly made up a more substantial proportion of gut contents in seven different species. For example, Shimose et al. (2010) reported that *R. laevis* (194–326 mm) were the dominant prey item by mass in the guts of shortbill spearfish *T. angustirostris* (n = 29) and large (≥ 140 cm) striped marlin *K. audax* (n = 22) in an open-ocean sampling area between Hawai'i and the North American continent. Specifically, *R. laevis* comprised 73 percent and 49 percent by mass, respectively. These results were however, not replicated for *K. audax* (of any size) caught in the near-continent and near-equator sampling areas (Shimose et al. 2010). Shimose et al. (2013) also reported that juvenile *R. laevis* (112–193 mm TL) comprised the dominant prey item in the gut of blue marlin, *Makaira nigricans* (n = 10) caught off Okinawa, Japan. Specifically, *R. laevis* had the highest index of relative importance (66 percent), as well as the highest frequency of occurrence (50 percent), relative number (54 percent) and relative mass (34 percent) of seven prey categories, in addition to plastic.

Juvenile *R. laevis* comprised a major component of gut contents of wahoo (*Acanthocybium solandri; ca.* 100–170 cm FL) in offshore areas northeast of the Hawai'ian archipelago (Oyafuso et al. 2016). This study used genetic barcoding of gut contents in addition to visual identification, revealing *R. laevis* in the gut of 14 percent of wahoo caught offshore (n = 175). The species constituted 24 percent of gut contents by mass. In contrast, *R. laevis* was only found in 0.5 percent of stomachs from wahoo caught nearshore (n = 135), contributing just 1.9 percent of ingested prey by mass. *Ma. lanceolatus* was negligible in offshore wahoo (0.6 percent of guts, 0.2 percent by mass) and absent altogether in nearshore wahoo samples (Oyafuso et al. 2016).

R. laevis was found in the gut of snake mackerel *Gempylus serpens* (66.5–140.5 cm FL; n = 47) sampled over several years in the central North Pacific (Choy et al. 2013). Of a diet mainly comprised of epipelagic fishes and epipelagic and mesopelagic squids, *R. laevis* was the single most abundant prey species both by number (19%) and weight (33%). No size or life stage indication was provided for *R. laevis* in this study, but given the size of the sampled predators (66.5–140.5 cm FL, mean 96.9 cm ± 15.4 SD; mean weight 1.3 kg ± 0.7 SD), the *R. laevis* prey had presumably not passed the juvenile stage.

Satoh et al. (2004) examined stomach content of several large pelagic piscivorous fishes and found molids to be the second and third most dominant prey family in the gut content of yellowfin tuna *Thunnus albacares* (n = 47) and Atlantic sailfish *Istiophorus platypterus* (n = 32), respectively. Specifically, molids contributed 5.8 percent and 11 percent by number, and 45.5 percent and 5.3 percent by mass of all prey items, respectively (Appendix 2). Satoh et al. (2004) noted that both *R. laevis* and *Mola* sp. (reported as *M. mola*) were found during the study, but provided no further information of relative numbers. This was the only dietary study unearthed that reported *Mola* sp. without accompanying size data or indication of their life stage (i.e., larva, *Molacanthus* stage, juvenile).

For *Ma. lanceolatus,* there are several accounts of its role as prey. Fitch (1950) reported that *Ma. lanceolatus* (6.4–51 mm, i.e., molacanthiforms) occurred commonly in the stomachs of yellowfin tuna taken in Hawai'ian waters (see Appendix 2 for details). Furthermore, 45 early *Molacanthus* stage *Ma. lanceolatus* (4.6–10.6 mm TL) were found in the stomach of a single skipjack tuna *K. pelamis* off Japan (T. Nakatsubo and G. Shinohara unpublished data), but this report was unusual among the discrete observations, where 1–7 (average of 2) molids were found per predator (Appendix 2). For *Mola* spp., reports of predation consisted almost exclusively of discrete observations (with the exception of Satoh et al. 2004), and these were all relatively small individuals in the *Molacanthus* or early juvenile stages (Appendix 2).

An anecdotal account of predation by sabertooth viperfish *Chauliodus sloani* on molid larvae (*R. laevis*) was given by Beebe (1934) alongside two illustrations (Beebe 1934, page IV and V; see Appendix 2). However, no information source was provided. Taken in the context of other questionable information in the narrative (e.g., "*...[R. laevis] are fully eight feet long [244 cm]...*") the account has limited scientific credibility [Note, the largest credible maximum length of *Ranzania* is 74 cm (Sawai et al. 2020 [Chapter 2]). We have found no other evidence in the literature or otherwise to corroborate predation on molids by viperfish, so we are reluctant to add *C. sloani* to the list of reported predators on early molid life stages at this time.

Predation by seabirds

Ma. lanceolatus and in particular, *R. laevis*, have been recorded in the diet of several seabird species investigated through regurgitate samples. These include black noddy *Anous minutus*, brown noddy *Anous stolidus*, Bulwer's petrel *Bulweria bulwerii*, Cory's shearwater *Calonectris diomedea*, great frigatebird *Fregata minor*, sooty tern *Onychoprion fuscatus*), red-tailed tropicbird *Phaethon rubricauda*, Laysan albatross *Phoebastria immutabilis* and red-footed booby *Sula sula* (Harrison et al. 1983, Seki and Harrison 1989, Wapp et al. 2017; see Appendix 2 for a review). Limited information is available on the size of the molid prey, but the general prey sizes reported in Harrison et al. (1983) indicated that the largest prey ingested whole by any of the above bird species was < 300 mm, with most species ingesting much smaller prey (Appendix 2). Further, the mass of the studied bird species (which had eaten molids) ranged from 85–133 g (black noddy) to 1,900–3,075 g (Laysan albatross) (Harrison et al. 1983, Appendix 2). Within this context *Ma. lanceolatus,* must have been, at the most, small juveniles. Conversely, it is possible that at least some ingested *R. laevis* were small adults (*R. laevis* weighs *ca.* 500 g at 300 mm TL; Smith et al. 2010), particularly as Laysan albatrosses may tear their prey prior to ingestion (Harrison et al. 1983). However, most ingested *R. laevis* were likely juveniles or even larvae.

The frequency of occurrence of molids in regurgitate samples of the seabirds in the reviewed studies was generally low (< 1.2% of samples, contributing < 0.6 percent by volume; Appendix 2), with a small number of exceptions. Red-tailed tropicbirds at the large Hawai'ian atoll French Frigate Shoals fed on juvenile *R. laevis* in the summer months (Harrison et al. 1983). During this time, Molidae (mainly *R. laevis*, combined with a small number of *Ma. lanceolatus*) ranked as the seventh most important prey by volume of 31 ranked prey groups (mainly at family level), and reportedly contributed to making the red-tailed tropicbird diet here potentially distinct from conspecifics found elsewhere throughout the wider Northwestern Hawai'ian Islands. The study did not quantify this statement, but reported that *R. laevis* was found in 5.2 percent of all red-tailed tropicbird samples from the wider study area (several islands in the Hawai'ian archipelago) across all sampled months, contributing to 4.3 percent by volume. Size was obtained for five *R. laevis* and these ranged from 123 to 140 mm TL (Harrison et al. 1983). *R. laevis* was also found in 3.1% of samples from red-footed booby, contributing to 1.5 percent by volume. In the same study, *R. laevis* was found in 2.2 percent of all regurgitate samples from Laysan albatross chicks (fed by their parents), but contributed only 0.9 percent by volume.

All of the seabirds discussed so far feed at the sea surface or in the upper few meters of the water column (Harrison et al. 1983). For example, the red-tailed tropicbird feeds by plunge-diving to depths of typically 2–4 m, with an ability to reach nearly 10 m (Le Correl 1987). The Laysan albatross obtains prey through surface seizing, during both daytime and nighttime (Conners et al. 2015). While little is known about the habitat and water column utilization of early life stages of molids, these findings indicate that at least some *R. laevis* and *Ma. lanceolatus* spend time near the sea surface.

More broadly, we found limited reports of molids listed as prey items in seabird diets, although our literature search was not exhaustive. Exceptions were the North Atlantic, where *R. laevis* was listed (without quantifying data) as a diet item in 673 regurgitates of Cory's shearwater (Alonso et al. 2013), and in 0.72% of 139 stomach samples of Bulwer's petrel (Wapp et al. 2017). It is unclear why the reports of molids in seabird diets are so localized to the Hawai'ian Islands, and particularly French Frigate Shoals.

Predation summary

In summary, evidence of predation on early molid life stages comprised mainly large, pelagic piscivorous fishes, and seabirds from the Hawai'ian archipelago. Notably, little evidence was found of predation on pre-molacanthiform *Mola* larvae (minimum prey size of 11 mm TL) and larva of *R. laevis* (minimum prey size of 24 mm PCBL). The smallest *Ma. lanceolatus* prey was 4.6 mm TL, close to the approximate size of transition into the *Molacanthus* stage at ca. 5 mm TL (Lyczkowski-Shulz 2003).

In light of the extremely high potential ova production of *Mola* spp. (Forsgren et al. 2020 [Chapter 6]), our review identified surprisingly few predation events on molids, particularly for *Ma. lanceolatus* and *Mola* spp. It is not clear if these genera are simply not widely preyed upon, if they are among the unidentified material commonly found in diet studies, or if future studies will reveal more widespread predation. It is also worth noting that despite a potential high fecundity of *M. mola*, it is unknown how many eggs are released per batch or across a total spawning season. It is also unknown how many eggs hatch and survive to become post-larvae and adults (Forsgren et al. 2020, Chapter 6) .

Future Directions and Remaining Questions

Further research is needed to describe egg and early developmental stages of *Mola* and *Masturus*, similar to what has been achieved for *R. laevis* (Leis 1977). These efforts would include genetic verification of the species identifications from the eggs and larvae. Such data will be invaluable in the development of a species-specific taxonomic guide for the early life stages of all Molidae species; particularly all species within the *Mola* genus. Additionally, further research is needed on larval diets and to track the ontogenetic presence of the swim bladder for larval *Masturus* and *Mola* spp. Further work is also required to describe the *Molocanthus* stage of *Mola* spp. and *Masturus* to better understand what, if any, survival advantage this transitional stage may provide for larger Molidae species, in contrast to the smaller *R. laevis* which lack this unique stage of development.

Closer inspection of long-term, global datasets (e.g., the NOAA fisheries COPEPOD (Coastal and Oceanic Plankton Ecology, Production and Observation Database, Global Plankton Database Project datasets), may reveal further evidence of sunfish spawning within a robust environmental framework. More targeted, depth stratified sampling during day and night times will also provide insights into larval behavior (e.g., diel or ontogenetic vertical migration). Greater care is also needed to cross collect and cross reference environmental parameters associated with larval occurrences (e.g., chlorophyll biomass, salinity, oxygen) coupled with concurrent plankton tows to identify co-occurring prey items. A more comprehensive database can provide essential information for habitat association models which could allow the identification of putative spawning areas and establishment of spatially defined sampling regimes. In such areas of high interest, autonomous underwater vehicles could provide useful tools for conducting targeted plankton sampling (Harvey et al. 2012). This approach, in turn, could help detect larval assemblages, record oceanic features in real time and even perform *in situ* genetic analyses (Zhang et al. 2019). As with many other areas of marine biology (e.g., Sequeira et al. 2019), a global portal for archiving and making data available on the occurrence of ocean sunfish larvae would have great utility, making existing records more accessible as well as helping to ensure that future records are preserved.

Acknowledgements

A special thanks to all parties listed in Appendix 2 who provided molid larval information including: K.E. Bemis, R. Bills, Tina Blanke, H. Carl, D. Catania, S. Charter, J. Cobb, T. Cullins, G. Duhamel, R. Feeney, B. Frable, L. Garrison, A. Graham, R. Hanel, A. Hay, D. Johnson, E. Kawamoto, A.

Knorrn, M. Kuroki (University of Tokyo Japan), B. Ludt, E. Malca, C. McMahan, H. Motomura, K. Mullen, B. Mundy, E. Nixon, M. Nogueria, P. Provoost, R. Robins, N. Schnell, T. Shimose, A. Suzumoto, D. Trimm, B. Watson, E. Weber, G. Zapfe, the Hakuho Maru Cruise KH-16-4, the Maria S. Merian cruise MSM41, the Walther Herwig cruise WH373, NOAA Southeast Fisheries Science Center Larval Fish Lab. Special thanks to all the people around the world curating fish collections and a huge debt of gratitude to D. Kohrs from the Stanford Library for helping find many obscure mola references and to O. Daly for her tireless copy editing.

References

Allain, V. 2003. Diet of mahi-mahi, wahoo and lancetfish in the western and central Pacific. 16th meeting of the standing committee on tuna and billfish (SCTB16) working paper BBRG-6. Oceanic Fisheries Programme, Secretariat of the Pacific Community, Noumea, New Caledonia.

Alonso, H., J.P. Granadeiro, J.A. Ramos and P. Catry. 2013. Use the backbone of your samples: fish vertebrae reduce biases associated with otoliths in seabird diet studies. J. Ornithol. 154: 883–886.

Ayala, D., L. Riemann and P. Munk. 2016. Species composition and diversity of fish larvae in the subtropical convergence zone of the Sargasso Sea from morphology and DNA barcoding. Fisheries Oceanography 25(1): 85–104.

Baensch, F. 2013. Hawai'i Larval Fish Project, Mola Larval Rearing. www.frankbaensch.com/marine-aquarium-fish-culture/my-research/slender- mola- culture/. Retrieved Jan 24 2020.

Beebe, W. 1934. A half mile down: strange creatures, beautiful and grotesque as figments of fancy reveal themselves at windows of the bathysphere, National Geographic 66(6): 661–704.

Bemis, K.E., J.C. Tyler, E.J. Hilton and W.E. Bemis. 2020. Overview of the anatomy of ocean sunfishes (Molidae: Tetraodontiformes). pp. 55–71. *In*: T.M. Thys, G.C. Hays and J.D.R. Houghton [eds.]. The Ocean Sunfishes: Evolution, Biology and Conservation, CRC Press. Boca Raton, FL, USA.

Borets, L.A. and A.S. Sokolovskij. 1978. Species composition of the ichthyoplankton of the Hawa'iian submarine ridge and the Emperor Seamounts. pp. 43–50. *In*: Izvestiya of the Pacific Ocean Scientific Research Institute for Fisheries and Oceanography (TINRO). Fisheries oceanography, hydrobiology, biology of fishes and other denizens of the Pacific Ocean. Vol 102. (English translation by W. G. Van Campen for the Southwest Fish. Cent. Honolulu Lab. Natl. Mar. Fish. Serv. NOAA Transl No. 105. 10 pp.

Brock, R.E. 1984. A contribution to the trophic biology of the blue marlin (*Makaria nigricans* Lacepede, 1802) in Hawaii. Pacific Science 38(2): 141–149.

Chanet, B., C. Guintard, T. Boisgard, M. Fusellier, C. Tavernier, E. Betti et al. 2012. Visceral anatomy of ocean sunfish (*Mola mola* L., 1758), Molidae, Tetraodontiformes) and angler (*Lophius piscatorius* L., 1758), Lophiidae, Lophiiformes) investigated by non-invasive imaging techniques. Comptes Rendus Biologies 335(12): 744–752.

Choy, C.A., E. Portner, M. Iwane and J.C. Drazen. 2013. Diets of five important predatory mesopelagic fishes of the central North Pacific. Marine Ecology Progress Series. 492: 169–184.

Conners, M.G., E.L. Hazen, D.P. Costa and S.A. Shaffer. 2015. Shadowed by scale: Subtle behavioral niche partitioning in two sympatric, tropical breeding albatross species. Movement Ecology 3: 28.

Dunning, J.B. 2008. CRC Handbook of avian body masses. Second edition. CRC Press. Boca Raton, FL USA. 655pp.

Feeney, R. 2019. LACM Vertebrate Collection. Version 18.7. Natural History Museum of Los Angeles County. http://ipt.vertnet.org:8080/ipt/resource.do?r=lacm_verts#versions. Accessed on 12-31-2019.

Fitch, J.E. 1950. Notes on some Pacific Fishes. Calif. Fish and Game. 36(2): 65–73.

Forsgren, K., R.S. McBride, T. Nakatsubo, T.M. Thys, C.D. Carson, E.K. Tholke et al. 2020. Reproductive biology of the ocean sunfishes. pp. 87–104. *In*: T.M. Thys, G.C. Hays and J.D.R. Houghton [eds.]. The Ocean Sunfishes: Evolution, Biology and Conservation, CRC Press. Boca Raton, FL, USA.

Fourmanoir, P. 1976. Formes post-larvaires et juvéniles de poissons côtiers pris au chalut pélagique dans le sub-ouest Pacifique. Cah. Pac. 19: 47–88.

Fraser-Brunner, A. 1951. The ocean sunfishes (family Molidae). Bull. Br. Mus. (Nat. Hist.) Zool. 1: 89–121.

Fricke, R., W.N. Eschmeyer and R. Van der Laan [eds.]. 2020. Eschmeyer's Catalog of Fishes: Genera, Species, References. http://researcharchive.calacademy.org/research/ichthyology/catalog/fishcatmain.asp). Accessed 05 Marc 2020.

Garrison, L. 2013. SEFSC Caribbean Survey 1995. Data downloaded from OBIS-SEAMAP (http://seamap.env.duke.edu/dataset/11) on 2015-10-17.

Greer, J.K. 1976. Activities of blue dolphins in the Atlantic Ocean. Tex. J. Sei. 27: 245.

Gudger, E.W. 1935. Some undescribed young of the pointed-tailed ocean sunfish, *Masturus lanceolatus*. Copeia. 1: 35–38.

Gudger, E.W. 1937. The natural history and geographical distribution of the pointed-tailed ocean sunfish (*Masturus lanceolatus*), with notes on the shape of the tail. Proceedings of the Zoological Society of London. 107A,3: 353–396.

Gudger, E.W. 1939. Three six-inch pointed-tailed ocean sunfish, *Masturus lanceolatus*, the largest post-larvae on record. Journal of the Elisha Mitchell Scientific Society 55(2): 305–313.

Günther, A.C.L.G. 1880. An Introduction to the Study of Fishes. R. and R. Clark, Edinburgh. 720 pp.

Harrison, C.S., T.S. Hida and M.P. Seki. 1983. Hawai'ian seabird feeding ecology. Wildlife Monographs 85: 3–71.

Harting, P. 1865. Notices zoologiques, anatomiques et histologiques, sur *l'Orthragoriscus ozodura*; suivies de considérations sur l'ostéogenèse des téléostieus en general. Amsterdam: CG Vanderpost.

Harvey, J.B.J, J.P. Ryan, R. Marin, C.M. Preston, N. Alvarado, C.A. Scholin et al. 2012. Robotic sampling, *in situ* monitoring and molecular detection of marine zooplankton. Journal of Experimental Marine Biology and Ecology 413: 60–70.

Hellenbrecht, L., M. Freese, J. Pohlmann, H. Westerberg, T. Blancke and R. Hanel. 2019. Larval distribution of the ocean sunfishes *Ranzania laevis* and *Masturus lanceolatus* (Tetraodontiformes: Molidae) in the Sargasso Sea subtropical convergence zone. Journal of Plankton Research 41: 595–608.

Johnson, G.D. and R. Britz. 2005. Leis' conundrum: Homology of the clavus of the ocean sunfishes. 2. Ontogeny of the median fins and axial skeleton of *Ranzania laevis* (Teleostei, Tetraodontiformes, Molidae). Journal of Morphology 266(1): 11–21.

Jordan, D.S. and B.W. Evermann. 1903. Descriptions of New Genera and Species of Fishes from the Hawai'ian Islands. Bull. U. S. Fish Comm. for 1902, XXII. 161–208.

Kang, M.J., H.J. Baek, D.W. Lee and J.H. Choi. 2015. Sexual maturity and spawning of ocean sunfish *Mola mola* in Korean waters. Korean Journal of Fisheries and Aquatic Sciences 48(5): 739–744. doi:10.5657/KFAS.2015.0739.

Kawakami, T., J. Aoyama and K. Tsukamoto. 2010. Morphology of pelagic fish eggs identified using mitochondrial DNA and their distribution in waters west of the Mariana Islands. Environmental Biology of Fishes 87: 221–235.

Kawakami, T. 2016. Species composition, distribution and embryonic development of fish eggs collected during the KH-16-4 research cruise. *In*: Kuroki, M., K. Tsukamoto and T. Otake [eds.]. Preliminary Report of the Hakuho Maru Cruise KH-16-4: Research on the spawning and migration ecology of the anguillid eels and the resource fluctuation mechanism in the South Pacific. Atmos. Ocean. Res. Inst. Univ. Tokyo. 223 p.

Kendall, A.W., Jr., E.H. Ahlstrom and H.G. Moser. 1984. Early life history stages of fishes and their characters. pp. 11–22. *In*: Ontogeny and Systematics of Fishes: based on an International Symposium Dedicated to the Memory of Elbert Halvor Ahlstrom.

King, J.E. 1951. Two juvenile pointed-tailed ocean sunfish, *Masturus lanceolatus*, from Hawai'ian waters. Pacific Science 5(1): 109.

Knorrn, A.H. 2019. Larval development and distribution of *Ranzania laevis* (Pennant, 1776) and *Masturus lanceolatus* (Liénard, 1840) across thermal frontal zones in the Sargasso Sea. Unpublished bachelor thesis. Johann Wolfgang Goethe University of Frankfurt, Germany.

Kuronuma, K. 1940. A young of ocean sunfish, *Mola mola*, taken from the stomach of *Germo germo*, and a specimen of *Masturus lanceolatus* as the second record from Japanese water. Bull. Biogeogr. Soc. Jap. 10(2): 25–28.

Lasker, R. 1981. Marine Fish Larvae: Morphology, Ecology, and Relation to Fisheries. University of Washington Press, Seattle, WA USA.

Le Correl, M. 1987. Diving depths of two tropical pelecaniformes: the red-tailed tropicbird and the red-footed Booby. The Condor 99(4): 1004–1007.

Leis, J.M. and J.M. Miller. 1976. Offshore distributional patterns of Hawai'ian fish larvae. Marine Biology 36(4): 359–367.

Leis, J.M. 1977. Development of the eggs and larvae of the slender mola, *Ranzania laevis* (Pisces, Molidae). Bulletin of Marine Science 27(3): 448–466.

Leis, J.M. 1984. Tetraodontoidei: Development. pp. 447–459. *In*: Ontogeny and systematics of fishes: based on an international symposium dedicated to the memory of Elbert Halvor Ahlstrom.

Lyczkowski-Shultz, J. 2003. Preliminary guide to the identification of the early life. *In*: W.J. Richards [eds.]. Early Stages of Atlantic Fishes: An Identification Guide for the Western Central North Atlantic. (August) NOAA Technical Memorandum NMFS-SEFSC-504, 10pp.

Marshall, K.N., J.T. Duffy-Anderson, E.J. Ward, S.C. Anderson, M.E. Hunsicker and B.C. Williams. 2019. Long-term trends in ichthyoplankton assemblage structure, biodiversity, and synchrony in the Gulf of Alaska and their relationships to climate. Prog. Oceanogr. 170: 134–45.

Martin, F.D. and G.E. Drewry. 1978. Family Molidae. In Development of fishes of the Mid-Atlantic Bight. An atlas of egg, larval and juvenile stages. VI. Stromateidae through Ogcocephalidae., U. S. Fish Wildl. Serv., Biol. Serv. Prog. FWS/OBS, 78/12: pp. 313–338.
Matthews, F.D., D.M. Damkaer, L.W. Knapp and B.B. Collette. 1977. Food of western North Atlantic tunas (*Thunnus*) and lancetfishes (*Alepisaurus*). NOAA Technical Report NMFS SSRF 706, 19 pp.
McCulloch, A.R. 1912. A description and figures of three specimens of Molacanthus from the central Pacific Ocean. The Proceedings of the Linnean Society of New South Wales. Vol XXXVII: 553–555, Pl LVIII-LIX.
Nakatsubo, T., M. Kawachi, N. Mano and H. Hirose. 2007a. Estimation of maturation in wild and captive ocean sunfish *Mola mola*. Aquaculture Sci. 55: 259–264.
Nakatsubo, T., M. Kawachi, N. Mano, H. Hirose. 2007b. Spawning period of ocean sunfish Mola mola in waters of the eastern Kanto region, Japan. Aquaculture Sci. 55(4): 613–618.
Nogueira, M.M., C.S. De Souza and P. Mafalda. 2012. The influence of abiotic and biotic factors on the composition of Tetraodontiformes larvae (Teleostei) along the Brazilian Northeast Exclusive Economic Zone (1°N–14°S). Pan-American Journal of Aquatic Sciences 7(1): 10–20.
Nyegaard, M., E. Sawai, N. Gemmell, J. Gillum, N.R. Loneragan, Y. Yamanoue et al. 2018. Hiding in broad daylight: molecular and morphological data reveal a new ocean sunfish species (Tetraodontiformes: Molidae) that has eluded recognition. Zoological Journal of the Linnean Society 182(3): 631–658.
Nyegaard, M., S. Andrzejaczek, C.S. Jenner and M.N.M. Jenner. 2019. Tiger shark predation on large ocean sunfishes (Family Molidae)—two Australian observations. Environmental Biology of Fishes 102(12): 1559–1567.
Oyafuso, Z.S., R.J. Toonen and E.C. Franklin. 2016. Temporal and spatial trends in prey composition of wahoo *Acanthocybium solandri*: A diet analysis from the central North Pacific Ocean using visual and DNA bar-coding techniques. Journal of Fish Biology 88: 1501–1523.
Parin, N.V. 1968. Ichthyofauna of the epipelagic zone. Academy of Sciences of the USSR, Institute of Oceanography. Translated from Russian (by M. Ravah. Edited by H. Mills). Jerusalem, Israel Program for Scientific Translations. U.S. Dept. of Commerce, Clearinghouse for Federal Scientific and Technical Information, Springfield, VA. USA.
Paris, C.B., L.M. Chérubin and R.K. Cowen. 2007. Surfing, spinning, or diving from reef to reef: effects on population connectivity. Mar. Ecol. Prog. Ser. 347: 285–300.doi: 10.3354/meps06985.
Perelman, J.N., K.N. Schmidt, I. Haro, I.R. Tibbetts and M.T. Zischke. 2017. Feeding dynamics, consumption rates and daily ration of wahoo *Acanthocybium solandri* in Indo-Pacific waters. Journal of Fish Biology 90(5): 1842–1860.
Perugia, A. 1889. Sui giovani dell' Orthagoriscus mola. Nota di Alberto Perugia. Annali del Museo civico di storia naturale di Genova. ser.2:v.27,1889: 365–368.
Putnam, F.W. 1870. On the young of *Orthagoriscus mola*. American Naturalist 4: 629–633.
Reuvens, C.L. 1895. Remarks on the genus *Orthragoriscus*. Notes from the Leyden Museum XVL: 128–130, Pl 5.
Richardson, D.E., J.K. Llopiz, C.M. Guigand and R.K. Cowen. 2010. Larval assemblages of large and medium-sized pelagic species in the Straits of Florida. Progress in Oceanography 86: 8–20.
Robinson, B.H. 1975. Observations on living juvenile specimens of the slender mola, *Ranzania laevis* (Pisces, Molidae). Pacific Science 29(1): 27–29.
Rodríguez, J.M., F. Alemany and A. García. 2017. A guide to the eggs and larvae of 100 common Western Mediterranean Sea bony fish species. FAO, Rome, Italy.
Sanzo, Li. 1919. Contributo alla conoscenza degli stadi larvali di Orthagoriscus Bl. [in Italian]. Mem. R. Comit. Talassogr. Ital 69. 7 pp.; pl. 1.
Sanzo, L. 1939. Rarissimi stadi larvali di Teleostei. Arch. Zool. Ital. 26: 121–151, 2 pls [Tavole 6 e 7]. [In Italian].
Satoh, K., K. Yokawa, H. Saito, H. Matsunaga, H. Okamoto and Y. Uozumi. 2004. Preliminary stomach contents analysis of pelagic fish collected by Shoyo-Maru 2002 research cruise in the Atlantic Ocean. Col. vol. Sci. Pap. ICCAT 56(3): 1096–1114.
Sawai, E., Y. Yamanoue, M. Nyegaard and Y. Sakai. 2018. Redescription of the bump-head sunfish *Mola alexandrini* (Ranzani 1839), senior synonym of *Mola ramsayi* (Giglioli 1883), with designation of a neotype for Mola mola (Linnaeus 1758) (Tetraodontiformes: Molidae). Ichthyol. Res. 65: 142–160.
Sawai, E., M. Nyegaard and Y. Yamanoue. 2020. Phylogeny, taxonomy and size records of ocean sunfishes. pp. 18–36. *In*: T.M. Thys, G.C. Hays and J.D.R. Houghton [eds.]. The Ocean Sunfishes: Evolution, Biology and Conservation, CRC Press. Boca Raton, FL, USA.
Schmidt, J. 1921a. New studies of sun-fishes made during the "Dana" Expedition, 1920. Nature. 107: 76–79.
Schmidt, J. 1921b. Contributions to the knowledge of the young of the sun-fishes (*Mola* and *Ranzania*). Meddelelser fra Kommissionen for havundersøgelser, Serie Fiskeri VI(6): 1–13.
Seki, M.P. and C.S. Harrison. 1989. Feeding ecology of two subtropical seabird species at French Frigate Shoals, Hawaii. Bulletin of Marine Science 45(1): 52–67.

Sequeira, A.M.M., G.C. Hays, D.W. Sims, V.M. Eguíluz, J. Rodriguez and M. Heupel et al. 2019. Overhauling ocean spatial planning to improve marine megafauna conservation. Frontiers in Marine Science 6: 639. doi: 10.3389/fmars.2019.00639.

Sherman, K. 1961. Occurrence of early developmental stages of the oblong ocean sunfish *Ranzania laevis* (Pennant) in the Central North Pacific. American Society of Ichthyologists and Herpetologists (ASIH) 1961: 467–470.

Shimose, T., H. Shono and K. Yokawa. 2006. Food and feeding habits of blue marlin, *Makaira nigricans*, around Yonaguni Island, southwestern Japan. Bulletin of Marine Science 79(3): 761–775.

Shimose, T., K. Yokawa and H. Saito. 2010. Habitat and food partitioning of billfishes (Xiphioidei). Journal of Fish Biology 76(10): 2418–2433.

Shimose, T., K. Yokawa and K. Tachihara. 2013. Occurrence of slender mola *Ranzania laevis* (Pennant, 1776) in stomachs of blue marlin *Makaira nigricans* Lacépède, 1802. Journal of Applied Ichthyology 29(5): 1160–1162.

Smith, J.A., A.G. Miskiewicz, L.E. Beckley, J.D. Everett, V. Garcia, C.A. Gray et al. 2018. A database of marine larval fish assemblages in Australian temperate and subtropical waters, Scientific Data 5: 180207.

Smith, K.A., M. Hammond and P.G. Close. 2010. Aggregation and stranding of elongate sunfish (*Ranzania laevis*) (Pisces: Molidae) (Pennant, 1776) on the southern coast of Western Australia. Journal of the Royal Society of Western Australia 93(4): 181–188.

Smith, L.J.B. 1949. The sea fishes of Southern Africa. With a message from his Excellency the Governor of Moçambique and a foreword by B.F.J. Schonland. Central News Agency, Cape Town pp. 788.

Sokolovskaya, T.S. and A.S. Sokolovskiy. 1975. New data on expansion of the area of reproduction of ocean sunfishes (Pisces, Molidae) in the Northwestern Part of the Pacific Ocean. Ocean. J. Ichthyol. 15: 675–678.

Southey, I. and P.G.H. Frost. 2013. Bulwer's petrel. *In*: C.M. Miskelly [ed.]. New Zealand Birds Online. www.nzbirdsonline.org.nz.

Steenstrup, J. and C. Lütken. 1898. Spolia Atlantica: bidrag til kundskab om klump-eller maanefiskene (Molidae). D Kgl Vidensk Selsk Skr; 6 Række, Naturvidenskabelig og mathemastisk Afd IX, 1: 102.

Trimm, D.L., J. Shenker, Amy Hirons and S. Mochrie. 2010. Occurrence of an ocean sunfish (*Mola mola*) larva in the Florida Current. Florida Scientist 2: 101–104.

Tyler, J.C. 1980. Osteology, phylogeny, and higher classification of the fishes of the order Plectognathi (Tetraodontiformes) (Vol. 434). US Department of Commerce, National Oceanic and Atmospheric Administration, National Marine Fisheries Service.

Waap, S., W.O.C. Symondson, J.P. Granadeiro, H. Alonso, C. Serra-Goncalves, M.P. Dias et al. 2017.The diet of a nocturnal pelagic predator, the Bulwer's petrel, across the lunar cycle. Sci. Rep. 7: 1384.

Wan, R.J. and R.Z. Zhang. 2005. Spatial distribution and morphological characters of the eggs and larvae of the slender mola *Ranzania laevis* from the tropical waters of the western Pacific Ocean. Acta Zoologica Sinica. 51(6): 1034–1043 [In Chinese with English abstract].

Watson, W. and J.M. Leis. 1974. Ichthyoplankton of Kaneohe Bay, Hawai'i: a one-year study of fish eggs and larvae. Univ. Hawai'i Sea Grant Prog. Tech. Rept. TR75-01: 1–178.

WEB Fish Picture Book. 2020. https://zukan.com/fish/leaf68953. Accessed 06 March 2020. [website in Japanese, original title: WEB魚図鑑].

Whitley, G. 1933. Sunfishes. Vict. Nat. XLIX: 207–213.

Yabe, H. 1950. Juvenile of the pointed-tailed ocean sunfish, *Masturus lanceolatus*. Bull. Jpn. Soc. Sci. Fish. 16: 40–42. [in Japanese with English abstract]

Young, J.W., M.J. Lansdell, R.A. Campbell, S.P. Cooper, F. Juanes and M.A. Guest. 2010. Feeding ecology and niche segregation in oceanic top predators off eastern Australia. Marine Biology 157(11): 2347–2368.

Zhang, Y., J.P. Ryan, B. Kieft, B.W. Hobson, R.T.S. McEwen, M.A. Godin et al. 2019. Targeted Sampling by Autonomous Underwater Vehicles. Front. Mar. Sci. 6: 415.

APPENDIX 1

Larval Records Sources

HISTORICAL/LITERATURE
Putnam 1870, Perugia 1889, Steenstrup and Lutken 1898, Jordan and Evermann 1903, Schmidt 1921b, Whitley 1933, Gudger 1939, Fitch 1950, King 1951, Sherman 1961, Watson and Leis 1974, Robinson 1975, Sokolovskaya and Sokolovskiy 1975, Leis 1977

Source	Contact	Institution	Address
Trimm et al. 2010	D. Trimm	Ecology and Environmental Inc.	Pensacola FL USA 32501
Nogueira et al. 2012	M. M. Nogueira	Universidade Federal da Bahia, Instituto de Biologia, Departamento de Zoologia, Laboratório de Plâncton	Av. Adhemar de Barros, Campus Ondina CEP 40170-290, Salvador, Bahia, Brazil
Shimose et al. 2013	T. Shimose	University of the Ryukyus, Faculty of Science, Laboratory of Fisheries Biology and Coral Reef Studies, Research Center for Subtropical Fisheries, Seikai National Fisheries Research Institute, Fisheries Research Agency	148–446, Fukai-Ohta, Ishigaki, Okinawa 907–0451, Japan
Hellenbrecht et al. 2019	M. Freese	1) University of Bergen, Department of Biological Sciences, and 2) Thuenen Institute of Fisheries Ecology	1) Thormøhlensgate 53 a/b, Bergen 5006, Norway. 2) Herwigstraße 31, Bremerhaven 27572, Germany
SEAMAP	G. Zapfe J. Cobb	Southeast Fisheries Science Center, Mississippi Laboratories	Pascagoula, MS Laboratory: Southeast Fisheries Science Center, 3209 Frederic Street Pascagoula, MS USA 39567
Huntsman Marine Science Center Fish Collection	N. D. Phillips	Huntsman Marine Science Center	1 Lower Campus Road, St. Andrews, New Brunswick E5B 2L7, Canada
SEAMAP	T. Cullins		
Australian Research Council (ARC)	ARC (via P. Provoost)	Australian Research Council (ARC)	11 Lancaster Pl, Australian Capital Territory 2609, Australia
TRAWL AND PLANKTON SURVEYS			
SEFSC Caribbean Survey 1995 (Garrison, L. 2013)	K. D. Mullen L. Garrison	NOAA Southeast Fisheries Science Center	Southeast Fisheries Science Center. 75 Virginia Beach Drive. Miami, FL, 33149 USA
ReviZEE Programa de Avaliação do Potencial Sustentável de Recursos Vivos na Zona Econômica Exclusiva [Program of Evaluation of the Sustainability of Living Resources in the Economic Zone]	P. Provoost	Comissão Interministerial para os Recursos do Mar	Esplanada dos Ministérios, Bloco B, Brasília – DF, CEP 70068-900

Contd. ...

...Contd.

Source	Contact	Institution	Address
ReviZEE	P. Provoost	International Oceanographic Data and Information Exchange" (IODE) of the Intergovernmental Oceanographic Commission (IOC) of UNESCO	Wandelaarkaai 7. Pakhuis 61 8400 Oostende Belgium
ReviZEE	P. Provoost	Institute of Marine Sciences (ICM)	ICM: Passeig Marítim de la Barceloneta, 37-49. E-08003 Barcelona (Spain)
SPURS-MIDAS (Project MIDAS-6 (AYA2010-22062-C05-01)		Spanish National Research Council (CSIC)	CSIC: Spanish National Research Council (CSIC), Serrano, 117, 28006 Madrid Spain
NOAA Pacific Islands Fisheries Science Center SE 1501 Cruise; Chief scientist Phoebe Woodworth-Jefcoats	E. Kawamoto	National Oceanica and Atmospheric Administration (NOAA)	1401 Constitution Ave NW, Washington, DC 20230, USA
FORCES NOAA Lab (NF-13-04)	E. Malca	NOAA; NMFS Miami Laboratory; University of South Florida; University of Miami; Cooperative Institute for Marine and Atmospheric Studies CIMAS-RSMAS NOAA Southeast Fisheries Science Center, Larval Fish Lab	NOAA Southeast Fisheries Science Center National Marine Fisheries Service, 75 Virginia Beach Dr, Key Biscayne, FL 33149, USA
		Instituto Nacional de Pesca; El Colegio de La Frontera Sur; Instituto Español de Oceanografía.	Calle del Corazón de María, 8 28002 Madrid Spain
FISH COLLECTIONS			
National Fish Collection	D. Johnson	Smithsonian Natural History Museum	National Mall, 10th St. & Constitution Ave. NW, Washington, DC 20560 USA
Australian National Fish Collection National Research Collections Australia	A. Graham	Commonwealth Scientific and industrial Research Organisation (CSIRO)	National Facilities and Collections, Castray Esplanade, Hobart Tas 7000, Australia; GPO Box 1538, Hobart, Tasmania 7001, Australia
National Museum of Nature and Science Tokyo Fish Collection	T. Nakatsubo G. Shinohara	ORIX Aquarium Corporation; National Museum of Nature and Science	Nippon Life Hamamatsucho Crea Tower 14F Tokyo, Japan
CAS Fish Collection	D. Catania	California Academy of Sciences (CAS)	Golden Gate Park, 55 Music Concourse Drive, San Francisco, CA USA 94118
LACM Vertebrate Collection	R. Feeney B. Ludt	Natural History Museum of Los Angeles Ichthyology Department	900 Exposition Blvd., Los Angeles, CA USA 90007
SIO Marine Vertebrate Collection	B. Frable	Scripps Institution of Oceanography (SIO)	9500 Gilman Drive, La Jolla, CA USA 90923
SAIAB National Fish Collection of South Africa	R. Bills	South African Institute for Aquatic Biodiversity (SAIAB)	Somerset Street, Grahamstown, 6139, South Africa
Bishop Museum Ichthyology Collection	A. Suzumoto	Bishop Museum, Hawai'i State Museum of Natural and Cultural History	1525 Bernice Street, Honolulu, HI USA 96817

Contd. ...

...Contd.

Source	Contact	Institution	Address
AMS Ichthyology Collection	A. Hay	Australian Museum (AMS)	1 William Street, Sydney NSW 2010, Australia
SWFSC Ichthyology Reference Archive	W. Watson E. Weber S. Charter	Southwest Fisheries Science Center (SWFSC)	8901 La Jolla Shores Drive La Jolla, CA USA 92037
UWFC Ichthyology Collection		Burke Museum Ichthyology Collection	4300 15th Ave NE, Seattle, WA USA
ZMUC Fish Collection	H. Carl	Natural History Museum of Denmark Zoological Museum (ZMUC)	Universitetsparken 15, 2100 København, Denmark
UH Fish Collection	J. Whitney	NOAA Pacific Islands Fisheries Science Center	1845 Wasp Blvd. Bldg 176 Honolulu, Hawai'i USA 96818
PIFSC Reference Collection	B. Mundy (retired)	NOAA Pacific Islands Fisheries Science Center	1845 Wasp Blvd. Honolulu, Hawai'i USA 96818
MNHN Fish and Larval Fish Collections	N. Schnell G. Duhamel	Muséum National d'Histoire naturelle (MNHN) Ichthyology	Muséum National d'Histoire Naturelle 57, rue Cuvier, Paris 75005, France

APPENDIX 2

Predation Table

Reports of predators on molids, from anecdotal narratives, discrete observations ('discrete obs') and diet studies. Length measurement abbreviations used: body length (BL), body weight (BW), fork length (FL), lower jaw fork length (LJFL), pre-clavus band length (PCBL) (here considered interchangeable with standard length, which is not used), total length (TL). Stomach content abbreviations used: percent of all prey items by number (%N), weight (%W) or volume (%Vol), frequency of occurrence in all stomach samples (%F). Also: standard deviation (SD), Standard Error (SE). The number (n) of predators in the diet studies exclude fish with empty stomachs.

Source	Predator info	Prey	Prey (quantity, size)	Date	Location	Reference
STOMIIDAE						
Saber-toothed Viperfish *Chauliodus sloani* (Bloch and Schneider 1801)						
Anecdotal	[Report is of uncertain scientific credibility]	*R. laevis*	*"...newly hatched young... have been taken in nets off [Bermuda] at a depth of a quarter of a mile, where they are the prey of the Viperfish and other predatory species..."*	1934	Bermuda	Beebe 1934

Contd. ...

...Contd.

Source	Predator info	Prey	Prey (quantity, size)	Date	Location	Reference
ALEPISAURIDAE						
Lancetfish *Alepisaurus* spp.						
Diet study	n = 114 39–100 cm SL	Molidae	0.5% W	2004–2006	E Australia	Young et al. 2010
	100–148 cm SL	Molidae	1.23% W			
Lancetfish *Alepisaurus ferox* Lowe 1833						
Diet study	n = 168	Molidae	0.6% F, 0.2% N, 2.6% W	23 Jun–16 Oct 2002	N Central Atlantic, tropical Atlantic	Satoh et al. 2004
Discrete obs.	–	*Ma. lanceolatus*	n = 1 *ca.* 30 mm TL	08 Apr 2007	–	ZMUC P2014848, P2014849
Diet study	n = 120 40–153 cm FL	*R. laevis*	0.83% F, 0.06% N, 2.10% W	2007–2012	Central N Pacific	Choy et al. 2013
CORYPHAENIDAE						
Dolphinfish *Coryphaena sp.*						
Discrete obs.	–	*Ma. lanceolatus*	n = 1 47.5 mm BL	Jul 1862	Near the Azores	Steenstrup and Lütken 1898, Schmidt 1921a, Gudger 1939, ZMUC specimen "Andrea Oct 62-7"
Discrete obs.	–	*Ma. lanceolatus*	n = 2 53 and ? mm PCBL	1880	W Sargasso Sea, E of Savannah, Georgia (USA)	Gudger 1939
Discrete obs.	–	*Ma. lanceolatus*	n = 4 35–50 mm BL	Mar 1882	Pensacola, Santa Rosa Island, Florida, Gulf of Mexico	Schmidt 1921a; Gudger 1939
Discrete obs.	–	*Ma. lanceolatus*	n = 3; 125, 127 130 mm PCBL	17 Apr 1938	New Caledonia	AMS I.43605-001
Discrete obs.	"Dolphin"[1], 110 cm FL	*Ma. lanceolatus*	n = 2 135, 129 mm TL/111.9, 107.2 PCBL	11 Apr 1950	W of Hawai'i	King 1951
Pompano mahi mahi *Coryphaena equiselis* Linnaeus 1758						
Discrete obs.	–	*Ma. lanceolatus*	n=3 "small"	March 1958	Central Atlantic (between Canary Is, Spain, and Antigua, West Indies)	Greer (1976), MSU specimens IC.1001 - IC.1003

Contd. ...

...*Contd.*

Source	Predator info	Prey	Prey (quantity, size)	Date	Location	Reference
Mahi mahi *Coryphaena hippurus* Linnaeus 1758						
Diet study	n = 51	Molidae	n = 1	–	W Pacific	Allain et al. 2003
Diet study	n = 95 50–100 cm	Molidae	0% W	1999, 2004–2006	E Australia	Young et al. 2010
	100–178 cm	Molidae	0.13% W ["...*dolphinfish... fed on a range of larval and smaller fishes*..."]			
GEMPYLIDAE						
Snake mackerel *Gempylus serpens* Cuvier 1829						
Diet study	n = 47 66.5–140.5 cm FL, mean 96.9 ± 15.4 SD	*R. laevis*	14.89% F, 18.75% N, 32.99% W	2007–2012	Central N Pacific	Choy et al. 2013
SCOMBRIDAE						
Wahoo *Acanthocybium solandri* (Cuvier 1832)						
Discrete obs.	–	*Ma. lanceolatus*	n = 3 9.5, 10, 13 mm ?BL	1911	Central Pacific	AMS I.12286 McCulloch 1912, Schmidt 1921a, Gudger 1939
Discrete obs.	–	*Ma. lanceolatus*	n = 7 22.5–77.5 mm BL	28 Jun 1949	Minamidaito Is., Japan	Yabe 1950
Diet study	n = 135 (nearshore)	*R. laevis*	1.71% F, 0.48% N, 1.9% W ["juveniles"]	Jun–Dec 2014	Central N. Pacific and Hawai'i	Oyafuso et al. 2016
	n = 175 (offshore)	*R. laevis*	20.96% F, 13.68% N, 23.72% W ["juveniles"]			
		Ma. lanceolatus	1.2% F, 0.58% N, 0.18% W			
Diet study	n = 254 80 – *ca.* 178 cm	Molidae	0.79% F, 0.53% N, 1.11% W [Overall teleost prey lengths: 77.5 ± 5.8 SE mm]	Nov 2008–Sep 2011	E Australia and Christmas Is.	Perelman et al. 2017
Skipjack tuna *Katsuwonus pelamis* (Linnaeus 1758)						
Discrete obs.	–	*Ma. lanceolatus*	n = 45 4.6–10.6 mm TL [length of claval filaments negligible]	Unknown	Japan	T. Nakatsubo and G. Shinohara *unpublished data*

Contd. ...

...Contd.

Source	Predator info	Prey	Prey (quantity, size)	Date	Location	Reference
Atlantic blue marlin *Makaira nigricans* (Lacépède 1802)						
Diet study	n = 27 Average 95 kg	*R. laevis*	n = 2	27–29 Jul 1981	Hawai'ian waters	Brock 1984
		Ma. lanceolatus	n = 0			
	n = 38 Average 114 kg	*R. laevis*	n = 0	16–20 Aug 1982		
		Ma. lanceolatus	n = 3 18 to 24 mm TL			
Diet study	n = 10 177–252 cm LJFL	*R. laevis*	50% F, 54% N, 34%W 112–193 mm PCBL	26–28 May, 17–18 Jun 2006	Off Okinawa, Japan	Shimose et al. 2013
Tuna *Thunnus* sp.						
Discrete obs.	–	*R. laevis*	n = 1 60 mm TL	Unknown	Atlantic Ocean	Harting 1865
Discrete obs.	"Bluefin tuna"	*Mola* sp.	n = 1 16 mm TL	21 Sep 1956	New South Wales, Australia	CSIRO B2355 2; Nyegaard et al. 2019
Discrete obs.	"Bluefin tuna"	*Mola* sp.	n = 1 32 mm TL	11 Nov 1963	New South Wales, Australia	CSIRO B2354; Nyegaard et al. 2019
Discrete obs.	"Yellowfin tuna"	*R. laevis*	n = 1 100 mm TL	Unknown	Off Kumejima, Okinawa, Japan	WEB fish picture book 2020
Albacore tuna *Thunnus alalunga* (Bonnaterre 1788)						
Discrete obs.	"*Germo germo* (Lacépède, 1801)"[2]	*Mola* sp.[2]	n = 1 53 mm TL	Aug 1939	Pacific coast of Honshyu, Japan	Kuronuma 1940
Diet study	n = 46 63–100 cm SL	Molidae	0.08% W	1997, 2004–2006	E Australia	Young et al. 2010
	100–113 cm SL	Molidae	0% W			
Yellowfin tuna *Thunnus albacares* (Bonnaterre 1788)						
Anecdotal account	–	*Ma. lanceolatus*	"*...found quite commonly in the stomachs of yellowfin tuna in Hawai'ian waters, nearly a hundred specimens of varying in size from ½ to 2 inches [6.4–51 mm] having been obtained from this source.*"	Unknown	Hawai'ian waters	Fitch 1950

Contd. ...

...*Contd.*

Source	Predator info	Prey	Prey (quantity, size)	Date	Location	Reference
Discrete obs.	–	*R. laevis*	n = 1 53 mm PCB	12 Sep 1957	Atlantic Ocean	NMNH 392798
Discrete obs.	–	*R. laevis*	n = 2 53, 62 mm PCBL	12 Sep 1957	Atlantic Ocean	NMNH 392801
Discrete obs.	–	*R. laevis*	n = 1 32 mm PCBL	12 Sep 1957	Atlantic Ocean	NMNH 392797
Discrete obs.	–	*R. laevis*	n = 1 38 mm PCBL	02 Oct 1957	Atlantic Ocean	NMNH 392799
Discrete obs.	–	*R. laevis*	n = 1 71 mm PCBL	02 Oct 1957	Atlantic Ocean	NMNH 392800
Discrete obs.	–	*R. laevis*	n = 1 24 mm PCBL	03 Oct 1957	Atlantic Ocean	NMNH 392794
Discrete obs.	–	*R. laevis*	n = 1 36 mm PCBL	03 Oct 1957	Atlantic Ocean	NMNH 392802
Discrete obs.	–	*R. laevis*	n = 3 42, 44, 51 mm PCBL	09 Oct 1957	Atlantic Ocean	NMNH 392803
Discrete obs.	–	Molidae	n = 2 15, 18 mm PCBL	03 Nov 1957	Atlantic Ocean	NMNH 392796
Diet study	n = 266 74–166 cm FL	*Mola* sp.[2]	n = 2	1957–1964	Western N Atlantic	Matthews et al. 1977
		R. laevis	n = 10			
		Molidae	n = 2 [No size information but "... *the majority of forage fishes were immature forms of midwater fishes and epipelagic* post-larvae."]			
Discrete obs.	–	*Mola* sp.[2]	n = 3	21 Apr 1960	Atlantic Ocean	NMNH 217452
Discrete obs.	–	*Mola* sp.	n = 2 11, 26 mm TL	08 Sep 1960	New South Wales, Australia	CSIRO B247; Nyegaard et al. 2019
Bigeye tuna *Thunnus obesus* (Lowe 1839)						
Discrete obs.	–	Molidae	n = 1 8.7 mm TL	Feb 1982	N. Hawai'ian Islands	T. Nakatsubo and G. Shinohara, *unpublished data*
Discrete obs.	–	*Ma. lanceolatus*	n = 5 5.4–11.4 mm TL	01 Jun 1975	Sea of Japan	T. Nakatsubo and G. Shinohara unpublished data

Contd. ...

...Contd.

Source	Predator info	Prey	Prey (quantity, size)	Date	Location	Reference
Discrete obs.	146 cm TL, 64 kg	Molidae	n = 1 8.7 mm TL	09 Feb 1982	N Pacific Ocean	T. Nakatsubo and G. Shinohara unpublished data
Diet study	n = 652 71–100 cm SL	Molidae	0.6%W	1994–2001, 2003–2006	E Australia	Young et al. 2010
	100–214 cm SL	Molidae	0.05%W			
Diet study	n = 47	Molidae	8.5% F, 5.8% N, 45.5% W	23 Jun–16 Oct 2002	N Central Atlantic and tropical Atlantic	Satoh et al. 2004
Southern Bluefin tuna *Thunnus maccoyii* (Castelnau 1872)						
Discrete obs.	–	*Mola* sp.	n = 4 20–23 mm TL	28 Oct 1958	New South Wales, Australia	CSIRO B306; Nyegaard et al. 2019
Atlantic Bluefin tuna *Thunnus thynnus* (Linnaeus 1758)						
Diet study	n = 38 58–232 cm FL	*Mola* sp.[2]	6% F [No size information but "*... the majority of forage fishes were immature forms of midwater fishes and epipelagic post-larvae....*"]	1957–1964	Western N Atlantic	Matthews et al. 1977
ISTIOPHORIDAE						
Atlantic sailfish *Istiophorus platypterus* (Shaw 1792)						
Diet study	"*I. albicans* Jolley (1977)"[3] n = 42 stomachs	Molidae	28.6% F, 11.2% N, 5.3% W	23 Jun–16 Oct 2002	N Central Atlantic and tropical Atlantic	Satoh et al. 2004
White marlin *Kajikia albida* (Poey 1860)						
Diet study	"*Tetrapturus albidus* Poey 1860"[3] n = 32 stomachs	Molidae	9.4% F, 2.7% N, 1.8% W	23 Jun–16 Oct 2002	N Central Atlantic and tropical Atlantic	Satoh et al. 2004
"Striped marlin" *Kajikia audax* (Philippi 1887)						
	"Striped marlin"[4]	Molidae	n = 1 *ca.* 80 mm PCBL	05 Nov 1963	*Ca.* 1,000 km off Baja California, Mexico	ZMUC specimen P.89122
Diet study	n = 16 (open ocean) 80–240 cm LJFL	*R. laevis*	50% F, 49% M [Overall range of *R. laevis* within study: 194–326 mm SL]	18 Sep–07 Nov 2004	E N Pacific	Shimose et al. 2010

Contd. ...

...Contd.

Source	Predator info	Prey	Prey (quantity, size)	Date	Location	Reference
Longbill spearfish *Tetrapturus pfluegeri* (Robins and DeSylva 1963)						
Diet study	n = 53	Molidae	9.4% F, 3.1% N, 1.1% W	23 Jun–16 Oct 2002	N Central Atlantic and tropical Atlantic	Satoh et al. 2004
Discrete obs.	35 lbs/*ca.* 16 kg	*R. laevis*	n = 1	18 Jan 2004	Caribbean Sea	NMNH 384696
Shortbill spearfish *Tetrapturus angustirostris* Tanaka 1915						
Diet study	n = 23 (open ocean) 153–177 cm LLJF	*R. laevis*	35% F, 75% M [Overall range of *R. laevis* within study: 194–326 mm SL]	18 Sep–07 Nov 2004	Eastern N Pacific	Shimose et al. 2010
Unidentified Actinopterygii						
Discrete obs.	"King snapper"	*Mola* sp.	n = 1 64 mm ?PCBL	21 Nov 1932	Near Fraser Is, Queensland, Australia	QM I4990
STERNIDAE						
Black noddy *Anous minutus* Boie 1844						
Diet study	n = 494 regurgitates 85–133 g BW	*R. laevis*	n = 9, 1.2% F, 0.4% Vol. [size range of all measurable prey items (n = 1,038): 5–167 mm, average 34 mm]	1978–80	Hawai'i	Harrison et al. 1983
Brown noddy *Anous stolidus* (Linnaeus 1758)						
Diet study	n = 354 regurgitates 153–275 g BW	*R. laevis*	n = 3, 0.8% F, 0.6% Vol. [size range of all measurable prey items (n = 460): 3–185 mm, average 48 mm]	Mar–Nov 1978–80	Hawai'i	Harrison et al. 1983
Sooty tern *Onychoprion fuscatus* (Linnaeus 1766)						
Diet study	n = 356 regurgitates 153–320 g BW	*R. laevis*	n = 11, 2.2% F, 1.0% Vol. [size range of all measurable prey items (n = 326): 1–120 mm, average 48 mm]	Mar–Sep 1978–1980	Hawai'i	Harrison et al. 1983
FREGATIDAE Great frigatebird *Fregata minor* (Gmelin 1789)						
Diet study	n = 284 regurgitates 1,060–1,950 g BW	*Ma. lanceolatus*	n = 5, 0.4% F, 0.1% Vol. [size range of all measurable prey items (n = 248): 12–272 mm, average 83 mm]	1978–80	Hawai'i	Harrison et al. 1983
		R. laevis	n = 2, 0.4% F, 0.1% Vol.			
PHAETHONTIDAE Red-tailed tropicbird *Phaethon rubricauda* Boddaert 1783						
Diet study	n = 270 regurgitates 540–750 g BW	*R. laevis*	n = 23; 5.2% F, 4.3% Vol. 123–140 mm SL (n = 5)	Aug–Sep 1978, May 1979, Jun 1980	Hawai'i	Harrison et al. 1983
		Ma. lanceolatus	n = 2, 0.7% F, 0.5% Vol. [size range of all measurable prey items (n = 169): 10–237 mm, average 101 mm]			

Contd. ...

...Contd.

Source	Predator info	Prey	Prey (quantity, size)	Date	Location	Reference
SULIDAE Red-footed booby *Sula sula* (Linnaeus 1766)						
Diet study	n = 360 regurgitates 905–1,400 g BW	*R. laevis*	n = 18, 3.1% F, 1.7% Vol. [size range of all measurable prey items (n = 550): 32–282 mm, average 88 mm]	1978–80	Hawai'i	Harrison et al. 1983
		Ma. lanceolatus	n = 2, 0.3% F, 0.1% Vol.			
Diet study	n = 456 regurgitates	*R. laevis*	n = 5, 0.7% F, 0.1% W [all present in austral spring and summer]	Jan 1981–Oct 1982	Hawai'i	Seki and Harrison 1989
DIOMEDEIDAE Laysan albatross *Phoebastria immutabilis* (Rothschild 1893)						
Diet study	n = 183 regurgitates 1,900–3,075 g BW	*R. laevis*	n = 12, 2.2% F, 0.9% Vol. [size range of all measurable prey items (n = 229): 8–150 mm, but this species tears prey]	Feb–July, 1978–80	Hawai'i	Harrison et al. 1983
		Molidae	n = 2, 1.1% F, 0.7% Vol.			
PROCELLARIIDAE						
Cory's Shearwater *Calonectris diomedea* (Scopoli 1769)[5]						
Diet study	n = 673 regurgitates	*R. laevis*	Noted as present, no quantitative data	Aug–Sep of 2008, 2009 and 2010	Selvagem Grande, North Atlantic	Alonso et al. 2013
Bulwer's petrel *Bulweria bulwerii* (Jardine and Selby, 1828)[6]						
Diet study	n = 139 stomach content samples	*R. laevis*	0.72% F	2012, 2013	Deserta Grande and Selvagem Grande, NE Atlantic	Waap et al. 2017

[1] Presumably Dolphinfish *Coryphaena* sp., alternatively a small marine mammal
[2] Reported as *Mola mola*
[3] Junior synonym
[4] Presumably *Kajikia audax*
[5] Adult body mass < 650 g (Dunning 2008)
[6] Adult body mass of 94 g (Southey and Frost 2013), or 78–130 g (Dunning 2008)

Chapter 8

Movements and Foraging Behavior of Ocean Sunfish

Lara L. Sousa,[1,*] *Itsumi Nakamura*[2] and *David W. Sims*[3,4]

Introduction

Ocean sunfishes (Molidae) are distributed widely throughout temperate and subtropical regions, with observation records from various sites in the Mediterranean, North and South Atlantic, Gulf of Mexico and Pacific Ocean, Indian Ocean and Australia (e.g., Silvani et al. 1999, Sims and Southall 2002, Petersen 2005, Houghton et al. 2006, Fulling et al. 2007, Abe et al. 2012, Garibaldi 2015, Phillips et al. 2015). Owing to their atypical morphology and characteristic basking behavior (see Pope et al. 2010, Nakamura et al. 2015), there was a historical perception that the species drifted passively (Pope et al. 2010). However, over recent years, these views have changed radically with strong empirical data revealing high performance burst swimming and long-distance movements against major current regimes (Cartamil and Lowe 2004, Sims et al. 2009b, Sousa et al. 2016a, Chang et al. 2019). These advancements in our understanding were facilitated by animal-borne sensors (biologging) and tracking technologies (telemetry). From this backdrop we consolidate our current understanding of long-range movements, together with the temporal and spatial drivers of habitat use and foraging behavior, both horizontally and vertically. We also explore a range of behavioral and physiological mechanisms, from thermal inertia to respiration, that allow ocean sunfishes to operate over broad spatial scales and into the ocean depths.

Large-Scale Horizontal Movements, Seasonality and Thermal Range

Historically, mark-recapture has been used to unravel the movements of aquatic species since the use of colored wool ribbons to document the return of salmon (*Salmo salar*) to natal rivers in the early 1600s (Kohler and Turner 2001). Over time, this technique has provided vital data on key life history

[1] Wildlife Conservation Research Unit, University of Oxford, Tubney, United Kingdom.

[2] Institute for East China Sea Research, Organization for Marine Science and Technology, Nagasaki University, 1551-7 Taira-machi, Nagasaki, Japan; Email: itsumi@nagasaki-u.ac.jp

[3] Marine Biological Association of the United Kingdom, The Laboratory, Citadel Hill, Plymouth PL1 2PB, UK.

[4] Ocean and Earth Science, University of Southampton, National Oceanography Centre Southampton, Waterfront Campus, Southampton, SO14 3ZH, UK; Email: dws@mba.ac.uk

* Corresponding author: lara.l.sousa@gmail.com

traits from growth and longevity (e.g., Pratt and Casey 1983) through to distribution and population dynamics (e.g., Holland et al. 2001, Kohler and Turner 2001, Queiroz et al. 2005). However, likely given the lack of commercial interest in sunfishes, such mark and recapture studies have not been employed widely across the group. Such approaches warrant consideration though, especially in areas where sunfish are caught routinely in discrete spatial areas over time (e.g., the set net tuna fishery in southern Portugal and northern Italy, Sims et al. 2009a,b, Sousa et al. 2016a,b,c, Phillips et al. 2018).

Satellite tracking of fish species started in 1984, with the remote tracking of a basking shark (*Cetorhinus maximus*), pioneering the biologging ecology advances (Priede 1984). Concurrent with studies of other fish species (e.g., Bluefin tuna, *Thunnus thynnus*, Block et al. 1998) the most significant advance in our understanding of sunfish movements to date followed the development of archival devices that could relay information via satellite (Hooker et al. 2007). Such animal-borne tags allow the track of an individual to be retrospectively reconstructed from light (i.e., day length and time of sunrise and sunset) and water temperature data. As communication between satellite transmitters and the satellites themselves is not possible through seawater, these data are relayed via the Argos satellite network when the device detaches from the animal (at a predetermined time) and floats to the surface (e.g., Southall et al. 2005, Sims et al. 2008, Hays 2008). The first deployment of these pop-off tags on a sunfish was on a *Mola mola* off San Diego by researchers at the Monterey Bay Aquarium in the year 2000 (T.M. Thys, personal communication), and the first published satellite tagging of a sharptail mola (*Masturus lanceolatus*) with pop-up tags happened two years later (Seitz et al. 2002). Since then, sunfish species have been satellite tracked worldwide (See Fig. 1 for a global map of published studies to date). *Mola mola* has received the most attention, constituting 81 percent (N = 57) of all tracked individuals (N = 71). Notably, all satellite tagging studies of sunfish species have been conducted in coastal regions; thus there may be a bias in the interpretation of movements and space-use of tracked individuals (see Nakano and Stevens 2008). However, a general seasonal pattern is evident in terms of broad-scale movements of these species. In the north-east Atlantic off the UK, a northerly movement of *M. mola* has been noted (Fig. 2A) in late winter/spring with movement to the south recorded separately in late summer, possibly representing a seasonal migration (Fig. 2B; Sousa et al. 2016a, Sims et al. 2009a, respectively). A similar seasonal pattern was detected in the western Pacific Ocean (Dewar et al. 2010) with sunfish moving to higher latitudes in the summer off Japan and the Kuril Islands while in the north-west Atlantic (Potter et al. 2010) sunfish moved southerly during winter movements from Massachusetts into the Gulf of Mexico. The seasonal shelf movements of sunfish in these three regions appear temporally correlated with oceanic productivity and underlying thermal regimes. Indeed, the seasonal appearance of sunfish at higher latitudes has been linked to seasonal increases in planktonic productivity (Sims et al. 2009a), which mirrors the movements of other gelatinous zooplankton feeders such as the leatherback (*Dermochelys coriacea*) (e.g., Houghton et al. 2006b, Witt et al. 2006, Hays et al. 2009). Likewise, satellite tracking studies in the North Atlantic and in North-West Pacific Ocean basins have shown population-level spatial changes linked to seasonal variations in sea surface temperature (SST) and/or forage availability (Sims et al. 2009a, Sims et al. 2009b, Dewar et al. 2010, Potter et al. 2010, Sousa et al. 2016a). Similarly, the seasonal movements of sunfish off Japan (Dewar et al. 2010) are consistent with the sunfish fisheries catches, which peaks twice in the year at the northwest part of Japan (Nakamura and Sato 2014).

During broad-scale movements, the thermal envelope occupied by the tracked sunfish in the north Atlantic peaked around 14–17°C (Sims et al. 2009, Potter et al. 2010), with *M. mola* spending 90 percent of the time in waters of 10–25°C (Sims et al. 2009). Furthermore, this thermal range matches the recorded sea surface temperatures for when sunfish catch rates peak in Japan (Nakamura and Sato 2014). Conversely, *M. alexandrini* were found to occupy warmer waters in general, with tracked individuals found within a thermal band of 20–25°C (Thys et al. 2016, Thys et al. 2017). These tracking data are broadly consistent with predictive global distribution models, which showed a strong correlation of species with sea surface temperature (Phillips et al. 2017).

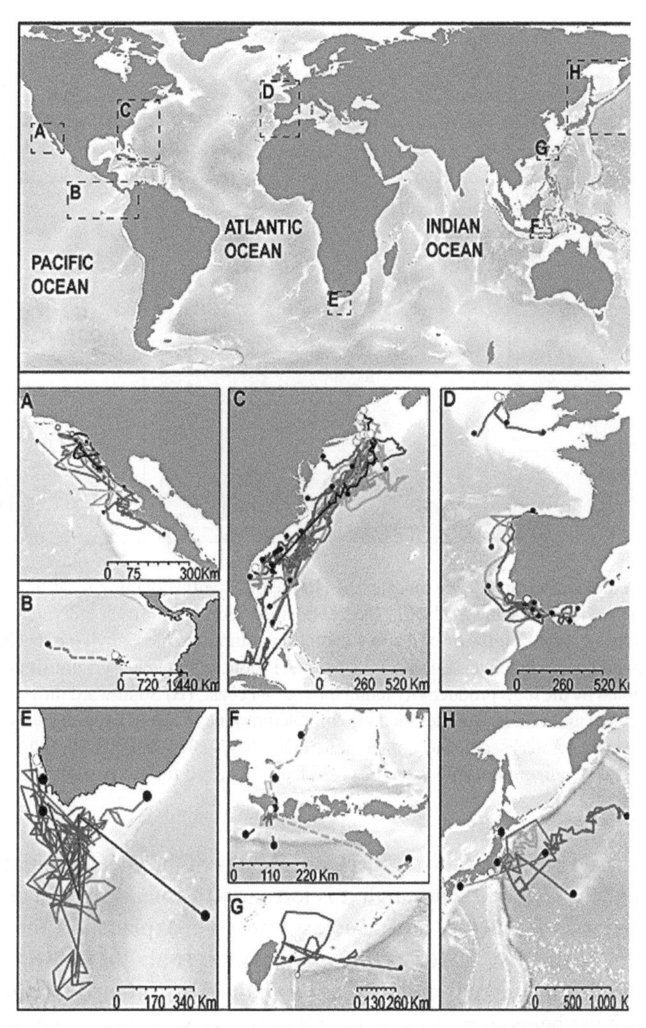

Figure 1. Movements of ocean sunfish revealed by satellite tracking. The top panel highlights the tracking regions (A–H). White dots denote transmitter attachment locations, black dots the last retrieved locations. (A) *M. mola* tracks from the NE Pacific (N = 9 Thys et al. 2015); (B) Galápagos *M. alexandrini* tracks (N = 1 Thys et al. 2017); (C) *M. mola* tracks from the NW Atlantic (N = 18 Potter et al. 2010); (D) *M. mola* tracks from the NE Atlantic (N = 18 Sims et al. 2009a,b, Sousa et al. 2016a); (E) *M. mola* tracks from South Africa (N = 4 Hays et al. 2009); (F) *M. alexandrini* tracks off Bali (N = 8 Nyegaard 2018 and N = 4 Thys et al. 2016); (G) *M. mola* (N = 1 [blue] and *M. alexandrini* N = 1 [red] tracks off Taiwan Chang et al. 2019); and (H) *M. mola* tracks from Japan (N = 7 Dewar et al. 2010).

Residency vs. Migration

Assessment of the *Mola* spp. distributions on a global scale suggested sunfishes are facultative seasonal migrants (Phillips et al. 2017). Aerial surveys and ship-based sunfish sightings have also been used to inform abundance estimates for sunfish in northern latitudes, for instance off the Irish and Welsh coasts (Houghton et al. 2006a, Breen et al. 2017), off British Columbia in Canada (Thys and Williams 2013) and Bali, Indonesia (Rob Williams, personal communication). Although smaller individuals were found year-round off Ireland, larger sunfish were only observed in summer (Breen et al. 2017). Off British Columbia, sightings of *M. mola* individuals 1 to 2 m in total length in spring and summer months also confirm the occurrence of larger individuals in temperate coastal waters during warmer months (Thys and Williams 2013). In general, the worldwide satellite tracking studies,

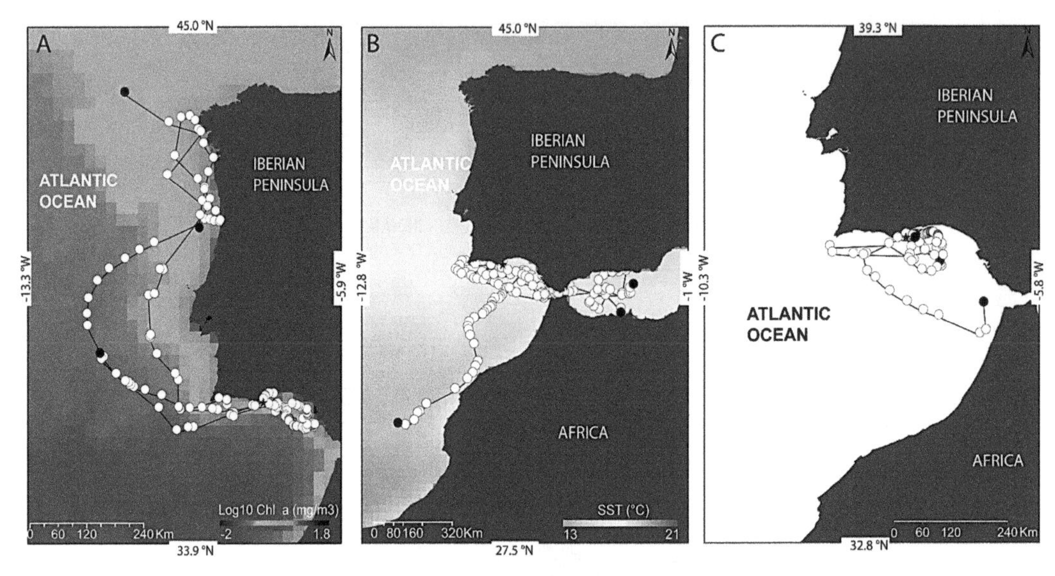

Figure 2. Satellite tracked movements of *Mola mola* in the Northeast Atlantic. Tagging location denoted by the star, white dots are the satellite retrieved locations and in black is the last position received for each individual sunfish. (A) Northwards movements in spring, early summer months overlaid on a cumulative summer chlorophyll *a* map (2007–2014 from MODIS Moderate Resolution Imaging Spectrometer, GlobColour level-3 Product 1/4 Degree). Sunfish appear to move along the upwelling boundary denoted by the high productivity along the Iberian coast; (B) Southward movements, likely followed the decreased water temperatures in northern latitudes, and into the Mediterranean Sea in autumn and winter months overlaid on a cumulative seasonal (autumn-winter) sea surface temperature map (2007–2014 AVHRR V2 NOAA Optimum Interpolation 1/4 Degree Daily Sea Surface Temperature Analysis); (C) Residency depicted for ocean sunfish of smaller sizes in the Gulf of Cadiz.

together with those based on opportunistic surface sightings of the species, point towards the notion of sunfish size-dependent residency, although it is too early to consider this a generalizable pattern. Nonetheless, maturation is frequently advanced as a possible explanation for partial migration in fishes, (Boyle 2008, Chapman et al. 2012a,b, Papastamatiou et al. 2013), which may hold true for sunfish. However, sunfish maturity is assumed to be attained at 1.40 m TL (Nakatsubo et al. 2007) (although this has not yet been confirmed across all regions), which is far larger than wide-ranging individuals tracked in the northeast Atlantic (e.g., Sousa et al. 2016a tracked a fish of 0.6 m travel 596.8 km before the pop-off tag detached). Thus, the drivers of sunfish residency and dispersal require further attention in future tracking studies, ideally from currently underrepresented regions.

Independent of size, long-distance movements of sunfish are clearly deliberate and often against the prevailing currents (Cartamil and Lowe 2004, Watanabe and Sato 2008, Sims et al. 2009, Thys et al. 2015, Chang et al. 2019), dispelling the myth of passive drifting once and for all. Such findings are consistent with data from biologging studies that show sunfish are highly motile swimmers (see Watanabe and Davenport 2020, Chapter 5) capable of burst swimming between 2.4 and 6.6 m/s (Watanabe and Sato 2008, Nakamura and Sato 2014, Nakamura et al. 2015, Thys et al. 2015).

Movements in Relation to Ocean Productivity

The long-term (> 90 days) GPS tracking of sunfish individuals in the north-east Atlantic, revealed intermittent periods of reduced movements in localized areas interspersed with faster, directional movements (Sims et al. 2009b). Such movements are logically driven by environmental suitability to satisfy the species intrinsic demands, such as feeding, which in turn is motivated by the abundance of prey (Hays et al. 2006, Sims et al. 2006). Similar to the leatherback turtle, sunfish also feed on gelatinous zooplankton (Hays et al. 2009, Pope et al. 2010), among other prey items (e.g., Harrod et al. 2013, Nakamura and Sato 2014, Sousa et al. 2016). In coastal waters, sunfish are often found in association with true jellyfish (Phylum Cnidaria, Class Scyphozoan) which, for the most part, have a metagenic life cycle whereby medusa (released as ephyrae) emerge from the asexual budding from polyps at the seabed (Lucas et al. 2014). This requirement of a hard substrate for polyp attachment means that scyphozoan jellyfish (with a few notable exceptions of directly developing species such as *Pelagia noctiluca*) are often more abundant in coastal seas (Arai 1997). Characteristically, medusae have a seasonal signal that correlates with increases in marine productivity during the spring and summer, followed by individuals often dying off in the autumn post-spawning (Houghton et al. 2007, Ceh et al. 2015). In line with this life history, it makes logical sense that ocean sunfish might be found in coastal seas during summer months to take advantage of the medusae, although when small (below 1 m TL) sunfish have a more mixed diet (Sousa et al. 2016b, Phillips et al. 2020, and see Phillips et al. 2020, Chapter 9). Beyond scyphozoans, sunfish may forage on more colonial gelata (e.g., salps and siphonophores) in the vicinity of pronounced upwelling, as seen off Japan (Nakamura et al. 2015). This strong association with frontal areas, was described for *M. mola* off the U.K., California, Iberia and Taiwan (Sims and Southall 2002, Thys et al. 2015, Sousa et al. 2016a,c, Chang et al. 2019) and also for the *M. alexandrini* in Bali, the Galápagos Islands and Taiwan (Thys et al. 2016, Thys et al. 2017, Chang et al. 2019), where tracked sunfish individuals were observed seeking out upwelling frontal regions.

Short-term, fine-scale tracks recorded using GPS tags integrated with environmental data (e.g., SST and SST frontal regions) have confirmed that frontal regions have a significant influence on foraging behavior (Sousa et al. 2016c). An Autonomous Underwater Vehicle (AUV) video recorded the path of the tracked sunfish. Its dynamic structure was described by a Lagrangian modeling approach which simulated the distribution of zooplankton in the region. This was linked to frontal areas that were positively correlated with sunfish areas of restricted search (ARS) behaviors (Sousa et al. 2016c). The ever-changing spatio-temporal patterns of gelatinous zooplankton make it challenging to measure prey densities consistently at the scale needed to rigorously investigate foraging behaviors of marine predators (Houghton et al. 2006b, Sims et al. 2006). To overcome this problem, studies of basking sharks and leatherback turtles in the North Atlantic have often used zooplankton prey fields gathered from Continuous Plankton Recorders (CPR) data (Sims et al. 2006, Witt et al. 2007). However, sunfish have been mostly tracked where CPR data are not collected routinely, with the added caveat that larger gelatinous species may be under-represented given the small (~ 2.5 cm) aperture of CPRs. Hence, different approaches are needed to further investigate predator-prey interactions of sunfishes.

Foraging at Depth

Sunfish across all ocean basins have been shown to exploit a broad range of depths, with *M. mola* occupying the water column from the surface to at least 844 m (Potter and Howell 2011). Perhaps more impressive are records for southern sunfish, *M. alexandrini*, off the Galápagos Islands, diving to 1,112 m (Thys et al. 2017). Similar to the horizontal movements, vertical distributions of sunfish likely reflect differences in prey distribution/abundance, bounded by thermal constraints (Dewar et al. 2010, Potter et al. 2010). Likewise, off Japan, vertical movements of *M. mola* were directly linked to seasonal differences in the thermal structuring of the water column (Nakamura and Sato 2014). Specifically, ocean sunfish equipped with multichannel data loggers (N = 12) remained nearer to the surface (> 20 m) in summer, shifting to more wide-ranging movements throughout the mixed layer (< 100 m) in autumn (Nakamura and Sato 2014). Strong stratification of the uppermost layers of the water column resulted in an increased vertical thermal gradient in the summer months, whereas the mixed layer extended over greater depths in autumn. Based on this evidence, it is feasible that ocean sunfish shift from predation on seasonally abundant surface medusa during the summer to vertically migrating gelatinous species found at depth during cooler months (Hays et al. 2009, Pope et al. 2010). This pattern mirrors similar behaviors observed in leatherback turtles which are thought to feed upon deep water, siphonophores, salps and pyrosomes more predominantly when medusae (class Scyphozoa) are scarce in surface waters (Houghton et al. 2009). Yet, off Iberia and Irish coasts, a wide range of different diving behaviors and depth use patterns were recorded and could not be linked to a specific water column structure (Sousa et al. 2016a) suggesting that such seasonal patterns are highly adaptive to local conditions.

The first evidence of diel patterns in sunfish vertical movements came from the sharptail sunfish (*M. lanceolatus*) in the Gulf of Mexico (Seitz 2002). In this study, the tracked individual performed deeper excursions in the daytime (250–1000 m in day vs. 100–250 m at night), albeit spending > 90 percent of the time in the top 200 m of the water column (Seitz 2002). Following on from this study, a similar diel pattern of depth use was described for *M. mola*, with individuals diving below the thermocline during daylight hours, whilst remaining in the surface mixed layer at night (Cartamil and Lowe 2004). These vertical movements are consistent with the classic pattern of normal diel vertical migration (nDVM; dusk ascent, dawn descent). Furthermore, nDVM has since been observed in sunfish worldwide, with individuals spending the majority of their daytime at depth, with vertical movements occurring between dawn and dusk, frequently diving to deeper waters (> 400 m), while resurfacing and using shallower waters during night-time (e.g., Hays et al. 2009, Sims et al. 2009a, Dewar et al. 2010, Potter and Howell 2011, Thys et al. 2015, Chang et al. 2019). This behavior has been attributed to a strategy of near continual feeding on vertically migrating prey (Hays et al. 2009, Sims et al. 2009a). However, despite the conspicuous nDVM, diving patterns of *M. mola* in the Atlantic were not restricted solely to this behavior (Potter and Howell 2011, Sousa et al. 2016a). Sunfish off the Iberian Peninsula in the northeast Atlantic performed reverse DVM (rDVM; dusk descent, dawn ascent) both in frontal regions and well-mixed waters, whereas nDVM and surface-oriented behaviors were detected in all the regions and independently of water column structure (Sousa et al. 2016a; Fig. 3). In addition, depth data revealed that the vertical range of sunfish varied with the size of the tracked individuals, with larger fish exhibiting significantly extended depth amplitudes and time spent in deeper layers (> 250 m), which may reflect tracking vertically migrating prey (Houghton et al. 2006a, Sims et al. 2009a). Altogether, this varied diel vertical activity may suggest a wider range of prey for sunfish or a preference for prey with more varied behavior at depth.

Overall, sunfish DVM is likely to be a foraging strategy (Cartamil and Lowe 2004, Sims et al. 2009a, Dewar et al. 2010), with fish potentially tracking the DVM of gelatinous prey in the water column. Importantly, current shear will disperse olfactory trails horizontally in the water column (Carey and Scharold 1990), indicating that diving sunfish will likely have a greater chance of encountering a prey (using olfaction) by moving vertically through the layers. For sunfish, there has been, to our knowledge, only one study that has complemented tracking technologies with the actual recording of feeding events (Nakamura et al. 2015), thus providing a direct link between movements and prey occurrences. In the latter study, sunfish were almost inactive during the night and fed exclusively during daylight hours, which indicated that the species' diel pattern was not related directly to diel vertical movements of their prey (Nakamura et al. 2015). They were diurnally active, possibly because they may need downwelling light from the surface to illuminate prey, which may not be possible when foraging in the dark. Yet, visual acuity of ocean sunfish, calculated from peak ganglion cell densities, is between 3.51 and 4.33 cycles per degree, comparable to adult sharks (2.8–3.7 cycles per degree; Kino et al. 2009) that are known to forage nocturnally. Moreover, ocean sunfish can find their prey at several hundred meters deep, indicating they are able to see in dark environments, or possibly homing in on pronounced olfactory cues.

From a theoretical perspective, the vertical movements of sunfish in the northeast Atlantic were analyzed and found to be consistent with an optimal foraging strategy, the so-called Lévy walk (Humphries et al. 2010). According to the Lévy foraging hypothesis (LF) (Viswanathan et al. 2008), searching is optimized if the animal adopts a Lévy walk when prey is sparse, whereas when in abundant prey regions, a Brownian (exponential) random search is sufficient (Humphries and Sims 2014). Despite some initial controversy (e.g., Viswanathan 1996, Viswanathan et al. 2000, Edwards et

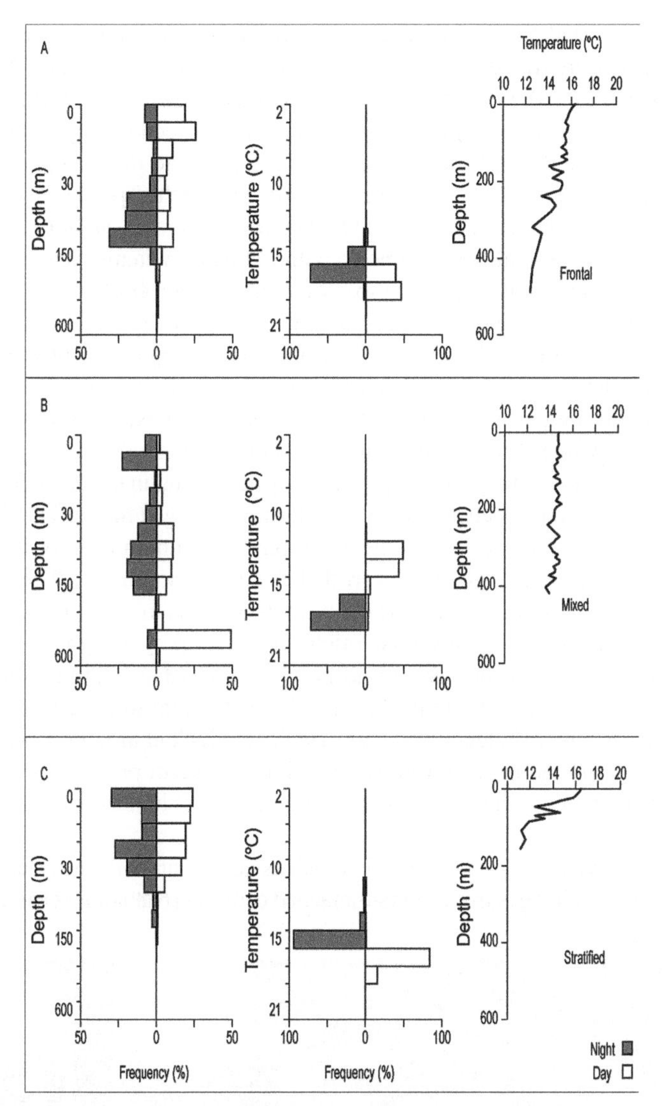

Figure 3. Sunfish (*Mola mola*) diel vertical movements recorded with pop-off satellite tag in the Northeast Atlantic (adapted from Sousa et al. 2016a). (A) reverse DVM in frontal waters; (B) normal DVM in mixed waters; (C) surface-oriented behavior in stratified waters. Left panel shows the time at depth occupied by sunfish; central panel the corresponding time at temperature (TAT) profiles; and right panel displays the water column structure.

al. 2007), empirical evidence for the occurrence of Lévy walks has increased over recent years (Sims et al. 2008, Humphries et al. 2010, Hays et al. 2012, Humphries et al. 2012, Ugland et al. 2014, Sims 2015, Humphries et al. 2016). Interestingly, and similar to other large marine predators such as sharks and tunas, sunfish vertical movements are consistent with a switching pattern between Lévy patterns of movement in less productive habitats (open waters) and Brownian-type motion in a high abundant prey region (continental-shelf rich in zooplankton area), as predicted by theory (Humphries et al. 2010). The latter study suggests that sunfish vertical movement patterns are adjusted to the resource distributions encountered in heterogeneous ocean habitats (Humphries et al. 2010).

Until recently, direct observations of sunfish feeding in deep water have not been possible and so theoretical movement models such as the optimal Lévy walk cannot presently be validated. The recent development of small, animal-borne data-logging devices such as accelerometers and cameras have provided new tools for studying difficult to observe behaviors such as deep-water foraging patterns and feeding itself (Aoki et al. 2012). In order to observe foraging behavior under natural conditions, animal-borne cameras were first deployed on *M. mola* by Nakamura and Sato (2014). The cameras were attached on small sunfish (49–54 cm TL) and recorded short-term movements (< 8 hours) but failed to identify obvious feeding events. However, the images recorded by small sunfish in coastal waters frequently recorded the seabed, and the stopovers in a head-down position while swimming near the seabed, revealed a possible benthic foraging strategy for this species. A year later, Nakamura et al. (2015) successfully characterized foraging behavior of large *M. mola* (> 105 cm TL) in deep water (max. 670 m) using animal-borne cameras with lights. Prey items consumed by sunfish were siphonophores (at least three species; *Praya* sp., *Nanomia* sp. and *Apolemia* sp.), scyphozoa (lion's mane *Cyanea capillata*) and ctenophore (non-identifiable species), with siphonophores being the most commonly observed (Fig. 4, 1–4 below). The obvious feeding events on these prey items were accompanied by deceleration of the sunfish when it encountered the prey, and acceleration as it began consuming it. Feeding events were mostly observed during excursions of sunfish to depths of 50–670 m during the daytime. The sunfish swam back and forth between the sea surface and the deep water, and most of possible feeding events appeared during staying in the deep. Its body temperature decreased during staying in the deep and recovered during staying at the surface (Fig. 4). Gelatinous plankton occurs at highest densities below the chlorophyll maximum layer or in the middle part of the main pycnocline (Ichikawa et al. 2006, Vereshchaka and Vinogradov 1999, Vinogradov 2005). Hence, ocean sunfish were thought to undertake deep excursions to target prey in these deep-water layers. As discussed previously, ocean sunfish might search for prey using vision (Pope et al. 2010), in which case, it is possible for the fish to find prey from a distance and accelerate to capture it. Some siphonophore species have bioluminescence that is thought to attract potential prey species (Davenport and Balazs 1991, Haddock and Case 1999, Haddock et al. 2005), which might, in turn, attract ocean sunfish to forage in deep water where bioluminescent prey can be sighted over longer distances. This acts to increase the sunfish search volume whilst in deep water and, presumably, its prey capture success.

Selective feeding events on parts of jellyfish such as gonads and oral arms leaving the bell intact were observed in the serial images from sunfish-attached cameras (Nakamura et al. 2015) (Fig. 4a, 4b).

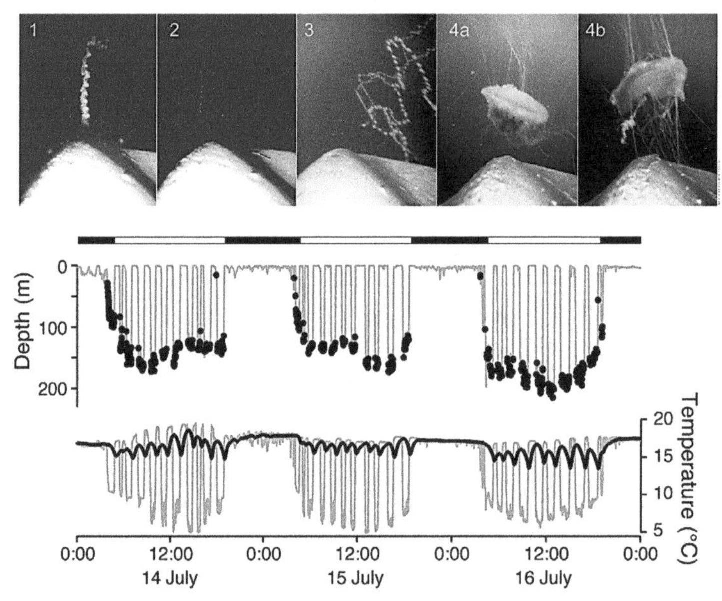

Figure 4. Top panel shows still frames taken from the cameras attached to sunfish recording feeding events (prey 1. *Praya* sp., 2. *Nanomia* sp., 3. *Apolemia* sp. 4. *Cyanea capillata* a: before, b: after). Bottom panel: Depth and temperature (grey: water, black: body) recordings with day/night-time highlighted on the top. Dots on the depth graph indicate each possible feeding event.

The gonads and oral arms of jellyfish have higher energetic value than the bell (Doyle et al. 2007), thus, it may be that sunfish exhibit selection for the most energy-rich parts whilst feeding. In fact, this selective feeding by sunfish may be an indication of higher abundances of gelata in deep water, when considering the rate of energy acquisition with time spent on a single prey (Hays 2015). In prey-rich environments, it may be more time-efficient to prey upon the nearby prey without consuming the whole body of the first prey, and thus limit the total amount that is ingested per prey item, and focus solely on the most energy-rich body parts. Taken together, the behavioral data from the camera studies further confirms that smaller sunfish likely forage on benthic prey and large sunfish mainly move into deep water where prey is more abundant, foraging principally on gelatinous zooplankton such as siphonophores. This non-obligatory gelatinous diet was brought to light again by stable isotope analysis (SIA) which showed a broader diet for smaller sunfish in the Mediterranean, with a similar proportion of pelagic and neritic prey (Syvaranta et al. 2012, Harrod et al. 2013). It should be noted however, that disparate but nonetheless valid, records of sunfish feeding on non-gelatinous prey existed long before this biochemical study (see Pope et al. 2010). On the other hand, in Japan, large sunfish feeding on siphonophores was often observed in cold water (< 12°C). The depth range of small sunfish was narrower than larger sunfish (Nakamura et al. 2015), as was also shown for the northeast Atlantic PSAT-tracked sunfish (Sousa et al. 2016a). A possible explanation might be that smaller sunfish are less able to withstand cold water and so are not able to feed on or prey on gelatinous zooplankton, like siphonophores, which inhabit the deeper water layers. In fact, coupling SIA and swimming stroke efficiency, Nakamura and Sato (2014) were able to confirm an ontogenetic shift in the diet, with smaller individuals feeding on benthic prey items whereas larger sunfish appeared more likely to feed on gelatinous prey (Nakamura and Sato 2014). In 2016, sunfish dietary habits were further examined using a molecular barcoding approach to identify consumed prey from stomach contents providing further support for the ontogenetic shift: whereas smaller sunfish consumed teleost fish, Hydrozoa, Malacostraca, Maxillopoda, Bivalvia, Gastropoda and Cephalopoda prey; subadults seemed to be more selective, feeding mainly on Malacostraca, Hydrozoan and Scyphozoa (Sousa et al. 2016b) (see Phillips et al. 2020, Chapter 9 for more details). In a recent review, Hays et al. (2018) suggested a direct link between a predator diet based on gelata and fasting endurance, given the long periods in between jellyfish encounters. Following this logic, it is possible that a sunfish limited fasting endurance dictates its recorded ontogenetic shift in diet, with gelatinous prey only found in adult *M. mola* individuals (Hays et al. 2020, Chapter 15).

Behavioral Thermoregulation

Being an ectothermic species, the body temperature of sunfish can be expected to be affected by ambient water temperature. Ectotherms regulate their body temperature by moving between different temperature environments (e.g., Dubois et al. 2009, Kiefer et al. 2007). Acoustic tracking of sunfish by Cartamil and Lowe (2004) suggested behavioral thermoregulation by showing a significant relationship between the maximum dive depth and the post-dive period spent in the surface mixed layer. Behavioral thermoregulation is commonly seen in many species of fishes, both endothermic (e.g., Pacific Bluefin tuna *Thunnus orientalis*, Kitagawa et al. 2001) and ectothermic (e.g., blue shark *Prionace glauca*; Carey and Scharold 1990), with all of these species alternating between deep excursions into cold water and 'rewarming' periods in the surface mixed layer. Moreover, ocean sunfish are often observed lying motionless at the sea surface, which has been interpreted as a 'basking' behavior (Pope et al. 2010). Although parasite removal by seabirds has been suggested to be the reason for this surfacing behavior (Abe et al. 2012), the permanence at the surface after every excursion to deep, cold water further supports the thermoregulatory hypothesis (Cartamil and Lowe 2004, Dewar et al. 2010).

To investigate putative behavioral thermoregulation behavior directly, Nakamura et al. (2015) measured the body temperature of the sunfish during foraging deep excursions. The body temperature decreased during the time that sunfish stayed in deep, cold water, but rebounded during the period they stayed at the surface after deep excursions. This supports the hypothesis that surfacing after deep dives into colder water is a behavioral thermoregulatory mechanism for recovering body temperature by remaining in warm surface water. In the latter study, sunfish repetitively explored the water column from the surface (16–22°C) to over 100 m depth (minimum 3°C at depth over 200 m) for foraging during the daytime. Before deep excursions, body temperature was measured and found to be similar to ambient water temperatures at the surface (16–21°C) and decreased by up to 6°C during deep excursions (Fig. 4). Importantly, body temperature intervals of the sunfish (12–21°C) were narrower than the range of ambient water temperature (3–22°C), and the minimum body temperature was higher than the minimum ambient water temperature. By estimating the whole-body heat-transfer coefficient using the heat-budget model (body temperature change by gaps between water temperature and body temperature), the authors found a smaller coefficient during cooling compared to the warming phase for sunfish. Importantly, these changes in body temperature were comparable to those of a dead sunfish which showed similar heat-transfer coefficients both during cooling and warming (Nakamura et al. 2015). Results suggest that sunfish might be able to physiologically regulate their heat transfer, which enables them to keep mean body temperature at 1–2°C warmer than average ambient water temperature, without heat production. Similar body temperature change has been observed in other ectothermic fish, such as blue shark, for which the cooling rate was smaller than the warming rate, enabling them to keep mean body temperature 4°C warmer than mean ambient water temperature (Carey and Scharold 1990). This suggests that heat loss might be a physical constraint for sunfish, which increase heat gain during warming. Moreover, the heat-transfer coefficient estimation indicated that larger fish have smaller heat-transfer coefficients during both cooling and warming phases, resulting in similar ratio of heat-transfer coefficients between warming and cooling. Consequently, larger sunfish are probably less likely to lose heat as quickly in cold water but require more time to recover their body temperature than smaller individuals.

Further insights into thermal ecology may also come from further studies into respiration. For example, fish use their gills to transfer oxygen from the surrounding water into their blood, and fish gills have between two and 48 times greater surface area than body surface area (Gray 1954). Gill surface area of ocean sunfish is similar to that of other marine fish with intermediate activity (Adeney and Hughes 1977). Large gill surface area of fish may work as a heat exchanger (Sorenson and Fromm 1976, Stevens and Sutterlin 1976). Changes in perfusion flow through the gill alter heat exchange by the gill (Sorenson and Fromm 1976), suggesting that the physiological regulation of heat is achieved partly by changing the blood flow through the gill. The circulation in the gill of ocean sunfish shows a difference from those of other marine fish in that afferent branchial arteries are double (Adeney and Hughes 1977), which might be related to the physiological regulation of efficient uptake of heat as well as oxygen, but more conclusive measurements on live sunfish are needed.

Advantage of Large Body Size

Another aspect for consideration of foraging ecology in a temperature limited environment is the advantage of large body size (Nakamura et al. 2015). In the latter, cycles of deep foraging and surface rewarming were suggested to be related to optimal foraging behavior of ocean sunfish, and examined the factors underpinning vertical movements in terms of foraging and thermoregulation. Larger body size provides larger thermal inertia and allows them to stay longer at cold depths. In fact, the whole-body heat-transfer coefficients, which is an indicator of the difficulty of changing body temperature, estimated from sunfish with variations in size showed a double logarithm relationship that became smaller as the body mass increased (Nakamura et al. 2015). In the following paragraph, the reason why the large body size increase foraging efficiency is explained.

A foraging-rewarming cycle consists of a deep excursion for foraging and subsequent surfacing for rewarming, and a deep excursion requires travelling between the surface and deep water. It is

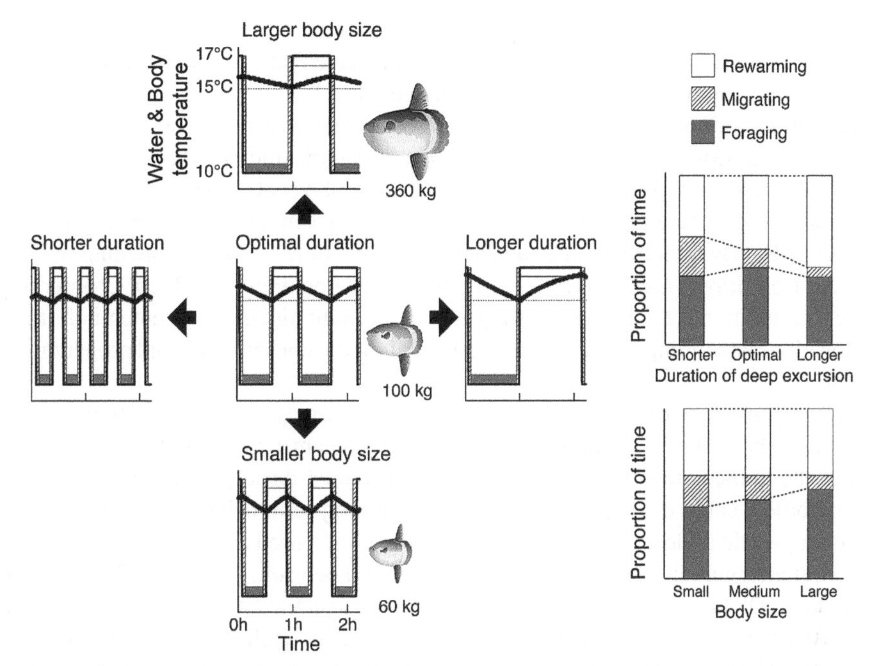

Figure 5. Schematic diagram of foraging/rewarming cycles of sunfish. Comparison of body temperature changes while diving between the sea surface at 17°C and feeding depth at 10°C, when the lower limit of body temperature is 15°C. Note the variation of duration (longer or shorter) of staying at the feeding ground with the body size (smaller or larger). Optimal duration provides maximum foraging time whereas sunfish of larger body size showed longer foraging time.

important for sunfish to increase the time spent in deep water in order to efficiently increase the number of encounters with prey. Changing the duration at the deep foraging area, the fluctuation range of body temperature becomes larger and the rate of time required to rewarm the body will increase as the fish stays longer in deep waters (Fig. 5). Thus, shortening the time of a deep excursion and frequent travelling between the sea surface and the deep water can reduce the body temperature fluctuation and the time required for body temperature recovery. However, frequent back and forth requires more travelling time and potentially decreases foraging time. In summary, a certain size sunfish has an efficient duration of deep excursions, and efficient durations were estimated to be longer in larger individuals because cooling speed was slower with increased body sizes. Nakamura et al. (2015) found that this efficient duration was longer in larger individuals and within ranges of the observed durations of deep excursions. Ultimately, it appeared that ocean sunfish swam between the sea surface and the deep water where their prey exists abundantly while regulating body temperature by vertical movements, perhaps such that the time for staying at a depth where their prey is abundant was longest.

Remaining Questions

Despite the broad coverage of *Mola* spp. movements in the various ocean basins, many questions still remain and further tracking studies are needed. For example, no tracking studies have been conducted in the Indian Ocean, Australia and New Zealand, or off South America, aside from the Galapagos. No one has ever tracked *Ranzania laevis* nor *Mola tecta.* To date, most tracked molid studies have been conducted in coastal waters despite the reported occasional sightings of sunfish species in open waters (e.g., longliners in the North Atlantic G. Mucientes personal communication; and community crowd-sourced data from oceansunfish.org). Are these open-water individuals lonely migrants who've lost their way or are they following major movement corridors we are unaware of given that the open ocean is so inaccessible and much less studied? Would tracking larger individuals or mark and recapture studies conducted in collaboration with either offshore fishermen or any groups that encounter large numbers of sunfish help disentangle this?

Another fruitful area of research would be to link reproduction activities with overall movements. Relatively little is known about molid spawning. In a landmark study off Japan, *M. mola* spawning period was found to extend from August to October (Nakatsubo et al. 2008). In the latter, researchers also found that mature females contained oocytes of distinct maturational stages suggesting that mature individuals spawn multiple times a year. In another study, *M. mola* spawning off Korea was found to occur from July to October (Kang et al. 2015) (see Forsgren et al. 2020, Chapter 6 for more details). Hence, additional long-term tracking of mature individuals could potentially reveal both spawning locations and timings, which will be essential for the conservation and proper management of this species.

Biologging provides crucial insights into otherwise largely inaccessible biological systems (Rutz and Hays 2009). While important species-specific dynamic traits, including spatial behavior in relation to the immediate environment, have been revealed through past tracking efforts, using finer-scale recorders could deepen our understanding of long-term behaviors. As an example, GPS tracking techniques that allow higher resolution tracking, increased accuracy and better correlation with behavioral information (e.g., Sims et al. 2009b) could be useful. On the other hand, electronic data storage tags (DSTs) store high-resolution data to non-volatile flash memory for up to 20 years (Sims et al. 2006), providing near-continuous temperature and depth recordings over entire seasons and years. New insights could also be gleaned by expanding studies using animal borne-cameras, with lights and body temperature probes. Nakamura et al. (2015) found that *M. mola* body temperatures decreased during deep excursions and recovered during surfacing periods and revealed, for the first-time, wild feeding bouts where ingested food could be identified. Repeating this protocol in other regions of the globe and relating findings to environmental factors would shed insight into the adaptive flexibility of this circumglobally group of fishes and their ability to respond to the increasing onslaught of human impacts such as climate pollution and overfishing.

Useful information is to be gleaned from additional simultaneous recording of both sunfish movement and behaviors (fine-scale) along with prey field recordings (previously attempted in Nakamura et al. 2015, Sousa et al. 2016b). Importantly, and for instance in relation to *M. mola*, smaller individuals should also be monitored given the size-related shift in the diet preferences in the species. Within this context, the coordinated inclusion of sunfishes and jelly blooms in statutory aerial survey programs (e.g., seabirds, cetaceans) would be an important step (e.g., Houghton et al. 2006a and b, Breen et al. 2017). Likewise, metanalytical studies of jelly blooms on a global scale could provide a phenomenal resource for understanding Molidae distribution on an oceanic scale, emphasizing why collaborations between gelata and sunfish researchers should always be encouraged.

Acknowledgements

L.L.S thanks Mr. Morikawa, Captain Alfredo and all the crew in M/V *Aragao* and M/V *Guentaru-maru*, Tunipex S.A., for the much-valued field support during the long tagging seasons off Southern Portugal. She also wants to thank Dr. Nuno Queiroz for all the guidance and knowledge transmitted that boosted the North Atlantic study of sunfish. Gonzalo Mucientes is also thanked for his support in the field. Funding for fieldwork in the North Atlantic was provided by the FCT through a Ph.D. Scholarship (SFRH/BD/68717/2010) to L.L.S. Lastly, we gratefully acknowledge funding from Oceanário de Lisboa to purchase two Argos-GPS tags. I.N. thanks the fishermen of the Funakoshi Bay Fisheries Cooperative, Otsuchi Fisheries Cooperative and the crew of the M/V *Koyo-maru* boat for the tagging studies in Japan. He would also like to thank Dr. Dhugal Lindsay for many suggestions for identifying gelata. He also thanks Grant-in-Aid for JSPS Fellows (12J07184). The tagging study in Japan was supported by Bio-Logging Science at the University of Tokyo (UTBLS), the Grants-in-Aid for Scientific Research from JSPS (25660152, 24241001), the Canon Foundation, Project Grand Maillet and the Cooperative Program of Atmosphere and Ocean Research Institute at the University of Tokyo. D.W.S. thanks G. Hays for provision of Fastloc GPS tags and for leading the tag deployments in Ireland. D.W.S. was supported by a Marine Biological Association (MBA) Senior Research Fellowship. Funding for satellite tags, field research and data analyses underpinning our published work presented in this chapter was provided by the UK Natural Environment Research Council (NERC) through the NERC Oceans 2025 Strategic Research Programme theme 6 (Science for Sustainable Marine Resources) in which D.W.S. was a principal investigator. Finally, the authors want to thank I. Potter, M. Nyegaard, T. Thys, G. Hays and C. Chang for sharing the sunfish tracked movements replotted in Fig. 1.

References

Abe, T., K. Sekiguchi, H. Onishi, K. Muramatsu and T. Kamito. 2012. Observations on a school of ocean sunfish and evidence for a symbiotic cleaning association with albatrosses. Mar. Biol. 159(5): 1173–1176.

Adeney, R.J. and G.M. Hughes. 1977. Some observation on the gills of the oceanic sunfish, *Mola mola*. J. Mar. Biol. Assoc. UK 57: 825–837.

Aoki, K., M. Amano, K. Mori, A. Kourogi, T. Kubodera and N. Miyazaki. 2012. Active hunting by deep-diving sperm whales: 3D dive profiles and maneuvers during bursts of speed. Mar. Ecol. Prog. Ser. 444: 289–301.

Arai, M.N. 1997. A functional biology of scyphozoa. London: Chapman and Hall.

Block, B.A., H. Dewar, T. Williams, E.D. Prince, C. Farwell and D. Fudge. 1998. Archival tagging of atlantic bluefin tuna (Thunnus thynnus thynnus). Mar. Technol. Soc. J. 32(1): 37–46.

Boyle, W.A. 2008. Partial migration in birds: tests of three hypotheses in a tropical lekking frugivore. J. Anim. Ecol. 77(6): 1122–1128.

Breen, P., A. Canadas, O.O. Cadhla, M. Mackey, M. Scheidat, S.C.V. Geelhoed et al. 2017. New insights into ocean sunfish (Mola mola) abundance and seasonal distribution in the northeast Atlantic. Sci. Rep. 7(1): 2025.

Carey, F.G. and J.V. Scharold. 1990. Movements of blue sharks (Prionace glauca) in depth and course. Mar. Biol. 106: 329–342.

Cartamil, D.P. and C.G. Lowe. 2004. Diel movement patterns of ocean sunfish Mola mola off southern California. Mar. Ecol. Prog. Ser. 266: 245–253.

Ceh, J., J. Gonzalez, A.S. Pacheco and J.M. Riascos. 2015. The elusive life cycle of scyphozoan jellyfish—metagenesis revisited. Sci. Rep., pp. 1–13. doi: 10.1038/srep12037.

Chapman, B.B., C. Skov, K. Hulthén, J. Brodersen, P.A. Nilsson, L.A. Hansson et al. 2012a. Partial migration in fishes: definitions, methodologies and taxonomic distribution. J. Fish Biol. 81(2): 479–499.

Chapman, B.B., K. Hulthén, J. Brodersen, P.A. Nilsson, C. Skov, L.A. Hansson et al. 2012b. Partial migration in fishes: causes and consequences. J. Fish Biol. 81(2): 456–478.

Ching-Tsun Chang, Shian-Jhong Lin, Wei-Chuan Chiang, Michael K. Musyl, Chi-Hin Lam, Hung-Hung Hsu et al. 2019. Horizontal and vertical movement patterns of sunfish off eastern Taiwan. Deep-Sea Res. Pt. II 104683, ISSN 0967-0645.

Davenport, J. and G.H. Balazs. 1991. 'Fiery bodies': are pyrosomas important items in the diet of leatherback turtles? Herpetol. Bull. 37: 33–38.

Dewar, H., T.M. Thys, S.L.H. Teo, C. Farwell, J. O'Sullivan, T. Tobayama et al. 2010. Satellite tracking the world's largest jelly predator, the ocean sunfish, Mola mola, in the Western Pacific. J. Exp. Mar. Biol. Ecol. 393: 32–42.

Doyle, T.K., J.D. Houghton, R. McDevitt, J. Davenport and G.C. Hays. 2007. The energy density of jellyfish: estimates from bomb-calorimetry and proximate-composition. J. Exp. Mar. Biol. Ecol. 343(2): 239–252.

Dubois, Y., G. Blouin-Demers, B. Shipley and D. Thomas. 2009. Thermoregulation and habitat selection in wood turtles Glyptemys insculpta: chasing the sun slowly. J. Anim. Ecol. 78(5): 1023–1032.

Edwards, A.M., R.A. Phillips, N.W. Watkins, M.P. Freeman, E.J. Murphy, V. Afanasyev et al. 2007. Revisiting Lévy flight search patterns of wandering albatrosses, bumblebees and deer. Nature 449(7165): 1044–1048.

Forsgren, K., R.S. McBride, T. Nakatsubo, T.M. Thys, C.D. Carson, E.K. Tholke et al. 2020. Reproductive biology of the ocean sunfishes. pp. 87–104. *In*: T.M. Thys, G.C. Hays and J.D.R Houghton [eds.]. The Ocean Sunfishes: Evolution, Biology and Conservation, CRC Press. Boca Raton, FL, USA.

Fulling, G.L., D. Fertl, K. Knight and W. Hoggard. 2007. Distribution of Molidae in the northern coast of Mexico. Gulf Caribb. Res. 19(2): 53–67.

Garibaldi, F. 2015. By-catch in the mesopelagic swordfish longline fishery in the Ligurian Sea (Western Mediterranean). Collect. Vol. Sci. Pap. ICCAT 71(3): 1495–1498.

Gray, I. 1954. Comparative study of the gill area of marine fishes. Biol. Bull. 107: 219–225.

Haddock, S.H.D. and J.F. Case. 1999. Bioluminescence spectra of shallow and deep-sea gelatinous zooplankton: ctenophores, medusae and siphonophores. Mar. Biol. 133: 571–582.

Haddock, S.H., C.W. Dunn, P.R. Pugh and C.E. Schnitzler. 2005. Bioluminescent and red-fluorescent lures in a deep-sea siphonophore. Science 309: 263–263.

Harrod, C., J. Syvaranta, L. Kubicek, V. Cappanera and J.D. Houghton. 2013. Reply to Logan & Dodge: 'stable isotopes challenge the perception of ocean sunfish Mola mola as obligate jellyfish predators'. J. Fish Biol. 82(1): 10–16.

Hays, G.C. 2008. Sea turtles: A review of some key recent discoveries and remaining questions. J. Exp. Mar. Biol. Ecol. 356: 1–7.

Hays, G.C., V.J. Hobson, J.D. Metcalfe, D. Righton and D.W. Sims. 2006. Flexible foraging movements of leatherback turtles across the North Atlantic Ocean. Ecology 87(10): 2647–2656.

Hays, G.C., M.R. Farquhar, P. Luschi, S.L.H. Teo and T.M. Thys. 2009. Vertical niche overlap by two ocean giants with similar diets: Ocean sunfish and leatherback turtles. J. Exp. Mar. Biol. Ecol. 370(1-2): 134–143.

Hays, G.C., T. Bastian, T.K. Doyle, S. Fossette, A.C. Gleiss, M.B. Gravenor et al. 2012. High activity and Levy searches: jellyfish can search the water column like fish. Proc. R. Soc. B 279(1728): 465–473.

Hays, G.C. 2015. New insights: animal-borne cameras and accelerometers reveal the secret lives of cryptic species. J. Anim. Ecol. 84(3): 587–589.

Hays, G.C., T.K. Doyle and J.D.R. Houghton. 2018. A Paradigm Shift in the Trophic Importance of Jellyfish? Trends Ecol. Evol. 33: 874–884.

Hays, G.C., J.D.R. Houghton, T.M. Thys, D.H. Adams, A.E. Ahuir-Baraja, J. Alvarez et al. 2020. Unresolved questions about ocean sunfishes, Molidae–a family comprising some of the world's largest teleosts. pp. 280–296. *In*: T.M. Thys, G.C. Hays and J.D.R. Houghton [eds.]. The Ocean Sunfishes: Evolution, Biology and Conservation, CRC Press. Boca Raton, FL, USA.

Holland, K.N., S.M. Kajiura, D.G. Itano and J.R. Sibert. 2001. Tagging techniques can elucidate the biology and exploitation of aggregated pelagic species. Am. Fish. Soc. Symp. 25: 211–218.

Hooker, S., M. Biuw, B. McConnell, P. Miller and C. Sparling. 2007. Bio-logging science: Logging and relaying physical and biological data using animal-attached tags. Deep-Sea Res. Pt. II 54(3-4): 177–182.

Houghton, J., T.K. Doyle, J. Davenport, G.C. Hays, J. Houghton, T.K. Doyle et al. 2006a. The ocean sunfish Mola mola: insights into distribution, abundance and behaviour in the Irish and Celtic Seas. J. Mar. Biol. Assoc. UK 86(5): 1237–1243.

Houghton, J.D., T.K. Doyle, M.W. Wilson, J. Davenport and G.C. Hays. 2006b. Jellyfish aggregations and leatherback turtle foraging patterns in a temperate coastal environment. Ecology 87(8): 1967–1972.

Houghton, J., T. Doyle, J. Davenport, M. Lilley, R. Wilson and G. Hays. 2007. Stranding events provide indirect insights into the seasonality and persistence of jellyfish medusae (Cnidaria : Scyphozoa). Hydrobiologia 589: 1–13.

Houghton, J.D.R., N. Liebsch, T.K. Doyle, A.C. Gleiss, M.K.S. Lilley, R.P. Wilson et al. 2009. Harnessing the Sun: testing a novel attachment method to record fine scale movements in ocean sunfish (Mola mola). pp. 229–242. *In*: J.L. Nielsen, H. Arriza-balaga, N. Fragoso, A. Hobday, M. Lutcavage and J. Sibert [eds.]. Tagging and tracking of marine animals with electronic devices, vol 9. Springer, Dordrecht, doi:10.1007/978-1-4020-9640-214.

Humphries, N.E., N. Queiroz, J.R.M. Dyer, N.G. Pade, M.K. Musyl, K.M. Schaefer et al. 2010. Environmental context explains Lévy and Brownian movement patterns of marine predators. Nature 465(7301): 1066–1069.

Humphries, N.E., H. Weimerskirch, N. Queiroz, E.J. Southall and D.W. Sims. 2012. Foraging success of biological Levy flights recorded *in situ*. Proc. Natl. Acad. Sci. USA 109(19): 7169–7174.

Humphries, N.E. and D.W. Sims. 2014. Optimal foraging strategies: Levy walks balance searching and patch exploitation under a very broad range of conditions. J. Theor. Biol. 358: 179–193.

Humphries, N.E., K.M. Schaefer, D.W. Fuller, G.E.M. Phillips, C. Wilding and D.W. Sims. 2016. Scale-dependent to scale-free: daily behavioural switching and optimized searching in a marine predator. Anim. Behav. 113: 189–201.

Ichikawa, T., K. Segawa and M. Terazaki. 2006. Estimation of Cnidaria and Ctenophora biomass and vertical distribution using the Video Plankton Recorder II (VPRII) in the meso- and epipelagic layers of the Oyashio and Transition zone off eastern Japan. Bull. Jpn. Soc. Fish Oceanogr. 70: 240–248.

Kang, M.J., H.J. Baek, D.W. Lee and J.H. Choi. 2015. Sexual maturity and spawning of ocean sunfish Mola mola in Korean Waters. Korean J. Fish. Aquat. Sci. 48(5): 739–744.

Kiefer, M.C., M. Van Sluys and C.F. Rocha. 2007. Thermoregulatory behaviour in Tropidurus torquatus (Squamata, Tropiduridae) from Brazilian coastal populations: an estimate of passive and active thermoregulation in lizards. Acta Zool. 88(1): 81–87.

Kino, M., T. Miayzaki, T. Iwami and J. Kohbara. 2009. Retinal topography of ganglion cells in immature ocean sunfish, *Mola mola*. Environ. Biol. Fishes 85: 33–38.

Kitagawa, T., H. Nakata, S. Kimura and S. Tsuji. 2001. Thermoconservation mechanisms inferred from peritoneal cavity temperature in free-swimming Pacific bluefin tuna Thunnus thynnus orientalis. Mar. Ecol. Prog. Ser. 220: 253–263.

Kohler, N.E. and P.A. Turner. 2001. Shark tagging: a review of conventional methods and studies. Environ. Biol. Fishes 60: 191–223.

Lucas, C.H., D. Jones, C. Hollyhead, R. Condon, C. Duarte, W. Graham et al. 2014. Gelatinous zooplankton biomass in the global oceans: Geographic variation and environmental drivers. Glob. Ecol. Biogeogr. 23: 701–714.

Nakamura, I. and K. Sato. 2014. Ontogenetic shift in foraging habit of ocean sunfish Mola mola from dietary and behavioral studies. Mar. Biol. 161(6): 1263–1273.

Nakamura, I., Y. Goto and K. Sato. 2015. Ocean sunfish rewarm at the surface after deep excursions to forage for siphonophores. J. Anim. Ecol. 84: 590–603.

Nakano, H. and J.D. Stevens. 2008. The biology and ecology of the blue shark, Prionace glauca. Sharks of the open ocean: biology, fisheries and conservation. M.D. Camhi, E.K. Pikitch and E.A. Babcock. Oxford, Blackwell Publishing: 140–151.

Nakatsubo, T., M. Kawachi, N. Mano and H. Hirose. 2007. Estimation of maturation in wild and captive ocean sunfish Mola mola. Aquacult. Sci. 55(2): 259–264.

Nakatsubo, T., M. Kawachi, N. Mano and H. Hirose. 2008. Spawning period of ocean sunfish Mola mola in waters of the eastern Kanto region, Japan. Aquacult. Sci. 55(4): 613–618.

Nyegaard, M. 2018. There be giants! The importance of taxonomic clarity of the large ocean sunfishes (genus Mola, Family Molidae) for assessing sunfish vulnerability to anthropogenic pressures. Doctor of Philosophy thesis, Murdoch University.

Papastamatiou, Y.P., C.G. Meyer, F. Carvalho, J.J. Dale, M.R. Hutchinson and K.N. Holland. 2013. Telemetry and random walk models reveal complex patterns of partial migration in a large marine predator. Ecology 94(11): 2595–606.

Petersen, S. 2005. Initial bycatch assessment: South Africa's domestic pelagic longline fishery, 2000–2003. BirdLife International (South Africa) Report. B. S. A. Seabird Conservation Programme. Phuket, Percy FitzPatrick Institute, University of Cape Town.

Phillips, N.D., C. Harrod, A.R. Gates, T.M. Thys and J.D. Houghton. 2015. Seeking the sun in deep, dark places: mesopelagic sightings of ocean sunfishes (Molidae). J. Fish Biol. 87(4): 1118–1126.

Phillips, N.D., N. Reid, T. Thys, C. Harrod, N.L. Payne, C.A. Morgan et al. 2017. Applying species distribution modelling to a data poor, pelagic fish complex: the ocean sunfishes. J. Biogeogr. 44(10): 2176–2187.

Phillips, N.D., L. Kubicek, N. Payne, C. Harrod, L. Eagling, C. Carson et al. 2018. Isometric growth in the world's largest bony fishes (Genus Mola)? Morphological insights from fisheries bycatch data. J. Morphol. 279(9): 1312–1320.

Phillips, N.D., E.A. Elliott Smith, S.D. Newsome, J.D.R. Houghton, C.D. Carson, J.C. Mangel et al. 2020. Bulk tissue and amino acid stable isotope analyses reveal global ontogenetic patterns in ocean sunfish trophic ecology and habitat use. Mar. Ecol. Prog. Ser. 633: 127–140.

Phillips, N.D., E.C. Pope, C. Harrod and J.D.R. Houghton. 2020. The diet and trophic role of ocean sunfishes. pp. 146–159. *In*: T.M. Thys, G.C. Hays and J.D.R. Houghton [eds.]. The Ocean Sunfishes: Evolution, Biology and Conservation, CRC Press. Boca Raton, FL, USA.

Pope, E.C., G.C. Hays, T.M. Thys, T.K. Doyle, D.W. Sims, N. Queiroz et al. 2010. The biology and ecology of the ocean sunfish Mola mola: a review of current knowledge and future research perspectives. Rev. Fish Biol. Fish. 20(4): 471–487.

Potter, I.F., B. Galuardi and W.H. Howell. 2010. Horizontal movement of ocean sunfish, Mola mola, in the northwest Atlantic. Mar. Biol. 158(3): 531–540.

Potter, I.F. and W.H. Howell. 2011. Vertical movement and behavior of the ocean sunfish, Mola mola, in the northwest Atlantic. J. Exp. Mar. Biol. Ecol. 396(2): 138–146.

Pratt, H.L. and J.G. Casey. 1983. Age and growth of the shortfin mako, Isurus oxyrinchus, using four methods. Can. J. Fish. Aquat. Sci. 40: 1944–1957.

Priede, I.G. 1984. A Basking shark (Cetorhinus maximus) tracked by satellite together with simultaneous remote sensing. Fish. Res. 2: 201–216.

Queiroz, N., F.P. Lima, A. Maia, P.A. Ribeiro, J.P.S. Correia and A.M. Santos. 2005. Movements of blue shark, Prionace glauca, in the north-east Atlantic based on mark-recapture data. J. Mar. Biol. Assoc. UK 85: 1107–1112.

Rutz, C. and G.C. Hays. 2009. New frontiers in biologging science. Biol. Lett. 5: 289–292.

Seitz, A.C., K.C. Weng, A.M. Boustany and B.A. Block. 2002. Behaviour of a sharptail mola in the Gulf of Mexico. J. Fish Biol. 60(6): 1597–1602.

Silvani, L., M. Gazo and A. Aguilar. 1999. Spanish driftnet and incidental catches in the western Mediterranean. Biol. Conserv. 90: 79–85.

Sims, D.W. and E.J. Southall. 2002. Occurrence of ocean sunfish, Mola mola near fronts in the western English Channel. J. Mar. Biol. Assoc. UK 82(5): 927–928.

Sims, D.W., M.J. Witt, A.J. Richardson, E.J. Southall and J.D. Metcalfe. 2006. Encounter success of free-ranging marine predator movements across a dynamic prey landscape. Proc. R. Soc. B 273(1591): 1195–1201.

Sims, D.W., E.J. Southall, N.E. Humphries, G.C. Hays, C.J. Bradshaw, J.W. Pitchford et al. 2008. Scaling laws of marine predator search behaviour. Nature 451(7182): 1098–1102.

Sims, D.W., N. Queiroz, T.K. Doyle, J.D.R. Houghton and G.C. Hays. 2009a. Satellite tracking of the world's largest bony fish, the ocean sunfish (*Mola mola* L.) in the North East Atlantic. J. Exp. Mar. Biol. Ecol. 370(1-2): 127–133.

Sims, D.W., N. Queiroz, N.E. Humphries, F.P. Lima and G.C. Hays. 2009b. Long-Term GPS tracking of Ocean sunfish Mola mola offers a new direction in fish monitoring. PLoS One 4(10): e7351.

Sims, D.W. 2015. Intrinsic Lévy behaviour in organisms–searching for a mechanism: Comment on Liberating Lévy walk research from the shackles of optimal foraging by A.M. Reynolds. Phys. Life Rev. 14: 111–114.

Sorenson, P.R. and P.O. Fromm. 1976. Heat transfer characteristics of isolated-perfused gills of rainbow trout. J. Comp. Physiol. B 112(3): 345–357.

Sousa, L.L. 2016. Behaviour, predator-prey and fisheries interactions of the Ocean sunfish (Mola mola) in the north-east Atlantic. Doctor in Philosophy, University of Southampton, eprints.soton.ac.uk/397413/.

Sousa, L.L., N. Queiroz, G. Mucientes, N.E. Humphries and D.W. Sims. 2016a. Environmental influence on the seasonal movements of satellite-tracked ocean sunfish Mola mola in the north-east Atlantic. Anim. Biotelem. 4(7): 1–19.

Sousa, L.L., R. Xavier, V. Costa, N.E. Humphries, C. Trueman, R. Rosa et al. 2016b. DNA barcoding identifies a cosmopolitan diet in the ocean sunfish. Sci. Rep. 6: 28762.

Sousa, L.L., F. Lopez-Castejon, J. Gilabert, P. Relvas, A. Couto, N. Queiroz et al. 2016c. Integrated monitoring of Mola mola behaviour in space and time. PLoS One 11(8): e0160404.

Southall, E.J., D. Sims, J.D. Metcalfe, J. Doyle, S. Fanshawe, C. Lacey et al. 2005. Spatial distribution patterns of basking sharks on the European shelf: preliminary comparison of satellite-tag geolocation, survey and public sightings data. J. Mar. Biol. Assoc. UK 85(05): 1083.

Stevens, E.D. and A.M. Sutterlin. 1976. Heat transfer between fish and ambient water. J. Exp. Biol. 65(1): 131–145.

Syvaranta, J., C. Harrod, L. Kubicek, V. Cappanera and J.D. Houghton. 2012. Stable isotopes challenge the perception of ocean sunfish Mola mola as obligate jellyfish predators. J. Fish Biol. 80(1): 225–231.

Thys, T. 2003. Tracking Ocean Sunfish, Mola mola with Pop-Up Satellite Archival Tags in California Waters. https://www.oceansunfish.org/MontereySanctuaryv11.html. Retrieved Feb 29 2020.

Thys, T. and R. Williams. 2013. Ocean sunfish in Canadian Pacific waters: Summer hotspot for a jelly-eating giant? OCEANS - San Diego, San Diego, CA: 1–5.

Thys, T., J.P. Ryan, K.C. Weng, M. Erdmann and J. Tresnati. 2016. Tracking a marine ecotourism star: movements of the short ocean sunfish Mola ramsayi in Nusa Penida, Bali, Indonesia. J. Mar. Biol. 2016: 1–6.

Thys, T.M., J.P. Ryan, H. Dewar, C.R. Perle, K. Lyons, J. O'Sullivan et al. 2015. Ecology of the ocean sunfish, Mola mola, in the southern California Current System. J. Exp. Mar. Biol. Ecol. 471: 64–76.

Thys, T.M., A.R. Hearn, K.C. Weng, J.P. Ryan and C. Peñaherrera-Palma. 2017. Satellite tracking and site fidelity of short ocean sunfish, Mola ramsayi, in the Galapagos Islands. J. Mar. Biol. 2017: 1–10.

Ugland, K.I., D.L. Aksnes, T.A. Klevjer, J. Titelman and S. Kaartvedt. 2014. Lévy night flights by the jellyfish Periphylla periphylla. Mar. Ecol. Prog. Ser. 513: 121–130.

Vereshchaka, A.L. and G.M. Vinogradov. 1999. Visual observations of the vertical distribution of plankton throughout the water column above Broken Spur vent field, Mid-Atlantic Ridge. Deep-sea Res. Pt I. 46: 1615–1632.

Vinogradov, G.M. 2005. Vertical distribution of macroplankton at the Charlie-Gibbs Fracture Zone (North Atlantic), as observed from the manned submersible "Mir-1". Mar. Biol. 146: 325–331.

Viswanathan, G.M. 1996. Levy flight search patterns of wandering albatrosses. Nature 381: 413–415.

Viswanathan, G.M., V. Afanasyev, S.V. Buldyrev, S. Havlin, M.G.E. da Luz, E.P. Raposo et al. 2000. Lévy flights in random searches. Physica A 282: 1–12.

Viswanathan, G.M., E.P. Raposo and M.G.E. da Luz. 2008. Lévy flights and superdiffusion in the context of biological encounters and random searches. Phys. Life Rev. 5: 133–150.

Watanabe, Y. and K. Sato. 2008. Functional Dorsoventral Symmetry in Relation to Lift-Based Swimming in the Ocean Sunfish Mola mola. PLoS ONE 3(10): e3446.

Watanabe, Y.Y. and J. Davenport. 2020. Locomotory systems and biomechanics of ocean sunfish. pp. 72–86. *In*: T.M. Thys, G.C. Hays and J.D.R. Houghton [eds.]. The Ocean Sunfishes: Evolution, Biology and Conservation, CRC Press. Boca Raton, FL, USA.

Witt, M.J., A. Broderick, D. Johns, C. Martin, R. Penrose, M. Hoogmoed et al. 2006. Prey landscapes help identify potential foraging habitat for leatherback turtles in the NE Atlantic. Mar. Ecol. Prog. Ser. 337: 231–243.

Witt, M.J., A.C. Broderick, D.J. Johns, C. Martin, R. Penrose, M.S. Hoogmoed et al. 2007. Prey landscapes help identify potential foraging habitats for leatherback turtles in the NE Atlantic. Mar. Ecol. Prog. Ser. 337: 231–244.

Chapter 9

The Diet and Trophic Role of Ocean Sunfishes

Natasha D. Phillips,[1,*] *Edward C. Pope*,[2] *Chris Harrod*[3] and *Jonathan D.R. Houghton*[4]

Introduction

Elucidating animal diets provides vital insight into their trophic position and role in ecosystem functioning. More precisely, understanding predator-prey interactions is an important step underpinning effective management and conservation efforts (McCauley et al. 2015). In 2010, an extensive review of the biology and ecology of ocean sunfishes (Pope et al. 2010), encouraged researchers to resolve long standing but fundamental questions regarding sunfish diet following the unearthing of overlooked studies that challenged the long-held idea of predation on gelatinous prey alone (e.g., Schmidt 1921, Norman and Fraser 1949, Fraser-Brunner 1951). A decade later, it seems timely to consider how our understanding has progressed, set against a backdrop of anthropogenic pressures that have reduced marine predator biomass by 90 percent over the last 50 years (Myers and Worm 2003, Bearzi et al. 2006, Young et al. 2014).

The most immediate challenge remains the translation of empirical field data into a format that can be used more effectively by fisheries and ecosystem modellers. Food web models for fishes are typically based on body size or mass, which is then used to investigate food chain coupling, prey size selection, predict productivity or fisheries impacts (Jennings et al. 2017). Such reliance on body mass as a predictor is mostly appropriate for the marine environment as the life histories of fishes are often not well understood, particularly in deep water or open ocean environments (Cushing 1975, Jennings et al. 2017). Moreover, marine food webs are strongly size structured with trophic roles driven as much by size as by taxonomy (Cushing 1975, Jennings et al. 2017). Whilst mass/length relationships have recently been published for ocean sunfishes (Phillips et al. 2018), linking these relationships to ontogenetic shifts in diet or trophic level presents a challenge beyond localized scales. Drawing inferences from closely related species as a proxy is unrealistic given that the sunfishes' closest relatives in the Tetraodontiformes include the families Tetraodontidae (pufferfishes) and Diodontidae (porcupine fishes), which are an order of magnitude smaller, and inhabit shallow water reef environments (Santini et al. 2013). More importantly, the assumption commonly used in these

[1] Queen's Marine Laboratory, 12-13 The Strand, Portaferry, BT22 1PF.

[2] Swansea University, Singleton Park, Swansea SA2 8PP.

[3] Instituto de Ciencias Naturales Alexander von Humboldt, Universidad de Antofagasta, Avenida Angamos 601, Antofagasta, Chile; Instituto Antofagasta, Universidad de Antofagasta, Avenida Angamos 601, Antofagasta, Chile.

[4] School of Biological Sciences, 19 Chlorine Gardens, Belfast, BT9 5DL.

* Corresponding author: nphillips01@qub.ac.uk

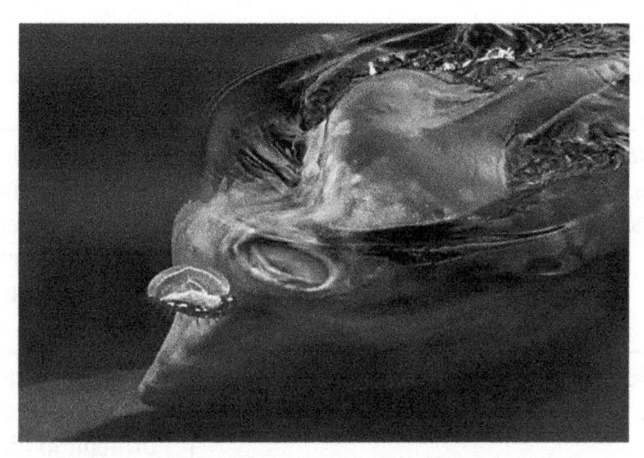

Figure 1. Ocean sunfish feeding on *Velella velella.* Photo credit, copyright and permission granted by Jodi Frediani.

models, that trophic level increases with body size, does not appear to hold true for ocean sunfishes (Phillips et al. 2020) see Fig. 1. In this review we explore the implications of these recent findings for sunfish ecology, and discuss how we might address the host of remaining questions.

Evidence of a Mixed Diet

The ocean sunfishes are highly charismatic and easily recognized across their global range. Despite their attributes, these fishes have frequently been cast in literature and popular media as inactive drifters of little ecological importance (see Pope et al. 2010). This narrow view of ocean sunfish ecology stems from their historical profiling as obligate 'gelativores', and an assumed peripheral role within marine food webs. However, descriptions of non-gelatinous prey items (identified from gut content analyses) have been noted from sunfish dissection studies since the 1890s, including littoral algae (Reuvens 1897, Barnard 1927), seagrass (Reuvens 1897) crustaceans, ophiuroids, molluscs, hydroids, (Schmidt 1921) and teleosts (Reuvens 1897, Schmidt 1921, Fraser-Brunner 1951). More recently, a pivotal study by Syväranta et al. (2012) reopened the discussion surrounding sunfish diet, using stable isotope analysis to support the view that Mediterranean sunfish consumed both gelatinous and neritic prey (Syväranta et al. 2012, Harrod et al. 2013). Further studies to assess sunfish diet in more detail rapidly followed using a variety of techniques (see Table 1 for overview) including gut content analysis, stable isotopes, DNA barcoding and animal-borne cameras (e.g., Harrod et al. 2013, Nakamura and Sato 2014, Sousa et al. 2016, Phillips et al. 2020). Taken together, this growing body of evidence demonstrates the gelatinous diet of smaller sunfishes (~ 1 m total length), is frequently supplemented by a wide range of invertebrate and vertebrate prey items (Harrod et al. 2013, Nakamura and Sato 2014, Sousa et al. 2016, Phillips et al. 2020; Figs. 2 and 3).

Trophic Role

Insights from historical dietary records and recent biomolecular analyses point unequivocally towards sunfish occupying a broader trophic role than previously realized (e.g., Sousa et al. 2016, Phillips et al. 2020; see Table 2 of known prey items).

More specifically, it appears that smaller sunfish (approx. < 1 m TL) adopt a mixed diet of benthic and pelagic prey (Syväranta et al. 2012, Nakamura and Sato 2014, Sousa et al. 2016, Phillips et al. 2020; Fig. 3). These smaller individuals often school seasonally in large numbers in shallow, coastal regions feeding widely within inshore food webs (e.g., Fraser-Brunner 1951, Syväranta et al. 2012, Harrod et al. 2013, Nakamura and Sato 2014, Sousa et al. 2016), with a substantial proportion of their diet comprised of benthic prey (see Fig. 2). However, as ocean sunfish (*Mola* spp.) grow larger, individuals appear to become more solitary,and are more commonly encountered offshore where their diet appears to shift towards primarily gelatinous zooplankton (e.g., Fraser-Brunner 1951, MacGinitie and MacGinitie 1968, Hooper et al. 1973, Nakamura and Sato 2014; Fig. 3).

Table 1. Pros and cons of key techniques used to investigate fish diet.

Method	Pros	Cons
Direct observation (inc. cameras)	• Provides a snapshot of consumed prey • Simple to conduct • Does not require dead specimens	• Restrictive (location/duration) • Identifies consumed, not assimilated prey • Prey identification difficult • Need to recover camera • Potential disturbance of normal behavior due to carrying a camera
Gut content and fecal analysis	• Provides a snapshot of consumed prey • Simple to conduct	• Typically requires dead specimens • Fragmented and partially digested prey difficult to identify • Identifies consumed not assimilated prey • Misrepresentation of different prey items, especially under representation of gelatinous animals
DNA of gut contents	• Provides a snapshot of consumed prey • Can identify fragmented or partially digested prey	• Requires dead specimens or difficult to collect fecal matter • Requires specialist equipment • Requires barcodes for prey taxa
Stable isotope analysis	• Identifies and quantifies consumed and assimilated prey from tissues • Does not require dead specimens • Can assess prey over differing time scales (i.e., recent days: blood, weeks: skin, months/years: bone)	• Requires specialist equipment • Longer time frame • Generally cannot identify prey to species
Inter-mandibular angle sensor (IMASEN; jaw sensor)	• Infers feeding • Does not require dead specimens	• Cannot identify or quantify prey type • Potential for over-estimation of prey encounters
Accelerometery	• Infers feeding or prey encounter events (deceleration and behavior) • Illustrates foraging behaviors • Does not require dead specimens	• Cannot identify or quantify prey type • Potential for misidentification of prey encounters
Dental wear patterns	• Infers feeding • May provide long term record of diet (see Bemis et al. 2020 [Chapter 4])	• Cannot identify or quantify prey type • Requires dead specimens • Cannot assess feeding on soft-bodied prey
Parasitology	• Studying parasites can elucidate predator/prey dynamics (for more detail, see Ahuir-Baraja 2020 [Chapter 10])	• Cannot identify or quantify prey type • Requires expertise • May require dead specimens
Fatty acid analysis	• Can be used as biochemical markers • Quantify broad seasonal changes • Assess dietary changes between broadly defined prey groups	• Turnover rate poorly known • Can be modified during digestion, deposition and metabolism • Cannot identify prey to species level

This apparent ontogenetic shift from a mixed, high-energy diet to predominantly low-energy density, gelatinous prey (Doyle et al. 2007, Phillips et al. 2020) initially posed an ecological conundrum for scientists as typically (although not in all cases), when body size increases, an animal's mouth gape and prey handling capabilities also increase which enables the capture and consumption of larger and more valuable prey (Werner and Gillian 1984). Although it is well known that mass specific metabolic rates decline with body size (e.g., Peters 1983, Freedman and Noakes 2002), the largest sunfishes' reliance on gelatinous prey appeared an unusual choice to meet the energetic demands of such a large fish. When we consider the mass-specific metabolic cost however, it appears that larger sunfishes may become more energetically efficient (Phillips et al. unpublished data) which may explain some of the sunfishes' apparent shift towards an increasingly gelatinous diet. However,

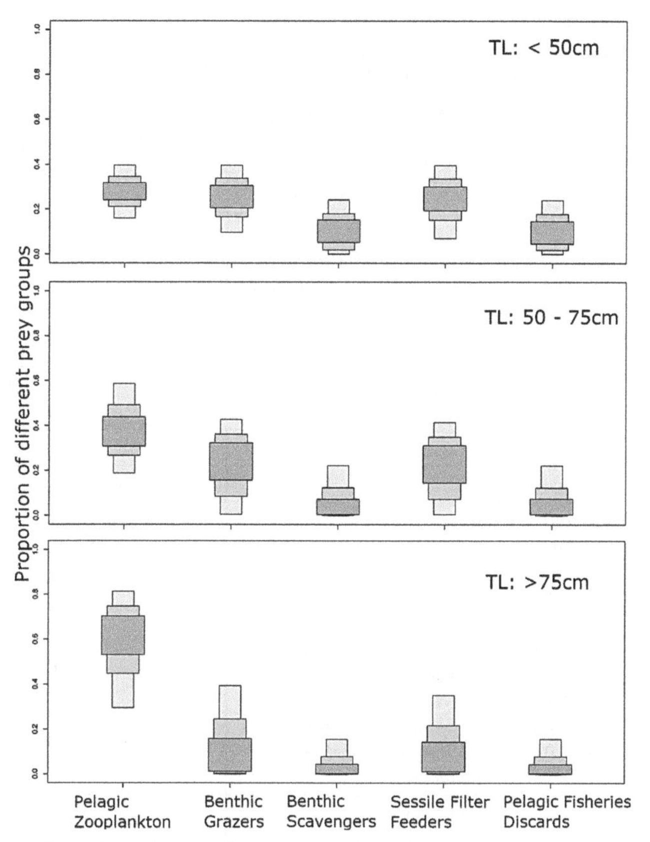

Figure 2. Estimated proportions of putative prey for ocean sunfish (*Mola* spp.) using stable isotope analysis (SIA) of tissue samples. Here SIAR mixing models (package: Stable Isotope Analysis in R) were used to estimate the diet of ocean sunfish between distinct size classes; (a) < 50 cm, b) 50 > 75 cm and (c) > 75 cm TL. The dietary proportions indicate the credibility intervals at 25, 75 and 95%. Putative prey categorized into the following functional groups: Pelagic Zooplankton, Benthic Grazers, Benthic Scavengers, Sessile Filter Feeders and Pelagic Fisheries Discards. Reproduced from Phillips et al. (2020).

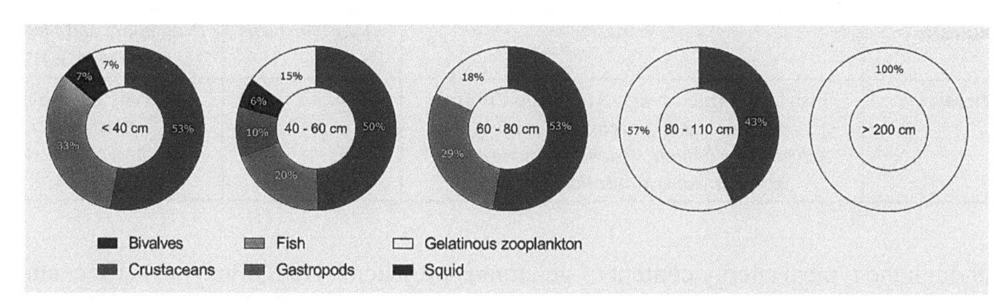

Figure 3. Smaller sunfish consume a mixed diet but display an ontogenetic shift towards the consumption of gelatinous animals as they grow. Percentage occurrence of different taxa in the stomachs of sunfish belonging to different size classes. Diet for animals < 110 cm from genetic analysis (NGS (Next Generation Sequencing) and cloning techniques) of digestive tracts of *M. mola* caught off southern Portugal (data from Sousa et al. 2016; 57 sunfish digestive tracts amplified and 33 successful sequences for identification *via* cloning, 36 samples successfully amplified for NGS). The dietary composition of animals > 200 cm comes from direct identification of stomach contents of adult ocean sunfish caught around Otsuchi Bay, Japan (data from Nakamura and Sato 2014; N = 2).

new research also suggests that many marine species, including birds, marine reptiles and many groups of fishes, incorporate gelatinous taxa in their diets and it appears that the nutritional value of gelatinous zooplankton has been significantly underestimated (e.g., Drazen and Sutton 2016, Briz et al. 2017, Phillips et al. 2017, Hays et al. 2018).

Table 2. Known Molidae prey items.

Prey taxa	Specific examples (species where known)	Molidae genus	References
Algae	Littoral algae; *Fucus vesiculosus*, Phaeophyceae filamentous macro-algae (*Ectocarpus* sp.), Bacillariophyceae epiphytic diatoms	*Ranzania, Mola, Masturus*	(Plancus 1746, Reuvens 1897, Barnard 1927, Bakenhaster 2016)
Seagrass	*Posidonia sinuosa, P. australis, Halodule wrightii, Zostera marina*	*Ranzania, Mola, Masturus*	(Reuvens 1897)
Cnidaria	Scyphozoa (*Aurelia* sp., *Cyanea capillata, Aequorea coerulescens, Chrysaora melanaster, Pelagia noctiluca*), hydrozoa (*Physalia physalis, Velella velella, Rosacea cymbiformis*), siphonophora (*Praya* sp., *Nanomia* sp., *Apolemia* sp., *Physophora hydrostatica, Sulculeolaria quadrivalvis*)	*Mola, Masturus*	(Schmidt 1921, Norman and Fraser 1949, MacGinitie and MacGinitie 1968, Nakamura and Sato 2014, Nakamura et al. 2015, Bakenhaster 2016, Sousa et al. 2016)
Ctenophora	*Beroe* sp., *Cydippida* sp.	*Mola*	(Nakamura et al. 2015)
Annelida	Benthic annelids	*Masturus*	(Yabe 1950)
Crustacea	Ostracods, calanoid and cyclopoid copepods (*Pleuromamma gracilis, Candacia* sp., *Oncacea* sp., *Candacia* sp.), shrimp (*Palaemonetes vulgaris, Pasiphaea sivado, Funchalia villosa, Thysanoessa gregaria, Jaxea nocturna, Meganyctiphanes norvegica*), crab (*Plagusia chabrus, Polybius henslowii, Liocarcinus holsatus, Goneplax rhomboides*), hyperiid amphipods (*Corophiidae* sp., *Phrosina semilunata*), sea lice (*Argulidae, Lepeophtheirus pollachiu*)	*Mola, Masturus, Ranzania*	(Donovan 1803, Schmidt 1921, Fitch 1969, Robison 1975, Nakamura and Sato 2014, Bakenhaster 2016, Sousa et al. 2016)
Echinodermata	Brittle stars	*Mola*	(Norman and Fraser 1949, Fraser-Brunner 1951)
Mollusca	Squid (*Ommastrephidae, Octopoteuthis* sp.), pteropods (*Clio pyramidata*), top snail, moon snail (*Naticidae* sp.) limpet, mussel (*Mytilus galloprovincialis*), oyster	*Mola, Masturus*	(Schmidt 1921, Sousa et al. 2016, Nyegaard et al. 2017)
Porifera	Triaxon sponge spicules	*Masturus*	(Yabe 1950)
Tunicata	Salps	*Mola, Masturus*	(Nakamura and Sato 2014, Briz et al. 2017)
Teleostei	Fish larvae (incl. *conger* sp.), anguilliforme sp., *Pleuronectes* sp., *Molva macrophthalma, Hygophum benoiti, Trachurus picturatus, Sparus aurata, Maurolicus muelleri*	*Ranzania, Mola*	(Reuvens 1897, Schmidt 1921, Fraser-Brunner 1951, Sousa et al. 2016)

Although the typical energy content of gelatinous prey items is relatively low (for example,the lion's mane, *Cyanea capillata*, has a gross energy density of 4.22 ± 0.98 kJ/g wet weight; mean, ± 1 standard deviation; Doyle et al. 2007), recent studies have shown that selective feeding in dense jelly blooms allows predators to exploit a highly abundant source of easily digestible prey (Davenport and Balazs 1991, Clarke et al. 1992, Doyle et al. 2007, Nakamura et al. 2015). Also, feeding on specific parts of gelatinous organisms, such as the energy-rich gonads and oral arms of scyphozoans (Doyle et al. 2007), provides a comparable energy gain to that from bait fish (Mourocq and Ridoux 2010). Although estimates for sunfish calorific intake are not yet available, the gonadal tissues of the lion's mane jelly have an energy density of 7.55 ± 1.88 kJ/g (mean ± 1 SD) wet weight and the oral arms, 7.98 ± 1.65 kJ/g (mean ± 1 SD) wet weight (Doyle et al. 2007); similar to the mean calorific value of whole fishes such as Atlantic mackerel, *Scomber scombrus* (7.9 ± 1.4 kJ/g mean ± 1 SD wet weight) and the European pilchard, *Sardina pilchardus* (8.7 ± 2.6 kJ/g mean ± 1 SD wet weight; Mourocq and Ridoux 2010). Famously, consumption of only gelatinous zooplankton is even able to provide sufficient calories for the giant leatherback turtle, *Dermochelys coricacea* (review by Hays 2008, Perrault 2014) that grow to a weight of > 900 kg (McClain et al. 2015).

It should also be noted that a gelatinous diet is not simply restricted to typical scyphozoa, with sunfishes consuming a broad variety of gelata including salps, ctenophores, siphonophores and pyrosomes (e.g., Potter and Howell 2011, Thys et al. 2015, Cardona et al. 2012, Nakamura et al. 2015). The total biomass of gelatinous animals in the mixed layer of the ocean has been estimated as 38.3 million tonnes (or Tg C; equivalent to 0.53 mg C m^3; Lucas et al. 2014), offering an abundant, widely available food source. Under ideal bloom conditions, such as those exploited by the 'boom and bust' feeding patterns of leatherback turtles, conditions can occur where a 300 kg turtle can reach its daily energetic requirements feeding on gelata weighing 4 g each, in less than four hours (Fossette et al. 2012). Consumption of gelata also appears to be linked to ectothermic animals because their low metabolic rates permit slow use of energy reserves and allow long fasting periods (review by Hays 2008; Hays et al. 2018; also see Hays et al. 2020 [Chapter 15]). The pelagic wandering behavior interspaced with deep water dives exhibited by large ocean sunfish is reminiscent of the behavior of leatherback turtles (Hays et al. 2009), suggesting a commonality in pelagic foraging strategies. However, whilst leatherbacks are obligate gelativores throughout their life history (Arai 2005, Houghton et al. 2006, Hays et al. 2009, Young et al. 2014), the sunfish's intriguing shift from higher to lower energy density prey as they grow raises some interesting questions. Quite simply, if larger sunfishes can survive on this broadly available, easy to capture prey, then why do the smaller sunfishes choose to feed on a mixed diet in coastal waters which presumably present stronger competition for resources?

Ontogenetic Shifts

Although much of the ocean sunfishes early life history (spawning grounds and larval ecology), is still unclear (see review Pope et al. 2010, Thys et al. 2020 [Chapter 7]), we do know that ocean sunfish can grow from a tiny larvae (~ 7 mm) to a 3.3 m long pelagic fish, a shift associated with a 10^8 fold increase in mass (e.g., Gudger 1936, Rowan 2006, Phillips et al. 2018). Smaller sunfish (~ 20–100 cm) are more commonly observed in nearshore environments; showing a possible bias towards land-based sightings but also potentially explained by the abundance of accessible, high energy density, benthic prey, gelatinous zooplankton and fisheries discards (see Table 2 and Fig. 2). A high energy diet may well be essential to fuel rapid growth in these smaller fishes (Nakatsubo and Hirose 2007, Pan et al. 2016) and to maximize net energetic gains (Phillips et al. unpublished data).The large schools of sunfish in these environments (see Fig. 4) may also function as protection from predators (Gladstone 1988, Fergusson et al. 2000, Nyegaard et al. 2019), although such a pronounced group strategy will also increase competition for resources. Indeed, the high number of sunfish in nearshore environments may actually drive the diversification of diet, such as opportunistic scavenging on fisheries discards (Phillips et al. 2020). Taken together, these lines of evidence suggest a complicated mix of factors may influence sunfish prey selection during such earlier life stages, extending beyond simple calorific explanations (see Bemis et al. 2020 [Chapter 4]).

As individual size and mass increase, sunfish develop a substantially thicker subcutaneous gelatinous layer (the hypodermis or capsule in Davenport et al. 2018), which is suggested to boost buoyancy (and decrease cost of transport), act as an exoskeleton (for attachment of the well-developed fin muscles) and increase tolerance to lower temperatures (e.g., Watanabe and Sato 2008, Nakamura et al. 2015, Potter et al. 2017, Davenport et al. 2018; for more detail on sunfish anatomy see Bemis et al. 2020 [Chapter 4]; Watanabe and Davenport 2020 [Chapter 5]). Larger fish (> 1 m) are also less tractable prey for pelagic predators (Johnsson 1993, Stamps 2007) and they may become more capable of travelling greater distances offshore and exploiting deeper prey fields (Davenport 1988, Angel and Pugh 2000, Houghton et al. 2008). Biologging studies and opportunistically obtained

Figure 4. School of smaller sunfishes (< 1 m total length) in Mediterranean coastal waters (note fishing nets in background). Photo credit Natasha Phillips.

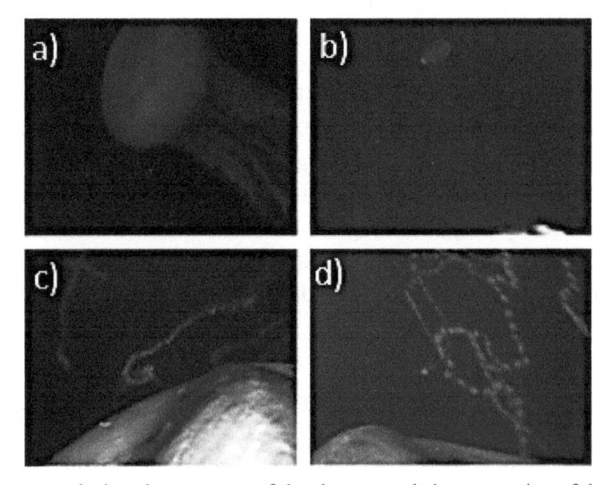

Figure 5. Animal-borne cameras deployed on ocean sunfishes have revealed consumption of deep water prey (reproduced with permission from Nakmura et al. 2015). Images show; (a) a lion's mane (*Cyanea capillata*), (b) a ctenophore (*Beroe* sp.), (c–d) siphonophore chains (*Praya* sp.).

camera footage have revealed the diving capabilites of larger specimens (total length 105–200 cm), which regularly swim to depths of > 200 m (Nakamura et al. 2015, Phillips et al. 2015). Animal-borne camera footage suggests such dives are discrete foraging trips to locate gelatinous prey (Nakamura et al. 2015; see Fig. 5). As medusae biomass for many species is limited in the open ocean by the bipartite lifecycle of scyphozoans (which typically require hard substrate for polyp attachment; Richardson et al. 2009), offshore prey items may contain more non-scyphozoan gelatinous groups such as salps, pyrosomes and siphonophores (Angel and Pugh 2000, Nakamura et al. 2015). In light of these findings, the gelatinous species found in deeper waters may require more extensive assessment as putative prey for sunfishes.

Ecological Role of Sunfishes

From these collective lines of evidence, the ecological role of sunfish is far broader than previously thought and appears to vary throughout ontogeny. The schools of smaller sunfish (typically < 1 m total length) are likely to have an important role in nearshore foodwebs, functioning as active predators and scavengers, possibly in sufficient numbers to influence top-down control on local benthic communities (see Figs. 2 and 3). This functional role may be disrupted when we consider the ongoing mass bycatch of these fishes in nearshore fisheries (see Fig. 6; Silvani et al. 1999, Cartamil and Lowe 2004, Petersen and McDonell 2007) and that the removal of large numbers of mesopredators could significantly impact local ecosystems (Myers et al. 2007, Heithaus et al. 2008, Baum and Worm 2009,

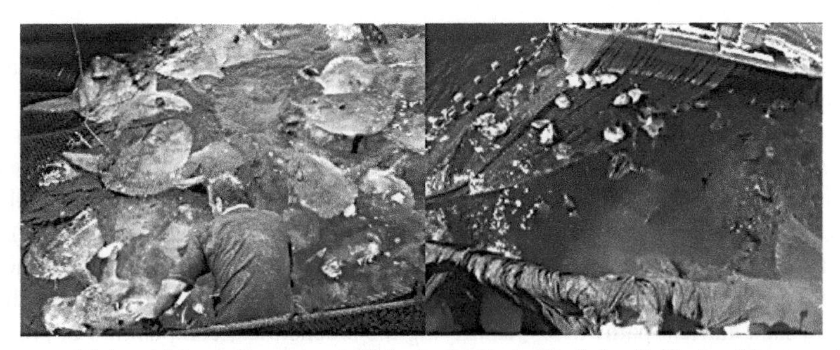

Figure 6. Ocean sunfish bycatch in small scale coastal fisheries (photo credits: Lukas Kubicek and Lawrence Eagling).

please see Nyegaard et al. 2020 [Chapter 12]). Although attempts have been made to estimate the volume of gelata consumed by sunfishes (e.g., Grémillet et al. 2017), so far these have been limited by model assumptions which could lead to significant over-estimations due to models assuming a single size of fish, population estimates based on seasonal sightings and a purely gelatinous diet throughout ontogeny. Future studies might like to include additional data in such models, such as fishes across a broad size range, factors associated with their ontogenetic shift in diet and regional population size estimates based on genetic analysis.

The larger sunfishes (> 1 m total length) may play a more pronounced role in pelagic ecosystems, acting as mass consumers of gelatinous zooplankton and transporting nutrients both vertically within the water column and horizontally across ocean basins (e.g., Sims et al. 2009a, Sims et al. 2009b, Cardona et al. 2012, Breen et al. 2017, Phillips et al. 2020). These larger fishes may also be at considerable risk from accidental fisheries capture (Sims et al. 2009b, Jing et al. 2015, Nyegaard et al. 2018). However, predicting the ecosystem impacts of removing these fish in significant numbers remains a challenge given a lack of data on rate processes, such as number and mass of prey ingested over time and realistic estimates of population size and structure in given areas. Gathering such data will not be simple, and will require close collaboration between geneticists and trophic ecologists on a broad geographic scale, but it is a clear aspiration deserving immediate attention. The knock-on effects of removing sunfishes in large numbers may also affect other marine animals that either directly prey upon them (sea lions, orca and large sharks, e.g., Nyegaard et al. 2019), indirectly feed on their parasites (sea birds and fish, Nyegaard et al. 2019) or scavenge stomach contents of dead individuals (observed in sea lions; Phillips personal communication; see review by Pope et al. 2010 for overview). However, knowledge of which species rely on the Molidae as prey are scarce, and the predators of molids will vary significantly across different life stages (e.g., mola larvae have been found in the stomachs of big fish such as mahi mahi, swordfish and tuna, Nyegaard et al. 2019, Thys et al. 2020 [Chapter 7]). The ichthyoplanktonic sunfish fry occupy a basal position in marine food chains (although this is higher than phytoplankton) but they can quickly outgrow the handling capabilities of many predators (e.g., Pan et al. 2016) and the largest sunfishes will be accessible to only large predators. Unfortunately, the early life history of the sunfishes is not well known, and so the importance of sunfishes as prey in marine ecosystems cannot yet be described. On occasion, anecdotal evidence arises, such as small *Ranzania* (< 30 cm) which have been noted in the stomach contents of sea birds including cormorants and Cory's Shearwaters (*Calonectris borealis*) (e.g., Alonso et al. 2013). Such evidence provides new insight into the trophic role of the Molidae, however further evidence is required to explore this facet of sunfish ecology.

Remaining Questions

Unravelling the diet of marine species is a complicated task, but comparing and combining the results of multiple techniques will reveal a more complete picture of sunfish diet and trophic role.

More specifically, we need to consider the qualitative dietary value of gelatinous prey beyond simple calorific content (Hays et al. 2018). The hypodermis, for example, is likely to require a significant collagen input (Davenport et al. 2018), but the synthesis of this structural protein or its possible dietary acquisition through gelatinous prey is unknown. The hypodermis becomes very thick in the large animals comprising a greater proportion of the individual's body weight (possibly tied to the need for increased buoyancy) and as such, the diet may require an increase in collagen as the animals grow (for further detail, please see Watanabe and Davenport 2020 [Chapter 5]). While it is plausible that gelatinous zooplankton can prove a significant source of collagen (some species of gelata contain up to 69 percent collagen by dry mass; Khong et al. 2015), further research is required to understand such nutrient pathways. Quantifying the biochemical composition of prey items may also answer whether the ocean sunfishes' ontogenetic dietary shift reflects changing nutritional requirements with increasing size. Current research has focused on sunfish (and prey items) in nearshore waters, however this approach may be missing a large part of their life history. Sunfishes are sighted throughout the mesopelagic zone (Phillips et al. 2015, Thys et al. 2017) and can even dive > 1000 m presumably in search of colonial gelatinous zooplankton (Thys et al. 2017, Nakamura et al. 2015). More data are required to explore the importance of these deeper water prey items, suggested to include siphonophores, ctenophores and pyrosomes (Potter and Howell 2011, Doyle et al. 2014, Nakamura et al. 2015) and the potential role of sunfishes in the advection of nutrients throughout the water column (Angel and Pugh 2000, Beckman 2012, Katija and Dabiri 2009). Prey taxa may also be inferred by dental wear patterns (e.g., Fahlkea et al. 2013) and closer examination of the pharyngeal teeth (see Bemis et al. 2020 [Chapter 4]). Finally, most dietary studies have been conducted in summer months so it is highly likely a seasonal element of the sunfishes' diet is currently being overlooked, particularly as these fishes are less commonly noticed in nearshore environments during the winter season, indicating possible differing habitat use and feeding strategies. It would also be of interest to assess the relative value of prey items alongside the energetic requirements of fishes ontogenetically. This may be achieved in future studies through accelerometry or metabolic rate measurements conducted under semi-wild conditions (e.g., the 'megaflume' used successfully to assess metabolic rate under controlled conditions such as set swim speeds, for large bodied sharks; Payne et al. 2015).

Future work should also shine a light on where and how sunfish transition between differing habitats, allowing us to minimize the current impacts of fisheries bycatch. This can be accomplished by increasing sampling, particularly of individuals from the open ocean and outside of summer months, and encouraging the collection and collaborative sharing of by caught sunfishes and their putative prey (such as in Phillips et al. 2018). Since both bulk stable isotope analysis and DNA barcoding require extensive background collection of potential prey items or risk constraining the modelling of predator trophic position, a collaborative approach will be vital in moving the field forwards.

Concluding Thoughts

Understanding the diet of consumers is a central theme in marine ecology. For the ocean sunfishes, recent research has revealed a hitherto unknown cosmopolitan diet in smaller individuals and shifts in their ecosystem role. Such knowledge is of particular importance given the ongoing pressures affecting these fish and the broader marine environment. However, despite this new understanding, further research is strongly recommended to derive generalizable rules on sunfish diet to inform trophic-level size models over broad temporal and spatial scales.

It is difficult to imagine the global ocean without this charismatic group of globally distributed predators with their intriguing life cycle and varied ecological role. However, current research suggests that anthropogenic influences are resulting in broad scale issues across the world's ocean with 33 percent of the world's marine fish stocks currently overfished and 60 percent fished at maximum sustainability (FAO 2018). As demands on oceanic resources intensify, further insights into the trophic role of all fishes are urgently needed for sustainable management (da Silva 2014). With the ever-increasing awareness of marine conservation issues, the sunfishes' charismatic nature may help to highlight the continuing efforts of conservation groups to understand the trophic role of many marine fishes and ensure a more sustainable future.

Acknowledgments

The authors would like to acknowledge and thank Lawrence Eagling and Olivia Daly who extensively proofread this chapter, alongside everyone who generously provided images to illustrate the text: Lukas Kubicek, Jodi Frediani, Lawrence Eagling and Itsumi Nakamura. Chris Harrod is supported by Núcleo Milenio INVASAL funded by Chile's Government program, Iniciativa Cientifica Milenio from the Ministerio de Economia, Fomento y Turismo. Natasha Phillips is supported by Sea Monitor, a project funded by the European Union's INTERREG VA Programme (Environment Theme), managed by the Special EU Programmes Body (SEUPB). Natasha Phillips' PhD thesis and data collection (elements of which are included in this chapter) was supported by the Fisheries Society of the British Isles, alongside grants for research and travel provided by: Flying Sharks, the Marine Institute, The Society of Experimental Biology, The Royal Society of Biology, The Alice McCosh Trust grant, The Sir Thomas Dixon Trust and The MacQuitty Trust.

References

Ahuir-Baraja, A. 2020. Parasites of the ocean sunfishes. *In*: T. Thys, J.D.R Houghton and G.C. Hays [eds.]. The Ocean Sunfishes: Evolution, Biology and Conservation, CRC Press Boca Raton FL USA.

Alonso, H., J. Granadeiro, J. Ramos and P. Catry. 2013. Use the backbone of your samples: Fish vertebrae reduces biases associated with otoliths in seabird diet studies. J. Ornithol. 154(2019): 883–886.

Angel, M.V. and P.R. Pugh. 2000. Quantification of diel vertical migration by macroplankton and micronektonic taxa in the northeast Atlantic. Hydrobiologia, 440, pp.161–179. Available at: http://eprints.soton.ac.uk/54572/.

Arai, M.N. 2005. Predation on pelagic coelenterates: A review. J. Mar. Biol. Assoc. UK 85(3): 523–536.

Bakenhaster, M.D. 2016. (Tetraodontiformes : Molidae) off Florida's Atlantic coast with discussion of historical context. Bull. Mar. Sci. 92(4): 497–511. Available at: file:///C:/Users/40144085/Downloads/Bakenhaster and Knight-Gray 2016_Molid diet.pdf.

Barnard, K.H. 1927. A monograph of the marine fishes of South Africa. Part II. Ann. S. Afr. Mus. 21: 19–1065. Available at: http://www.biodiversitylibrary.org/page/40934782#page/13/ mode/1up.

Baum, J.K. and B. Worm. 2009. Cascading top-down effects of changing oceanic predator abundances. J. Anim. Ecol. 78(4): 699–714.

Bearzi, G., E. Politi, S. Agazzi and A. Azzellino. 2006. Prey depletion caused by overfishing and the decline of marine megafauna in eastern Ionian Sea coastal waters (central Mediterranean). Biol. Conserv. 127(4): 373–382.

Beckman, D. 2012. Classification and biology of marine organisms. In Marine Environmental Biology and Conservation. Jones and Bartlett, pp. 41–43.

Bemis, K.E., J.C. Tyler, E.J. Hilton and W.E. Bemis. 2020. Overview of the anatomy of ocean sunfishes (Molidae: Tetraodontiformes). pp. 55–71. *In*: T. Thys, J.D.R. Houghton and G.C. Hays [eds.]. The Ocean Sunfishes: Evolution, Biology and Conservation, CRC Press Boca Raton FL USA.

Breen, P., O. Canadas, O. Cadhla, M. Mackey, M. Scheidat, S. Geelhoed et al. 2017. New insights into ocean sunfish (*Mola mola*) abundance and seasonal distribution in the northeast Atlantic. Sci. Rep. 7: 2025.

Briz, L.D., F. Sanchez and N. Mari. 2017. Gelatinous zooplankton (ctenophores, salps and medusae): an important food resource of fishes in the temperate SW Atlantic Ocean. Mar. Biol. Res. 13(6): 630–644.

Cardona, L., I. Quevedo, A. Borrell and A. Aguilar. 2012. Massive consumption of gelatinous plankton by mediterranean apex predators. PLoS ONE 7(3).

Cartamil, D.P. and C.G. Lowe. 2004. Diel movement patterns of ocean sunfish *Mola mola* off southern California. Mar. Ecol. Prog. Ser. 266: 245–253.

Clarke, Holmes, J. and D.J. Gore. 1992. Proximate and elemental composition of gelatinous zooplankton from the Southern Ocean. J. Exp. Mar. Biol. Ecol. 155(1): 55–68.

Cushing, D. 1975. Marine ecology and fisheries. Cambridge: Cambridge University Press.

da Silva, J. 2014. FAO: The state of world fisheries and aquaculture report. Available at: http://scholar.google.com/scholar?hl=enandbtnG=Searchandq=intitle:THE+STATE+OF+WORLD+FISHERIES+AND+AQUACULTURE#0.

Davenport, J. 1988. Do diving leatherbacks pursue glowing jelly? Herpetol. Bull. 24: 20–21.

Davenport, J. and G.H. Balazs. 1991. Fiery Bodies -are pyrosomes an important component of the diet of leatherback turtles? Analysis (37): 33–38.

Davenport, J., N. Phillips, E. Cotter, L. Eagling and J. Houghton. 2018. The locomotor system of the ocean sunfish *Mola mola* (L.): role of gelatinous exoskeleton, horizontal septum, muscles and tendons. J. Anat. 223(3): 347–357. Available at: http://doi.wiley.com/10.1111/joa.12842.

Donovan, F.L.S. 1803. The Natural History of British Fishes: Including Scientific and General Descriptions of the Most Interesting Species and an Extensive Selection of Accurately Finished Coloured Plates, Taken Entirely from Original Drawings, Purposely Made from the Specimen. London.

Doyle, T.K., J. Houghton, R. McDevitt, J. Davenport and G. Hays. 2007. The energy density of jellyfish: Estimates from bomb-calorimetry and proximate-composition. J. Exp. Mar. Biol. Ecol. 343(2): 239–252.

Doyle, T.K., J.Y. Georges and J.D.R. Houghton. 2014. A leatherback turtle's guide to jellyfish in the North East Atlantic. Munibe Monogr. Nat. Ser. 1: 15–21.

Drazen, J.C. and T.T. Sutton. 2016. Dining in the Deep: The Feeding Ecology of Deep-Sea Fishes. Annu. Rev. Mar. Sci. 9: 337–366.

Fahlke, M., K. Bastl, G. Semprebon and P. Gingerich. 2013. Paleoecology of archaeocete whales throughout the Eocene: Dietary adaptations revealed by microwear analysis. Palaeogeogr. Palaeoclimatol. Palaeoecol. 386: 690–701.

FAO. 2018. The State of World Fisheries and Aquaculture 2018—Meeting the sustainable development goals.

Fergusson, I.K., L.J.V. Compagno and M.A. Marks. 2000. Predation by white sharks *Carcharodon carcharias* (Chondrichthyes: Lamnidae) upon chelonians, with new records from the mediterranean sea and a first record of the ocean surfish *Mola mola* (Osteichthyes: Molidae) as stomach contents. Environ. Biol. Fishes 58(4): 447–453.

Fitch, J.E. 1969. A second record of the slender Mola, *Ranzania laevis* (Pennant), from California. Bull. S. Calif. Acad. Sci. 68: 115–188.

Fossette, S., A.C. Gleiss, J. Cassey, A. Lewis and G. Hays. 2012. Does prey size matter? Novel observations of feeding in the leatherback turtle (*Dermochelys coriacea*) allow a test of predator—prey size relationships. Biol. Lett. 8(3): 351–4.

Fraser-Brunner, A. 1951. The ocean sunfishes (Family Molidae). Bull. Br. Mus. Nat. Hist. 1(6): 87–121. Available at: http://archive.org/stream/cbarchive_52333_theoceansunfishfamilymolidae1950/theoceansunfishfamilymolidae1950_djvu.txt.

Freedman, J.A. and D.L.G. Noakes. 2002. Why are there no really big bony fishes? A point-of-view on maximum body size in teleosts and elasmobranchs. Rev. Fish Biol. Fish. 12: 403–416. Available at: http://download.springer.com/static/pdf/829/art%25253A10.1023%25252FA%25253A1025365210414.pdf?auth66=1416412953_1cef5c809fd1544042e2303034378a0dandext=.pdf.

Gladstone, W. 1988. Killer Whale Feeding Observed Underwater. J. Mammal. 69(3): 629–630. Available at: https://academic.oup.com/jmammal/article-lookup/doi/10.2307/1381360.

Grémillet, D., C. White, M. Authier, G. Doremus, V. Ridoux and E. Pettex. 2017. Ocean sunfish as indicators for the "rise of slime". Curr. Biol. 27(23): R1263–R1264. doi: 10.1016/j.cub.2017.09.027.

Gudger, E.W. 1936. From atom to colossus. Bull. Am. Mus. Nat. Hist. 38: 26–30. Available at: https://archive.org/stream/naturalhistory38newy/naturalhistory38newy_djvu.txt.

Harrod, C., J. Syvaranta, L. Kubicek, V. Cappenera and J. Houghton. 2013. Reply to Logan and Dodge: Stable isotopes challenge the perception of ocean sunfish *Mola mola* as obligate jellyfish predators. J. Fish Biol. 82(1): 10–6. Available at: http://www.ncbi.nlm.nih.gov/pubmed/23331134.

Hays, G.C. 2008. Sea turtles: A review of some key recent discoveries and remaining questions. J. Exp. Mar. Biol. Ecol. 356: 1–7.

Hays, G.C., M. Farquhar, P. Luschi, S. Teo and T. Thys. 2009. Vertical niche overlap by two ocean giants with similar diets: Ocean sunfish and leatherback turtles. J. Exp. Mar. Biol. Ecol. 370(1-2): 134–143. Available at: http://linkinghub.elsevier.com/retrieve/pii/S0022098108006096.

Hays, G.C., T.K. Doyle and J.D.R. Houghton. 2018. A Paradigm Shift in the Trophic Importance of Jellyfish? Trends Ecol. Evol. 33(11): 874–884.

Hays, G.C., J.D.R. Houghton, T.M. Thys, D. Adams, A. Ahuir-Barqia, J. Alvarez et al. 2020. Unresolved questions about the Ocean Sunfishes, Molidae—a family comprising some of the world's largest teleosts. pp. 280–296. *In*: T. Thys, J.D.R. Houghton and G.C. Hays [eds.]. The Ocean Sunfishes: Evolution, Biology and Conservation, CRC Press Boca Raton FL USA.

Heithaus, M.R., F. Wirsing and B. Worm. 2008. Predicting ecological consequences of marine top predator declines. Trends Ecol. Evol. 23(4): 202–210.

Hooper, S.N., M. Paradis and R.G. Ackman. 1973. Distribution of trans 6 hexa decenoic acid 7 methyl 7 hexa decenoic acid and common fatty acids in lipids of the ocean sunfish. Lipids 8(9): 509–516.

Houghton, J.D.R., T. Doyle, M. Wilson, J. Davenport and G. Hays. 2006. Jellyfish aggregations and leatherback turtle foraging patterns in a temperate coastal environment. Ecology 87(8): 1967–1972.

Houghton, J.D.R., T.K. Doyle, J. Davenport, R. Wilson and G. Hays. 2008. The role of infrequent and extraordinary deep dives in leatherback turtles (*Dermochelys coriacea*). J. Exp. Biol. 211: 2566–2575.

Jennings, S., K.H. Andersen and J.L. Blanchard. 2017. Size-based analysis of aquatic food webs. In Marine Ecology and Fisheries.

Jing, L., G. Zapfe, K.-T. Shao, J.L. Leis, K. Matsuura, G. Hardy et al. 2015. IUCN Red list of threatened species. IUCN Red List. Available at: www.iucnredlist.org.

Johnsson, J.I. 1993. Big and brave: size selection affects foraging under risk of predation in juvenile rainbow trout, *Oncorhynchus mykiss*. Anim. Behav. 45: 1219–1225.

Katija, K. and J.O. Dabiri. 2009. A viscosity-enhanced mechanism for biogenic ocean mixing. Nat. Lett. 460: 624–627. doi:10.1038/nature08207.

Khong, N.M.H., F. Yusogg, J. Bakar, M. Basri, M. Ismail, K. Chan et al. 2015. Nutritional composition and total collagen content of three commercially important edible jellyfish. Food Chem. 196. Available at: http://dx.doi.org/10.1016/j.foodchem.2015.09.094.

Lucas, C.H., D. Jones, C. Hollyhead, R. Condon, C. Duarte, W. Graham et al. 2014. Gelatinous zooplankton biomass in the global oceans : Geographic variation and environmental drivers. Glob. Ecol. Biogeogr. 23(November 2017): 701–714.

MacGinitie, G. and N. MacGinitie. 1968. Natural history of marine animals. 2nd edn., New York: McGraw-Hill Book Company.

McCauley, D.J., M. Pinsky, S. Palumbi, J. Estes, F. Joyce and R. Warner. 2015. Marine defaunation: Animal loss in the global ocean. Science 347(6219).

McClain, C.R., M. Balk, M. Benfield, T. Branch, C. Chen, J. Cosgrove et al. 2015. Sizing ocean giants: patterns of intraspecific size variation in marine megafauna. PeerJ. 3, p.e715. Available at: https://peerj.com/articles/715%255Cnhttp://www.pubmedcentral.nih.gov/articlerender.fcgi?artid=4304853andtool=pmcentrezandrendertype=abstract.

Mourocq, E. and V. Ridoux. 2010. Proximate composition and energy content of forage species from the Bay of Biscay : high- or low-quality food? ICES J. Mar. Sci. 67(5): 909–915.

Myers, R.A. and B. Worm. 2003. Rapid worldwide depletion of predatory fish communities. Nature 423(6937): pp. 280–283. Available at: http://www.nature.com/doifinder/10.1038/nature01610.

Myers, R.A., J. Baum, T. Shepherd, S. Powers and C. Peterson. 2007. Cascading effects of the loss of predatory sharks from a coastal ocean. Science (March), pp. 1846–1850.

Nakamura, I. and K. Sato. 2014. Ontogenetic shift in foraging habit of ocean sunfish *Mola mola* from dietary and behavioral studies. Mar. Biol. 161(6): 1263–1273. Available at: http://link.springer.com/article/10.1007/s00227-014-2416-8/fulltext.html.

Nakamura, I., Y. Goto and K. Sato. 2015. Ocean sunfish rewarm at the surface after deep excursions to forage for siphonophores. J. Anim. Ecol. 84(3): 590–603. Available at: http://doi.wiley.com/10.1111/1365-2656.12346.

Nakatsubo, T. and H. Hirose. 2007. Growth of captive ocean sunfish *Mola mol*a. Aquacult. Sci. 55(3): 403–407.

Norman, J.R. and F.C. Fraser. 1949. Field book of giant fishes. G. P. Putnam, ed., New York.

Nyegaard, M., N. Loneragan and M.B. Santos. 2017. Squid predation by slender sunfish *Ranzania laevis* (Molidae). J. Fish Biol. 90(6): 2480–2487.

Nyegaard, M., N. Loneragan, S. Hall, J. Andrew, E. Sawai and M. Nyegaard. 2018. Giant jelly eaters on the line: species distribution and bycatch of three dominant sunfishes in the Southwest Pacific. Estuar. Coast. Shelf Sci. 207: 1–15. Available at: https://doi.org/10.1016/j.ecss.2018.03.017.

Nyegaard, M., S. Andrzejaczek, C.S. Jenner and M.-N. Jenner. 2019. Tiger shark predation on large ocean sunfishes (Family Molidae)—two Australian observations. Environ. Biol. Fish. 102: 1559–1567. https://doi.org/10.1007/s10641-019-00926-y.

Nyegaard, M., S. Garcia Barcelona, N.D. Phillips and E. Sawai. 2020. Fisheries interactions, distribution modeling and conservation issues of the ocean sunfishes. *In*: T. Thys, J.D.R. Houghton and G.C. Hays [eds.]. The Ocean Sunfishes: Evolution, Biology and Conservation, CRC Press Boca Raton FL USA.

Pan, H., H. Yu, V. Ravi, C. Li, A. Lee, M. Lian et al. 2016. The genome of the largest bony fish, ocean sunfish (*Mola mola*), provides insights into its fast growth rate. GigaScience 5(1): 1–12. Available at: http://dx.doi.org/10.1186/s13742-016-0144-3.

Payne, N.L., E. Snelling, R. Fitzpatrick, J. Seymour, R. Courtney, A. Barnett et al. 2015. A new method for resolving uncertainty of energy requirements in large water breathers : the 'mega-flume' seagoing swim-tunnel respirometer. Methods Ecol. Evol. 6: 668–677.

Perrault, J.R. 2014. Mercury and selenium ingestion rates of Atlantic leatherback sea turtles (*Dermochelys coriacea*): a cause for concern in this species? Mar. Environ. Res. 99: 160–9. Available at: http://www.ncbi.nlm.nih.gov/pubmed/24853722.

Peters, R.H. 1983. The Ecological Implications of Body Size. Cambridge University Press, Cambridge.

Petersen, S. and Z. McDonell. 2007. A bycatch assessment of the Cape horse mackerel *Trachurus trachurus capensis* mid- water trawl fishery off South Africa. Birdlife/WWF Responsible Fisheries Programme Report 2002–2005.

Phillips, N.D., C. Harrod, A. Gates, T. Thys and J. Houghton. 2015. Seeking the sun in deep, dark places: Mesopelagic sightings of ocean sunfishes (Molidae). J. Fish Biol. 87(4): 1118–1126.

Phillips, N., L. Eagling, C. Harrod, V. Cappenera, J. Houghton and N. Reid. 2017. Quacks snack on smacks: mallard ducks (*Anas platyrhynchos*) observed feeding on hydrozoans (*Velella velella*). Plankton Benthos Res. 12(2): 143–144. Available at: https://www.jstage.jst.go.jp/article/pbr/12/2/12_P120207/_article.

Phillips, N.D., L. Kubicek, N. Payne, C. Harrod, L. Eagling and C. Carson. 2018. Isometric growth in the world's largest bony fishes (Genus Mola)? Morphological insights from fisheries bycatch data. J. Morphol. 279(9).

Phillips, N.D., E.A.E. Smith, S.D. Newsome, J.D.R. Houghton, C.D. Carson, J.C. Mangel et al. 2020. Bulk tissue and amino acid stable isotope analyses reveal global ontogenetic patterns in ocean sunfish trophic ecology and habitat use. Mar. Ecol. Prog. Ser. 633: 127–140.

Phillips, N.D., L. Eagling, L. Kubicek, N. Payne, C. Harrod, C. Carson et al. Swimmingly efficient! Linking routine metabolic costs to counter-intuitive dietary shifts in ocean sunfishes (Genus Mola). Unpublished data.

Plancus, J.A. 1746. De Mola pisce. Commentarii de Bononiensi Scientiarum et Artium Instituto Atque Academia tom. Bologna: Bononiae.

Pope, E.C., G. Hays, T. Thys, T. Doyle, D. Sims, N. Queiroz et al. 2010. The biology and ecology of the ocean sunfish *Mola mola*: A review of current knowledge and future research perspectives. Rev. Fish Biol. Fish. 20(4): 471–487.

Potter, I.F. and W.H. Howell. 2011. Vertical movement and behavior of the ocean sunfish, *Mola mola*, in the northwest Atlantic. J. Exp. Mar. Biol. Ecol. 396(2): 138–146. Available at: http://linkinghub.elsevier.com/retrieve/pii/S0022098110004193.

Potter, M.C., D.C. Wiggert and H. Bassem. 2017. Mechanics of Fluids 5th Edition, Boston, USA: Cengage Learning. Available at: https://books.google.co.uk/books?id=U3gcCgAAQBAJandpg=PA356andlpg=PA356anddq=Reynolds+number+between+10%255E5+and+10%255E6,+the+drag+coefficient+takes+a+sudden+dipandsource=blandots=Y2SnB9I242andsig=Xs3eMnmZtGJF3UrOYpbPrbufKQ4andhl=enandsa=Xandved=0ahUKEwjC5L6zwvvSAhVoLcAKHZ8K.

Reuvens, C.L. 1897. *Orthragoriscus nasus* Ranz. on the Dutch coast. Notes Leyden Mus. 18: 209–212.

Richardson, A.J., A. Bakun, G. Hays and M. Gibbons. 2009. The jellyfish joyride: causes, consequences and management responses to a more gelatinous future. Trends Ecol. Evol. 24(6): 312–322.

Robison, B.H. 1975. Observations on living juvenile specimens of the slender mola *Ranzania laevis* (Pisces, Molidae). Pac. Sci. 29(27–29).

Rowan, J. 2006. Tropical sunfish visitor as big as a car. The New Zealand Herald APN News & Media.

Santini, F., M. Nguyen, L. Sorenson, T. Waltzek, J. Alfaro, J. Eastman et al. 2013. Do habitat shifts drive diversification in teleost fishes? An example from the pufferfishes (Tetraodontidae). J. Evol. Biol. 26(5): 1003–1018.

Schmidt, J. 1921. New studies of sun-fishes made during the "Dana" Expedition, 1920. Nature 107: 76–79.

Silvani, L., M. Gazo and A. Aguilar. 1999. Spanish driftnet fishing and incidental catches in the western Mediterranean. Biol. Conserv. 90(1): 79–85.

Sims, D.W., N. Queiroz, N. Humphries, F. Lima and G. Hays. 2009a. Long-term GPS tracking of ocean sunfish *Mola mola* offers a new direction in fish monitoring. PLoS ONE 4(10): 1–6.

Sims, D.W., N. Queiroz, T. Doyle, J. Houghton and G. Hays. 2009b. Satellite tracking of the world's largest bony fish, the ocean sunfish (*Mola mola* L.) in the North East Atlantic. J. Exp. Mar. Biol. Ecol. 370(1-2): 127–133. Available at: http://linkinghub.elsevier.com/retrieve/pii/S0022098108006114.

Sousa, L.L., R. Xavier, V. Costa, N. Humphries, C. Trueman, R. Rosa et al. 2016. DNA barcoding identifies a cosmopolitan diet in the ocean sunfish. Sci. Rep. 6(June): 28762. Available at: http://www.pubmedcentral.nih.gov/articlerender.fcgi?artid=4931451andtool=pmcentrezandrendertype=abstract.

Stamps, J.A. 2007. Growth-mortality tradeoffs and personality traits in animals. Ecol. Lett. 10(5): 355–363.

Syväranta, J., C. Harrod, L. Kubicek, V. Cappenera and J. Houghton. 2012. Stable isotopes challenge the perception of ocean sunfish *Mola mola* as obligate jellyfish predators. J. Fish Biol. 80(1): 225–31. Available at: http://www.ncbi.nlm.nih.gov/pubmed/22220901.

Thys, T.M., J. Ryan, H. Dewar, C. Perle, K. Lyons, J. O'Sullivan et al. 2015. Ecology of the Ocean Sunfish, *Mola mola*, in the southern California Current System. J. Exp. Mar. Biol. Ecol. 471: 64–76. Available at: http://linkinghub.elsevier.com/retrieve/pii/S0022098115001239.

Thys, T.M., A. Hearn, K. Weng, J. Ryan and C. Penaherrera-Palma. 2017. Satellite Tracking and Site Fidelity of Short Ocean Sunfish, *Mola ramsayi*, in the Galapagos Islands. J. Mar. Biol., pp. 1–10.

Thys, T., M. Nyegaard, J.L. Whitney, J. Ryan, I. Potter, T. Nakatsubo et al. 2020. Ocean sunfish larvae: detections, identification and predation. pp. 105–128. *In*: T. Thys, J.D.R Houghton and G.C. Hays [eds.]. The Ocean Sunfishes: Evolution, Biology and Conservation, CRC Press Boca Raton FL USA.

Watanabe, Y. and K. Sato. 2008. Functional dorsoventral symmetry in relation to lift-based swimming in the ocean sunfish *Mola mola*. PLoS ONE 3(10): e3446. Available at: http://www.pubmedcentral.nih.gov/articlerender.fcgi?artid=2562982andtool=pmcentrezandrendertype=abstract.

Watanabe and Davenport. 2020. Locomotory systems and biomechanics of the ocean sunfish. pp. 72–86. *In*: T. Thys, J.D.R. Houghton and G.C. Hays [eds.]. The Ocean Sunfishes: Evolution, Biology and Conservation, CRC Press Boca Raton FL USA.

Werner, E.E. and J.F. Gillian. 1984. The ontogenetic niche and species interactions in size-structured populations. Annu. Rev. Ecol. Evol. Syst. 15: 393–425.

Yabe, H. 1950. Juvenile of the pointed-tailed ocean sunfish, *Masturus lanceolatus*. Bull. Jap. Soc. Sci. Fish. 16(2): 40–42.

Young, J.W., B. Hunt, T. Cook, J. Llopiz, E. Hazen, H. Pethybridge et al. 2014. The trophodynamics of marine top predators: Current knowledge, recent advances and challenges. Deep Sea Res. Part II Top. Stud. Oceanogr. 113: 170–187. Available at: http://dx.doi.org/10.1016/j.dsr2.2014.05.015.

Chapter 10

Parasites of the Ocean Sunfishes

Ana E. Ahuir-Baraja

Introduction

The strong relationship between parasites and their hosts offers important insights into the latter's phylogeny, ecology, biogeography and population regulation (Campbell et al. 1980, Brooks and McLennan 1993, Tompkins and Begon 1999, Hoberg and Klassen 2002, Moore 2002). Furthermore, parasitic communities are integral elements within marine food webs (Lafferty et al. 2008, Lafferty 2013, Thieltges et al. 2013) and help to maintain equilibrium, integrity and stability (e.g., Hudson et al. 2006, Lafferty 2013, de Azevedo and Abdallah 2016). For example, some parasitic species can alter the fecundity rate of their hosts (Lafferty and Kuris 2009, Gilardoni et al. 2012, Labaude et al. 2015), whilst others can modify host behavior to facilitate predation by a subsequent host (Thomas et al. 1998). Moreover, the study of the parasites can assist with the identification of fish populations and stock differentiation via information on host origin, migration and/or life history (e.g., Lester 1990, Thomas et al. 1996, MacKenzie 2004, Timi 2007, Lester and MacKenzie 2009, MacKenzie and Abaunza 2014, Mattiucci et al. 2015). Indeed, as nearly all phyla are subjected to some form of parasitism (Poulin and Morand 2000), opportunities for new discoveries abound. However, there is an imbalance in our knowledge of fish parasitism, with studies biased towards species of high commercial value (Lloret et al. 2012, Lafferty 2013). The ocean sunfishes exemplify this problem clearly, as the family Molidae includes some of the most heavily parasitized fish worldwide (Dollfus 1946, Fraser-Brunner 1951, Love and Moser 1983, Ahuir-Baraja et al. 2017, de Figueiredo et al. 2017). However, our knowledge of how these parasites affect the biology or behavior of molids is limited compared to other teleosts.

Poulin and Morand (2000) described a parasite as an organism that lives on, or in its host during a significant part of its life. Parasites can be categorized into two primary groups: microparasites (e.g., viruses, bacteria, protists) and macroparasites (e.g., helminths and arthropods, described as metazoans hereafter) (Price 1980, Anderson 1993, Viney and Cable 2011). Further distinctions can be made between ectoparasites (living mostly on the outside of their hosts, e.g., on the skin or gills) and endoparasites (living inside their hosts). Ectoparasites are mainly monoxenous (i.e., living on one type of host during its life cycle), with direct life cycles involving one species of host where

Department of Animal Production and Health, Public Veterinary Health and Food Science and Technology, Faculty of Veterinary Science, Universidad Cardenal Herrera-CEU, CEU Universities, Alfara del Patriarca, Valencia. P.O. Box 46115, Spain; Marine Zoology Unit, Cavanilles Institute of Biodiversity and Evolutionary Biology, Science Park, University of Valencia. P.O. Box 22085, Valencia 46071, Spain.

Email: ana.ahuir@uchceu.es, baraja@uv.es

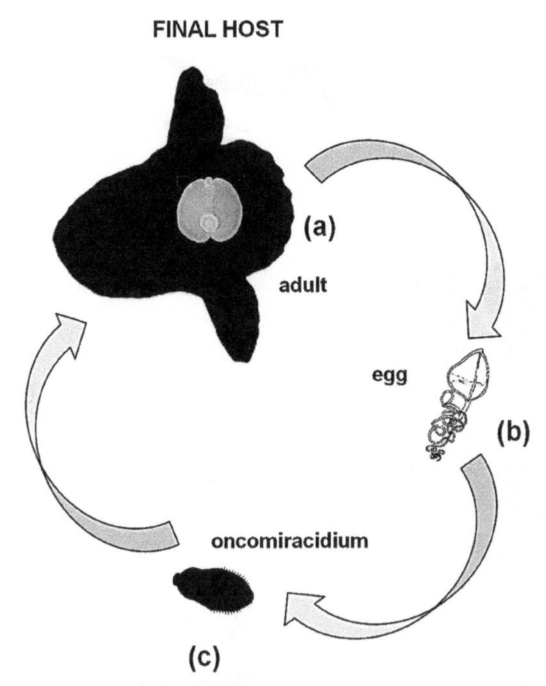

Figure 1. Schematic representation of the life cycle of *Capsala martinieri* (Monogenea) in *Mola mola* as an example of ectoparasitic direct life cycle. (A) Adults living on the host are attached to the skin. (B) Eggs (drawing from Volodymyr et al. 2015) released by adults can fix to the same host or to another *M. mola*. (C) Oncomiracidium (ciliated larval stage) hatched from an egg swims to attach itself to the host where it will develop into its adult form (Based on Klimpel et al. 2019).

larval/immature stages attach and develop to adult stages (Fig. 1; Kearn 1998, Cordero del Campillo 2000, Rohde 2005). Examples of ectoparasites include protists, monogeneans, copepods, isopods, and hirudineans.

By contrast, endoparasites are typically heteroxenous (living in more than one type of host, e.g., intermediate host and final host, during its life cycle). They display indirect life cycles in which the adult specimens reproduce in the final host, whilst the larval/immature stages need to parasitize and develop in intermediate hosts from different species, until parasitizing their final host where they can start the life cycle (Fig. 2). Trematodes, cestodes, nematodes and acanthocephalans are all examples of endoparasites (Kearn 1998, Cordero del Campillo 2000, Rohde 2005). Parasites can also be classified as specialists, infecting a narrow host range or as generalists, infecting a great host range (Kearn 1998, Lajeunesse and Forbes 2002). Lastly, the term "paratenic hosts" refers to hosts that can take part in the life cycle of the parasite (typically as a penultimate phase) in which the parasites neither grow nor mature (Bush et al. 2001).

From this vantage point, the parasitic fauna of the ocean sunfishes is described. A checklist of parasites is given and contextualized alongside our current knowledge of the Molidae, and promising lines of future investigation are offered.

Previous Studies on Parasites of Ocean Sunfishes

Ocean sunfish are famous for carrying a large parasite load and numerous studies have attempted to provide comprehensive overviews with a focus primarily on *Mola mola* (e.g., Dollfus 1946, Hillis and O'Riordan 1960, Threlfall 1967, Love and Moser 1983, Villalba and Fernández 1985, Aureli 2004, Gustinelli et al. 2006, Ahuir-Baraja 2012, Radujković and Šundić 2014, Ahuir-Baraja et al. 2017, de Figueiredo et al. 2017). While some studies concentrate on one species or taxonomical group of parasites (e.g., Hewitt 1968, Bray and Gibson 1977), others focus on the parasites' morphology and

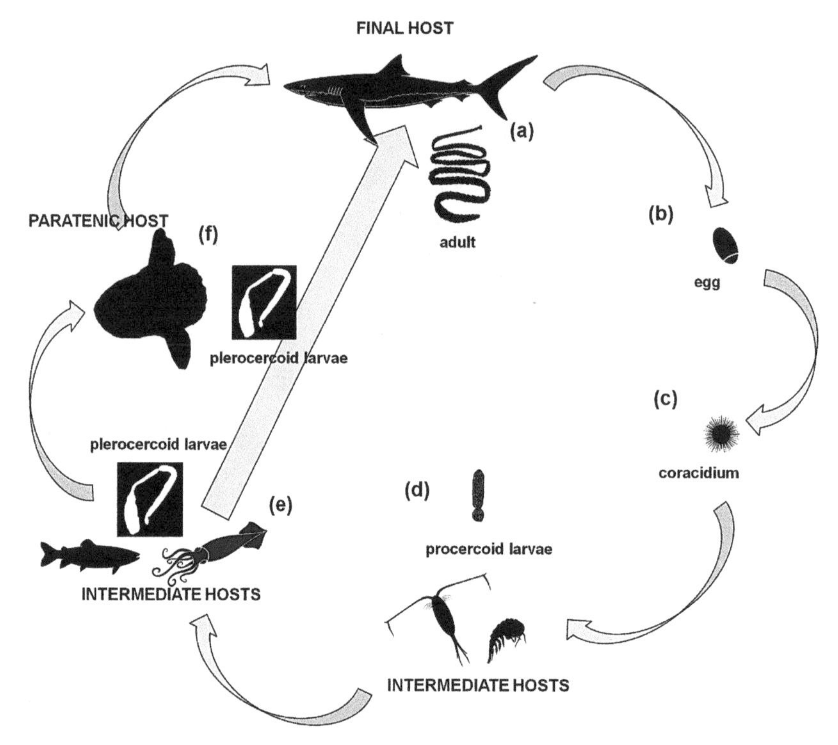

Figure 2. A hypothetical life cycle of *Molicola horridus* (Cestoda) in *Mola mola* as an example of an endoparasite. (A) Adults of *M. horridus* are depicted within stomach or spiral valves of sharks. (B) Eggs are released into the water column via host feces. (C) Coracidium (first larval stage) emerge from the egg. (D) The first intermediate host ingests the coracidium which develops into a procercoid (Chervy 2002). (E) A second intermediate host allows the procercoid to develop into the next larval stage, the plerocercoid. The second intermediate host could be directly eaten by the final host or (F) by *M. mola* acting as a paratenic host. In turn, *M. mola* could be eaten by the final host where the parasite could reach the adult stage (Based on Klimpel et al. 2019).

the affected host organs (e.g., Dollfus 1946, Hillis and O'Riordan 1960, McCann 1961, Threlfall 1967, Cooper et al. 1982, Aureli 2004, Gustinelli et al. 2006). Typically, these studies include only one or at best a few host molids due to isolated stranding events and sporadic bycatch (Noble and Noble 1966a,b, Hewitt 1968, Gibson and Bray 1979, Aureli 2004, Barreiros and Teves 2005, Gustinelli et al. 2006, Pastore 2009, Fernández et al. 2016).

Other studies have reported on epibionts like the pelagic gooseneck barnacle, *Lepas (Anatifa) anatifera* (Linnaeus, 1758), (Crustacea, Lepadidae), which live harmlessly on mola skin (Barreiros and Teves 2005). Also occuring on the skin is the spider crab *Acanthonyx petiverii* H. Milne Edwards, 1834 (Crustacea, Epialtidae) (cited as *A. scutiformis* (Dana, 1851) in the checklist from de Figueiredo et al. 2017). Commensals, such as remoras, *Echeneis naucrates* Linnaeus, 1758 (Perciformes, Echeneidae), have also been reported within the gill chamber of both *M. mola* and *Masturus lanceolatus* (Liénard, 1840) (Schwartz and Lindquist 1987). Some epibionts use other parasites as attachment substrata, e.g., goose barnacles, *Conchoderma virgatum* Spengler, 1789 (Crustacea, Lepadidae), attached to parasitic copepods in *M. mola* (Balakrishnan 1969, Cooper et al. 1982).

Captive specimens can have different arrays of parasites. Sato et al. (1982) reported several bacterial infections in different *M. mola* organs (e.g., kidney, liver or gills) housed at the Matsushima Aquarium in Japan. There is also a report of infection with *Enteromyxum leei* (Diamant, Lom and Dyková, 1994), (Myxozoa), (cited as *Myxidium leei* Diamant, Lom and Dyková, 1994), in aquarium-reared *M. mola* from L'Aquàrium in Barcelona (Spain) (Padrós et al. 2001). Of particular concern are some species of metazoan parasites on *M. mola* that have been reported to survive quarantine procedures at the Oceanogràfic in Valencia (Spain) (Ahuir-Baraja 2012; Table 1). One such species is *Accacoelium contortum* (Rudolphi, 1819) Looss, 1899 (Trematoda) which can severely alter tissues and cause an inflammatory response in the gills and pharynx (Ahuir-Baraja et al. 2015b). This capacity, along with its resistance to antihelmintic procedures, renders this species a prime pathogen for captive *M. mola* (Ahuir-Baraja et al. 2015b). Another species that survives the quarantine period is the cestode *Molicola horridus* (Goodsir, 1841) Dollfus, 1935, the plerocercoid stage of which lives in the liver, kidney and dorsal muscles where it induces inflammatory responses, occupying in some cases up to 28 percent of the liver surface (Ahuir-Baraja 2012). A pronounced inflammatory response was also produced by two ectoparasitic species, *Capsala martinieri* Bosc, 1811 (Monogenea) and *Philorthagoriscus serratus* (Krøyer, 1863) (Copepoda), collected from the dermis and hypodermis of *M. mola* (Logan and Odense 1974). However, given the great diversity of parasites associated with wild molids (Dollfus 1946, Fraser-Brunner 1951, Love and Moser 1983, Ahuir-Baraja et al. 2017, de Figueiredo et al. 2017), the pathology of many more parasite species remains to be investigated.

Parasitic Faunal Groups from Molids

From wild individuals, a total of 73 different species of mainly metazoan parasites have been recorded in *Mola mola* (N = 66 species), *M. alexandrini* (Ranzani, 1839) (17), *Masturus lanceolatus* (5) and *Ranzania laevis* (Pennant, 1776) (3) (Table 1). To date, no references for *Mola tecta* Nyegaard et al. 2017 are included since no parasite faunal studies have been published for this species.

Protists

Three species of protists have been reported in *M. mola*: *Entamoeba molae* Noble and Noble, 1966, *Chilomastix* sp. and *Monocercomonas molae* Noble and Noble, 1966 (Table 1), all found in the hindgut of the same individual from Santa Barbara, California (USA). There was no evidence of pathogenicity caused by these protists (Noble and Noble 1966a) although fishes with diarrhea seem to have more amoebic parasites, such as *E. molae* (Noble and Noble 1966a). Protists generally need one or several hosts to complete their life cycles (Klimpel et al. 2019) and are the only microparasites reported in free-ranging molids.

Trematodes

Trematodes are the most predominant parasitic taxa found in molids (N = 23 species; Table 1). The presence of the family Accacoeliidae Odhner, 1911 is highlighted, with 10 different species (mainly molid specific), located in different organs such as digestive system or gills (Table 1). Bray and Gibson (1977) explored two possible mechanisms of trophic transmission for the accacoeliids. The first involves the mola feeding on known prey items such as ctenophores and cnidarians, with the second involving fish that play host to accacoeliid larval stages (metacercaria) (Fraser-Brunner 1951, Syväranta et al. 2012, Nakamura and Sato 2014). Amongst these, ctenophores and hydromedusae boast the greatest occurrence of trematode infection (Purcell and Arai 2001, Pope et al. 2010, Syväranta et al. 2012, Bakenhaster and Knight-Gray 2016, Busch and Klimpel 2016, Kondo et al. 2016). Other trematode species belong to the Acanthocolpidae Lühe, 1906 (one species), Didymozoidae Monticelli, 1888 (seven species) (Fig. 3), Lepocreadiidae Odhner, 1905 (two species) and Macroderoididae McMullen, 1937 (one species) families (Table 1). Related to family Didymozoidae, its metacercariae are found typically in cephalopods, its intermediate hosts (Busch and Klimpel 2016), suggesting that such species may also feature in the molid diet.

It is noteworthy that all these reported trematodes are usually found in their adult form, suggesting that molids serve as the definitive host for them. The one exception for trematodes is the Acanthocolpidae family which may use molids as an intermediate host before reaching their adult stage in their final hosts, e.g., sharks (Bray and Gibson 2003).

Table 1. Checklist of the parasitefauna from the Molidae Bonaparte, 1835 family. The grade of specificity for the parasite species is indicated as follows: G, generalist, being reported in great host range; Mm, *Mola mola* (Linnaeus, 1758) specialist, being only reported in *M. mola*; Ma, *M. alexandrini* (Ranzani, 1839) specialist, being only reported in *M. alexandrini*; Mal, *Masturus lanceolatus* (Liénard, 1840) specialists, being only reported in *Ma. lanceolatus*; Rl, *Ranzania laevis* (Pennant, 1776) specialists, being only reported in *R. laevis*; and MS, molid specialist, being reported in different species of molids. The location of the parasites, the locality/ies where the parasites have been cited, and the reference/s for these reports are also included. "N.d." means no data available, "N" means north, "S" means south, and "NE" means northeast.

Parasite	Specificity	Location	Locality	Reference
PROTISTS				
Phylum Amoebozoa Lühe, 1913				
Family Entamoebidae Chatton, 1925				
Entamoeba molae Noble and Noble, 1966	Mm	Hindgut	N Pacific	Noble and Noble 1966
Phylum Metamonada Grassé, 1952				
Family Chilomastigidae Cavalier-Smith, 2013				
Chilomastix sp.	G (Mm)	Hindgut	N Pacific	Noble and Noble 1966
Family Monocercomonadidae Kirby, 1944				
Monocercomonas molae Noble and Noble, 1966	Mm	Hindgut	N Pacific	Noble and Noble 1966
Phylum Platyhelminths Gegenbaur, 1859				
TREMATODA				
Family Acanthocolpidae Lühe, 1906				
Stephanostomum valdeinflatum (Stossich, 1883) Linton, 1940 (*taxon inquirendum*) [Syn. *S. baccatum* (Nicoll, 1907) Manter, 1934]	G (Mm)	Digestive system	N Atlantic	Linton 1940
Family Accacoeliidae Odhner, 1911				
Accacoelium contortum (Rudolphi, 1819) Looss, 1899*[Syn.: *Distomum contornum* Rudolphi, 1819]	MS (Mm, Ma)	Gills, pharynx, stomach	Adriatic and Mediterranean Seas, France*, N Atlantic and S Pacific	Olsson 1867, Monticelli 1893, Linton 1898 and 1940, Johnston 1909, Dollfus 1946*, Threlfall 1967, Timon-David and Musso 1971, Bray and Gibson 1977, Cooper et al. 1982, Love and Moser 1983, Villalba and Fernández 1985, Aureli 2004, Gibson et al. 2005, Gustinelli et al. 2006, Radujković and Šundić 2014, Ahuir-Baraja et al. 2015b, Ahuir-Baraja et al. 2017 [Linton 1898]
Accacladium serpentulus Odhner, 1928 [Syn.: [1]*Accacladium nematulum* Noble & Noble, 1937; [2]*Accacladium serpentulum* Odhner, 1928; [3]*Distomum nigroflavum* Rudolphi, 1819]	G[a] (Mm, Ma)	Digestive system, rectum	Mediterranean Sea, France*, N and S Pacific and N Atlantic	Linton 1940, Dollfus 1946*, Dawes 1947, Bray and Gibson 1977, Villalba and Fernández 1985, Ahuir-Baraja 2012, Fernández et al. 2016, Ahuir-Baraja et al. 2017 [[1]Noble and Noble 1937, [1]Threlfall 1967, [1]Timon-David and Musso 1971, [2]Odhner 1928, [2]Yamaguti 1934a, [2]Linton 1940 and [3]1898]

Table 1 contd. ...

... *Table 1 contd.*

Parasite	Specificity	Location	Locality	Reference
Accacladocoelium alveolatum Robinson, 1934 [Syn.: *Guschanskiana alveolata* (Robinson, 1934) Skrjabin, 1959	Mm	Digestive system	Atlantic and Pacific	Robinson 1934, Manter 1960, Thulin 1973, Bray and Gibson 1977, Love and Moser 1983 [Skrjabin 1959, Tantalean et al. 1992]
Accacladocoelium macrocotyle (Diesing, 1858) Robinson, 1934 [Syn.: [1]*Distoma* sp. Bellingham, 1844; [2]*Distomum macrocotyle* Diesing, 1858; [3]*Podocotyle macrocotyle* (Diesing, 1858) Stossich, 1898]	G[b] (Mm, Ma)	Intestine, rectum	Adriatic and Mediterranean Seas, Atlantic and Pacific	Olsson 1867, Monticelli 1893, Linton 1898 and 1940, Guiart 1938, Dawes 1947, Pratt and McCauley 1961, Threlfall 1967, Timon-David and Musso 1971, Bray and Gibson 1977, Villalba and Fernández 1985, Rivera and Sarmiento 1992, Luque and Oliva 1993, Gaevskaya 2002, Aureli 2004, Gustinelli et al. 2006, Radujković and Šundić 2014, Ahuir-Baraja et al. 2015a, Fernández et al. 2016, Ahuir-Baraja et al. 2017 [[1]Bellingham 1844, [2]Linton 1898, [3]Stossich 1898]
Accacladocoelium nigroflavum (Rudolphi, 1819) Robinson, 1934 [Syn. [1]*Distoma nigroflavum* Rudolphi, 1819, [2]*Echinostoma nigroflavum* (Rudolphi, 1819) Barbagallo & Drago, 1903]	MS (Mm, Ma)	Stomach, intestine	Mediterranean Sea and Atlantic	Dawes 1947, Timon-David and Musso 1971, Bray and Gibson 1977, Ahuir-Baraja et al. 2015a, Ahuir-Baraja et al. 2017 [[1]Barbagallo and Drago 1903, [2]Rudolphi 1819]
Accacladocoelium petasiporum Odhner, 1928	MS (Mm, Ma)	Intestine	Adriatic and Mediterranean Seas, France* and N Atlantic	Odhner 1928, Dollfus 1946*, Dawes 1947, Thulin 1973, Bray and Gibson 1977, Gaevskaya 2002, Ahuir-Baraja 2012, Radujković and Šundić 2014, Ahuir-Baraja et al. 2017
Rhynchopharynx paradoxa Odhner, 1928	MS (Mm, Ma)	Digestive system	Adriatic and Mediterranean Seas, France*, N Atlantic and Pacific	Odhner 1928, Yamaguti 1934a, Manter 1934 and 1960, Dollfus 1946*, Bray and Gibson 1977, Radujković and Šundić 2014, Ahuir-Baraja et al. 2017
Odhnerium calyptocotyle (Monticelli, 1893) Yamaguti, 1934[c] [Syn.: [1]*Distoma foliatum* Linton, 1898; [2]*D. nigroflavum* Rudolphi, 1819; [3]*D. fragile* Linton, 1901; [4]*Mneiodhneria calyptrocotyle* (Monticelli, 1891) Dollfus, 1935; [5]*M. foliata* (Linton, 1898) Dollfus, 1935; [6]*Orophocotyle foliata* (Linton, 1898) Looss, 1902; [7]*O. planci* (Stossich, 1899) Loos, 1902; [8]*Accacoelium foliatum* (Linton, 1898) Stafford, 1904; [9]*Caballeriana lagodovsky* Skrjabin & Guschanskaja, 1959; [10]*Distomum calyptrocotyle* Monticelli, 1893]	MS[c] (Mm, Ma)	Intestine	Adriatic and Mediterranean Seas, Pacific and Atlantic	Yamaguti 1934a, Lloyd 1938, Manter 1934, Montgomery 1957, McCann 1961, Bray and Gibson 1977, Love and Moser 1983, Villalba and Fernández 1985, Gibson et al. 2005, Ahuir-Baraja 2012, Radujković and Šundić 2014 [[1]Linton 1898 and [6]1940, [2]Olsson 1867, [3]Threlfall 1967, [4]Dollfus 1935, [4]Timon-David and Musso 1971, [5]Guiart 1938, [7]Dawes 1947, [8]Stafford 1904, [9]Skrjabin and Guschanskaja 1959, [10]Monticelli 1893]
Orophocotyle divergens Loos, 1902	Rl	n.d.	Mediterranean Sea	Looss 1902

Table 1 contd. ...

... Table 1 contd.

Parasite	Specificity	Location	Locality	Reference
O. planci (Stossich, 1899) Looss, 1902 [Syn. *Podocotyle planci* Stossich, 1899]	Rl	n.d.	Mediterranean Sea	Yamaguti 1971 [Stossich 1899]
Family Didymozoidae (Monticelli, 1888)				
Didymozoidae gen. n. sp. n**	Ma	Subcutaneous tissues	Mediterranean Sea	Ahuir-Baraja et al. 2017
Didymozoon molae (Rudolphi, 1819)	Mm	Dorsal muscles	Atlantic and France*	Dollfus 1946*, Pozdnyakov 1996
Didymozoon taenioides Monticelli, 1883 (*taxon inquirendum*) [Syn. *Monostoma molae* Rudolphi, 1819; *Didymozoum taenioides* Monticelli, 1888]	Mm	Dorsal muscle	Adriatic, Ligurian, Tyrrhenian and Mediterranean Seas and NE Atlantic	Bona et al. 1995, Pozdnyakov 1996, Ortis and Paggi 2008, Radujković and Šundić 2014
Nematobothrium benedeni (Monticelli, 1893) (*taxon inquirendum*) [Syn. *Koellikeria benedenii* (Monticelli, 1893)]	Mm	n.d.	Atlantic	Pozdnyakov 1996 [Pozdnyakov 1996]
Nematobothrium molae (Maclaren, 1904) [Syn. *Benedenozoum molae* (Maclaren, 1904) Ishii, 1935]	Mm	Gills	Mediterranean Sea, France*, Atlantic and N Pacific	Maclaren 1904, Dollfus 1946*, Dawes 1947, Timon-David and Musso 1971, Noble 1975 [Ishii 1935]
Nematobibothrioides histoidii Noble, 1974	Mm	Connective tissue	Atlantic and N Pacific	Noble 1975, Thulin 1980
Koellikeria filicollis (Rudolphi, 1819)	G (Mm)	Intestine	S Pacific	Nicoll 1915
Gonapodasmius squamata (Pozdnyakov, 1993)*	Mm	Gills and hyoid gills	N Pacific and Mediterranean Sea	Pozdnyakov 1994, Ahuir-Baraja, 2012
Family Hemiruridae Looss, 1899				
Lecithochirium sp.	Ma[d]	Oral cavity, digestive system	S Atlantic	Mendez-Ahid et al. 2009[d]
Family Macroderoididae McMullen, 1937				
Cirkennedya porlockensis Gibson & Bray, 1979	Mm	Intestine	N Atlantic	Gibson and Bray 1979
Family Lepocreadiidae Odhner, 1905				
Dihemistephanus lydiae (Stossich, 1896) Looss 1901 [Syn.: [1]*Echinostoma lydiae*; [2]*Dihemistephanus fragilis* (Linton, 1900) Stafford 1904; [3]*Distomum fragile* (Linton, 1900)]	MS (Mm, Ma)	Intestine	Adriatic and Mediterranean Seas, France*, Atlantic and S Pacific	Linton 1900, Nicoll 1909 and 1915, Dollfus 1946*, Manter 1960, Peters 1960, Yamaguti 1968, Bray and Gibson 1977, Gibson 1996, Bray and Gibson 1997, Ahuir-Baraja 2012, Radujković and Šundić 2014, Ahuir-Baraja et al. 2017 [[1]Stossich 1896, [2]Stafford 1904, [3]Threlfall 1967]

Table 1 contd. ...

... *Table 1 contd.*

Parasite	Specificity	Location	Locality	Reference
Dihemistephanus fragilis (Linton, 1900) Yamaguti, 1971 [Syn. [1]*Stenocollum fragile* (Linton, 1900) Stafford, 1904, [2]*Distomum fragile* (Linton, 1900)]	Mm	Intestine	N Atlantic, France*	Dollfus 1946* [[1]Stafford 1904, [2]Linton 1900]
MONOGENEA				
Family Capsalidae Carus, 1863				
Capsala martinieri Bosc, 1811 [Syn.: [1]*C. cephala* (Risso, 1826); [2]*C. cutaneum* Guiart, 1938; [3]*C. grimaldi* Guiart, 1938; [4]*C. maculata* Martiniere 1787; [5]*C. molae* (Blanchard, 1847) Johnston, 1929; [6]*Tricotyla cutanea,* Guiart, 1938; [7]*T. molae* (Blanchard, 1847) Guiart, 1938; [8]*Tristomella grimaldii* Guiart, 1938; [9]*Tristomum rudolphianum* Diesing, 1850]	G^e (Mm, Ma, Mal)	Body surface	Atlantic, Pacific, Mediterranean and Adriatic Seas and France*	Stossich 1896, Dollfus 1946*, Sproston 1946, Brinkmann 1952, Threlfall 1967, Logan and Odense 1974, Cooper et al. 1982, Love and Moser 1983, Villalba and Fernández 1985, Radujković and Euzet 1989, Euzet et al. 1993, Lamothe-Argumedo 1997, Kohn and Cohen 1998, Egorova 2000, Kohn and Paiva 2000, Olson and Littlewood 2002, Brito 2003, Simková et al. 2003, Hernández-Orts et al. 2009, Riera et al. 2010, Ahuir-Baraja 2012, Cohen et al. 2013, Radujković and Šundić 2014 [[1]Yamaguti 1963b, [2,3]Dawes 1947, [4]Timon-David and Musso 1971, [4]Dollfus 1946*, [5]Linton 1940 and [9]1898, [6,7,8]Guiart 1938, [7]Robinson 1934, [7]Draper 1941, [7]Price 1962]
Capsala pelamydis (Taschenberg, 1878) Johnston, 1929 [Syn.: *Caballerocotyla pelamydis* (Taschenberg, 1878)]	G (Mm)	Body surface	N Pacific	Dawes 1947, Love and Moser 1983
Tristoma papillosum Diesing, 1836 [unaccepted synonym of *T. coccineum* Cuvier, 1817]	G (Mm)	Body surface	France* and N Pacific	Dollfus 1946*, Love and Moser 1983
CESTODA				
Family Gymnorhynchidae Dollfus, 1935				
Molicola horridus (Goodsir, 1841) Dollfus, 1935* [Syn.: [1]*Gymnorhynchus horridus* Goodsir, 1841; [2]*Rhynchobothrium* sp. (Linton, 1899) Linton, 1909; [3]*Tetrarhynchus elongatus* Wagener, 1901; [4]*Tetrarhynchus reptans* Wag.]	G (Mm, Ma)	Liver, kidney, dorsal muscles	Mediterranean Sea, S Pacific and Atlantic	Robinson 1959, McCann 1961, Villalba and Fernández 1985, Andersen 1987, Schwartz and Lindquist 1987, Bates 1990, Aureli 2004, Gustinelli et al. 2006, Ahuir-Baraja 2012, Fernández et al. 2016, Ahuir-Baraja et al. 2017 [[1]Goodsir 1841, [2]Threlfall 1967, [2]Bates 1990, [3]Linton 1924, [3]Threlfall 1967, [4]Johnston 1909]
Family Lacistorhynchidae Guiart, 1927				
Floriceps saccatus Cuvier, 1817	G (Mm)	Liver, mesentery, encysted in the walls of the digestive system	Mediterranean Sea, France* and N Atlantic	Guiart 1935, Dollfus 1946*, Hillis and O'Riordan 1960. Timon-David and Musso 1971, Andersen 1987, Bates 1990, Ahuir-Baraja 2012

Table 1 contd. ...

... *Table 1 contd.*

Parasite	Specificity	Location	Locality	Reference
Family Pterobothriidae Pintner, 1931				
Pterobothrium sp.	G (Mm)	Liver	Mediterranean Sea	Aureli 2004, Gustinelli et al. 2006
Family Taeniidae Ludwig, 1886				
Cysticercus sp.	G (Mm)	Intestine	S Pacific	Johnston 1909
Family Tentaculariidae Poche, 1926				
Nybelinia sp.[f]	G (Ma)	Intestine	S Pacific	Villalba and Fernández 1985
Tetraphyllidea *incertae sedis* van Beneden, 1849				
Tetraphyllidean plerocercoids**	G (Mm)	Digestive system	Mediterranean Sea	Ahuir-Baraja 2012
Family Triaenophoridae Lönnber, 1889				
Anchistrocephalus microcephalus (Rudolphi, 1819) [Syn.: [1]*Bothriocephalus microcephalus* Rudolphi, 1819; [2]*Bothriocephalus monorchis* Linstow, 1903]	MS (Mm, Ma)	Intestine	Adriatic and Mediterranean Seas France*, Atlantic and Pacific	Robinson 1934, Yamaguti 1934b, Dollfus 1946*, Robinson 1959, Hillis and O'Riordan 1960, McCann 1961, Threlfall 1967, Timon-David and Musso 1971, Kennedy and Andersen 1982, Villalba and Fernández 1985, Schwartz and Lindquist 1987, Johnston 1909, Shimazu et al. 1996, Olson et al. 2001, Aureli 2004, Gustinelli et al. 2006, Ahuir-Baraja 2012, Radujković and Šundić 2014, Fernández et al. 2016, Ahuir-Baraja et al. 2017 [[1]Hartwich and Kilias 1992, [2]Dollfus 1946*]
Fistulicola plicatus (Rudolphi, 1819) Lühe, 1899	G (Mm)	Intestine	N Atlantic	Linton 1941
Rhynchobotrium sp. (*genus inquirendum*)	G (Mm)	Intestine wall	N Atlantic	Threlfall 1967
Order Trypanorhyncha[g]	G (Mal)	Liver	S Atlantic	Araújo et al. 2010[g]
Phylum Acanthocephala Kolhreuther, 1771				
Family Echinorhynchidae Cobbold, 1879				
Echinorhynchus acus Rudolphi, 1902 [unaccepted synonym of *Echinorhynchus gadi* Zoega in Muller, 1776]	G (Mm)	Gills[h]	N Atlantic, France*	Linton 1901, Dollfus 1946*, Love and Moser 1983
Family Polymorphydae Meyer, 1931				
Bolbosoma capitatum (von Linstow, 1880) Porta, 1908**	G (Mm)	Encysted in digestive system wall	Mediterranean Sea	Ahuir-Baraja 2012
Phylum Nematoda Rudolphi, 1808				
Family Anisakidae Skrjabin et Karokhin, 1945				

Table 1 contd. ...

... Table 1 contd.

Parasite	Specificity	Location	Locality	Reference
Anisakis sp.	G (Mm)	Encapsulated in mesentery and viscera	Atlantic and S Pacific	Hewitt and Hine 1972, Fernández et al. 2016
Anisakis sp. type I (sensu Berland, 1961)[**]	G (Mm, Ma)	Encapsulated in digestive system wall	Mediterranean Sea	Ahuir-Baraja 2012, Ahuir-Baraja et al. 2017
Anisakis sp. type II (sensu Berland, 1961)[**]	G (Mm)	Encapsulated in digestive system wall	Mediterranean Sea	Ahuir-Baraja 2012
Family Ascarididae Baird, 1853				
Ascaris ranzania (M. Stossich 1893) (*incertae sedis*)[i]	Rl	n.d.	Adriatic Sea	Stossich, 1883
Family Cucullanidae Cobbold, 1864				
Cucullanus orthagorisci (Rudolphi, 1819) [Syn.: *Ascaris orthragorisci* Rudolphi, 1819]	Mm	Digestive system	Mediterranean Sea and France*	Rudolphi 1819 [Dollfus 1946*, Bruce et al. 1994]
Family Cystidicolidae Skrjabin, 1946				
Ascarophis sp.[**]	G (Mm)	Encapsulated in digestive system wall	Mediterranean Sea	Ahuir-Baraja 2012
Agamonema sp. (larval name, *incertae sedis*)[j]	G (Mm)	Encysted in intestine	N Atlantic	Linton 1901, Dollfus 1946*
Phylum Arthropoda Latreille, 1829				
BRANCHIURA				
Family Argulidae, Leach 1819				
Argulus scutiformis Thiele, 1900	G (Mm)	Body surface	N Pacific and France*	Tokioka 1936, Dollfus 1946*, Yamaguti 1963a
COPEPODA				
Family Caligidae Burmeister, 1835				
Caligus elongatus von Nordmann, 1832	G (Mm)	Body surface	Mediterranean Sea and N Atlantic	Olsson 1869, Wilson 1905, Parker 1969
C. rapax Milne Edwards, 1840	G (Mm)	n.d.	North Sea	Hamond 1969
Caligus bonito bonito Wilson, 1905[**]	G (Mm)	Body surface	Mediterranean Sea	Ahuir-Baraja 2012
C. pelamydis Krøyer, 1863[**]	G (Mm)	Body surface	Mediterranean Sea	Ahuir-Baraja 2012
Caligus sp.[**]	Mm	Body surface	Mediterranean Sea	Ahuir-Baraja 2012
Lepeophtheirus formosanus Ho & Lin, 2010	Mm	Body surface	N Pacific	Ho and Lin 2010
L. hastatus Shiino, 1960 [Syn.: *L. molae* Heegaard, 1962]	Mm	Body surface	Pacific and Atlantic	Shiino 1960 [Heegard 1962, Hewitt 1964]

Table 1 contd. ...

... *Table 1 contd.*

Parasite	Specificity	Location	Locality	Reference
L. nordmanni (Milne Edwards, 1840) [Syn.: *L. insignis* Wilson, 1908]	G[k] (Ma, Mm, Mal)	Body surface, bucal cavity	Mediterranean Sea, France*, Pacific and Atlantic	Milne Edwards 1840, Baird 1850, Heller 1866, Scott 1901, Wilson 1905 and 1932, Robinson 1934, Dollfus 1946*, Stuardo 1958, Hillis and O'Riordan 1960, Hewitt 1964 and 1971, Villalba and Fernández 1985, Oldewage 1993, Díaz 2000, Aureli 2004, Dippenaar 2005, Gustinelli et al. 2006, Ahuir-Baraja 2012 [Wilson 1907 and 1932, Barnard 1948, Shiino 1957 and 1960, Threlfall 1967, Kensley and Grindley 1973]
Family Dichelesthiidae Milne Edwards, 1840				
Anthosoma crassum (Abildgaard, 1794)	G (Mm)	Bucal cavity, operculum, fins	Mediterranean Sea and France*	Dollfus 1946*, Timon-David and Musso 1971
Family Cecropidae Dana, 1849				
Cecrops latreillii Leach, 1816 [Syn.: *Cecrops exiguus* Wilson, 1923]	G[l] (Mm, Ma, Mal)	Gills	Mediterranean Sea France*, Atlantic and Pacific	Wilson 1932, Dollfus 1946*, Barnard 1955, Hillis and O'Riordan 1960, McCann 1961, Threlfall 1967, Hewitt 1968, Timon-David and Musso 1971, Kensley and Grindley 1973, Grabda 1973, Bakke 1981, Cooper et al. 1982, Villalba and Fernández 1985, Raibaut et al. 1998, Díaz 2000, Aureli 2004, Dippenaar 2005, Gustinelli et al. 2006, Pastore 2009, Ahuir-Baraja 2012, Fernández et al 2016 [Shiino 1965, Kensley and Grindley 1973]
Philorthragoriscus serratus (Krøyer, 1863) [Syn. *Dinematura serrata* Krøyer, 1863]	MS (Mm, Mal)	Body surface	Mediterranean Sea, Atlantic, N Pacific and France*	Wilson 1932, Dollfus 1946*, Barnard 1948 and 1955, Shiino 1960, Threlfall 1967, Kensley and Grindley 1973, Logan and Odense 1973, Cooper et al. 1982, Raibaut et al. 1998 [KrØyer 1863]
Orthagoriscicola muricatus (Krøyer, 1837) [Syn. [1]*Orthagoriscicola wilsoni* Schuurm-Stekhoven Jr & Schuurm-Stekhoven, 1956; [2]*Laemargus muricatus* Krøyer, 1837]	Mm	Gills	Mediterranean Sea, France* and Atlantic	Wilson 1907, Dollfus 1946*, Barnard 1955, Kensley and Grindley 1973, Cooper et al. 1982, Raibaut et al. 1998 [[1]Threlfall 1967, [1,2]Walter and Boxshall 2019, [2]Evans 1906]
Family Lernaeopodidae Milne Edwards, 1840				
Lernaeopoda bidiscalis Kane, 1892	G (Mm)	Body surface	France*	Dollfus 1946*
Family Pandaridae Milne Edwards, 1840				
Dinemoura latifolia (Steenstrup & Lütken, 1861) [Syn. *Dinematura latifolia* Steenstrup & Lütken, 1861]	G (Mm)	n.d.	S Atlantic	Kensley and Grindley 1973, Oldewage 1992

Table 1 contd. ...

... *Table 1 contd.*

Parasite	Specificity	Location	Locality	Reference
Echthrogaleus coleoptratus (Guérin-Méneville, 1837)	G (Mm)	Body surface	France*	Dollfus 1946*
Nogaus sp. n.••	Mm	Body surface	Mediterranean Sea	Ahuir-Baraja 2012
Pandarus bicolor Leach, 1816	G (Mm)		France*	Dollfus 1946*
Family Pennellidae Burmeister, 1816				
Pennella filosa (L.) [Syn.: [1]*P. crassicornis* Streenstrup and Lutken 1861; [2]*P. orthagorisci* Wright E.P., 1870; [3]*P. rubra* Brian, 1906]	G (Mm)	Body surface	Mediterranean Sea, France*, Atlantic and Pacific	Dollfus 1946*, Atria 1967, Cooper et al. 1982, Hogans 1987, Raibaut et al. 1998, Ahuir-Baraja 2012 [[1]Yamaguti 1963a, [2,3]Wilson 1932, [3]Barnard 1955, [3]Pastore 2009]
Family Trebiidae Wison C. B., 1905				
Trebius sp.	G (Mm)	Gills	France*	Dollfus 1946*
ISOPODA				
Family Cymothoidae Leach, 1818				
Ceratothoa steindachneri Koelbel, 1879	G (Mm)	Oral cavity	Mediterranean Sea	Aureli 2004, Gustinelli et al. 2006
Nerocila japonica Schioedte & Meinert, 1881	G (Mm)	Body surface	N Atlantic	Hata et al. 2017
Nerocila orbignyi (Guérin-Méneville, 1832)	G (Mm)	Body surface	France*	Dollfus 1946*
Nerocila macleayi Miers 1884 (*nomen nudum*)	G (Mm)	Body surface	France*	Dollfus 1946*
Family Gnathiidae Harger, 1880				
Gnathia sp.••	G (Mm)	Body surface	Mediterranean Sea	Ahuir-Baraja 2012

* In Dollfus (1946) it is not indicated if it is French Mediterranean shore or Atlantic shore.

[a] Pérez-del Olmo et al. (2006) reported the presence of *Accacladium serpentulus* in *Boops boops* (L.) (Sparidae) from NE Atlantic.

[b] Reported in *Lophius piscatorius* (L.) (Lophiidae) as *Distomum macrocotyle* Diesing, 1858 (WoRMS 2019).

[c] Reported in *Beroe ovata* Bruguière, 1789 (Ctenophora) as larval stage (WoRMS 2019).

[d] In Mendez-Ahid et al. (2009) there is a misidentification of *Mola mola*. The analyzed specimen is *Masturus lanceolatus*. In this work, the authors also reported plerocerci from the order Trypanorhyncha in "...bucal cavity, digestive system, liver, muscle and heart..." but without indicating genus or species.

[e] *Capsala martinieri* is also reported in *Hemigymnus melapterus* (Bloch, 1791) (Labridae) (Gibson et al. 2005).

[f] For *Nybelinia* sp. the authors refers to two different morphotypes reported in *M. alexandrini* (Villalba and Fernández 1985).

[g] Araújo et al. (2010) reported for *Ma. lanceolatus* in liver "...numerous cysts from nematodes and cestodes from the order Trypanorhyncha..." not indicating genus or species.

[h] *Echinorhynchus acus* is reported in gills, according to the comments of Dollfus (1946) for Linton's (1901) report. This is a not common location for acanthocephalans, which are usually located in digestive system (Klimpel et al. 2019). This unusual location for parasites commonly located in digestive system is considered as doubtful in other parasitic groups as trematodes (Gibson 2002).

[i] As *Agamonema ranzianae* in Stossich (1893).

[j] For *Agamonema* sp. Dollfus (1946) refers to a little larvae found by Linton (1901) insufficient for a correct identification of the genus. We suggest keeping it as *incertae sedis*.

[k] *L. nordmanni* classified as "generalist" but it is cited in *Mola mola*, *M. alexandrini*, *Masturus lanceolatus* and *Scophthalmus maximus* (Linnaeus, 1758) (Scophthalmidae) (WoRMS 2019).

[l] *Cecrops latreilli* classified as "generalist" but it is cited in *Mola mola*, *M. alexandrini*, *Masturus lanceolatus* and *Thunnus thynnus* (Linnaeus, 1758) (Scombridae) (WoRMS 2019).

• Parasitic species that survived the quarantine procedures at the *Oceanogràfic* of Valencia (Spain) (Ahuir-Baraja 2012).

•• Parasitic species reported for the first time in *Mola mola* or *M. alexandrini* from the Mediterranean Sea (Ahuir-Baraja 2012, Ahuir-Baraja et al. 2017).

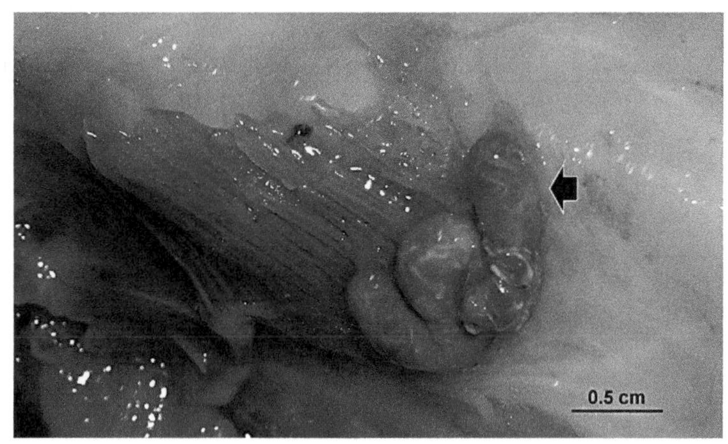

Figure 3. *Gonapodasmius squamata* (Trematoda) (black arrow) located on a structure known as the hyoid gills (Gregory and Raven 1934) of *Mola mola* from the Mediterranean Sea.

Monogeneans

Three generalist species of monogeneans (e.g., Silas 1962, Pozdnyakov 1990, Hendrix 1994, Gibson et al. 2005) have been reported on the body surface of molids: *Capsala martinieri* Bosc, 1811 (Fig. 4), *Capsala pelamydis* (Taschenberg, 1878) Johnston, 1929 and *Tristoma papillosum* Diesing, 1836 (unaccepted synonym of *T. coccineum* Cuvier, 1817) (Table 1). All three species have been reported for *M. mola*. *C. martinieri* has been also reported in *M. alexandrini* and *Ma. lanceolatus* (Table 1). However, it is important to note that monogeneans, and specifically *C. martinieri*, have had numerous synonymizations (Table 1). Furthermore, as monogeneans are ectoparasitic, they may drop off during handling and avoid being recorded, so our current understanding may under represent their use of *Mola* spp. as hosts.

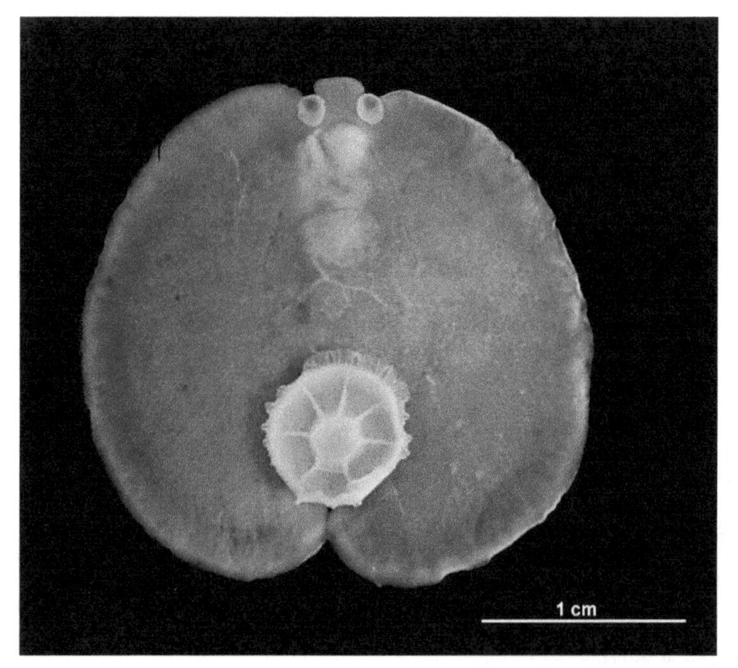

Figure 4. Isolated specimen of *Capsala martinieri* (Monogenea) from the skin of *Mola mola*, from the Mediterranean Sea.

Cestodes

Knowledge of the life cycles of marine cestodes is scarce (Marcogliese 1995, Klimpel et al. 2019) with limited information about their intermediate and final hosts. In molids, 10 species have been recorded, nine from *M. mola*, three from *M. alexandrini* and one, identified to order level, in *Ma. lanceolatus*. Seven species include only larval stages (plerocerci) encysted in different organs, with molids acting as intermediate or paratenic hosts, e.g., *Molicola horridus*, which is located in the liver (Fig. 5), kidney and muscle; Table 1). Likewise, *Floriceps saccatus* Cuvier, 1817 is found in the liver and mesentery, while *Pterobothrium* sp. has only been found in the liver of *M. mola*. In contrast, *Nybelinia* sp., *Cysticercus* sp. and *Rhynchobothrium* sp. are reported in the intestines of *M. alexandrini*. All these species are generalist parasites and have been reported as larval stages in a wide range of teleosts who act as intermediate hosts (e.g., plerocerci of *Floriceps saccatus* in swordfish, *Xiphias gladius* [Linnaeus 1758; Love and Moser (1983)]. Finally, tetraphyllidean plerocerci have been described (with different morphotypes) in the digestive system of *M. mola* (Ahuir-Baraja 2012). These larvae are found in cetaceans and other marine mammals in different viscera (Agustí et al. 2005, Aznar et al. 2007, Oliveira et al. 2011) and are common in other teleosts, molluscs and crustaceans (Stunkard 1977, Jensen and Bullard 2010, Ferrer-Maza et al. 2015).

Adult specimens of all these cestodes mentioned above are located in stomach and spiral valves of sharks and rays which act as final hosts (Dollfus 1942, Heinz and Dailey 1974, Love and Moser 1983, Butler 1987, Jensen and Bullard 2010, Mendez and Galván-Magana 2016). For example, adult stages of *M. horridus* are reported in shortfin mako, *Isurus oxyrinchus* Rafinesque, 1810 and blue shark, *Prionace glauca* (Linnaeus, 1758) (Dollfus 1942, Heinz and Dailey 1974, Arru et al. 1991, Gaevskaya and Kovaljova 1991, Palm 2004). For the blue shark, small molids are reported as prey (Garibaldi and Orsi Relini 2000, Pope et al. 2010). Larger molids (up to 3.1 m length) (Carwardine 1995, Roach 2003) are known prey for orcas, *Orcinus orca* (Linnaeus, 1758), white sharks, *Carcharodon carcharias* (Linnaeus, 1758), sea lions, *Zalophus californianus* (Lesson, 1828), and tiger sharks, *Galeocerdo cuvier* (Péron and Lesueur, 1822) (Gladstone 1988, Fergusson et al. 2000, Ryan and Holmes 2012, Nyegaard et al. 2019).

The other two species of cestodes reported in molids are *Anchistrocephalus microcephalus* (Rudolphi 1819) and *Fistulicola plicatus* (Rudolphi, 1819) Lühe, 1899 (Table 1). These cestodes are only recorded in their adult form, suggesting that sunfish are the final host. The larval stages of both species are found in zooplankton (e.g., copepods; Dollfus 1923 and 1964, Riser 1956, Mudry and Dailey 1971, Busch et al. 2012) and in a broad range of teleosts (Beveridge et al. 2017, Klimpel et al. 2019) acting as intermediate or paratenic hosts who may be part of the molids' diet.

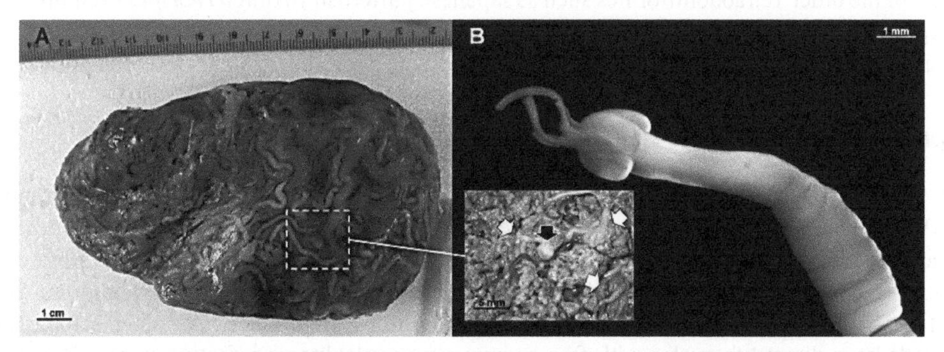

Figure 5. The liver of a *Mola mola* collected from the Mediterranean Sea, infected by *Molicola horridus* (Cestoda). (A) Isolated liver that has been highly parasitized by *M. horridus*. (B) Detail of the anterior region of the plerocerci of *M. horridus* from the liver (inset: detail of the blastocyst (black arrow) and body elongation (white arrow) of the plerocerci of *M. horridus* in the liver). Scale bars: a, 1 cm; b, 1 mm; inset, 5 mm.

Acanthocephalans

To date, only two species of acanthocephalans have been reported in *M. mola*: adult stages of *Echinorhynchus gadi* (Zoega in Müller, 1776) (reported as *E. acus* Rudolphi, 1902) located in gills; and cystacanth (larval stage) of *Bolbosoma capitatum* (von Linstow, 1880) Porta, 1908 encysted in the digestive system wall. The latter species was reported in a *M. mola* for the first time in the Mediterranean Sea (Table 1). For *Echinorhynchus gadi*, sunfish appear to be final hosts, and intermediate hosts for *B. capitatum*. Typically, acanthocephalans use different species of benthic amphipods or ostracods (Crustacea) as intermediate hosts (Marcogliese 1995). Cystacanth stages are quite unspecific and are found in many families of marine teleosts (Measures 1992). However, records of larval acanthocephalans in zooplankton are scarce and reported only for euphausiids (Crustacea) (Marcogliese 1995).

Nematodes

Larval stages of seven nematode species have been reported in molids. Three of these were reported for the first time in *M. mola* in the Mediterranean Sea: *Anisakis* sp. Larvae 3 Type I (sensu Berland, 1961), *Anisakis* sp. Larvae 3 Type II (sensu Berland, 1961) and *Ascarophis* sp., all within the digestive system wall. *Anisakis* sp. Larvae 3 Type I has also been recorded in *M. alexandrini* in the Mediterranean Sea (Ahuir-Baraja et al. 2017). The other species of nematodes reported in sunfish are *Cucullanus orthagorisci* (Rudolphi, 1819) (as *Ascaris orthragorisci* Rudolphi, 1819), and *Agamonema* sp. in digestive system and mesenteries. *Ascaris ranzania* (Stossich, 1893) is the only species reported from *Ranzania laevis*, however its specific location is not stated (Table 1).

Usually, nematodes have four larval stages that can infect a great range of paratenic hosts, with molids most likely acting as intermediate hosts (McClelland et al. 1990, Nagasawa 1990, Marcogliese 1995, Petter and Radujković 1989, Busch and Klimpel 2016). The final hosts of these marine nematodes are typically marine mammals (pinnipeds and cetaceans) and other piscivorous fish (Marcogliese 1995, Mattiucci and Nascetti 2007, Busch and Klimpel 2016), although the trophic linkages to sunfish in this case, are not clear.

Branchiurans

There is only one species of the subclass Branchiura Thorell, 1864 reported for *M. mola*, *Argulus scutiformis* Thiele, 1900. This generalist ectoparasite has been reported in other teleosts, including species of the order Tetraodontiformes such as Japanese pufferfish *Takifugu rubripes* (Temminck and Schlegel, 1850) (Leong and Colorni 2002, Nagasawa 2009; Table 1). As with monogeneans, these parasites have direct life cycles, requiring no intermediate hosts during their development.

Copepods

With 19 species reported to date, copepods are the second most common taxon found in molids. All 19 have been reported for *M. mola*, three for *Ma. lanceolatus* and two in *M. alexandrini* (Table 1; Fig. 6; see also Ahuir-Baraja 2012). From the listed species *Caligus* sp., *Caligus bonito bonito* Wilson, 1905, *C. pelamydis* (Krøyer, 1863) and *Nogaus* sp. n. have been reported for the first time in *M. mola* from the Mediterranean Sea (Ahuir-Baraja 2012). As other ectoparasites, copepods have direct life cycles with free-swimming, zooplanktonic infective stages (Boxshall et al. 2005).

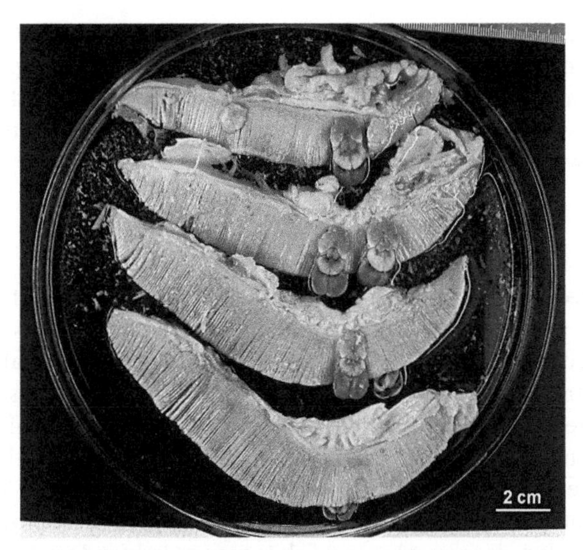

Figure 6. Isolated gills of *Mola mola* collected from an individual in the Mediterranean Sea, parasitized by several specimens of *Cecrops latreilli* (Copepoda). Scale bar: 2 cm.

Isopods

Five species of isopods have been reported in *M. mola*; all of them generalist parasites. Four species, reported as adult stages, are *Ceratothoa steindachneri* Koelbel, 1879, *Nerocila japonica* Schioedte and Meinert, 1881, *N. orbignyi* (Guérin-Méneville, 1832) Schioedte and Meinert, 1881 and *N. macleayi* White, 1843 (Table 1). Praniza larvae of the genus *Gnathia* Leach, 1814 is the fifth species reported (Ahuir-Baraja 2012). Like monogeneans or copepods, recording the diversity and abundance of such species during necropsy can be problematic owing to their weak host fixation. This issue is apparent for gnathids (Smit and Davies 2004) which have free-swimming infective larval stages that must find hosts in order to develop to the adult stage (Mladineo 2003, Smit et al. 2003, Smit and Davies 2004).

Symbiotic Cleaning Associations

The basking behavior in molids has been described in a number of studies (e.g., Cartamil and Lowe 2004, Watanabe and Sato 2008, Houghton et al. 2009, Abe and Sekiguchi 2012). One partial explanation for such behavior includes parasite removal. There is some evidence to corroborate this theory with observations of albatrosses apparently feeding on sunfish ectoparasites (Abe et al. 2012, Abe and Sekiguchi 2012). Specifically, *Phoebastria immutabilis* (Rothschild, 1893) (Laysan albatross) and *P. nigripes* (Audubon, 1839) (black-footed albatross), in western North Pacific Ocean, have been observed picking off copepods from the genus Pennella Oken, 1815. Furthermore, Abe and Sekiguchi (2012) observed scars related to the removal of other specimens of *Pennella* sp. and injuries caused by the albatrosses (or other seabirds) when removing parasites, suggesting that such behavior may be common. This assertion is supported by reports of *Pennella* sp. from albatross stomachs in Hawaii (Harrison et al. 1988).

However, such cleaning events are not restricted to the sea surface. For example, fish species such as butterfly fish, *Heniochus diphreutes* Jordan, 1903, angelfish, *Pomacanthus imperator* (Bloch, 1787) or wrasses such as *Labroides dimidiatus* (Valenciennes, 1839) or *Coris julis* (L.), have been observed feeding on the ectoparasites of *M. ramsayi* (now known as *M. alexandrini*) (see Sawai et al. 2020 [Chapter 2], Thys et al. 2020 [Chapter 14]) in Indonesia and the Galápagos Islands, Ecuador (Konow 2006, Thys et al. 2016, Thys et al. 2017) and *M. mola* in western Mediterranean (Vasco-Rodrigues and Cabrera 2015). During these specific events, the fish appeared to be feeding on copepods from the genera *Lepeophtheirus* von Nordmann, 1832 or *Caligus* Müller O.F., 1785 and monogeneans from the genera *Capsala* Bosc, 1811 or *Tristoma* Cuvier, 1817, whilst avoiding direct feeding on the mucus or skin of the host (Poulin and Grutter 1996). In this way the cleaners get direct access to readily ingestible prey whilst reducing the parasitic load of the host (most of which are hematophagous) which inevitably could improve host health.

Future Research

While progress has been made identifying the myriad numbers and types of parasites living in and around molids, more research is needed into the direct or indirect effects of parasites on molid health. Protocols to conduct parasitological studies on *Mola tecta* should also be implemented to gain a better understanding of how attachment locations (e.g., liver or digestive tract) and parasite feeding behaviors affect growth, locomotion, buoyancy and survivorship of sunfish hosts. A more thorough examination of behavior at cleaning stations could offer additional insight. To date, such behavior has been reported mainly for *M. alexandrini* (Konow 2006, Thys et al. 2016, Thys et al. 2017). Coordinated efforts through citizen science initiatives may provide the mechanism to gather such data, particularly in areas where sunfish occur regularly and are of high tourist value (Thys et al. 2016, Thys et al. 2020 [Chapter 14]). Collecting movement data via acoustic telemetry arrays at known sunfish cleaning stations would also shed light on how much time is spent getting cleaned. Considering the allocated time and frequency of visits to cleaning stations, such services likely provide health benefits or improved survivorship for ocean sunfishes. Further comparative studies of parasitic species abundance and diversity in juvenile and adult molids can lend insight into a number of other areas of interest including ontogenetic shifts in sunfish diet (e.g., Nakamura and Sato 2014, Sousa et al. 2016. Phillips et al. 2020 [Chapter 9]). It is also noteworthy that juvenile sunfish swim in schools (Abe et al. 2012), which may influence transmission of ectoparasites during early life history stages. Additionally studying specialist parasites can be a valuable tool for identifying molid species adding to existing morphological and molecular identification methods (Caldera et al. 2020 [Chapter 3]). For example the trematode *Accacladocoelium alveolatum* (Robinson, 1934) (Accacoeliidae) is only found on *M. mola* (Table 1). Such studies can also be useful in the identification of fish populations (e.g., Lester 1990, Blaylock et al. 2003, MacKenzie 2004, Timi 2007, Lester and MacKenzie 2009, MacKenzie and Abaunza 2014, Mattiucci et al. 2015). Likewise, established protocols for the collection of parasite data in areas of high bycatch need to be developed and disseminated widely so that such information becomes interwoven more routinely into overall studies of sunfish ecology.

Acknowledgements

I am thankful to M. Dunphy-Collis for going over the English of the manuscript. I am indebted to the editors, Jonathan Houghton and Tierney Thys, for all the valuable and helpful comments and suggestions to improve the quality of the chapter. I am very grateful to my friends and colleagues J.A. Raga, F.E. Montero, J. Aznar, M. Fernández, C. Blanco, A. Repullés-Albelda, F.J. Palacios-Abella, N. Fraija-Fernández and A. Pérez del Olmo (University of Valencia), F. Padrós (Autonomous University of Barcelona) and F. de la Gándara (Spanish Institute of Oceanography, IEO) for their useful help in fish sampling and necropsy and in the parasite collection and identification. I would also like to express my gratitude to my friends and colleagues L. Kubicek, E. Sawai (Hiroshima University), Y. Yamanoue (The University of Tokyo) and N. Phillips (Queen's University Belfast), for sharing their wide sunfish knowledge and sage advice. I am grateful for the collaboration of the Biology and Veterinary staffs of the Oceanogràfic for their help in sampling aquarium-reared fish. I received a PhD student grant from the Ministry of Science, Innovation and Universities (MICINN) of the Spanish Government.

References

Abe, T. and K. Sekiguchi. 2012. Why does the ocean sunfish bask? Commun. Integr. Biol. 5: 395–398.

Abe, T., K. Sekiguchi, H. Onishi, K. Muramatsu and T. Kamito. 2012. Observations on a school of ocean sunfish and evidence for a symbiotic cleaning association with albatrosses. Mar. Biol. 159: 1173–1176.

Agustí, C., F.J. Aznar and J.A. Raga. 2005. Tetraphyllidean plerocercoids from western Mediterranean cetaceans and other marine mammals around the world: a comprehensive morphological analysis. J. Parasitol. 91: 83–92.

Ahuir-Baraja, A.E. 2012. Parasitological study of ocean sunfish, *Mola mola* (L.) in the Western Mediterranean. Ph.D. Thesis, University of Valencia, Valencia, Spain. [in Spanish].

Ahuir-Baraja, A.E., N. Fraija-Fernández, J.A. Raga and F.E. Montero. 2015a. Molecular and morphological differentiation of two similar species of Accacoeliidae (Digenea): *Accacladocoelium macrocotyle* and *A. nigroflavum* from sunfish, *Mola mola*. J. Parasitol. 101: 231–235.

Ahuir-Baraja, A.E., F. Padrós, J.F. Palacios-Abella, J.A. Raga and F.E. Montero. 2015b. *Accacoelium contortum* (Trematoda: Accacoeliidae) a trematode living as a monogenean: morphological and pathological implications. Parasit. Vectors 8: 540.

Ahuir-Baraja, A.E., Y. Yamanoue and L. Kubicek. 2017. First confirmed record of *Mola* sp. A in the western Mediterranean Sea: morphological, molecular and parasitological findings. J. Fish Biol. 90: 1133–1141.

Andersen, K. 1987. SEM Observations on plerocercus larvae of *Floriceps saccatus* Cuvier, 1817 and *Molicola horridus* (Goodsir, 1841) (Cestoda; Trypanorhyncha) from sunfish (*Mola mola*). Fauna Norv. Ser. A. 8: 25–28.

Anderson, R.M. 1993. Epidemiology. pp. 75–116. *In*: F.E.G. Cox [ed.]. Modern Parasitology. Blackwell, Oxford, UK.

Araújo, M.E., E.C. Silva-Falcão, P.D. Falcão, V.M. Marques and I.R. Joca. 2010. Stranding of *Masturus lanceolatus* (Actinopterygii: Molidae) in the estuary of the Una River, Pernambuco, Brazil: natural and anthropogenic causes. Marin. Biodivers. Rec. 3: 1–5.

Arru, E., G. Garippa and E. Sanna. 1991. *Molicola horridus* in *Luvaris imperialis* and *Mola mola*. Boll. Soc. Ital. Patol. Ittica. 5: 92–96.

Atria, G. 1967. Un ectoparásito del pez luna (*Mola ramsayi*, Giglioli). *Pennella* cf. filosa L. (Crustacea, Copepoda). Not. Mens. Mus. Nac. Hist. Nat. Chile. 131: 3–5.

Aureli, G. 2004. Indagine parasitologica in *Mola mola* spiaggiati lungo le coste italiane. Corso di Laurea in Acquacoltura ed Ittiopatologia. University of Bologna, Bologna, Italy.

Aznar, F.J., C. Agustí, D.T.J. Littlewood, J.A. Raga and P.D. Olson. 2007. Insight into the role of cetaceans in the life cycle of the tetraphyllideans (Platyhelminthes: Cestoda). Int. J. Parasitol. 37: 243–255.

Baird, W. 1850. The Natural History of the British Entomostraca. Vol. 2. Ray Society. London, UK.

Bakenhaster, M.D. and J.S. Knight-Gray. 2016. New diet data for *Mola mola* and *Masturus lanceolatus* (Tetraodontiformes: Molidae) off Florida's Atlantic coast with discussion of historical context. B. Mar. Sci. 92: 497–511.

Bakke, T.A. 1981. Mussel epizootic on sunfish parasite. Fauna. 34: 77–79.

Balakrishnan, K.P. 1969. Observations on the occurrence of *Conchoderma virgatum* (Spengler) (Cirripedia) on *Diodon hystrix* Linnaeus (Pisces). Crustaceana 16: 101–103.

Barbagallo, P. and U. Drago. 1903. Primo contributo all della fauna elmintologia dei pesci della Sicilia orientale. Archs. Parasit. 7: 408–427.

Barnard, K.H. 1948. New records and descriptions of new species of parasitic Copepoda from South Africa. Ann. Mag. Nat. Hist. 12: 242–254.

Barnard, K.H. 1955. South African parasitic Copepoda. Ann. South African Mus. 41: 223–312.

Barreiros, J.P. and M. Teves. 2005. The sunfish *Mola mola* as an attachment surface for the lepadid cirriped *Lepas anatifera* a previously unreported association. Aqua 10: 1–4.

Bates, R.M. 1990. A checklist of the Trypanorhyncha (Platyhelminthes: Cestoda) of the world (1935–1985) (No. 1). National Museum of Wales.

Bellingham, O'B. 1844. Catalogue of Irish entozoa, with observations. An. Mag. Nat. Hist. 13: 422–430.

Berland, B. 1961. Nematodes from some Norwegian marine fishes. Sarsia 2: 1–50.

Beveridge, I. T.H. Cribb and S.C. Cutmore. 2017. Larval trypanorhynch cestodes in teleost fish from Moreton Bay, Queensland. Mar. Freshwater Res. 68: 2123–2133.

Blaylock, R.B., L. Margolis and J.C. Holmes. 2003. The use of parasites in discriminating stocks of Pacific halibut (*Hippoglossus stenolepis*) in the northeast Pacific. Fish. B-NOAA 101: 1–9.

Bona, F., E. Buriola, S. Cerioni, P. Orecchia and L. Paggi. 1995. Digenea. pp. 1–31. *In*: A. Minelli, S. Ruffo and S. La Posta [eds.]. Checkist delle Specie della Fauna Italiana, Fascicolo 5. Edizioni Calderini, Bologna, Italy.

Boxshall, G., R. Lester, M.J. Grygier, J.T. Hoeg, H. Glenner, J.D. Shields et al. 2005. Crustacean parasites. pp. 123–169. *In*: K. Rodhe [ed.]. Marine Parasitology. CSIRO Publishing, Collingwood, Australia.

Bray, R.A. and I.D. Gibson. 1977. The Accacoeliidae (Digenea) of fishes from the north-east Atlantic. Bull. Br. Mus. Nat. Hist. 31: 53–99.
Bray, R.A. and I.D. Gibson. 1997. The Lepocreadiidae Odhner, 1905 (Digenea) of fishes from the north-east Atlantic: summary paper, with keys and checklists. Syst. Parasitol. 36: 223–228.
Bray, R.A. and I.D. Gibson. 2003. The digeneans of elasmobranchs—distribution and evolutionary significance. pp. 67–96. *In*: C. Combes and J. Jourdanes [eds.]. Taxonomie écologie et évolution des métazoaires parasites—Taxonomy ecology and evolution of metazoan parasites. Livre hommage à Louis Euzet. PUP, Perpignan, France.
Brinkmann, A. 1952. Some Chilean monogenetic trematodes. Reports of the Lund University, Chile Expedition 1948–1949. Lunds Universitets Årsskrift: 1–26.
Brito, J.L. 2003. Nuevos registros de *Balistes polylepis* (Balistidae), *Sphoeroides lobatus* (Tetraodontidae), *Mola mola* y *M. ramsayi* (Molidae) en San Antonio, Chile (Pisces, Tetraodontiformes). Invest. Mar. 31: 77–83.
Brooks, D.R. and D.A. McLennan. 1993. Parascript. Parasites and the Language of Evolution. Smithsonian Institution Press, Washington D.C., USA.
Bruce, N.L., R.D. Adlard and L.R.G. Cannon. 1994. Synoptic checklist of ascaridoid parasites (Nematoda) from fish hosts. Invertebr. Taxon. 8: 583–674.
Bush, A.O., J.C. Fernández, G.W. Esch and R.J. Seed. 2001. Parasitism (The Diversity and Ecology of Animal Parasites). Cambridge University Press, Cambridge, USA.
Busch, M.W. and S. Klimpel. 2016. Marine intermediate hosts. pp. 1–4. *In*: H. Mehlhorn [ed.]. Encyclopedia of Parasitology. Springer, Heidelberg, Berlin, Germany.
Busch, M.W., T. Kuhn, J. Münster and S. Klimpel. 2012. Marine crustaceans as potential hosts and vectors for metazoan parasites. pp. 329–360. *In*: H. Mehlhorn [ed.]. Arthropods as Vectors of Emerging Diseases. Parasitology Research Monographs. Vol. 3. Springer, Heidelberg, Berlin, Germany.
Butler, S.A. 1987. Taxonomy of some Tetraphyllidean cestodes from elasmobranch fishes. Aust. J. Zool. 35: 343–371.
Caldera, E., J.L. Whitney, T.M. Thys, E. Ostalé-Valriberas and L. Kubicek. 2020. Genetic insights regarding the taxonomy, phylogeography and evolution of the ocean sunfishes (Molidae: Tetraodontiformes). pp. 37–54. *In*: T.M. Thys, G.C. Hays and J.D.R. Houghton [eds.]. The Ocean Sunfishes: Evolution, Biology and Conservation, CRC Press. Boca Raton, FL, USA.
Campbell, R.A., R.L. Haedrich and T.A. Munroe. 1980. Parasitism and ecological relationships among deep-sea benthic fishes. Mar. Biol. 57: 301–313.
Cartamil, D.P. and C.G. Lowe. 2004. Diel movement patterns of ocean sunfish *Mola m*ola off southern California. Mar. Ecol. Prog. Ser. 266: 245–253.
Carwardine, M. 1995. The Guinness Book of Animal Records, Middlesex, Guinness Publishing.
Chervy, L. 2002. The terminology of larval cestodes or metacestodes. Systemat. Parasitol. 52: 1–33.
Cohen, S.C., M.C. Justo and A. Kohn. 2013. South American Monogenoidea parasites of fishes, amphibians and reptiles. Conselho Nacional de Desenvolvimento Cientifico e Tecnologico (CNPq).
Cooper, T., D. McGrath and B. O'Connor. 1982. Species associated with a sunfish *Mola mola* (L.) from the west coast of Ireland. Ir. Nat. J. 20: 382–383.
Cordero del Campillo, M. 2000. El parasitismo y otras asociaciones biológicas. Parásitos hospedadores. pp. 22–38. *In*: M. Cordero del Campillo and F.A. Rojo Vázquez [eds.]. Parasitología Veterinaria. McGraw-Hill Interamericana, Spain.
Dawes, B. 1947. The Trematoda of British fishes. Ray Society (N° 131), London, UK.
de Azevedo, R.K. and V.D. Abdallah. 2016. Chapter 10. Fish parasites and their use as environmental research indicators. pp. 163–177. *In*: C. Gheler-Costa, M.C. Lyra-Jorge and L.M. Verdade [eds.]. Biodiversity in Agricultural Landscapes of Southeastern Brazil. De Gruyter Open, Warsaw, Poland.
de Figueiredo, N.C., J.T.A.X. de Lima, C.I. Freitas and C.G. da Silva. 2017. Checklist dos parasitos do peixe Lua (*Mola mola*: Molidae) no mundo. PUBVET 12: 1–9.
Díaz, O. 2000. Copépodos ectoparásitos del pez luna *Mola mola* (Giglioli, 1883) (Pisces: Molidae) en el Golfo de Cariaco, Venezuela. Bol. Inst. Ocean. Universidad de Oriente, Cumaná 39: 11–17.
Dippenaar, S.M. 2005. Reported siphonostomatoid copepods parasitic on marine fishes of southern Africa. Crustaceana 77: 1281–1328.
Dollfus, R.P.H. 1923. Enumeration des cestodes du plancton et des invertebres marins. Ann. Parasitol. 1: 363–394.
Dollfus, R.P.H. 1935. Sur quelques parasites de poissons récoltés à Castiglione (Algerie). Bulletin de la Station d'Aquiculture et de Pêche Castiglione 1933: 197–279.
Dollfus, R.P.H. 1942. Etudes critiques sur les Tétrarhynques du Muséum de Paris. Ann. Mus. Hist. Nat. Paris. 19: 1–466.
Dollfus, R.P.H. 1946. Essai de catalogue des parasites poisson-lune *Mola mola* (L. 1758) et autres Molidae. Ann. Soc. Sci. Nat. Charente-Marit. 7: 69–76.

Dollfus, R.P.H. 1964. Enumération des Cestodes du plancton et des Invertébrés marins—(6e Contribution). Ann. Parasitol. 3: 329–379.

Draper, H. 1941. A record of trematode parasites from *Mola mola* and *Raniceps raninus* (Linn.). Parasitology 33: 209–210.

Egorova, T.P. 2000. Occurrence of monogeneans of the subfamily Capsalinae (Capsalidae)—parasites of marine fishes. Parazitologiya. 34: 111–117.

Euzet, L., C. Combes and A. Caro. 1993. A check list of monogenea of Mediterranean fish. Proc. 2nd Intern. Symp. Monog. France: 5–8.

Evans, W. 1906. *Laemargus muricatus* Kröy, on a sunfish captured in the Firth of Forth. Ann. Scot. Nat. Hist. Edinburgh 57: 57.

Fergusson, I.K., L.J. Compagno and M.A. Marks. 2000. Predation by white sharks *Carcharodon carcharias* (Chondrichthyes: Lamnidae) upon chelonians, with new records from the Mediterranean Sea and a first record of the ocean sunfish *Mola mola* (Osteichthyes: Molidae) as stomach contents. Environ. Biol. Fishes. 58: 447–453.

Fernández, I., C. Oyarzún, A. Valenzuela, C. Burgos, V. Guaquín and V. Campos. 2016. Parásitos del pez luna *Mola mola* (Pisces: Molidae). Primer registro en aguas de la costa centro sur de Chile. Gayana 80: 192–197.

Ferrer-Maza, D., M. Muñoz, J. Lloret, E. Faliex, S. Vila and P. Sasal. 2015. Health and reproduction of red mullet, *Mullus barbatus*, in the western Mediterranean Sea. Hydrobiologia 753: 189–204.

Fraser-Brunner, A. 1951. The ocean sunfishes (family Molidae). Bull. Br. Mus. Nat. Hist. Zool. 1: 89–121.

Gaevskaya, A.V. and A. Kovaljova. 1991. Handbook on the diseases and parasites of food fish from Atlantic Ocean. Ed. IBSS, Kaliningrad, Ukraine.

Gaevskaya, A.V. 2002. New data on trematodes of the families Opecoelidae and Accacoeliidae from fishes in the Atlantic Ocean and its seas. Parazitologiya 36: 219–223 [in Russian].

Garibaldi, F. and L. Orsi Relini. 2000. Abbondanza estiva, struttura di taglia e nicchia alimentare della verdesca, *Prionace glauca*, nel santuario pelagico del Mar Ligure. Biol. Mar. Mediterr. 7: 324–333.

Gibson, D.I. and R.A. Bray. 1979. *Cirkennedya porlockensis*, a new genus and species of digenean from the sunfish *Mola mola* (L.). J. Helminthol. 53: 245–250.

Gibson, D.I. 1996. Trematoda. *In*: L. Margolis and Z. Kabata [eds.]. Guide to the Parasites of Fishes of Canada. Part IV. Canadian Special Publication of Fisheries and Aquatic Sciences 124. National Research Council of Canada.

Gibson, D.I., A. Jones and R.A. Bray. 2002. Keys to the Trematoda, volume I. CAB International and Natural History Museum, Wallingford and London, UK.

Gibson, D.I., R.A. Bray and E.A. Harris. (Compilers). 2005. Host-Parasite Database of the Natural History Museum, London. URL. Available from http://www.nhm.ac.uk/research-curation/scientific-resources/taxonomy-systematics/host-parasites/.

Gilardoni, C., C. Ituarte and F. Cremonte. 2012. Castrating effects of trematode larvae on the reproductive success of a highly parasitized population of *Crepipatella dilatata* (Caenogastropoda) in Argentina. Mar. Biol. 159: 2259–2267.

Gladstone, W. 1988. Killer whale feeding observed underwater. J. Mammal. 69: 629–630.

Goodsir, J. 1841. On *Gymnorhynchus horridus* a new cestoid entozoon. Edinburgh Philos. J. 31: 9–12.

Grabda, J. 1973. Contribution to knowledge of biology of *Cecrops latreilli* Leach, 1816 (Caligoida: Cecropidae) the parasite of the ocean sunfish *Mola mola* (L.). Acta Ichthyol. Piscat. 3: 61–74.

Gregory, W.K. and H.C. Raven. 1934. Notes on the anatomy and relationships of the ocean sunfish (*Mola mola*). Copeia 4: 145–151.

Guiart, J. 1935. Le véritable *Floriceps saccatus* de Cuvier n'est pas la larve géante de Tétrarhynque vivant dans le foie du Môle (*Mola mola*). Bull. Inst. Océanograph. 666: 1–15.

Guiart, J. 1938. Trematodes parasites provenant des campagnes scientifiques de S.A.S. le Prince Albert I de Mónaco (1886–1912). Resultats Campagnes Scient. Albert I Prince Mónaco. Fasc. C: 1–75.

Gustinelli, A., G. Nardini, G. Aureli, M. Trentini, M. Affronte and M.L. Fioravanti. 2006. Parasitofauna of *Mola mola* (Linnaeus, 1758) from Italian sea. Biol. Mar. Mediterr. 13: 872–876.

Hamond, R. 1969. The copepods parasitic on Norfolk marine fishes. Trans. Norfolk Norwich Nat. Soc. 21: 229–243.

Harrison, C.S., T.S. Hida and M.P. Seki. 1988. Hawaiian seabird feeding ecology. Wildl. Monogr. 85: 3–71.

Hartwich, G. and I. Kilias. 1992. Die Typen der Cercomeromorphae (Plathelminthes) des Zoologischen Museum in Berlin. Berlin Mitt. Zool. Mus. 68: 209–248.

Hata, H., A. Sogabe, S. Tada, R. Nishimoto, R. Nakano, N. Kohya, et al. 2017. Molecular phylogeny of obligate fish parasites of the family Cymothoidae (Isopoda, Crustacea): evolution of the attachment mode to host fish and the habitat shift from saline water to freshwater. Mar. Biol. 164: 105.

Heegard, P. 1962. Parasitic copepod from Australian waters. Rec. Aust. Mus. 25: 149–233.

Heinz, M.L. and M.D. Dailey. 1974. The Trypanorhyncha (Cestoda) of elasmobranch fishes from southern California and northern Mexico. Proc. Helminthol. Soc. Wash. 41: 161–169.

Heller, C. 1866. Carcinologische beiträge zur fauna der Adriatischen meeres. Verh. Zool - bot. Ges. Wien. 16: 723–760.
Hendrix, S.S. 1994. Marine flora and fauna of the eastern Unites States. Platyhelminthes: Monogenea. NOAA Technical Report NMFS, Seattle 121: 1–106.
Hernández-Orts, J.S. A.E. Ahuir-Baraja, N.A. García, E.A. Crespo, J.A. Raga and F.E. Montero. 2009. New geographical records of *Capsala martinieri* (Capsalidae) from the Argentinean Patagonia and the Spanish Mediterranean: Puerto Madryn's wins the first prize at the pumpkin fair. Proc. 6th Int. Symp. Monog. South Africa 43.
Hewitt, G.C. 1964. The occurrence of *Lepeophtheirus insignis* Wilson (Copepoda parasitica) in New Zealand waters and its relationship to *L. molae* Heegard. T. Roy. Soc. NZ. Zool. 4: 153–155.
Hewitt, G.C. 1968. *Cecrops latreilli* Leach (Cecropidae, Copepoda) on *Mola mola* in New Zealand waters. Rec. Dom. Mus. 6: 49–59.
Hewitt, G.C. 1971. Species of *Lepeophtheirus* (Copepoda, Caligidae) recorded from the ocean sunfish (*Mola mola*) and their implications for the caligid genus *Dentigryps*. J. Fish. Res. Board Can. 28: 323–334.
Hewitt, G.C. and P.M. Hine. 1972. Checklist of parasites of New Zealand fishes and of their hosts. N.Z. J. Mar. Freshwater Res. 6: 69–114.
Hillis, J.P. and C.E. O'Riordan. 1960. Parasites of a sunfish, *Mola mola*, from the Irish coast. Ir. Nat. J. 13: 123–124.
Ho, J.S. and C.L. Lin. 2010. Three more unrecorded sea lice (Copepoda, Caligidae) parasitic on marine fishes collected off Tai-dong, Taiwan. Crustaceana 83: 1261–1277.
Hoberg, E.P. and G.J. Klassen. 2002. Revealing the faunal tapestry: co-evolution and historical biogeography of hosts and parasites in marine systems. Parasitology 124: 3–22.
Hogans, W.E. 1987. Description of *Pennella filosa* L. (Copepoda: Pennellidae) on the ocean sunfish (*Mola mola* L.) in the Bay of Fundy. B. Mar. Sci. 40: 59–62.
Houghton, J.D.R., N. Liebsch, T.K. Doyle, A. Gleiss, M.K.S. Lilley, R.P. Wilson et al. 2009. Harnessing the sun: testing a novel attachment method to record fine scale movements in Ocean Sunfish (*Mola mola*). pp. 229–242. *In*: J.L. Nielsen, H. Arrizabalaga, N. Fragoso, A. Hobday, M. Lutcavage and J. Sibert [eds.]. Tagging and Tracking of Marine Animals with Electronic Devices. Vol 9. Springer, Dordrecht, Netherlands.
Hudson, P.J., A.P. Dobson and K.D. Lafferty. 2006. Is a healthy ecosystem one that is rich in parasites? Trends Ecol. Evol. 21: 381–385.
Ishii, N. 1935. Studies on the family Didymozoidae (Monticelli, 188). Jpn. J. Zool. 6: 279–335.
Jensen, K. and S.A. Bullard. 2010. Characterization of a diversity of tetraphyllidean and rhinebothriidean cestode larval types, with comments on host associations and life-cycles. Int. J. Parasitol. 40: 889–910.
Johnston, T.H. 1909. Notes on Australian Entozoa. N°. 1. Rec. Aust. Mus. 7: 329–344.
Kearn, G.C. 1998. Parasitism and the platyhelminths. Chapman & Hall Ltd, London, UK.
Kennedy, C.R. and K.I. Andersen. 1982. Redescription of *Anchistrocephalus microcephalus* (Rudolphi) (Cestoda, Pseudophyllidea) from the sunfish *Mola mola*. Zool. Scr. 11: 101–105.
Kensley, B. and J.R. Grindley. 1973. South African parasitic copepoda. Ann. South African Mus. 62: 69–130.
Klimpel S., T. Kuhn, J. Münster, D.D. Dörge, R. Klapper and J. Kochmann. 2019. Parasitic groups. pp. 29–76. *In*: S. Klimpel, T. Kuhn, J. Münster, D.D. Dörge, R. Klapper and J. Kochmann [eds.]. Parasites of Marine Fish and Cephalopods. Springer, Cham, Germany.
Kohn, A. and S.C. Cohen. 1998. South American Monogenea—list of species, hosts and geographical distribution. Int. J. Parasitol. 28: 1517–1554.
Kohn, A. and M.P. Paiva. 2000. Fishes parasitized by Monogenea in South America. Metazoan parasites in the tropics: a systematic and ecological perspective. Univ. Nac. Auton. Mex. 25–60.
Kondo, Y., S. Ohtsuka, T. Hirabayashi, S. Okada, N.O. Ogawa, N. Ohkouchi et al. 2016. Seasonal changes in infection with trematode species utilizing jellyfish as hosts: evidence of transmission to definitive host fish via medusivory. Parasite. 23: 1–14.
Konow. N., R. Fitzpatrick and A. Barnett. 2006. Adult emperor angelfish (*Pomacanthus imperator*) clean giant sunfishes (*Mola mola*) at Nusa Lembongan, Indonesia. Coral Reefs 25: 208. doi:10.1007/s00338-006-0086-9.
Krøyer, H. 1863. Bidrag til Kundskab om Snyltekrebsene. Naturhistorisk Tidsskrift 2: 75–426.
Labaude, S., T. Rigaud and F. Cézilly. 2015. Host manipulation in the face of environmental changes: ecological consequence. Int. J. Parasitol. Parasites Wildl. 4: 442–451.
Lafferty, K.D., S. Allesina, M. Arim, C.J. Briggs, G. DeLeo, A.P. Dobson et al. 2008. Parasites in food webs: the ultimate missing links. Ecol. Lett. 11: 533–546.
Lafferty, K.D. and A.M. Kuris. 2009. Parasitic castration: the evolution and ecology of body snatchers. Trends Parasitol. 25: 564–572.
Lafferty, K.D. 2013. Parasites in marine food webs. Bull. Mar. Sci. 89: 123–134.
Lajeunesse, M.J. and M.R. Forbes. 2002. Host range and local parasite adaptation. Proc. Royal Soc. B. 269: 703–710.

Lamothe-Argumedo, R. 1997. Nuevo arreglo taxonómico de la subfamilia Capsalinae (Monogenea: Capsalinae), clave para los géneros y dos combinaciones nuevas. An. Inst. Bio. UNAM, Zool. Ser. 68: 207–223.

Leong, T.S. and A. Colorni. 2002. Infection diseases of warm water fish in marine and brackish waters. pp. 193–230. *In*: P.T.K. Woo, D.W. Bruno and L.H. Susan Lim [eds.]. Diseases and Disorders of Finfish in Cage Culture. CABI Publishing, Oxford, UK.

Lester, R.J.G. 1990. Reappraisal of the use of parasites for fish stock identification. Mar. Freshwater Res. 41: 855–864.

Lester, R.J.G. and K. MacKenzie. 2009. The use and abuse of parasites as stock markers for fish. Fish. Res. 97: 1–2.

Linton, E. 1898. Notes on trematode parasites of fishes. Proc. U.S. National Mus. 20: 507–548.

Linton, E. 1900. Fish parasites collected at Woods Hole in 1898. U.S. Fish Comm. Bull. for 1898: 267–304.

Linton, E. 1901. Parasites of fishes of the Woods Hole region. U.S. Fish Comm. Bull. for 1899: 405–495.

Linton, E. 1924. Notes on cestode parasites of sharks and skates. Proc. U.S. National Mus. 64: 1–114.

Linton, E. 1940. Trematodes from fishes mainly from the Woods Hole region, Massachusetts. Proc. U.S. National Mus. 88: 1–172.

Linton, E. 1941. Cestode parasites of teleost fishes of the Woods Hole region, Massachusetts. Proc. U.S. National Mus. 90: 417–442.

Lloret, J., E. Faliex, G.E. Shulman, J.A. Raga, P. Sasal, M. Muñoz et al. 2012. Fish health and fisheries, implications for stock assessment and management: the Mediterranean example. Rev. Fish. Sci. 20: 165–180.

Lloyd, L.C. 1938. Some digenetic trematodes from Puget Sound fish. J. Parasitol. 24: 103–133.

Logan, V.H. and P.H. Odense. 1974. The integument of the ocean sunfish (*Mola mola* L.) (Plectognathi) with observations on the lesions from two ectoparasites, *Capsala martinierei* (Trematoda) and *Philorthagoriscus serratus* (Copepoda). Can. J. Zool. 52: 1039–1045.

Looss, A. 1902. Zur Kenntnis der Trematodenfauna des Triester Hafens. I. Ueber die Gattung *Orophocotyle* n. g. Centralblatt für Bakteriologie, Parasitenkunde und Infektionskrankheiten un Hygeine 31: 637–644.

Love, M. and M. Moser. 1983. Technical Report NMFS SSRF-777. A Checklist of Parasites of California. Oregon and Washington Marine and Estuarine Fishes. US Department of Commerce, Washington D.C., USA.

Luque, J. and M. Oliva. 1993. Trematodes of marine fishes from the Peruvian faunistic province (Peru and Chile), with description of *Lecithochirium callaoensis* n. sp. and new records. Rev. Biol. Mar. 28: 271–286.

MacKenzie, K. 2004. Parasites as biological tags for marine fish populations. Biologist 51: 86–90.

MacKenzie, K. and P. Abaunza. 2014. Parasites as biological tags. pp. 185–203. *In*: S.X. Cadrin, L.A. Kerr and S. Mariani [eds.]. Stock Identification Methods: Applications in Fishery Science. 2nd edn. Academic Press, Elsevier, New York, USA.

Maclaren, N. 1904. Beitrage zur Kenntnis einiger Trematoden (*Diplectanum aequans* Wagener und *Nemathobothrium molae* n. sp.). Jena Zoologischen Naturwissenschaftliche 38: 573–618.

Manter, H.W. 1934. Some digenetic trematodes from deep-water fishes of Tortugas, Florida. Pap. Tortugas Lab. Carnegie Inst. Washington 28: 257–345.

Manter, H.W. 1960. Some additional Digenea (Trematoda) from New Zealand fishes. Caballero Jubilee 328: 197–201.

Marcogliese, D.J. 1995. The role of zooplankton in the transmission of helminth parasites to fish. Rev. Fish Biol. Fisher. 5: 336–371.

Mattiucci, S. and G. Nascetti. 2007. Genetic diversity and infection levels of anisakid nematodes parasitic in fish and marine mammals from Boreal and Austral hemispheres. Vet. Parasitol. 148: 43–57.

Mattiucci, S., R. Cimmaruta, P. Cipriani, P. Abaunza, B. Bellisario and G. Nascetti. 2015. Integrating *Anisakis* spp. parasites data and host genetic structure in the frame of a holistic approach for stock identification of selected Mediterranean Sea fish species. Parasitology 142: 90–108.

McCann, Ch. 1961. The sunfish, *Mola mola* (L.) in New Zealand waters. Rec. Domin. Mus. 4: 7–20.

McClelland, G., R. Misra and D.J. Martell. 1990. Larval anisakine nematodes in various fish species from Sable Island Bank and vicinity. Bull. Fish. Aquat. Sci. 222: 83–118.

Measures, L.N. 1992. *Bolbosoma turbinella* (Acanthocephala) in a blue whale, *Balaenoptera musculus*, stranded in the St. Lawrence Estuary, Quebec. J. Helmint. Soc. Wash. 59: 206–211.

Mendez-Ahid, S.M., K.D. Filgueira, Z.A. Araújo de Souza Fonseca, B. Soto-Blanco and M.F. de Oliveira. 2009. Ocorrência de parasitismo em *Mola mola* (Linnaeus, 1758) por metazoários no litoral do rio grande do norte, Brasil. Acta Vet. Bras. 3: 43–47.

Mendez, O. and F. Galván-Magana. 2016. Cestodes of the blue shark, *Prionace glauca* (Linnaeus 1758), (Carcharhiniformes: Carcharhinidae), off the west coast of Baja California Sur, Mexico. Zootaxa 4085: 438–444.

Milne Edwards, H. 1840. Histoire Naturelle des Crustacés, Comprenant l'Anatomie, la Physiologie et la Classification de ces Animaux. Vol. 3. Roret, Paris, France.

Mladineo, I. 2003. Life cycle of *Ceratothoa oestroides*, a cymothoid isopod parasite from sea bass *Dicentrarchus labrax* and sea bream *Sparus aurata*. Dis. Aquat. Organ. 57: 97–101.

Montgomery, W.R. 1957. Studies on digenetic trematodes from marine fishes of Lajolla, California. T. Am. Microsc. Soc. 76: 13–36.
Monticelli, F.S. 1893. Studii sui trematodi endoparassiti: Primo contributo di osservazioni sui distomidi. Vol. 3. G. Fischer.
Moore, J. 2002. Parasites and the Behavior of Animals. Oxford University Press. Oxford, UK.
Mudry, D.R. and M.D. Dailey. 1971. Postembryonic development of certain tetraphyllidean and trypanorhynchan cestodes with a possible alternative life cycle for the order Trypanorhyncha. Can. J. Zool. 49: 1249–1253.
Nagasawa, K. 1990. The life cycle of *Anisakis simplex*: a review. pp. 31–40. *In*: H. Ishikura and K. Kikuchi [eds.]. Intestinal Anisakiasis in Japan. Infected Fish, Sero-Immunological Diagnosis, and Prevention. Springer, Tokyo, Japan.
Nagasawa, K. 2009. Synopsis of branchiurans of the genus *Argulus* (Crustacea, Argulidae), ectoparasites of freshwater and marine fishes, in Japan (1900–2009). Bull. Biogeograph. Soc. Japan 64: 135–148 [in Japanese].
Nakamura, I. and K. Sato. 2014. Ontogenetic shift in foraging habit of ocean sunfish *Mola mola* from dietary and behavioral studies. Mar. Biol. 161: 1263–1273.
Nicoll, W. 1909. A contribution towards a knowledge of the entozoan of British marine fishes. Part II. Ann. Mag. Nat. Hist. 4: 1–25.
Nicoll, W. 1915. A list of the trematode parasites of British marine fishes. Parasitology 7: 339–378.
Noble, E.R. and G.A. Noble. 1937. *Accacladium nematulum* n. sp., a trematode from the sunfish *Mola mola*. T. Am. Microsc. Soc. 56: 55–60.
Noble, E.R. and G.A. Noble. 1966. Amebic parasites of fishes. J. Eukaryot. Microbiol. 13: 478–480.
Noble, G.A. and E.R. Noble. 1966. *Monocercomonas molae* n. sp., a flagellate from the Sunfish *Mola mola*. J. Eukaryot. Microbiol. 13: 257–259.
Noble, G.A. 1975. Description of *Nematobibothrioides histoidii* (Noble, 1974) (Trematoda: Didymozoidae) and comparison with other genera. J. Parasitol. 61: 224–227.
Nyegaard, M., S. Andrzejaczek, C.S. Jenner and M.N.M. Jenner. 2019. Tiger shark predation on large ocean sunfishes (Family Molidae)–two Australian observations. Environ. Biol. Fish. 102: 1559–1567.
Odhner, T. 1928. *Rhynchopharynx paradoxa* n. g. n. sp., nebst revision der Accacoeliiden von *Orthagoriscus mola*. Zool. Anz. 38: 513–531.
Oldewage, W.H. 1992. Aspects of the fine structure of female *Dinemoura latifolia* Steenstrup & Lütken, 1861 (Copepoda: Pandaridae): an SEM study. South African J. Zool. 27: 6–10.
Oldewage, W.H. 1993. Three species of piscine parasitic copepods from southern African coastal waters. South African J. Zool. 28: 113–121.
Oliveira, J.B., J.A. Morales, R.C. González-Barrientos, J. Hernández-Gamboa and G. Hernández-Mora. 2011. Parasites of cetaceans stranded on the Pacific coast of Costa Rica. Vet. Parasitol. 182: 319–328.
Olson, P.D., D.T.J. Littlewood, R.A. Bray and J. Mariaux. 2001. Interrelationships and evolution of the tapeworms (Platyhelminthes: Cestoda). Mol. Phylogenet. Evol. 19: 443–467.
Olson, P.D. and D.T.J. Littlewood. 2002. Phylogenetics of the Monogenea—evidence from a medley of molecules. Int. J. Parasitol. 32: 233–244.
Olsson, P. 1867. Entozoa, iakttagna hos Skandinaviska hafsfiskar: Platyelminthes I. Berling.
Olsson, P. 1869. Prodromus faunae copepodorum parasitantium Scandinaviae. Acta Univ. Lundensis 5: 1–49.
Ortis, M. and L. Paggi. 2008. Digenei. pp. 140–149. *In*: G. Rellini [ed.]. Checklist della Flora e della Fauna dei Mari Italiani. Prima Parte. Biologia Marina Mediterranea, 15 (suppl. 1). Ministero dell'ambiente e della tutela del territorio e del mare.
Padrós, F., O. Palenzuela, C. Hispano, O. Tosas, C. Zarza, S. Crespo et al. 2001. *Myxidium leei* (Myxozoa) infections in aquarium-reared Mediterranean fish species. Dis. Aquat. Organ. 47: 57–62.
Palm, H.W. 2004. The Trypanorhyncha Diesing, 1863. PKSPL-IPB Press, Bogor, Indonesia.
Parker, R.R. 1969. Validity of the binomen *Caligus elongatus* for a common parasitic copepod formerly misidentified with *Caligus rapax*. J. Fish. Res. Board Can. 26: 1013–1035.
Pastore, M.A. 2009. Necropsy of an ocean sunfish stranded along the Taranto coast (Apulian, south Italy). Mar. Biodivers. Rec. 2.
Pérez-del Olmo, A., D.I. Gibson, M. Fernández, O. Sanisidro, J.A. Raga and A. Kostadinova. 2006. Descriptions of *Wardula bartolii* n. sp. (Digenea: Mesometridae) and three newly recorded accidental parasites of *Boops boops* L. (Sparidae) in the NE Atlantic. Syst. Parasitol. 63: 97–107.
Peters, L.E. 1960. The systematic position of the genus *Dihemistephanus* Loos, 1901 (Trematoda: Dinegea), with the redescription of *D. lydiae* (Stossich, 1896) from the South Pacific. P. Helm. Soc. Wash. 27: 134–138.
Petter, A.J. and B.M. Radujković. 1989. Parasites of marine fishes from Montenegro: nematodes. Acta Adriat. 30: 195–236.

Phillips, N.D., E.C. Pope, C. Harrod and J.D.R. Houghton. 2020. The diet and trophic role of ocean sunfishes. pp. 146–159. *In*: T.M. Thys, G.C. Hays and J.D.R. Houghton [eds.]. The Ocean Sunfishes: Evolution, Biology and Conservation, CRC Press. Boca Raton, FL, USA.

Pope, E.C., G.C. Hays, T.M. Thys, T.K., Doyle, D.W. Sims, N. Queiroz et al. 2010. The biology and ecology of the ocean sunfish *Mola mola*: a review of current knowledge and future research perspectives. Rev. Fish Biol. Fisher. 20: 471–487.

Poulin, R. and A.S. Grutter. 1996. Cleaning symbioses: proximate and adaptive explanations. BioScience 46: 512–517.

Poulin, R. and S. Morand. 2000. The diversity of parasites. Q. Rev. Biol. 75: 277–293.

Pozdnyakov, S.E. 1990. Helminths of Scombrid-like Fishes of the World's Oceans. DVO AN SSSR, Vladivostok, Russia [in Russian].

Pozdnyakov, S.E. 1994. *Gonapodasmius squamata* sp. n. (Trematoda: Didymozoata) a parasite of the moon fish in the north-eastern subtropical zone of the Pacific Ocean. Parazitologiâ 28: 73–75 [in Russian].

Pozdnyakov, S.E. 1996. Trematodes suborder Didymozoata. Tikhookeanskii Nauchno-Issledovatel'Skii Rybokhozyaistvennyi Tsentr, Vladivostok, Russia.

Pratt, I. and J.E. McCauley. 1961. Trematodes of the Pacific Northwest: An annotated catalog. Oregon State University Press, Corvallis, Oregon, USA.

Price, E.W. 1962. A description of *Tricotyla molae* (Blanchard), with a discussion of the monogenetic trematodes of the sunfish (*Mola mola*, L.). J. Parasitol. 48: 748–751.

Price, P.W. 1980. Evolutionary Biology of Parasites. Princeton University Press, Princeton, USA.

Purcell, J.E. and M.N. Arai. 2001. Interactions of pelagic cnidarians and ctenophores with fish: a review. Hydrobiologia 451: 27–44.

Radujković, B.M. and L. Euzet. 1989. Parasites des poissons marins du Montenegro: Monogènes. Acta Adriat. 30: 51–135.

Radujković, B.M. and D. Šundić. 2014. Parasitic flatworms (Platyhelminthes: Monogenea, Digenea, Cestoda) of fishes from the Adriatic Sea. Nat. Montenegrina 13: 7–280.

Raibaut, A., C. Combes and F. Benoit. 1998. Analysis of the parasitic copepod species richness among Mediterranean fish. J. Marine Syst. 15: 185–206.

Riera, R., L. Moro and M. Carrillo. 2010. Primera cita para Canarias de *Capsala martinieri* Bosc, 1811 (Monogenea: Capsalidae: Capsalinae), ectoparásito del pejeluna de cola (*Masturus lanceolatus*). Rev. Acad. Canar. Cienc. 22: 85–90.

Riser, N.W. 1956. Early larval stages of two cestodes from elasmobranch fishes. Proc. Helminthol. Soc. Wash. 23: 120–124.

Rivera, G. and L. Sarmiento. Helmintos parásitos de *Mola mola* (L.) "pez sol", en Perú. 1992. Bol. Lima (Perú) 14: 71–73.

Roach, J. 2003. World's heaviest bony fish discovered? National Geographic News, http://news.nationalgeographic.com/news/2003/05/0513_030513_sunfish.html.

Robinson, E.S. 1959. Records of cestodes from marine fishes of New Zealand. Trans. R. Soc. N.Z. 86: 143–153.

Robinson, V.C. 1934. A new species of Accacoeliid trematode (*Accacladocoelium alveolatum* n. sp.) from the intestine of a sun-fish (*Orthagoriscus mola* Bosch). Parasitology 26: 346–351.

Rohde, K. 2005. Marine parasitology. Csiro publishing, Collingwood, Australia.

Rudolphi, C.A. 1819. Entozoorum synopsis: cui accedunt mantissa duplex et indices locupletissimi. Rücker.

Ryan, C. and J.M.C. Holmes. 2012. Killer whale *Orcinus orca* predation on sunfish *Mola mola*. Mar. Biodiver. 5.

Sato, N., N. Yamame and T. Kawamura. 1982. Systemic *Citrobacter freundii* infection among sunfish *Mola mola* in Matsushima Aquarium. B. Jpn. Soc. Sci. Fish. 48: 1551–1557.

Sawai, E., Y. Yamanoue, M. Nyegaard. 2020. Phylogeny, taxonomy and size records of the ocean sunfishes. pp. 18–36. *In*: T.M. Thys, G.C. Hays and J.D.R. Houghton [eds.]. The Ocean Sunfishes: Evolution, Biology and Conservation, CRC Press. Boca Raton, FL, USA.

Schwartz, F.J. and D.G. Lindquist. 1987. Observations on *Mola* basking behavior, parasites, echeneidid associations, and body-organ weight relationships. J. Elisha. Mitchell. Sci. Soc. 103: 14–20.

Scott, A. 1901. On the fish parasites, *Lepeohtheirus* and *Lernea*. Rep. Lancs. Sea Fish. Labs. 9: 63–94.

Shiino, S.M. 1957. Copepods parasitic on Japanese fishes. 13. Parasitic copepods collected off Kesennuma, Miyagi Prefecture. Rep. Fac. Fish. Pref. U. Mie. 2: 359–375.

Shiino, S.M. 1960. Copepods parasitic on fishes collected on the coast of Province Shima, Japan. Rep. Fac. Fish. Pref. U. Mie. 3: 471–500.

Shiino, S.M. 1965. On *Cecrops exiguus* Wilson found in Japan. Rep. Fac. Fish. Pref. U. Mie. 5: 381–390.

Shimazu, T., J. Araki and S. Kamegai. 1996. Further notes on the Platyhelminth parasites reported by Yoshimasa Ozaki, 1923–1966, with a list of helminth parasite specimens deposited in the Department of Zoology, University Museum, University of Tokyo, Tokyo. J. Nagano Prefec. Colleg. 51: 11–15.

Silas, E.G. 1962. Parasites of scombroid fishes. Part I. Monogenetic trematodes, digenetic trematodes, and cestodes. Proc. Symp. Scombroid Fish. India 3: 799–875.
Simková, A., L. Plaisance, I. Matejusová, S. Morand and O. Verneau. 2003. Phylogenetic relationships of the Dactylogyridae Bychowsky, 1933 (Monogenea: Dactylogyridea): the need for the systematic revision of the Ancyrocephalinae Bychowsky, 1937. Syst. Parasitol. 54: 1–11.
Skrjabin, K.I. 1959. On the position of the trematode *Accacladocoelium alveolatum* Robinson, 1934 in the suborder Hemiurata. Trudy Gelmintologicheskoi Laboratorii 9: 278–279 [in Russian].
Skrjabin, K.I. and L.H. Guschanskaja. 1959. Suborder Hemiurata (Markevitsch, 1951) Skrjabin et Guschanskaja, 1954. Family Accacoeliidae Looss, 1912. Osnovy Trematodologii 16: 99–183 [in Russian].
Smit, N.J., L. Basson and J.G. Van As. 2003. Life cycle of the temporary fish parasite, *Gnathia africana* (Crustacea: Isopoda: Gnathiidae). Folia Parasitol. 50: 135–142.
Smit, N.J. and A.J. Davies. 2004. The curious life-style of the parasitic stages of Gnathiid isopods. Adv. Parasitol. 58: 289–391.
Sousa, L.L., R. Xavier, V. Costa, N.E. Humphries, C. Trueman, R. Rosa et al. 2016. DNA barcoding identifies a cosmopolitan diet in the ocean sunfish. Sci. Rep. 6: 28762.
Sproston, N. 1946. A synopsis of the Monogenetic Trematodes. Trans. Zool. Soc. London 25: 185–600.
Stafford, J. 1904. Trematodes from Canadian fishes. Zool. Anz. 27: 481–495.
Stossich, M. 1883. Brani di elmintologia tergestina. Serie prima. Boll. Soc. Adriat. Sci. Nat. Trieste 8: 111–121.
Stossich, M. 1896. Elminti trovati in un *Orthagoriscus mola*. Boll. Soc. Adriat. Sci. Nat. Trieste 17: 189–191.
Stossich, M. 1898. Saggio di una fauna elmintologia di Trieste e provincie contermini. Programma della Civica Scuola Reale Superiore 1–162.
Stossich, M. 1899. Appunti di elmintologia. Boll. Soc. Adriat. Sci. Nat. Trieste 19: 1–6.
Stuardo, J. 1958. *Lepeophtheirus ornatus* (Milne Edwards) a synonym of *L. nordmanii* (Milne Edwards) (Copepoda: Caligidae). Univ. Bergen. 8: 3–11.
Stunkard, H.W. 1977. Studies on Tetraphyllidean and Tetrarhynchidean metacestodes from squids taken on the New England coast. Biol. Bull. 153: 387–412.
Syväranta, J., C. Harrods, L. Kubicek, V. Cappanera and J.D.R. Houghton. 2012. Stable isotopes challenge the perception of ocean sunfish *Mola mola* as obligate jellyfish predators. J. Fish Biol. 80: 225–231.
Tantalean, V.M., B.L. Sarmiento and P.A. Huiza. 1992. Digeneos (Trematoda) del Perú. Bol. Lima 80: 47–84.
Thieltges, D.W., P.A. Amundsen, R.F. Hechinger, P.T. Johnson, K.D. Lafferty, K.N. Mouritsen et al. 2013. Parasites as prey in aquatic food webs: implications for predator infection and parasite transmission. Oikos 122: 1473–1482.
Thomas, F., O. Verneau, T. De Meeûs and F. Renaud. 1996. Parasites as to host evolutionary prints: Insights into host evolution from parasitological data. Int. J. Parasitol. 26: 677–686.
Thomas, F., F. Renaud, T. De Meeûs and R. Poulin. 1998. Manipulation of host behavior by parasites: ecosystem engineering in the intertidal zone? Proc. Royal Soc. B. 265: 1091–1096.
Threlfall, W. 1967. Some parasites recovered from the ocean sunfish, *Mola mola* (L.), in Newfoundland. Can. Field Nat. 81: 168–172.
Thulin, J. 1973. Some parasites in a sun-fish caught in Gothenburg harbour. Zool. Revy. 35: 82–84.
Thulin, J. 1980. Redescription of *Nematobibothrioides histoidii* Noble, 1974 (Digenea: Didymozooidea). Z. Parasitenkund. 63: 213–219.
Thys, T.M. and R. Williams. 2013. Ocean sunfish in Canadian Pacific waters: Summer hotspot for a jelly-eating giant? Oceans-San Diego (IEEE): 1–5.
Thys, T.M., J.P. Ryan, K.C. Weng, M. Erdmann and J. Tresnati. 2016. Tracking a marine ecotourism star: movements of the short ocean sunfish *Mola ramsayi* in Nusa Penida, Bali, Indonesia. J. Mar. Biol. doi.org/10.1155/2016/8750193.
Thys, T.M., A.R. Hearn, K.C. Weng, J.P. Ryan and C. Peñaherrera-Palma. 2017. Satellite tracking and site fidelity of short ocean sunfish, *Mola ramsayi*, in the Galápagos Islands. J. Mar. Biol. doi.org/10.1155/2017/7097965.
Timi, J.T. 2007. Parasites as biological tags for stock discrimination in marine fish from South American Atlantic waters. J. Helminthol. 81: 107–111.
Timon-David, P. and J.J. Musso. 1971. Digenetic trematodes of the sunfish (*Mola-mola*) in the gulf of Marseille (Accacoeliidae, Didymozoidae). Ann. Parasitol. Hum. Comp. 46: 233–256 [in French].
Tokioka, T. 1936. Preliminary report on Argulidae found in Japan. Annot. Zool. Japon. 15: 334–341 [in Japanese].
Tompkins, D.M. and M. Begon. 1999. Parasites can regulate wildlife populations. Parasitol. Today 15: 311–313.
Vasco-Rodrigues, N. and P.M. Cabrera. 2015. *Coris julis* cleaning a *Mola mola*, a previously unreported association. Cybium 39: 315–316.
Villalba, S.C. and B.J. Fernández. 1985. Parásitos de *Mola ramsayi* (Giglioli, 1883) (Pisces: Molidae) en Chile. Bol. Soc. Biol. Concepc. 56: 71–78.
Viney, M. and J. Cable. 2011. Macroparasite life histories. Curr. Biol. 21: R767–R774.

Volodymyr, K.M., S. Al-Jufaili, R. Khalfan and N.A.M. Al-Mazrooei. 2015. Marine parasites as an object and as a factor in the problem of invasive species in marine ecosystems: reflections on the topic. J. Biodivers. Endanger. Species. 3: 154. doi: 10.4172/2332-2543.1000153.

Walter, T.C. and G. Boxshall. 2019. World of Copepods database. Accessed at www.marinespecies.org/copepoda%20on%201019-01-31.

Watanabe, Y. and K. Sato. 2008. Functional dorsoventral symmetry in relation to lift-based swimming in the ocean sunfish *Mola mola*. PLoS ONE 3: e3446. doi.org/10.1371/journal.pone.0003446.

Wilson, C.B. 1905. North American parasitic copepods belonging to the family Caligidae. Part 1. The Caliginae. Proc. U.S. National Mus. 28: 479–672.

Wilson, C.B. 1907. North American parasitic copepods belonging to the family Caligidae. Parts 3 and 4. A revision of the Pandarinae and the Cecropinae. Proc. U.S. Nat. Mus. 33: 323–490.

Wilson, C.B. 1932. Copepods of the Woods Hole region, Massachusetts. U.S. Nat. Mus. 158: 490–941.

WoRMS Editorial Board. 2019. World Register of Marine Species. Available from http://www.marinespecies.org at VLIZ. Accessed 2019-03-06. doi:10.14284/170.

Yamaguti, S. 1934a. Studies on the helminth fauna of Japan. II. Trematodes of fishes. I. Jpn. J. Zool. 5: 249–541.

Yamaguti, S. 1934b. Studies on the helminth fauna of Japan. IV. Cestodes of fishes. Jpn. J. Zool. 6: 1–112.

Yamaguti, S. 1963a. Parasitic Copepoda and Branchiura of Fishes. John Wiley and Sons, Inc., New York, USA.

Yamaguti, S. 1963b. Systema Helminthum. Vol. IV, Monogenea and Aspidocotylea. John Wiley and Sons, Inc., New York, USA.

Yamaguti, S. 1968. Monogenetic Trematodes of Hawaiian Fishes. University of Hawaii Press, Honolulu, Hawaii.

Yamaguti, S. 1971. Synopsis of the Digenetic Trematodes of Vertebrates.Vols. I and II. Keigaku Publ. Co. Tokyo, Japan.

Chapter 11

Biotoxins, Trace Elements, and Microplastics in the Ocean Sunfishes (Molidae)

Miguel Baptista,[1,*] *Cátia Figueiredo,*[2,3] *Clara Lopes,*[3] *Pedro Reis Costa,*[3,4] *Jessica Dutton,*[5] *Douglas H. Adams,*[6] *Rui Rosa*[1] and *Joana Raimundo*[3]

Introduction

Any substance may become a contaminant when it occurs at a significantly higher concentration than the natural background level in a particular area or organism (GESAMP 2015). Some contaminants exist naturally in the wild, while others are introduced or released by human activities. An anthropogenic contaminant can become a pollutant when there is clear evidence of its harmful effects for ecosystems and life forms (GESAMP 2015). Man-made contaminants are ubiquitously distributed in the marine environment and enter the ocean through riverine runoff, the atmosphere, or through direct discharge in nearshore and offshore waters. Since most anthropogenic contaminants have a land-based source, greater levels are often found in coastal areas compared to the open ocean.

One major concern is the persistence of marine contaminants and their subsequent accumulation in seawater, sediments and biota. Additionally, many can bioaccumulate with the potential to biomagnify along marine food webs. Thus, the tissues of marine top predators can contain contaminants in concentrations several orders of magnitude greater than those found in lower trophic level organisms (Chen et al. 2000, Metian et al. 2013, Lucia et al. 2014), manifesting in reduced health (e.g., Kannan et al. 2000). Owing to public health concerns, most contaminant studies in marine fish are conducted on economically important species such as flounders, herrings, salmon, groupers, and tunas (e.g., Baechler et al. 2019, Dural et al. 2007, Lusher et al. 2017, Raimundo et al. 2013, Adams 2004,

[1] MARE—Marine and Environmental Sciences Centre, Laboratório Marítimo da Guia, Faculdade de Ciências, Universidade de Lisboa, Av. Nossa Senhora do Cabo, 939, 2750-374 Cascais, Portugal; Email: rarosa@fc.ul.pt

[2] MARE—Marine and Environmental Sciences Centre, Faculdade de Ciências, Universidade de Lisboa, Campo Grande 016, 1600-548 Lisbon, Portugal; Email: cafigueiredo@fc.ul.pt

[3] Division of Oceanography and Marine Environment, IPMA—Portuguese Institute of the Sea and Atmosphere, Av. Doutor Alfredo Magalhães Ramalho, 6, 1495-165 Algés, Portugal; Email: catia.figueiredo@ipma.pt; clara.lopes@ipma.pt;prcosta@ipma.pt; jraimundo@ipma.pt

[4] CCMAR—Centre of Marine Sciences, University of Algarve, Campus of Gambelas, 8005-139 Faro, Portugal; Email: prcosta@ipma.pt

[5] Department of Biology, Texas State University, Aquatic Station, 601 University Drive, San Marcos, TX 78666, USA; Email: jdutton@txstate.edu

[6] Florida Fish and Wildlife Conservation Commission, Fish and Wildlife Research Institute, 1220 Prospect Avenue, No. 285, Melbourne, FL 32901, USA; Email: doug.adams@myfwc.com

* Corresponding author: msbaptista@fc.ul.pt

Taniyama et al. 2002). Fish with low commercial value, like the ocean sunfishes (Molidae), have received considerably less attention. The subsequent paucity of fisheries data, coupled with their patchy distribution (Kenney 1996, Breen et al. 2017, Grémillet et al. 2017) make ocean sunfishes challenging study subjects. However, beyond commercial interest they are far from peripheral given their broad distribution and diverse roles in temperate and tropical food webs (Pope et al. 2010, Sousa et al. 2020 [Chapter 8] and Phillips et al. 2020 [Chapter 9]). From this perspective, and with a focus on *Mola* spp., this chapter reviews all available information on three types of contaminants known to occur in the ocean sunfishes; biotoxins, trace elements and microplastics.

Biotoxins

Fish of the order Tetraodontiformes are generally associated with the potent neurotoxins tetrodotoxin (whose name directly derives from their order name; TTX) and palytoxin (PTX). TTX blocks voltage-gated sodium (Na) channels, affecting the regular Na^+ current in mammalian excitable neuronal cells which causes paralysis. In severe cases, this can lead to respiratory failure and fatalities (Rodriguez et al. 2008, Walker et al. 2012). TTX is mostly known in pufferfish (Tetraodontidae) but it is also found in many marine vertebrates and invertebrates, freshwater fishes and terrestrial vertebrates (amphibians) (Bane et al. 2014, Hanifin 2010). The biogenic origin of TTX remains under debate but is most commonly attributed to symbiotic bacteria (Chau et al. 2011, Lago et al. 2015 and references therein; Magarlamov et al. 2017). Nevertheless, trophic transfer may play a role in TTX dynamics, with the presence of this toxin in shellfish associated with blooms of the benthic microalgae *Prorocentrum minutum* (Vlamis et al. 2015). By contrast, PTX exhibits complex mechanisms that involve the blockage of the Na^+/K^+-ATPase pump in mammals but may also act as a hemolysin and hence destroy erythrocytes (Patocka et al. 2015). This biotoxin was first detected in the zoanthid *Palythoa* sp. (Moore and Scheuer 1971) and later in various organisms such as soft corals, mussels, sponges, crustaceans and polychaetes living in close association with zoanthid colonies. This grouping also extended to echinoderms and fishes feeding on *Palythoa* (Gleibs and Mebs 1999). PTX is potentially produced by dinoflagellates of the genus *Ostreopsis* (Usami et al. 1995), which are widely distributed throughout tropical and subtropical seas (Gallitelli et al. 2005).

To date, only two studies have investigated the presence of TTX in ocean sunfish species (Saito et al. 1991, Huang et al. 2011), while only one study investigated the presence of PTX (Huang et al. 2011). No TTX was found during a mouse biological assay on the acidic extract of *Mola* spp. liver, and muscle of *Masturus lanceolatus* (Liénard 1840) and *Ranzania laevis* (Pennant 1776) (Saito et al. 1991, Huang et al. 2011). More sensitive chemical methods, based on hydrophilic interaction liquid chromatography tandem mass spectrometry (HILIC-MS/MS), are currently being used to evaluate the presence of TTX in *Mola* spp. from the Portuguese coast. Preliminary results indicate that no toxin is present in the three examined tissues (i.e., liver, white muscle and red muscle; Baptista et al. unpublished data).

PTX or PTX-like compounds may have been responsible for a food poisoning event resulting from the ingestion of *Ma. lanceolatus* in Taiwan (Huang et al. 2011). Two persons, a 40-year old mother and her 16-year old daughter, suffered breast pain, muscle pain, tympanites, vomiting and tachycardia 30 minutes after eating *Ma. lanceolatus* muscle (Huang et al. 2011). The symptoms of the patients were similar to those of PTX and a hemolytic assay suggested the presence of PTX-like compounds. In case of future occurrences, the employment of more selective detection methods such as fluorescence polarization (Alfonso et al. 2012) would be valuable to confirm the presence/absence of PTX or PTX-like compounds.

While the presence of either PTX or TTX has not been confirmed in any ocean sunfish species (Molidae) to date, erring on the side of caution, EU regulators have adopted a conservative approach to trading fish potentially containing toxins. Thus fish from families Tetraodontidae, Molidae and Diodontidae currently cannot be placed in EU markets (European Commission 2004).

Trace Elements

Trace elements can be critical contaminants in aquatic environments (Alquezar et al. 2006, Yilmaz and Doğan 2008), accumulating, like biotoxins, within food webs (Chen et al. 2000, Lucia et al. 2014). Fish occupying upper trophic levels are particularly susceptible to elemental bioaccumulation. Trace elements are considered essential when they serve a biological function, e.g., Co, Cr, Cu, Fe, Mn, Ni, Se, Zn, or nonessential when no biological function is evident, e.g., Ag, As, Cd, Hg, Pb, Sn, V, (See Table 1 for element abbreviations.) (Watanabe et al. 1997, Wood et al. 2012a,b). While nonessential trace elements can be toxic at low concentrations, essential trace elements typically display sub-lethal and lethal effects at high concentrations (Rainbow 1985). Either way, the presence of these elements above a certain threshold level can disrupt fish physiology and reproductive success and may increase fish mortality (Dunier 1996, Authman et al. 2015).

Trace element levels have been evaluated in numerous fish species from different habitats, regions and trophic levels (e.g., Jureša and Blanuša 2003, Alibabić et al. 2007, Yilmaz and Doğan 2008, Metian et al. 2013, Maulvault et al. 2015, Raimundo et al. 2015, Mataba et al. 2016, Kaleshkumar et al. 2017). However, due to public health concerns, research has mostly focused on economically important fishes (e.g., Dural et al. 2007, Ersoy and Çelik 2009, Pedro et al. 2013, Raimundo et al. 2013, Alkan et al. 2016). Nonetheless, these studies revealed several factors that influence elemental levels in fish tissues: sex (Protasowicki and Morsy 1993, Shakweer and Abbas 2005, Costa and Hartz 2009), size and therefore age (Al-Yousuf et al. 2000, Jureša and Blanuša 2003, Kojadinovic et al. 2007), diet (Kojadinovic et al. 2007, Dutton and Fisher 2011), habitat use (Walker 1976, Jureša and Blanuša 2003) and season (Dural et al. 2007, Yilmaz and Doğan 2008, Ersoy and Çelik 2009).

Trace Elements in Ocean Sunfishes

Ocean sunfishes have limited commercial value worldwide (Silvani et al. 1999, Pope et al. 2010), with targeted fisheries occurring mainly in East Asia, namely Japan, Korea and Taiwan (Sagara and Ozawa 2002, Liu et al. 2009, Lee et al. 2013, Nyegaard 2020 [Chapter 12]). As such, finite research has been conducted on elemental levels in their tissues. To date, seven published works and one unpublished screening by Dr. Tierney Thys and Sea Studios Foundation (Adams et al. 2003, Kumar et al. 2004, Zaccaroni et al. 2008, Perrault et al. 2014, Dutton et al. 2016, Baptista 2019, Baptista et al. 2019) evaluated elemental concentrations in *Mola* spp. tissues. With respect to other ocean sunfish species, only one elemental composition study has been conducted for *Ma. lanceolatus* (Dutton et al. 2016) (Supplemental Table 1).

Sex-Related Differences in the Elemental Composition of Juvenile *Mola* spp. (Adapted from Baptista 2019)

Baptista (2019) found sex-related differences in the trace element composition of immature *Mola* spp. [nine females (F) and 11 males (M)], 37.5–85.5 cm total length (TL) (Kang et al. 2015), from southern Portugal. Inter-sex comparison (Wilcoxon rank sum tests; $p < 0.05$) of tissues (brain, gills, hypodermis, gonads, spleen, liver, white muscle and red muscle) revealed differences in trace element composition solely in gonadal tissues. These differences likely reflect biochemical and structural differences between ovaries and testes, with the former displaying higher levels of Mn, Cu and Zn. By contrast, testes exhibited greater concentration of V. Since V is mainly regarded as toxic for fish (Holdway and Sprague 1979, Gravenmier et al. 2005, Zaki et al. 2010), there may be no physiological purpose underlying the observed difference in V concentration among different gonadal tissues. Quite simply, the greater values of V found in testes may derive from its organotropism (i.e., special affinity for particular tissues). In the ovaries the greater accumulation of the essential elements Mn, Cu and Zn (Watanabe et al. 1997, Chanda et al. 2015), may reflect their importance in embryonic development.

Indeed, a reduction in Mn and Zn in fish eggs leads to a decrease in successful embryogenesis and hatchability (Takeuchi et al. 1981). Accordingly, Satoh et al. (1987) hypothesized the existence of a link between egg elemental content and successful embryonic development, with Mn and Zn appearing particularly important. In this way, the biochemistry of ovaries may simply promote the accumulation of those elements. Indeed, the greater concentration of Zn in ovaries when compared to testes appears common (Shakweer and Abbas 2005, Rajkowska and Protasowicki 2013) and a similar trend may occur with Mn (Rajkowska and Protasowicki 2013). With regard to Cu, however, although higher levels were observed in ovaries of *Mola* spp., greater concentrations are found in ovaries or testes depending on fish species (Protasowicki and Morsy 1993, Shakweer and Abbas 2005, Rajkowska and Protasowicki 2013). Regardless, if certain essential elements play a role in embryonic development, it is logical that they would accumulate in ovaries, with species-specific differences potentially occurring as a result of different physiologies and reproductive strategies.

Elemental Distribution Within the Tissues of Juvenile *Mola* spp. (Adapted from Baptista 2019)

Zinc (Zn) is often present in high concentrations in teleost tissues (Kojadinovic et al. 2007, Bashir et al. 2012, Raimundo et al. 2013, 2015) serving as a cofactor for a number of enzymes and as an integral component of many metalloenzymes (e.g., alkaline phosphatase; Watanabe et al. 1997, Chanda et al. 2015). Unsurprisingly, Zn was found in higher levels than other elements in almost all body tissues of juvenile *Mola* spp. (Figs. 1, 2, 3 and 4), with the highest levels being found in the ovaries and gills (Fig. 4; Table 1). Given that Zn can regulate growth during early life history, its notable presence in the ovaries is logical (Ogino and Yang 1978, Li and Huang 2015). Likewise, the high concentration of Zn in the gills, alludes to its role in metabolic processes (Watanabe et al. 1997, Chanda et al. 2015).

Arsenic (As) was the second most abundant trace element in *Mola* spp. (Baptista 2019). In marine fish, levels of As depend on diet and trophic level (Rudneva 2013). More specifically, those feeding on crustaceans and algae appear to accumulate greater amounts of this trace element than typically piscivorous species (Rudneva 2013). Given that juvenile *Mola* spp. are known to prey on crustaceans (Pope et al. 2010 and references therein, Sousa et al. 2016a, Phillips et al. 2020 [Chapter 9]) and algae (Bakenhaster and Knight-Gray 2016), a likely pathway for bioaccumulation emerges.

Juvenile *Mola* spp. revealed considerable inter-individual variability in the concentration of each element in any given tissue, particularly in the liver (Figs. 1, 2, 3 and 4) (Baptista 2019). Such differences could be linked to variations in diet and elemental composition of seawater (Kraal et al. 1995, Ptachynski et al. 2002, Rosseland et al. 2007). Since juvenile ocean sunfish tissues were all sampled from a single site during the same month, the observed variability is likely not derived from differences in waterborne elemental concentration. Concomitantly, it was argued that dietary differences between individuals provided the most parsimonious origin for the inter-individual variability in elemental concentration. Although diet was not examined explicitly, juvenile *Mola mola* exhibit a generalist diet composed of pelagic and benthic prey (Syväranta et al. 2012, Harrod et al. 2013, Sousa et al. 2016a, Phillips et al. 2020 [Chapter 9]), including teleosts, crustaceans, bivalves, cephalopods, gastropods and gelatinous zooplankton (Sousa et al. 2016a). As sediments serve as a deposit of trace elements, benthic-feeding fish usually display higher levels of trace elements when compared to pelagic feeders (Yi et al. 2011). Such a diverse diet composed of both benthic and pelagic prey may translate into large variations in elemental intake and, ultimately, tissue levels (Baptista et al. 2020). Notably, the key role of the liver in the digestion, storage and detoxification of trace elements (Metian et al. 2013, Raimundo et al. 2015) may explain why inter-individual variability in elemental concentration was more evident in this tissue (Baptista 2019).

Overall, greater trace element load was found in the metabolically active tissues of the liver and gills (Fig. 5; Baptista 2019). This result aligns with the function, morphological configuration and biochemistry of those tissues. While playing a key role in the detoxification and storage of trace

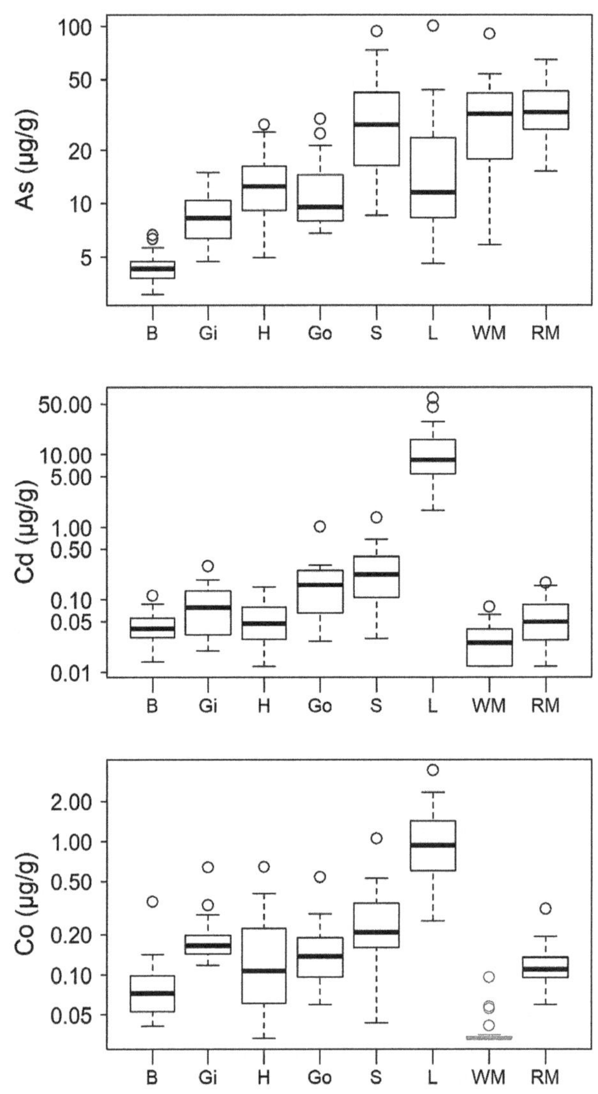

Figure 1. Concentration (µg/g, dry weight) of arsenic (As), cadmium (Cd) and cobalt (Co) in tissues of *Mola* spp. (n = 20) from southern Portugal. Box-plots show median, 25 and 75 percentiles, whiskers indicate the range. Note the y-axis logarithmic scale in plots. B, Brain; Gi, Gills; H, Hypodermis; Go, Gonads; S, Spleen; L, Liver; WM, White muscle; RM, Red muscle. The grey color in the boxplot for Co in white muscle is used to indicate that the data set comprised more than 30 percent of values below the detection limit (0.033 µg/g). Adapted from Baptista (2019).

elements (Metian et al. 2013, Raimundo et al. 2015), the liver is bound to display considerable trace element levels. Accordingly, this tissue exhibited the greatest concentration of Co, Cu and Cd (Figs. 1 and 2; Table 1; Baptista 2019). Conversely, the gills are in immediate contact with the surrounding water and can accumulate high levels of waterborne elements (Kraal et al. 1995, Kalay and Canli 2000, Ricketts et al. 2015), particularly those with greater affinity for binding sites (e.g., Na^+ and Ca^{2+} transport sites), such as Ni, Zn and Pb (Niyogi and Wood 2004). As such, this tissue exhibited the highest levels of Zn and Ni (together with the ovaries and gonads, respectively), and the greater median value of Pb (Figs. 3 and 4; Table 1; Baptista 2019). Additionally, the gills exhibited the greatest concentration of Mn (i.e., one order of magnitude greater than in any other tissue; Fig. 2; Table 1; Baptista 2019). The considerable levels of Mn in gills likely result from the abundance of

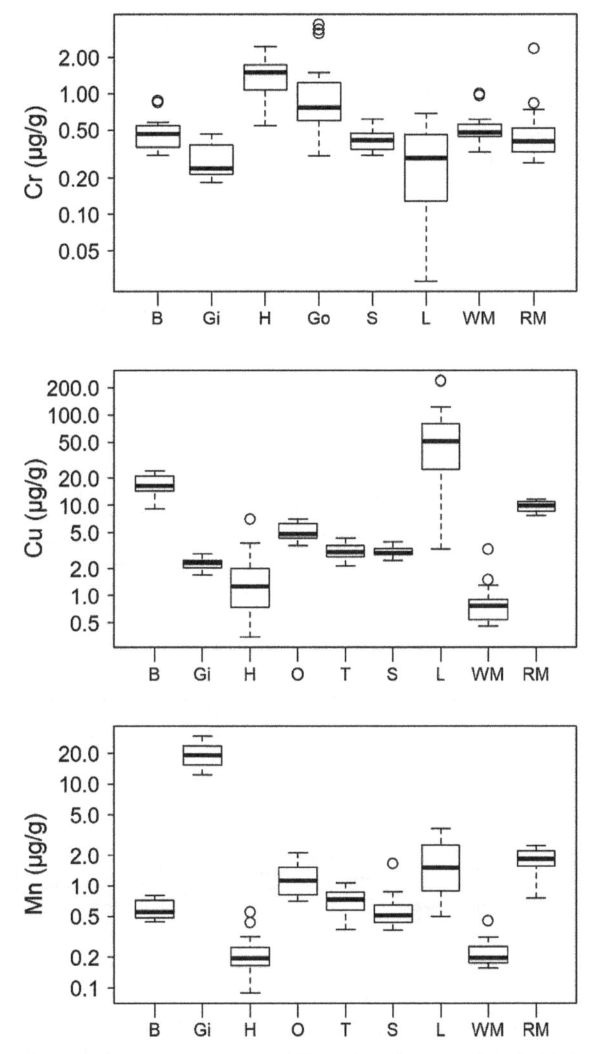

Figure 2. Concentration (μg/g, dry weight) of chromium (Cr), copper (Cu) and manganese (Mn) in tissues of *Mola* spp. (n = 20) from southern Portugal. Box-plots show median, 25 and 75 percentiles, whiskers indicate the range. Note the y-axis logarithmic scale in plots. B, Brain; Gi, Gills; H, Hypodermis; O, Ovaries; T, Testes; S, Spleen; L, Liver; WM, White muscle; RM, Red muscle. Adapted from Baptista (2019).

mitochondria in gill tissue (Lin and Sung 2003, Kaneko et al. 2008) and the close association between those organelles and Mn (reviewed by Gunter et al. 2009, Naranuntarat et al. 2009).

A generalized pattern of lower elemental load was found in white muscle, brain and hypodermis when compared to other tissues (Figs. 1, 2, 3, 4 and 5; Table 1; Baptista 2019). Muscle tissue commonly exhibits low elemental levels (Protasowicki and Morsy 1993, Szarek-Gwiazda et al. 2006, Raimundo et al. 2015). Few studies have examined trace element composition of fish brain tissue, yet, low elemental accumulation should occur as it is protected from generalized contamination by the blood/brain barrier (Sloman et al. 2003). Still, the second greatest levels of Cu and greater median concentrations of Mn, Ni and Cr were found in this tissue (Figs. 2 and 3; Table 1; Baptista 2019), leading on to the brain via the blood/brain barrier (i.e., Cu, Choi and Zheng 2009) or otherwise (e.g., via the olfactory nerves; Gottofrey and Tjälve 1991, Tjälve and Henriksson 1999, Lushchak et al. 2009). The low elemental concentrations in the hypodermis may indicate a minimal function in body metabolic processes (Baptista 2019). Curiously, however, the hypodermis displayed the greatest

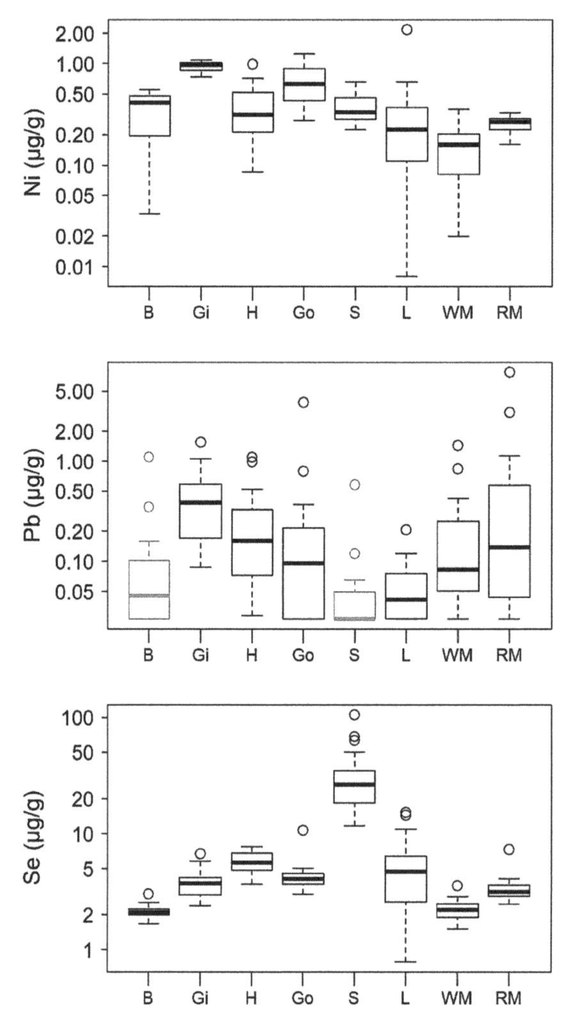

Figure 3. Concentration (µg/g, dry weight) of nickel (Ni), lead (Pb) and selenium (Se) in tissues of *Mola* spp. (n = 20) from southern Portugal. Box-plots show median, 25 and 75 percentiles, whiskers indicate the range. Note the y-axis logarithmic scale in plots. B, Brain; Gi, Gills; H, Hypodermis; Go, Gonads; S, Spleen; L, Liver; WM, White muscle; RM, Red muscle. The grey color in the boxplots for Pb in brain and spleen are used to indicate that each of the data sets comprised more than 30 percent of values below the detection limit (0.026 µg/g). Adapted from Baptista (2019).

median value of Cr, when compared to other tissues, which indicates either a high affinity of Cr for hypodermis or a low elimination ability of the tissue with respect to that element (Baptista 2019).

Interestingly, white and red muscles in ocean sunfish differed considerably in terms of elemental composition (Baptista 2019). Red muscle exhibited greater concentrations of Mn, Ni, Cu, Zn and Se and there is strong indication that V and Co may also be found in greater concentration (Figs. 1, 2, 3 and 4; Table 1). Concomitantly, red muscle showed a greater overall elemental load than white muscle (Fig. 5; Baptista 2019). Both tissues power fin movement in *Mola* spp. (Watanabe and Davenport 2020 [Chapter 5]), yet they play different roles in locomotion, with white muscle being responsible for fast vigorous movements whilst red muscle predominates during slow continuous swimming (reviewed by Mosse and Hudson 1977), and cruising (Davenport et al. 2018). Red muscle is well-adapted for aerobic metabolism using fat primarily as fuel, whereas white muscle is adapted for anaerobic metabolism using mostly glycogen (George 1962). As such, red muscle contains more mitochondria and greater fat content and lipase activity than white muscle (George 1962). Thus, while heightened levels of Mn may derive from a greater mitochondria presence, Mn, Cu and

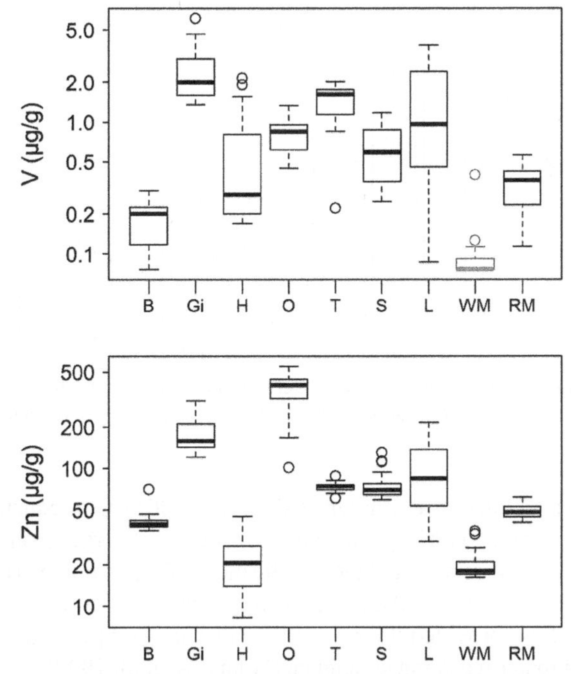

Figure 4. Concentration (µg/g, dry weight) of vanadium (V) and zinc (Zn) in tissues of *Mola* spp. (n = 20) from southern Portugal. Box-plots show median, 25 and 75 percentiles, whiskers indicate the range. Note the y-axis logarithmic scale in plots. B, Brain; Gi, Gills; H, Hypodermis; O, Ovaries; T, Testes; S, Spleen; L, Liver; WM, White muscle; RM, Red muscle. The grey color in the boxplot for V in white muscle is used to indicate that the data set comprised more than 30 percent of values below the detection limit (0.076 µg/g). Adapted from Baptista (2019).

Table 1. Differences in trace element composition among *Mola* spp. (n = 20) tissues from southern Portugal (Kruskal-Wallis rank sum tests and pairwise comparisons using Wilcoxon rank sum tests with Bonferroni correction; $p < 0.05$). Adapted from Baptista (2019).

Trace element	Significant result
As (arsenic)	Lowest concentration in Brain
Cd (cadmium)	Greatest concentration in Liver
Co (cobalt)	Greatest concentration in Liver
	Lowest concentration in Brain
Cu (copper)	Greatest concentration in Liver
	Second greatest concentration in Brain
	Lowest concentration in Gills, Hypodermis, White muscle
	Red muscle > White muscle
Mn (manganese)	Greatest concentration in Gills
	Lowest concentration in Hypodermis, White muscle
	Red muscle > White muscle
Ni (nickel)	Greatest concentration in Gills, Gonads
	Red muscle > White muscle
Se (selenium)	Greatest concentration in Spleen
	Lowest concentration in Brain and White muscle
	Red muscle > White muscle
V (vanadium)	Lowest concentration in Brain
Zn (zinc)	Greatest concentration in Gills, Ovary
	Lowest concentration in Hypodermis and White muscle
	Red muscle > White muscle

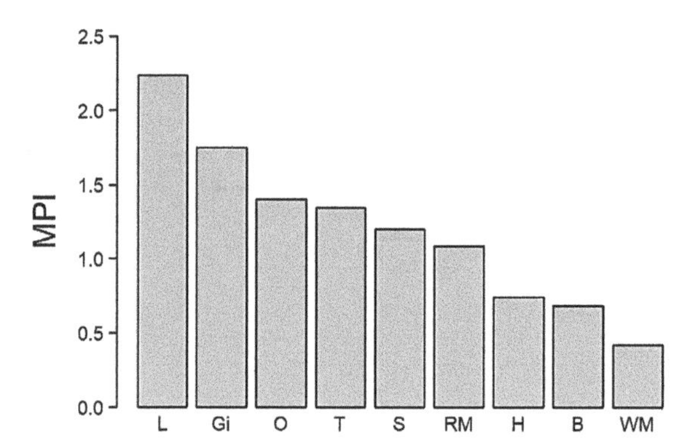

Figure 5. Metal pollution index (MPI) of the examined tissues of *Mola* spp. (n = 20) from Portugal. B, Brain; Gi, Gills; H, Hypodermis; O, Ovaries; T, Testes; S, Spleen; L, Liver; WM, White muscle; RM, Red muscle. Adapted from Baptista (2019).

Zn are involved in lipase production and activity (Maia et al. 2001). Consequently, the requirement of red muscle for those elements should be greater and the observed divergence in elemental composition between both tissues should be related to their specific metabolic physiology (Baptista 2019). Finally, the occurrence of higher Se levels in red muscle when compared to white muscle may be related with the also higher levels of Ni, provided the role of the latter in lipid peroxidation (Stohs and Bagchi 1995) and that of Se in oxidative damage defence (Watanabe et al. 1997).

Relationship Between Body Size and Trace Element Concentration in Juvenile *Mola* spp. (Adapted from Baptista et al. 2019)

Body size has been noted to have an effect in the elemental concentration of fish (Marcovecchio et al. 1991, Al-Yousuf et al. 2000, Yi and Zhang 2012, Ayas and Köşker 2018). In *Mola* spp., the existence and type of effect (increase or decrease with fish size) was not always observed depending on tissue/trace element examined (Table 2; Baptista et al. 2019). Mainly, elemental levels tended to scale inversely with body-size (Fig. 6; Table 2) with a few notable exceptions (Fig. 7; Table 2). Similarly contrasting results have been observed in various fish species from different regions of the world. The existence, or type of relationship, between elemental concentration and body size was found to vary depending on tissue, trace element and species (e.g., Marcovecchio et al. 1991, Al-Yousuf et al. 2000, Szefer et al. 2003, Farkas et al. 2008, Yi and Zhang 2012, Ayas and Köşker 2018, Yeltekin and Oğuz 2018).

Mola spp. exhibit fast growth rates in captivity of 0.02–0.49 kg a day (Howard and Nakatsubo, personal communication; Pope et al. 2010 and references therein). Indeed, Nakatsubo and Hirose

Table 2. Summary of results of generalized linear models (GLMs) evaluating the relationships between trace element concentrations in body tissues of *Mola* spp. from southern Portugal and body size (total length, TL) and the interaction between body size and season. Spring (April-May; n = 31), autumn (September-November; n = 26). ↓, decreases with; ↑, increases with.

	Gills	Hypodermis	Liver	White muscle	Red muscle
As	No effect	No effect	↑ TL	↓ TL	No effect
Cd	↓ TL	↓ TL	No effect	↓ TL	↓ TL
Co	↓ TL	↓ TL	No effect	-	↓ TL
Cu	↓ TL	↓ TL	No effect	No effect	↓ TL (Autumn)
Pb	No effect	No effect	-	↓ TL	↓ TL (Spring)
Zn	↑ TL	↓ TL (Spring)	No effect	No effect	↑ TL

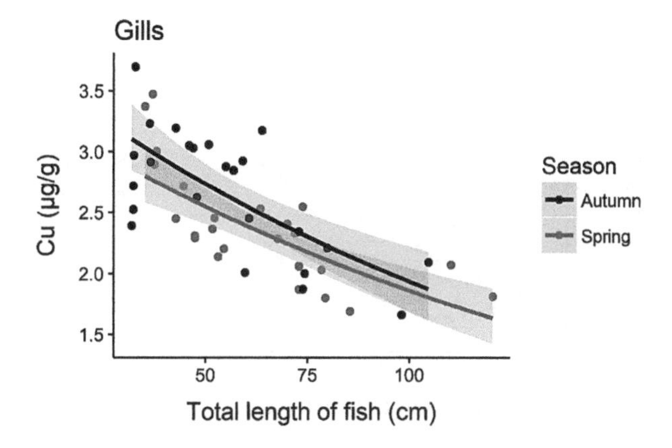

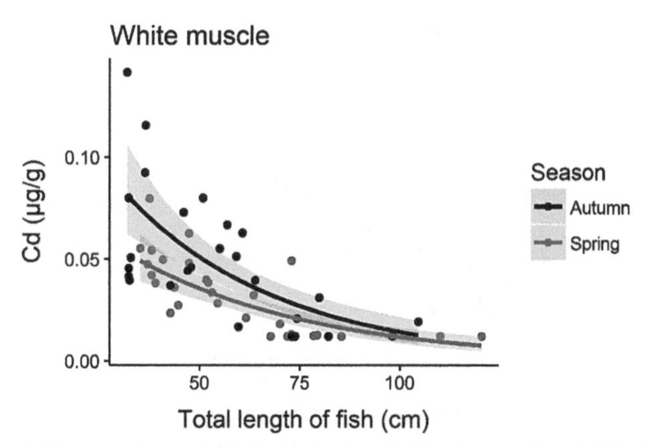

Figure 6. Body size related differences in concentration (µg/g, dry weight) of copper (Cu) in gills (n = 49) and cadmium (Cd) in white muscle (n = 56) of *Mola* spp. from southern Portugal in spring (April–May) and autumn (September–November). Regression lines and corresponding 95% confidence intervals (shaded areas) of generalized linear models are shown.

(2007) observed an average growth of 40 cm TL per year in captive juveniles whilst in the size range of the specimens studied by Baptista et al. (2019; i.e., 32–120 cm TL). Such growth is naturally accompanied by a fast increase in tissue volume. Certain tissues may grow at a faster rate than the rate at which trace elements get incorporated and hence lead to a decrease of elemental concentrations as the fish increases in size. Nonetheless, different trends in elemental levels relative to fish size were observed depending on tissue and trace element (Table 2; Baptista et al. 2019). As considerable inter-individual variability in the concentration of each element in any given tissue was found in juvenile *Mola* spp. (discussed previously in Baptista et al. 2019) it may obscure decreasing trends for some tissues/trace elements. Likewise, the specific function and volumetric increase of tissues over time, combined with the organotropism of trace elements may also account for absence of decreasing trends (Baptista et al. 2019). Given the role of the liver in storage and detoxification of trace elements (Metian et al. 2013, Raimundo et al. 2015), an ontogenetic increase in elemental levels would be expected, as shown in other fish species (Al-Yousuf et al. 2000, Ayas and Köşker 2018). However, the rate of volumetric increase in liver of *Mola* spp. over time may result in an overall accumulation of trace elements that does not manifest as an increase in concentration (Baptista et al. 2019).

Levels of the essential trace element Zn were found to increase with fish size in both the gills and red muscle (Baptista et al. 2019). This element is involved in several metabolic pathways, serving as a cofactor of a number of enzymes and being an integral part of many metalloenzymes (Watanabe et al. 1997, Chanda et al. 2015). Thus, it is possible that higher levels of Zn in the red muscle of larger

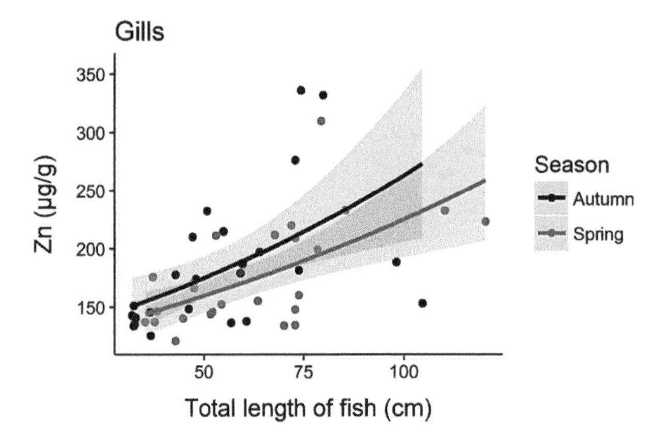

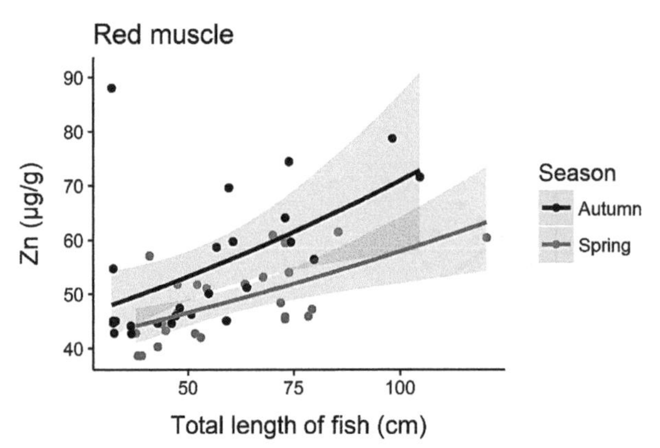

Figure 7. Body size-related differences in concentration (μg/g dry weight) of zinc (Zn) in gills (n = 49) and red muscle (n = 50) of *Mola* spp. from southern Portugal in spring (April–May) and autumn (September–November). Regression lines and corresponding 95% confidence intervals (shaded areas) of generalized linear models are shown. Adapted from Baptista et al. (2019).

fish are related to greater metabolic demands. Red muscle samples were obtained from the anal fin musculature (Baptista et al. 2019) which plays a key role in the swimming activity of ocean sunfish. Watanabe and Sato (2008) reported a remarkable decrease in the aspect ratio of both dorsal and anal fins with fish growth, indicating a decrease in the swimming efficiency of larger *Mola* spp. when compared to smaller specimens. Interestingly, Phillips et al. (2018) found a decrease in the aspect ratio of the anal fin but not in the dorsal fin. Regardless, such a decrease in swimming efficiency in larger animals may require compensation in terms of anal fin musculature performance through enhanced metabolic activity and a greater requirement of Zn when compared to smaller fish (Baptista et al. 2019). Concomitantly, as red muscle relies on aerobic metabolism (George 1962), greater respiratory activity should occur to fuel the enhanced metabolism of that tissue in larger animals. Such greater respiratory activity likely translates into greater metabolic activity in the gills and consequently, greater levels of Zn (Baptista et al. 2019).

Within a seasonal context, Baptista et al. (2019) revealed that Zn levels in the hypodermis, and Pb in red muscle, decreased with fish size only in spring while the concentration of Cu in red muscle decreased with size solely in autumn (Table 2). These results indicated that body size effects may be confounded by seasonal factors, such as differences in the rate of uptake or elimination of trace elements (Baptista et al. 2019).

Relationship Between Season and Trace Element Concentration in Juvenile *Mola* spp. (Adapted from Baptista et al. 2019)

The elemental composition of fish varies seasonally (Dural et al. 2007, Yilmaz and Doğan 2008, Ersoy and Çelik 2009). In *Mola* spp. collected from southern Portugal, seasonal (spring [April–May] vs autumn [September–November]) differences in trace element concentration were observed, albeit depending on tissue/trace element examined (Table 3; Baptista et al. 2019). Greater elemental concentration was found in spring or autumn depending on the tissue/trace element examined (Baptista et al. 2019). Similar results of greater elemental levels, regardless of season and depending on the tissue/trace element examined, have been reported for other fish species (Dural et al. 2007, Ersoy and Çelik 2009). Regardless, when a seasonal effect was evident, elemental levels were higher in the autumn in the majority of cases.

This seasonal pattern was most obvious in liver tissue for both individual trace elements and overall elemental content (Figs. 8 and 9; Table 3; Baptista et al. 2019). Contrastingly, the hypodermis mainly exhibited greater trace element concentrations during the spring. Seasonal differences in trace element concentration were less common in gills, white muscle and red muscle (Baptista et al. 2019). These divergent findings are somewhat intuitive given that different tissues display different turnover rates. Liver tissue, for instance, exhibits a faster turnover rate than muscle (Fauconneau and Arnal 1985, de la Higuera et al. 1999), and thus the latter should reflect elemental uptake over a longer time frame (Kojadinovic et al. 2007).

Higher autumnal concentrations in liver trace elements are indicative of a greater elemental uptake on a relatively recent period from late summer through autumn (Baptista et al. 2019).[1] Such elemental uptake in fish can occur via superior elemental load in dietary items and/or surrounding water (Kraal et al. 1995, Rosseland et al. 2007). Indeed, access to different prey items and exposure to different waterborne elements can affect trace element levels in fish (Kojadinovic et al. 2007, Dutton and Fisher 2011, Ricketts et al. 2015), with the latter noticeable in a relatively short period of time in the gills (i.e., days to weeks; Kraal et al. 1995, Kalay and Canli 2000, Ricketts et al. 2015). The overall lack of a seasonal effect in the concentration of trace elements in the gills indicated a similarity in waterborne elemental concentrations between spring and autumn at the study site. Thus, the greater trace element levels in liver tissue in the autumn most likely arose from differences in dietary uptake (Baptista et al. 2019). Conversely, the infrequency of seasonal effects in white and red muscle tissues may reflect slow turnover rates that may shroud changes in short/medium-term elemental uptake (Baptista et al. 2019).

Table 3. Summary of results of generalized linear models (GLMs) evaluating the relationships between trace element concentrations in body tissues of juvenile *Mola* spp. from southern Portugal and season. Spring (April–May; n = 31), autumn (September–November; n = 26). >, greater than.

	Gills	Hypodermis	Liver	White muscle	Red muscle
As	Autumn > Spring	Autumn > Spring	Autumn > Spring	No effect	No effect
Cd	No effect	No effect	Autumn > Spring	Autumn > Spring	No effect
Co	No effect	No effect	Autumn > Spring	-	No effect
Cu	No effect	Spring > Autumn	Autumn > Spring	No effect	Autumn > Spring
Pb	No effect	Spring > Autumn	-	Spring > Autumn	Spring > Autumn
Zn	No effect	Spring > Autumn	Autumn > Spring	No effect	Autumn > Spring

[1] While changes in the elemental concentration of the liver are noticeable in a relatively short time scale, on the order of weeks-months, it is difficult to pinpoint the exact moment when the actual increase in elemental uptake occurred and hence we must consider it occurred sometime during the end of summer and autumn months.

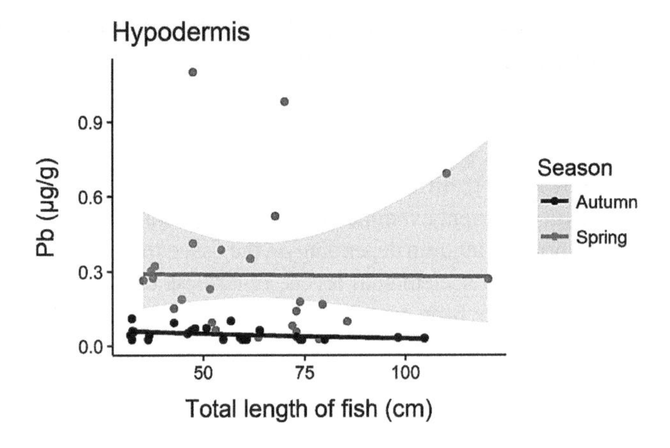

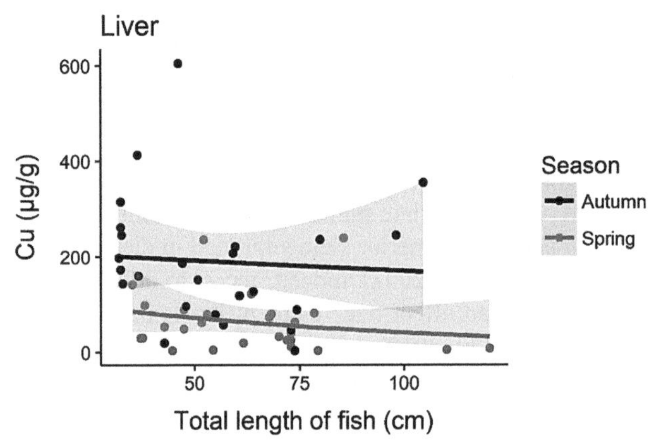

Figure 8. Concentration (µg/g dry weight) of lead in hypodermis (n = 51) and copper (Cu) in the liver (n = 52) of juvenile *Mola* spp. from southern Portugal in spring (April–May) and autumn (September–November). Regression lines and corresponding 95% confidence intervals (shaded areas) of generalized linear models are shown. Adapted from Baptista et al. (2019).

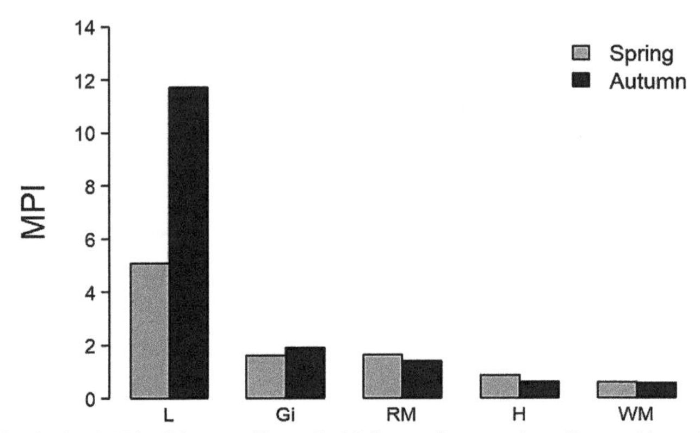

Figure 9. Metal pollution index (MPI) of tissues of juvenile *Mola* spp. from southern Portugal in spring (April–May; n = 31) and autumn (September–November; n = 26). Adapted from Baptista et al. (2019).

Seasonal differences in elemental concentrations of ocean sunfish may reflect habitat shifts. Baptista et al. (2019) studied *Mola* spp. occurring in waters off southern Portugal where two annual abundance peaks occur (spring and autumn; Baptista et al. 2018). These seasonal abundance peaks are likely related to latitudinal movements in which fish occupying the region during late winter and spring move northward along the west coast of the Iberian Peninsula and then switch into a southward movement in summer-autumn (Sims et al. 2009, Sousa et al. 2016b). Baptista et al. (2019) hypothesized that the increase in elemental load in liver tissue of *Mola* spp. between spring and autumn might arise from an increase in dietary elemental uptake at the end of the summer/autumn, although the exact mechanisms by which this might occur require further investigation.

It is also noteworthy that the hypodermis of ocean sunfish captured in Portugal typically exhibited greater elemental concentrations in the spring (Baptista et al. 2019). Lacking any vasculature it is likely that the hypodermis has a slow turnover rate (Watanabe and Sato 2008, Davenport et al. 2018, Watanabe and Davenport 2020 [Chapter 5]). Subsequently, the observed seasonal differences may indicate long-term differences in dietary or waterborne exposure to trace elements, providing the possibility for studying long-term differences in trace element uptake (Baptista et al. 2019).

Trace Elements in *Masturus lanceolatus*

Information on the concentration of trace elements in sharptail sunfish (*Ma. lanceolatus*) is provided by Dutton et al. (2016). This study reported the concentration of 14 trace elements in six tissues (muscle, liver, kidney, spleen, gill, testes) collected from five *Ma. lanceolatus* stranded in the surf zone on the Atlantic coast of Florida between February 2010 and August 2014 (Supplemental Table 1). The five specimens (four males and one female) were between 41.9 and 250.1 cm TL. Samples were collected following the methods described in Adams and Sonne (2013), dried at 60°C for 48 hours, and the concentration of trace elements determined using microwave acid digestion and Inductively Coupled Plasma Mass Spectrometry (ICP-MS) analysis following EPA Method 6020A (U.S. EPA 1998). To allow for conversion between dry weight and wet weight, the percentage moisture content was 85 percent for muscle, 77 percent for gill, 66 percent for liver, 75 percent for kidney and spleen, and 84 percent for testes.

Overall, essential trace elements were found at higher concentrations than non-essential trace elements (Supplemental T 1). The essential trace elements, Fe, Zn, Se, and Cu, had the highest mean concentrations, whereas the nonessential elements, As, Cd, and Hg, had the highest mean concentrations. For the majority of trace elements, the mean concentration was highest in the liver, kidney, and spleen due to their role in filtration, excretion, as well as storage and detoxification of contaminants, and lowest in the muscle. The mean (± SD) Mn, Cr, and Zn concentration was highest in the gill (50.7 ± 8.63, 0.556 ± 0.297, and 420 ± 79.2 μg/g, respectively) and As was highest in the testes (21.1 ± 20.6 μg/g). Silver measured in the liver of all five specimens, was BDL (below detection limit; < 0.010 μg/g) in the muscle of all specimens and was only at a measurable concentration in the kidney, spleen, gill, and testes of one specimen, a 250.1 cm TL male. Lead was detected only in the spleen and gills of all specimens. The low concentration of Hg in *Ma. lanceolatus* can most likely be attributed to their generally low trophic position (Phillips et al. 2020 [Chapter 9]).

Selenium is known to have an antagonistic relationship with Hg, and it has been proposed that a Se:Hg molar ratio > 1:1 may have a protective effect against Hg toxicity, due to Se being present in molar excess (Ralston and Raymond 2010). All tissues from all five specimens examined in this study had a Se:Hg molar ratio > 1:1 (mean ± SD; muscle: 15.2 ± 7.5; liver: 41.4 ± 25.0; kidney: 84.1 ± 35.4; gill: 143 ± ND; testes: 69.6 ± ND; ND = not determined) indicating that Se may have a protective effect against Hg toxicity in *Ma. lanceolatus.*

Comparison of Elemental Composition Among Ocean Sunfishes

The comparison of trace element concentrations in tissues from *Mola* spp. from different geographical locations (i.e., Florida, Dutton et al. 2016; Mediterranean Sea, Zacarroni et al. 2008; Portugal, Baptista 2019, Baptista et al. 2019) reveals some generalities. Overall, both essential and non-essential trace

Table 4. Comparison between trace element concentrations in tissues of *Mola* spp. from different geographical areas. MD, Mediterranean Sea; PT, southern waters of Portugal; FL, waters off eastern Florida.

			Hypodermis	Gills	Kidney	Liver	Ovary	Spleen	White muscle
Trace elements	**essential**	**Co**		PT > FL					PT > FL
		Cr	MD > PT	MD > FL > PT	MD > FL	PT > MD >FL	MD > PT > FL	M/PT > FL	MD > PT/FL
		Cu	MD > PT	MD > FL/PT	MD > FL	MD > FL > PT	MD > FL > PT		PT > MD > FL
		Mn		FL > PT					PT > FL
		Ni	MD > PT	MD > PT > FL	MD > FL	MD > PT > FL	PT > FL	PT > M > FL	MD > PT > FL
		Se	MD > PT	FL > PT > MD	FL > MD	PT > FL/MD	MD/FL > PT	M > FL > PT	MD > PT > FL
	non-essential	**As**	MD > PT	PT > FL > MD		PT > M > FL	MD > FL > PT		
		Cd	MD > PT	MD > PT > FL		MD > FL > PT	MD > PT/ FL		PT > MD > FL
		Hg			FL > MD	PT > FL			

Table 5. Comparison between trace element concentrations in tissues of *Mola* spp. and *Masturus lanceolatus* from waters off eastern Florida.

			Gills	Kidney	Liver	Spleen	White muscle
Trace elements	**essential**	Cr	*Mola* > *Masturus*	*Masturus* > *Mola*			
		Cu			*Mola* > *Masturus*	*Mola* > *Masturus*	
		Fe		*Masturus* > *Mola*	*Masturus* > *Mola*	*Masturus* > *Mola*	
		Mn					
		Ni				*Masturus* > *Mola*	
		Se	*Mola* > *Masturus*				
		Zn			*Masturus* > *Mola*	*Mola* > *Masturus*	
	non-essential	As	*Mola* > *Masturus*	*Mola* > *Masturus*		*Mola* > *Masturus*	
		Cd					*Masturus* > *Mola*
		Hg				*Masturus* > *Mola*	*Masturus* > *Mola*
		Pb				*Masturus* > *Mola*	*Mola* > *Masturus*
ratio		Se/Hg		*Mola* > *Masturus*			

elements were found in greater levels in Mediterranean specimens (Table 4). The clearer trends of greatest concentrations in Mediterranean specimens were found in the hypodermis and ovary. However, in the liver and white muscle, the greatest elemental concentrations were found in Mediterranean or Portuguese specimens depending on the trace element examined. Interestingly, in the kidney, Florida specimens exhibit greater concentrations of both Se and Hg than Mediterranean ones. This result is somewhat unexpected since the Mediterranean is recognized as a heavily polluted water body (Boyle et al. 1985).

Conversely, comparison of *Mola* spp. and *Ma. lanceolatus* from the same geographical location (i.e., Florida) is possible with the data provided by Dutton et al. (2016). These data provide some insight into phylogenetic-related differences in elemental dynamic (Table 5). A trend of greater elemental concentration in *Mola* spp. appears to occur in gill tissue. With regard to the concentration of specific elements, *Ma. lanceolatus* appears to exhibit greater values of Fe across tissues whereas *Mola* spp. seem to contain higher levels of As.

Microplastics in Ocean Sunfish—A First Report

Plastics are one of the most popular materials used worldwide due to their physicochemical properties, including endurance, light weight, durability and low manufacturing costs (Zobkov and Esiukova 2018). One main concern however, is the pollution caused by plastic debris, including those pieces not visible to the naked eye, referred to as microplastics (Andrady 2011). Microplastics (MP) are defined as any synthetic solid particle or polymeric matrix, with regular or irregular shape and with size ranging from 1 μm to 5 mm, of either primary or secondary manufacturing origin, which is insoluble in water (Frias and Nash 2019). Over the last decade, research investment into MP as a novel pollutant has seen a major increase on a global scale (Frias and Nash 2019).

Microplastics have been reported in the marine biotic and abiotic compartments, from the sea surface, water column, deep sea, coastal sediments to marine organisms (e.g., Karlsson et al. 2017, Andrady et al. 2017). Consumed either through direct ingestion, or indirectly by the ingestion of a prey item that is already contaminated (Nelms et al. 2018, Wright et al. 2013), MP are finding their way into numerous species. In the wake of this emergent concern, academic attention to the impact of MP has increased in recent years (e.g., Avio et al. 2015, Rochman et al. 2014, Wright et al. 2013). Recently, Nyegaard et al. (2018) reported a 3 × 5 mm polystyrene ball being found in the digestive tract of a *M. tecta*, supporting the need to evaluate the presence of MP in Molidae.

Fifty-three stomach contents from *Mola* spp. specimens were collected in the southern waters of Portugal, off Olhão, for MP analysis as described in Lopes et al. (unpublished data). All potential MP were visualized and photographed using a stereomicroscope LEICA S9i (Leica Microsystems GmbH, Wetzlar, Germany) with an integrated IC80 HD camera. Particles were categorized by color (black, blue, transparent, white, red, green, and other) and measured at their largest cross section using the ImageJ software and categorized according to the following size classes (≤ 0.5 mm; 0.5–1 mm; 1–2 mm; 2–3 mm; > 3 mm). In addition, fibers were distinguished from another type of MP shape. For all procedures, in the field and laboratory, contamination was accounted based on the report by Bessa et al. (2019). Particles were analysed by Fourier Transform Infrared Spectroscopy (FTIR) using a PerkinElmer Spotlight 200i, equipped with a mercury cadmium telluride array detector (MCT) cooled by liquid nitrogen. Spectra were collected in μ-ATR mode, 4000–600 cm^{-1}, with a resolution of 4 cm^{-1} and four scans. Spectra were compared with a database of custom references and only polymers matching reference spectra for > 70 percent were considered.

To our knowledge, this is the first comprehensive study on MP in *Mola* spp. Forty-two out of the 53 *Mola* spp. specimens sampled (79 percent) presented potential microplastics in their stomach contents, with a median of one microparticle per individual, ranging from 0 to 11 microplastics per sample. The number of ingested MP registered in this study is similar to the ones reported for various fish species from the Portuguese coast and estuaries (Bessa et al. 2018b, Neves et al. 2015). All MP were characterized according to the shape, color, length and polymer type. Notably, from a total of 119 ingested MP, fibers and fragments were equally identified (50:50 percent). These results are not in line with other studies where fibers are often the most common synthetic microparticles observed in fish (e.g., Neves et al. 2015, Bessa et al. 2018a and references therein). One possible explanation for this result is the shift in the feeding habits of *Mola* spp. In fact, individuals in the same size range as those sampled in this study may switch from coastal benthic and pelagic prey species to more oceanic pelagic prey as they grow larger (Nakamura and Sato 2014, Sousa et al. 2016a). In general, benthic species ingest more fibers than pelagic ones, while pelagics have a higher content of fragments (Neves et al. 2015). The size classes, separated according to changes in feeding habits (Sousa et al. 2016a), did not affect significantly ($p > 0.05$) the number or type of particles observed in the stomach contents of *Mola* spp., and for this reason data were treated all together.

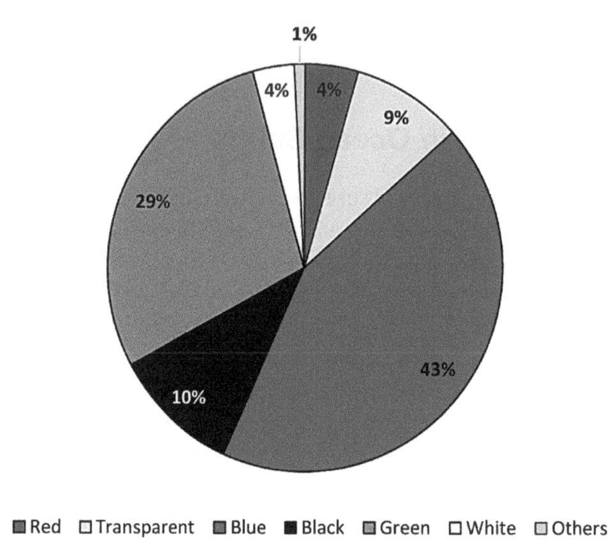

Figure 10. Percentage (%) of the different colors registered in the microplastics extracted from the stomach contents of *Mola* spp. (n = 42) by-caught in the southern waters of Portugal.

Microplastics had different colors, with blue being the most representative (43 percent) followed by green (29 percent), black (10 percent), transparent (9 percent), red and white (4 percent) and other colors (1 percent) (Fig. 10). An abundance of blue microparticles has already been described for other fish species from the Portuguese coastal area and estuaries (Bessa et al. 2018b, Neves et al. 2015). Color is considered to influence the MP ingestion, due to prey resemblance (Wright et al. 2013). However, as described below, the overall sizes of the MP identified in this study were very small and hence a selective ingestion by *Mola* spp. specimens is unlikely. Alternatively, ingested MP may result from food-web transfer or through water intake, rather than active ingestion deriving from resemblance to prey items.

Microplastics size ranged from 0.032 to 16 mm, median size of 0.57 mm, being smaller particles the most representative ones (< 0.3 mm, 43%) (Fig. 11). Although it was not confirmed in the present study, such small MP may potentially transfer to vital tissues/organs with adverse effects (Green et al. 2019, Ding et al. 2018).

No seasonal difference was noted in MP presence in the stomach contents of *Mola* spp. Differently, sex appears to significantly affect ($p < 0.05$) MP ingestion, with females presenting an enhanced number of MP in comparison to males (Fig. 12). The number of males and females sampled in this study was equal, meaning that the comparison among sexes is not biased. To the best of our knowledge, there is no evidence showing that male and female *Mola* spp. exhibit differences in feeding habit, food selection or habitat selection. Regardless, the potential existence of such sex-related differences does not appear to make sense on a biological or ecological perspective when considering that studied specimens were immature and hence reproductive-related divergent behaviors were not likely to exist. It is possible that the obtained results derive from sampling bias and that upon an even larger sampling effort, such sex-related difference is not verified.

A total of 25 MP (21 percent of the total particles observed), nine fibers and 16 fragments, were identified using FTIR by comparing their chemical spectra with reference spectra from library databases and comparison of characteristics bands with reference literature. Results showed that fibers are composed of rayon, cellulose and polyacrylic. The predominance of rayon, a semi-synthetic cellulose-based polymer, is in agreement with results published by Frias et al. (2016) regarding sediments samples from the Portuguese southern coast. One fragment matched with polyethylene (PET) reference spectrum (99 percent) PET is one of the world's most commonly produced plastic (PlasticsEurope 2018). Four colored fragments were identified as paint containing montmorillonite

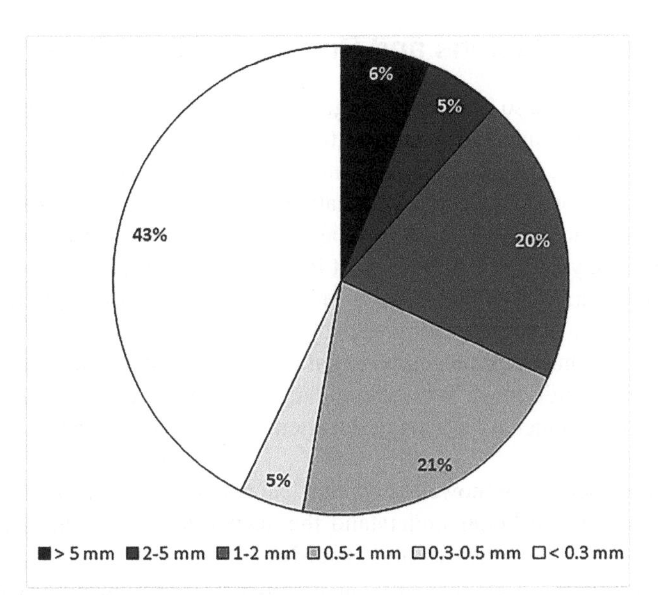

Figure 11. Percentage (%) of the different size classes (< 0.3, 0.3–0.5, 0.5–1, 1–2, 2–5 and > 5 mm) registered for the microplastics extracted from the stomach contents of *Mola* spp. (n = 42) by-caught in the southern waters of Portugal.

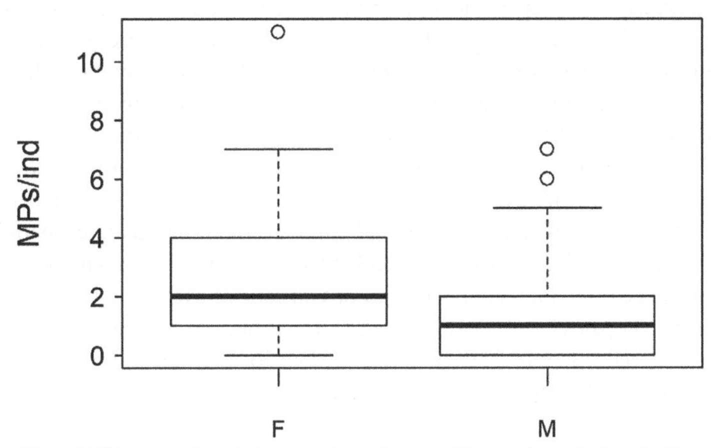

Figure 12. Median, 25 and 75% percentile, minimum and maximum, of the number of microplastics per individual (MPs/ind) extracted from the stomach contents of females (F; n = 26) and males (M; n = 15 of *Mola* spp. by-caught in the southern waters of Portugal, off Olhão.

and six fragments as acrylic acid. The remaining five MP fragments did not show a significant fit with any of the materials within the FTIR spectra libraries. However, all of those were blue or green particles with spectrum similarity with acrylic based polymers.

Overall findings indicate that *Mola* spp. are vulnerable to microplastic ingestion due to the ubiquitous presence of these anthropogenic particles in the southern Portuguese coast. Although there are no available data on the effects of microplastics in these fish, there is evidence suggesting their uptake has deleterious effects of both physical and chemical nature. Studies on several fish species have shown that ingestion of microplastics may cause physical damage in the digestive tract, inflammatory responses and nutritional disturbances (Jovanović 2017). Additionally, both the additives incorporated in plastic composition and pollutants adsorbed on the surface of microparticles may cause adverse effects to the fish (Avio et al. 2015, Rochman et al. 2014, Wright et al. 2013). The potential threats to *Mola* spp. resulting from the ingestion of microplastics should be investigated in future research studies.

Conclusions and Remaining Questions

As our human footprint grows, so too does the urgency to understand the impacts of contaminants on marine ecosystems and their inhabitants such as the Molidae. Studies on ocean sunfish, namely on *Mola* spp. from the Atlantic, provide a baseline understanding of trace element dynamics, however data from other world regions are needed. Information on the presence of biotoxins in ocean sunfish is scarce with only two studies and a preliminary analysis having been performed. Tetrodotoxin was never found in these fish yet PTX may be present in *Ma. lanceolatus*. Further research is needed to clarify the potential occurrence of biotoxins in Molidae. Regarding microplastics, only one study has explored dietary intake of these contaminants by *Mola* spp. (in southern Portugal). Seventy-nine percent of examined specimens presented potential MPs in their stomach contents, with a median of one microparticle per individual. While providing an insight into MP contamination in ocean sunfish, this study was conducted on a single location and further research on other geographical areas is recommended.

In summary, the present work showed an overall deficit of information concerning contamination of ocean sunfish. In order to better understand the accumulation dynamics and the impact of contaminants in ocean sunfish, further research is needed. Studies should be conducted addressing the presence and effects of toxins, trace elements, and man-made contaminants, such as the halogenated organic pollutants (HOPs; e.g., polychlorinated biphenyl (PCBs), hexachlorobenzene (HCB), and dichlorodiphenyltrichloroethanes (DDTs) and their metabolites) and those of emerging concern (e.g., rare earth elements, platinum group elements, pharmaceuticals, personal care products, and endocrine disrupting compounds) in tissues of these fishes. While assessing the presence of contaminants is relatively straightforward, performing ecotoxicological studies to understand the effects of such contaminants on organisms is more challenging. Such studies are generally conducted in a laboratory under controlled conditions. Due to the sheer size of specimens, such studies on ocean sunfish are perhaps unfeasible. Maybe focusing on more manageable tetraodontiformes such as pufferfishes could prove useful as phylogenetic closeness may allow for insights into the effects of contaminant accumulation in Molidae. Future studies should take into account ontogenetic, spatial, and temporal factors which are bound to affect the exposure and subsequent uptake of contaminants. Information on contaminant presence (e.g., elements in calcareous structures such as bones, otoliths, scales, and fin rays or epidermal biopsies) may also be used to track habitat use and the movement patterns of these migratory animals. The use of contaminant data for that specific purpose, although often challenging, could be useful effective population management.

Acknowledgements

The authors wish to thank Brian Jackson for the ICP-MS analysis of the Florida samples presented in Dutton et al. (2016), Dr. Tierney Thys (California Academy of Sciences) and Sea Studios Foundation for sharing trace element data of a *Mola mola* specimen from Monterey Bay (California) and Dr. Annalisa Zaccaroni for providing weight data on *Mola* spp. specimens from the Mediterranean.

The research presented in Baptista (2019) and Baptista et al. (2018, 2019) was funded by the Portuguese Foundation for Science and Technology (FCT) through the Ph.D. scholarship awarded to M. Baptista (SFRH/BD/88175/2012), the Postdoctoral Grant awarded to J. Raimundo (SFRH/BPD/91498/2012) and the FCT Investigator Fellowship awarded to R. Rosa (IF/01373/2013), and the strategic project UID/MAR/04292/2013 granted to MARE. The research presented in Baptista (2019) and Baptista et al. (2018, 2019) was also supported by Programa Operacional Pesca 2007–2013—Promar through the project MOLA (31-03-05-FEP-0037) awarded to R. Rosa. The research presented in Dutton et al. (2016) was funded by Texas State University.

Supplemental Table 1. Concentration (µg/g) of essential and non-essential trace elements and Se:Hg ratios (mean ± standard deviation) in different tissues of *Mola* spp. and *Masturus lanceolatus*.*indicates that in the cases where values of trace elements were below detection limit (BDL), the detection limit value was used for calculation of the mean concentration. For Baptista (2019) and Baptista et al. (2019), detection limits are 0.033 for Co and Mn, 0.028 for Cr, 0.348 for Cu, 0.008 for Ni, 0.232 for Se, 0.078 for Zn, 0.057 for As, 0.012 for Cd, 0.026 for Pb and 0.076 for V. For Dutton et al. (2016), detection limits are 0.010 for Ag, 0.002 for Cd, 0.026 for Cr, 0.033 for Hg, and 0.015 for Pb. For Perrault et al. (2014), detection limits are 0.093 for Cr and 0.925 for Pb. For Thys (2007), detection limits are 0.020 for Cr, 0.010 for Ni, 0.050 for Se, 0.050 for As, 0.030 for Be, 0.010 for Cd, 0.050 for Hg and 0.100 for Pb. For Zaccaroni et al. (2008), no detection limits are available. **Ag only detected in one *M. lanceolatus* (250.1 cm TL male). †trace element concentrations were converted from wet weight to dry weight using the following conversion factors. For Adams et al. (2003) the conversion factor used for white muscle was 0.140. For Perrault et al. (2014) the conversion factor used for the liver was 0.540. For Zaccaroni et al. (2008) the conversion factors used were 0.363 for white muscle, 0.398 for gills, 0.497 for ovary, 0.556 for spleen, 0.260 for kidney, 0.577 for liver and 0.068 for hypodermis. For Kumar et al. (2004) and Thys (2017) the conversion factor used for white muscle was 0.360. Dry weight concentrations were calculated by dividing the wet weight concentrations by the tissue-specific conversion factor. ND = not determined due to n = 1.

	Tissue	Brain	Fat	Hypodermis	Hypodermis	Gills	Gills	Gills
	Species	*Mola* spp.	*Mola* spp.	*Mola* spp.	*Mola* spp.	*Mola* spp.	*Mola* spp.	*Mola* spp.
	Ref.	*Baptista et al. (2019)*	*Zaccaroni et al. (2008)*	*Baptista et al. (2019)*	*Zaccaroni et al. (2008)*	*Baptista et al. (2019)*	*Dutton et al. (2016)*	*Zaccaroni et al. (2008)*
	Water body	northeast Atlantic	north-central Mediterranean	northeast Atlantic	north-central Mediterranean	northeast Atlantic	northwest Atlantic	north-central Mediterranean
	Location	southern Portugal	northeastern Italy	southern Portugal	northeastern Italy	southern Portugal	southeastern USA (Florida)	northeastern Italy
	Units	µg/g dry weight	µg/g wet weight	µg/g dry weight	µg/g dry weigh†	µg/g dry weight	µg/g dry weight	µg/g dry weight†
	TL (cm)	31.8–120	284	31.8–120	219–284	31.8–120	149	274–284
	n	49	1	51	2	49	1	2
Trace elements – essential	Co	0.121 ± 0.525*	–	0.176 ± 0.156*	–	0.270 ± 0.200	0.086	–
	Cr	0.478 ± 0.153*	0.226	0.491 ± 0.323	11.6 ± 12.4	0.271 ± 0.131	1.15	7.24 ± 3.05
	Cu	16.7 ± 7.67*	1.32	9.55 ± 1.83*	13.9 ± 16.4	2.51 ± 0.499	2.77	8.13 ± 2.09
	Fe	–	1.85	–	–	–	188	–
	Mn	0.855 ± 0.461*	–	1.28 ± 0.578	–	10.0 ± 9.81	42.2	–
	Ni	0.426 ± 0.332*	0.049	0.279 ± 0.169*	1.47	0.924 ± 0.172	0.245	1.47
	Se	2.18 ± 0.520*	0.470	3.40 ± 1.21	26.2 ± 22.9	3.85 ± 1.01	7.32	0.729 ± 0.126
	Zn	41.7 ± 11.0*	–	51.0 ± 12.7	–	180 ± 51.7	271	–
Trace elements – non-essential	Ag	–	–	–	–	–	BDL	–
	As	5.02 ± 2.01*	20.8	33.5 ± 16.4	179 ± 208	13.1 ± 9.28	8.97	3.61 ± 1.92
	Be	–	–	–	–	–	–	–
	Cd	0.063 ± 0.041*	0.008	0.096 ± 0.140*	1.09 ± 0.294	0.133 ± 0.114	0.065	0.309 ± 0.118
	Hg	–	0.021	–	–	–	BDL	–
	Pb	0.129 ± 0.324*	BDL	0.468 ± 1.21*	0.471	0.410 ± 0.286	0.192	1.05 ± 0.379
	V	0.486 ± 0.577*	–	0.236 ± 0.173	–	2.88 ± 2.06	–	–
ratio	Se:Hg	–	–	–	–	–	–	–

Supplemental Table 1 contd. ...

...Supplemental Table 1 contd.

	Tissue	Gills	Heart	Kidney	Kidney	Kidney	Liver	Liver
	Species	*Masturus lanceolatus*	*Mola* spp.	*Mola* spp.	*Mola* spp.	*Masturus lanceolatus*	*Mola* spp.	*Mola* spp.
	Ref.	*Dutton et al. (2016)*	*Dutton et al. (2016)*	*Dutton et al. (2016)*	*Zaccaroni et al. (2008)*	*Dutton et al. (2016)*	*Baptista et al. (2019)*	*Dutton et al. (2016)*
	Water body	northwest Atlantic	northwest Atlantic	northwest Atlantic	north-central Mediterranean	northwest Atlantic	northeast Atlantic	northwest Atlantic
	Location	southeastern USA (Florida)	southeastern USA (Florida)	southeastern USA (Florida)	northeastern Italy	southeastern USA (Florida)	southern Portugal	southeastern USA (Florida)
	Units	µg/g dry weight	µg/g dry weight	µg/g dry weight	µg/g dry weight†	µg/g dry weight	µg/g dry weight	µg/g dry weight
	TL (cm)	41.9–250.1	156	156	219–284	41.9–250.1	31.8–120	156
	n	3	1	1	3	4	52	1
Trace elements – essential	**Co**	0.082 ± 0.043	0.187	0.387	–	0.636 ± 0.219	0.406 ± 0.782	0.251
	Cr	0.556 ± 0.297	0.090	0.049	1.47 ± 2.63	0.279 ± 0.265	0.393 ± 0.103*	0.025
	Cu	1.31 ± 0.632	7.94	3.60	29.2 ± 29.8	4.01 ± 1.24	3.23 ± 0.980	21.1
	Fe	160 ± 123	118	353	360 ± 350	594 ± 158	–	240
	Mn	50.7 ± 8.63	1.05	4.66	–	3.00 ± 0.911	0.678 ± 0.320	1.10
	Ni	0.226 ± 0.149	0.166	0.183	2.69	0.509 ± 0.359	0.389 ± 0.220*	0.070
	Se	2.60 ± 1.27	10.6	24.2	7.30 ± 1.44	21.3 ± 14.0	31.7 ± 21.6	3.61
	Zn	420 ± 79.2	160	89.6	–	107 ± 32.9	83.9 ± 47.0	77.9
Trace elements – non-essential	**Ag**	0.009 ± ND**	BDL	BDL	–	0.022 ± ND**	–	0.428
	As	5.07 ± 2.63	28.3	22.7	18.1 ± 7.60	9.01 ± 1.75	55.3 ± 53.0	18.1
	Be	–	–	–	–	–	–	–
	Cd	0.088 ± 0.043	0.167	2.14	6.09 ± 6.25	2.32 ± 1.27*	0.410 ± 0.355	2.11
	Hg	0.056 ± 0.032*	0.169	0.293	0.177	0.809 ± 0.660	0.871 ± 0.857 n = 39	0.057
	Pb	0.247 ± 0.123	BDL	0.041	BDL	0.045 ± 0.037*	0.047 ± 0.080*	BDL
	V	-	-	–	–	–	0.939 ± 0.832	–
ratio	**Se:Hg**	142 ± 66.1	160	209	–	84.1 ± 35.4	23.9 ± 10.7 n = 39	161

	Tissue	Liver	Liver	Liver	Ovary	Ovary	Ovary	Red Muscle
	Species Ref.	*Mola* spp. *Perrault et al. (2014)*	*Mola* spp. *Zaccaroni et al. (2008)*	*Masturus lanceolatus* *Dutton et al. (2016)*	*Mola* spp. *Baptista et al. (2019)*	*Mola* spp. *Dutton et al. (2016)*	*Mola* spp. *Zaccaroni et al. (2008)*	*Mola* spp. *Baptista et al. (2019)*
	Water body	northwest Atlantic	north-central Mediterranean	northwest Atlantic	northeast Atlantic	northwest Atlantic	north-central Mediterranean	northeast Atlantic
	Location	southeastern USA (Florida)	northeastern Italy	southeastern USA (Florida)	southern Portugal	southeastern USA (Florida)	northeastern Italy	southern Portugal
	Units	µg/g dry weight†	µg/g dry weight†	µg/g dry weight	µg/g dry weight	µg/g dry weight	µg/g dry weight†	µg/g dry weight
	TL (cm)	140	274–284	41.9–250.1	32.2–120	156	219–284	31.8–120
	n	1	2	4	24	1	3	50
Trace elements – essential	Co	0.481	–	0.597 ± 0.343	0.224 ± 0.240	0.133	–	0.046 ± 0.024
	Cr	BDL	0.255 ± 0.090	0.039 ± 0.023*	0.926 ± 0.930	0.159	1.86 ± 2.66	0.591 ± 0.619
	Cu	5.19	125 ± 54.2	9.26 ± 1.04	5.47 ± 2.27	9.07	19.4 ± 10.4	0.853 ± 0.394
	Fe	–	–	2104 ± 1676	–	55.1	–	–
	Mn	1.57	–	1.95 ± 0.924	1.09 ± 0.396	1.70	–	0.219 ± 0.082
	Ni	–	0.549 ± 0.731	0.155 ± 0.113	0.726 ± 0.572	0.042	7.75	0.302 ± 0.683
	Se	2.98	1.02 ± 3.98	7.98 ± 8.34	5.07 ± 1.46	10.8	12.2 ± 15.4	2.27 ± 0.619
	Zn	87.0	–	137 ± 12.4	306 ± 179	277	–	21.4 ± 4.5
Trace elements – non-essential	Ag	–	–	0.314 ± 0.106	–	BDL	–	–
	As	2.78	28.5 ± 15.9	8.62 ± 2.44	16.4 ± 18.0	28.4	42.0 ± 47.8	34.6 ± 24.5
	Be	–	–	–	–	–	–	–
	Cd	6.48	23.9 ± 21.2	8.17 ± 5.15	0.242 ± 0.252	0.113	1.06 ± 1.07	0.040 ± 0.027*
	Hg	–	–	0.574 ± 0.454*	–	0.047	–	0.661 ± 0.473 n = 45
	Pb	BDL	BDL	0.062 ± 0.057*	0.297 ± 0.778*	BDL	0.265	0.276 ± 0.651*
	V	–	–	–	1.10 ± 0.784	–	–	0.127 ± 0.104*
ratio	Se:Hg	–	–	41.4 ± 25.0	–	583	–	19.5 ± 9.77 n = 45

Supplemental Table 1 contd. ...

...Supplemental Table 1 contd.

	Tissue	Spleen	Spleen	Spleen	Spleen	Testes	Testes	White Muscle
	Species	*Mola* spp.	*Mola* spp.	*Mola* spp.	*Masturus lanceolatus*	*Mola* spp.	*Masturus lanceolatus*	*Mola* spp.
	Ref.	*Baptista et al. (2019)*	*Dutton et al. (2016)*	*Zaccaroni et al. (2008)*	*Dutton et al. (2016)*	*Baptista et al. (2019)*	*Dutton et al. (2016)*	*Adams et al. (2003)*
	Water body	northeast Atlantic	northwest Atlantic	north-central Mediterranean	northwest Atlantic	northeast Atlantic	northwest Atlantic	northwest Atlantic
	Location	southern Portugal	southeastern USA (Florida)	northeastern Italy	southeastern USA (Florida)	southern Portugal	southeastern USA (Florida)	southeastern USA (Florida)
	Units	µg/g dry weight	µg/g dry weight	µg/g dry weight†	µg/g dry weight	µg/g dry weight	µg/g dry weight	µg/g dry weight†
	TL (cm)	31.8–120.3	156	219–284	41.9–250.1	31.8–110.1	41.9–250.1	174
	n	50	1	2	3	26	3	1
Trace elements – essential	**Co**	0.233 ± 0.215	0.596	–	0.449 ± 0.117	0.242 ± 0.194*	0.091 ± 0.047	–
	Cr	1.03 ± 0.792	0.042	0.353 ± 0.092	0.070 ± 0.045	1.13 ± 0.641	0.128 ± 0.056	–
	Cu	4.51 ± 1.97	3.62	3.92 ± 0.640	2.30 ± 0.313	3.62 ± 1.06	1.84 ± 1.08	–
	Fe	–	1189	–	8747 ± 7793	–	171 ± 88.7	–
	Mn	0.893 ± 0.373	1.16	–	0.869 ± 0.122	0.716 ± 0.247	1.13 ± 0.561	–
	Ni	0.999 ± 1.09	0.081	0.567 ± 0.293	0.277 ± 0.090	1.25 ± 1.37	0.104 ± 0.016	–
	Se	5.07 ± 1.85	48.6	121 ± 151	57.9 ± 19.8	5.07 ± 2.17	5.47 ± 2.95	–
	Zn	185 ± 170	82.4	–	59.3 ± 9.95	73.1 ± 18.9	65.1 ± 46.4	–
Trace elements – non-essential	**Ag**	–	BDL	–	0.012 ± ND**	–	0.011 ± ND**	–
	As	16.3 ± 14.0	28.2	38.5 ± 28.4	18.0 ± 0.859	16.1 ± 9.25	21.1 ± 20.6	–
	Be	–	–	–	–	–	–	–
	Cd	0.245 ± 0.221	0.648	0.993 ± 1.04	0.649 ± 0.313	0.248 ± 0.192	0.351 ± 0.085	–
	Hg	–	0.117	–	0.979 ± 0.454	–	0.184 ± 0.131*	0.143
	Pb	0.501 ± 1.71*	0.031	0.014	0.231 ± 0.126	0.690 ± 2.3*	0.024 ± 0.016*	–
	V	1.64 ± 1.80	–	–	–	2.12 ± 2.29	–	–
ratio	**Se:Hg**	–	1054	–	155 ± 23.7	–	69.6 ± 10.0	–

	Tissue	White Muscle	White Muscle	White Muscle	White Muscle	White Muscle	White Muscle
	Species	*Mola* spp.	*Mola* spp.	*Mola* spp.	*Mola* spp.	*Mola* spp.	*Masturus lanceolatus*
	Ref.	*Baptista et al. (2019)*	*Dutton et al. (2016)*	*Kumar et al. (2004)*	*Thys (2017)*	*Zaccaroni et al. (2008)*	*Dutton et al. (2016)*
	Water body	northeast Atlantic	northwest Atlantic	southwest Pacific	northeast Pacific	north-central Mediterranean	northeast Atlantic
	Location	southern Portugal	southeastern USA (Florida)	Fiji islands	southwestern USA (California)	northeastern Italy	southeastern USA (Florida)
	Units	µg/g dry weight	µg/g dry weight	µg/g dry weight†	µg/g dry weight†	µg/g dry weight†	µg/g dry weight
	TL (cm)	31.8–120.3	149–156	–	–	219–274	41.9–250.1
	n	56	2	5	1	2	5
Trace elements – essential	Co	2.58 ± 3.14*	0.016 ± 0.008	–	–	–	0.057 ± 0.047
	Cr	0.484 ± 1.01	0.109 ± 0.012	–	BDL	3.20 ± 3.73	0.208 ± 0.158
	Cu	133 ± 137	0.641 ± 0.090	–	1.76	14.9 ± 9.21	1.13 ± 0.573
	Fe	–	12.5 ± 10.8	–	–	–	17.2 ± 19.8
	Mn	2.38 ± 1.27	0.309 ± 0.090	–	–	–	0.507 ± 0.428
	Ni	0.527 ± 0.918	0.084 ± 0.006	–	BDL	1.64 ± 2.13	0.134 ± 0.072
	Se	7.44 ± 4.95	2.99 ± 1.94	–	BDL	13.5 ± 18.2	2.90 ± 0.313
	Zn	126 ± 57.1	23.9 ± 1.08	–	19.8	–	38.8 ± 21.2
Trace elements – non-essential	Ag	–	BDL	–	2.38	–	BDL
	As	31.5 ± 38.1	14.5 ± 6.04	–	BDL	46.9 ± 57.6	16.6 ± 8.63
	Be	–	–	–	BDL	–	–
	Cd	24.3 ± 22.3*	0.016 ± 0.002	–	BDL	10.8 ± 15.0	0.046 ± 0.024
	Hg	0.402 ± 0.314 n = 39	0.167 ± 0.080	2.00 ± 0.190	BDL	–	0.467 ± 0.325*
	Pb	0.080 ± 0.092*	0.217 ± 0.286*	–	BDL	4.74 *	0.027 ± 0.017*
	V	2.4 ± 3.2*	–	–	–	–	–
ratio	Se:Hg	20.6 ± 15.7 n = 39	59.7 ± 58.3	–	–	–	15.2 ± 7.46

References

Adams, D.H., R.H. McMichael, Jr. and G.E. Henderson. 2003. Mercury levels in marine and estuarine fishes of Florida 1989–2001. Florida Marine Research Institute Technical Report TR-9. 2nd ed. rev. 57 p.

Adams, D.H. 2004. Total mercury levels in tunas from offshore waters of the Florida Atlantic coast. Mar. Pollut. Bull. 49: 659–667.

Adams, D.H. and C. Sonne. 2013. Mercury and histopathology of the vulnerable goliath grouper, *Epinephelus itajara*, in U.S. waters: A multi-tissue approach. Environ. Res. 126: 254–263.

Alfonso, A., A. Fernández-Araujo, C. Alfonso, B. Caramés, A. Tobio, M.C. Louzao et al. 2012. Palytoxin detection and quantification using the fluorescence polarization technique. Anal. Biochem. 424: 64–70.

Alibabić, V., N. Vahcić and M. Bajramović. 2007. Bioaccumulation of metals in fish of Salmonidae family and the impact on fish meat quality. Environ. Monit. Assess. 131: 349–364.

Alkan, N., A. Alkan, K. Gedik and A. Fisher. 2016. Assessment of metal concentrations in commercially important fish species in Black Sea. Toxicol. Ind. Health 32: 447–456.

Alquezar, R., S.J. Markich and D.J. Booth. 2006. Metal accumulation in the smooth toadfish, *Tetractenos glaber*, in estuaries around Sydney, Australia. Environ. Pollut. 142: 123–131.

Al-Yousuf, M.H., M.S. El-Shahawi and S.M. Al-Ghais. 2000. Trace metals in liver, skin and muscle of *Lethrinus lentjan* fish species in relation to body length and sex. Sci. Total Environ. 256: 87–94.

Andrady, A.L. 2011. Microplastics in the marine environment. Mar. Pollut. Bull. 62: 1596–1605.

Andrady, A.L. 2017. The plastic in microplastics: a review. Mar. Pollut. Bull. 119(1): 12–22.

Ashoka, S., B.M. Peake, G. Bremner and K.J. Hageman. 2011. Distribution of trace metals in a ling (*Genypterus blacodes*) fish fillet. Food Chem. 125: 402–409.

Authman, M.M., M.S. Zaki, E.A. Khallaf and H.H. Abbas. 2015. Use of fish as bio indicator of the effects of heavy metals pollution. J. Aquac. Res. Dev. 6: 328.

Avio, C.G., S. Gorbi, M. Milan, M. Benedetti, D. Fattorini, G. D'Errico et al. 2015. Pollutants bioavailability and toxicological risk from microplastics to marine mussels. Environ. Pollut. 198: 211–222.

Ayas, D. and A.R. Köşker. 2018. The effects of age and individual size on metal accumulation of *Lagocephalus sceleratus* (Gmelin, 1789) from Mersin Bay, Turkey. Nat. Eng. Sci. 3: 45–53.

Baechler, B.R., C.D. Stienbarger, D.A. Horn, J. Joseph, A.R. Taylor, E.F. Granek et al. 2019. Microplastic occurrence and effects in commercially harvested North American finfish and shellfish: Current knowledge and future directions. Limnol. Oceanogr.: Lett. 5: 113–136.

Bakenhaster, M. and J. Knight-Gray. 2016. New diet data for *Mola mola* and *Masturus lanceolatus* (Tetraodontiformes: Molidae) off Florida's Atlantic coast with discussion of historical context. Bull. Mar. Sci. 92: 497–511.

Bane, V., M. Lehane, M. Dikshit, A. O'Riordan and A. Furey. 2014. Tetrodotoxin: Chemistry. toxicity. source. distribution and detection. Toxins 6(2): 693–755.

Baptista, M., A. Couto, J.R. Paula, J. Raimundo, N. Queiroz and R. Rosa. 2018. Seasonal variations in the abundance and body size distribution of the ocean sunfish *Mola mola* in coastal waters off southern Portugal. J. Mar. Biolog. Assoc. UK, 1–7.

Baptista, M. 2019. Bio-ecology and elemental composition of a giant—the ocean sunfish *Mola mola*. Doctoral Dissertation, Faculdade de Ciências, Universidade de Lisboa, Lisbon, Portugal.

Baptista, M., O. Azevedo, C. Figueiredo, J.R. Paula, M.T. Santos, N. Queiroz et al. 2019. Body size and season influence elemental composition of tissues in ocean sunfish Mola mola juveniles. Chemosphere 223: 714–722.

Bashir, F.A., M. Shuhaimi-Othman and A.G. Mazlan. 2012. Evaluation of trace metal levels in tissues of two commercial fish species in Kapar and Mersing Coastal Waters, Peninsular Malaysia. J. Environ. Public Health 2012: 352309.

Bentur, Y., J. Ashkar, Y. Lurie, Y. Levy, Z.S. Azzam, M. Litmanovich et al. 2008. Lessepsian migration and tetrodotoxin poisoning due to Lagocephalus sceleratus in the eastern Mediterranean. Toxicon 52: 964–968.

Bernardini, I., F. Garibaldi, L. Canesi, M. Cristina and M. Baini. 2018. First data on plastic ingestion by blue sharks (Prionace glauca) from the Ligurian Sea (North-Western Mediterranean Sea). Mar. Pollut. Bull. 135: 303–310.

Bessa, F., P. Barr, M. Neto, P.G.L. Frias, V. Otero and P. Sobral. 2018a. Proceedings of the International Conference on Microplastic Pollution in the Mediterranean Sea.

Bessa, F., P. Barría, J.M. Neto, J.P.G.L. Frias, V. Otero, P. Sobral et al. 2018b. Occurrence of microplastics in commercial fish from a natural estuarine environment. Mar. Pollut. Bull. 128: 575–584.

Bessa, F., J. Frias, T. Knögel, A. Lusher, J. Antunes, J.M. Andrade et al. 2019. Harmonized protocol for monitoring microplastics in biota. JPI-Oceans BASEMAN Proj. 29.

Boyle, E.A., S.D. Chapnick, X.X. Bai and A. Spivack. 1985. Trace metal enrichments in the Mediterranean Sea. Earth Planet. Sci. Lett. 74: 405–419.

Breen, P., A. Cañadas, O.Ó. Cadhla, M. Mackey, M. Scheidat, S.C.V. Geelhoed et al. 2017. New insights into ocean sunfish (*Mola mola*) abundance and seasonal distribution in the northeast Atlantic. Sci. Rep. 7: 1–9.

Chanda, S., B.N. Paul, K. Ghosh and S.S. Giri. 2015. Dietary essentiality of trace minerals in aquaculture-A Review. Agric. Rev. 36: 100–112.

Chau, R., J.A. Kalaitzis and B.A. Neilan. 2011. On the origins and biosynthesis of tetrodotoxin. Aquat. Toxicol. 104: 61–72.

Chen, C.Y., R.S. Stemberger, B. Klaue, J.D. Blum, P.C. Pickhardt and C.L. Folt. 2000. Accumulation of heavy metals in food web components across a gradient of lakes. Limnol. Oceanogr. 45: 1525–1536.

Choi, B.-S. and W. Zheng. 2009. Copper Transport to the Brain by the Blood-Brain Barrier and Blood-CSF Barrier. Brain Res. 1248: 14–21.

Clarke, A. and N.M. Johnston. 1999. Scaling of metabolic rate with body mass and temperature in teleost fish. J. Anim. Ecol. 68: 893–905.

Costa, S.D.C. and S.M. Hartz. 2009. Evaluation of trace metals (cadmium, chromium, copper and zinc) in tissues of a commercially important fish (*Leporinus obtusidens*) from Guaíba Lake, Southern Brazil. Braz. Arch. Biol. Technol. 52: 241–250.

Davenport, J., N.D. Phillips, E. Cotter, L.E. Eagling and J.D.R. Houghton. 2018. The locomotor system of the ocean sunfish *Mola mola* (L.): role of hypodermisous exoskeleton, horizontal septum, muscles and tendons. J. Anat. Epub ahead of print.

de la Higuera, M., H. Akharbach, M.C. Hidalgo, J. Peragon, J.A. Lupiáñez and M. García-Gallego. 1999. Liver and white muscle protein turnover rates in the European eel (*Anguilla anguilla*): effects of dietary protein quality. Aquaculture 179: 203–216.

Dunier, M. 1996. Water pollution and immunosuppression of freshwater fish. Ital. J. Zool. 63: 303–309.

Dural, M., M.Z.L. Göksu and A.A. Özak. 2007. Investigation of heavy metal levels in economically important fish species captured from the Tuzla lagoon. Food Chem. 102: 415–421.

Dutton, J. and N.S. Fisher. 2011. Bioaccumulation of As, Cd, Cr, Hg (II), and MeHg in killi fish (*Fundulus heteroclitus*) from amphipod and worm prey. Sci. Total Environ. 409: 3438–3447.

Dutton, J., B.P. Jackson and D.H. Adams. 2016. Tissue distribution of essential and nonessential trace elements in ocean sunfish (*Mola mola*) and sharptail mola (*M. lanceolata*). Society of Environmental Toxicology and Chemistry 7th World Congress/North America 37th Annual Meeting. Orlando, FL

Ersoy, B. and M. Çelik. 2009. Essential elements and contaminants in tissues of commercial pelagic fish from the Eastern Mediterranean Sea. J. Sci. Food Agric. 89: 1615–1621.

European Commission 2004. Commission Regulation (EC) No 854/2004 of the European Parliament and of the Council of 29 April 2004 laying down specific rules for the organisation of official controls on products of animal origin intended for human consumption. OJEU L226: 83–127.

Farkas, A., J. Salánki and I. Varanka. 2008. Heavy metal concentrations in fish of Lake Balaton. Lakes Reserv. Res. Manag. 5: 271–279.

Fauconneau, B. and M. Arnal. 1985. *In vivo* protein synthesis in different tissues and the whole body of rainbow trout (*Salmo gairdnerii* R.). Influence of environmental temperature. Comp. Biochem. Physiol. A 82: 179–187.

Frias, J.P.G.L., J. Gago, V. Otero and P. Sobral. 2016. Microplastics in coastal sediments from Southern Portuguese shelf waters. Mar. Environ. Res. 114: 24–30.

Frias, J.P.G.L. and R. Nash. 2019. Microplastics: Finding a consensus on the definition. Mar. Pollut. Bull. 138: 145–147.

Gallitelli, M., N. Ungaro, L.M. Addante, V. Procacci, N.G. Silver and C. Sabbà. 2005. Respiratory illness as a reaction to tropical algal blooms occurring in a temperate climate. JAMA 293: 2599–2600.

GESAMP. 2015. Sources, fate and effects of microplastics in the marine environment: a global assessment. P.J. Kershaw [ed.]. (IMO/FAO/UNESCO-IOC/UNIDO/WMO/IAEA/UN/UNEP/UNDP Joint Group of Experts on the Scientific Aspects of Marine Environmental Protection). Rep. Stud. GESAMP No. 90, 96 p.

George, J.C. 1962. A Histophysiological study of the red and white muscles of the mackerel. Am. Midl. Nat. 68: 487–494.

Gleibs, S. and D. Mebs. 1999. Distribution and sequestration of palytoxin in coral reef animals. Toxicon 37: 1521–1527.

Gottofrey, J. and H. Tjälve. 1991. Axonal transport of cadmium in the olfactory nerve of the pike. Pharmacol. Toxicol. 69: 242–252.

Gravenmier, J.J., D.W. Johnston and W.R. Arnold. 2005. Acute toxicity of vanadium to the threespine stickleback, *Gasterosteus aculeatus*. Environ. Toxicol. 20: 18–22.

Grémillet, D., C.R. White, M. Authier, G. Dorémus, V. Ridoux and E. Pettex. 2017. Ocean sunfish as indicators for the 'rise of slime'. Curr. Biol. 27: R1263–R1264.

Gunter, T.E., C.E. Gavin and K.K. Gunter. 2009. The Case for Manganese Interaction with Mitochondria. Neurotoxicology 30: 727–729.
Hanifin, C.T. 2010. The chemical and evolutionary ecology of tetrodotoxin (TTX) toxicity in terrestrial vertebrates. Mar. Drugs 8: 577–593.
Harrod, C., J. Syväranta, L. Kubicek, V. Cappanera and J.D.R. Houghton. 2013. Reply to Logan and Dodge: 'Stable isotopes challenge the perception of ocean sunfish *Mola mola* as obligate jellyfish predators'. J. Fish Biol. 82: 10–16.
Holdway, D.A. and J.B. Sprague. 1979. Chronic toxicity of vanadium to flagfish. Water Res. 13: 905–910.
Huang, K., S. Liu, Y. Huang, K. Huang and D. Hwang. 2011. Food poisoning caused by sunfish *Masturus lanceolatus* in Taiwan. J. Food Drug Anal. 19: 191–196
Jovanović, B. 2017. Ingestion of microplastics by fish and its potential consequences from a physical perspective. Integr. Environ. Assess. Manag. 13(3): 510–515.
Jureša, D. and M. Blanuša. 2003. Mercury, arsenic, lead and cadmium in fish and shellfish from the Adriatic Sea. Food Addit. Contam. 20: 241–246.
Kalay, M. and M. Canli. 2000. Elimination of Essential (Cu, Zn) and Non-essential (Cd, Pb) Metals from Tissues of a Freshwater Fish *Tilapia zilli*. Turk. J. Zool. 24: 429–436.
Kaleshkumar, K., R. Rajaram, P. Dhinesh and A. Ganeshkumar. 2017. First report on distribution of heavy metals and proximate analysis in marine edible puffer fishes collected from Gulf of Mannar Marine Biosphere Reserve, South India. Toxicol. Rep. 4: 319–327.
Kaneko, T., S. Watanabe and K.M. Lee. 2008. Functional Morphology of Mitochondrion-Rich Cells in Euryhaline and Stenohaline Teleosts. Aqua-BioSci. Monogr. 1: 1–62.
Kang, M.J., H.J. Baek, D.W. Lee and J.H. Choi. 2015. Sexual Maturity and Spawning of Ocean Sunfish *Mola mola* in Korean Waters. Korean J. Fish. Aquat. Sci. 48: 739–744.
Kannan, K., A. Blankenship, P. Jones and J. Giesy. 2000. Toxicity reference values for the toxic effects of polychlorinated biphenyls to aquatic mammals. Hum. Ecol. Risk Assess. 6: 181–201.
Karlsson, T.M., A.D. Vethaak, B.C. Almroth, F. Ariese, M. van Velzen, M. Hassellöv et al. 2017. Screening for microplastics in sediment, water, marine invertebrates and fish: Method development and microplastic accumulation. Mar. Pollut. Bull. 122: 403–408.
Kenney, R. 1996. Preliminary assessment of competition for prey between leatherback sea turtles and ocean sunfish in northeast shelf waters. pp. 144–147. *In*: J.A. Keinath, D.E. Barnard, J.A. Musick and B.A. Bell [eds.]. Fifteenth Annual Symposium on Sea Turtle Biology and Conservation. National Marine Fisheries Service, Miami, FL.
Kojadinovic, J., M. Potier, M. Le Corre, R.P. Cosson and P. Bustamante. 2007. Bioaccumulation of trace elements in pelagic fish from the Western Indian Ocean. Environ. Pollut. 146: 548–566.
Kraal, M., M. Kraak, C. de Groot and C. Davids. 1995. Uptake and Tissue Distribution of Dietary and Aqueous Cadmium by Carp (*Cyprinus carpio*). Ecotoxicol. Environ. Saf. 31: 179–183.
Kumar, M., B. Aalbersberg and L. Mosley. 2004. Mercury Levels in Fijian Seafoods and Potential Health Implications. Suva, Fiji.
Lago, J., L.P. Rodríguez, L. Blanco, J.M. Vieites and A.G. Cabado. 2015. Tetrodotoxin, an extremely potent marine neurotoxin: distribution, toxicity, origin and therapeutical uses. Mar. Drugs 13: 6384–6406.
Lee, D., J. Choi and K. Choi. 2013. Catch distribution of ocean sunfish *Mola mola* off Korean Waters. Korean J. Fish. Aquat. Sci. 46: 851–855.
Li, M.-R. and C.-H. Huang. 2015. Effect of dietary zinc level on growth, enzyme activity and body trace elements of hybrid tilapia, *Oreochromis niloticus* × *O. aureus*, fed soya bean meal-based diets. Aquac. Nutr. 22: 1–8.
Lin, H. and W. Sung. 2003. The distribution of mitochondria-rich cells in the gills of air-breathing fishes. Physiol. Biochem. Zool. 76: 215–228.
Linnaeus, C. 1758. Tomus I. Systema Naturae per Regna Tria Naturae, Secundum Classes, Ordines, Genera, Species, Cum Characteribus, Differentiis, Synonymis, Locis. Holmiae (Laurentii Salvii).
Liu, K.-M., M.-L. Lee, S.-J. Joung and Y.-C. Chang. 2009. Age and growth estimates of the sharptail mola, *Masturus lanceolatus*, in waters of eastern Taiwan. Fish. Res. 95: 154–160.
Lucia, M., P. Bocher, M. Chambosse, P. Delaporte and P. Bustamante. 2014. Trace element accumulation in relation to trophic niches of shorebirds using intertidal mudflats. J. Sea Res. 92: 134–143.
Lushchak, O.V, O.I. Kubrak, I.M. Torous, T.Y. Nazarchuk, K.B. Storey and V.I. Lushchak. 2009. Trivalent chromium induces oxidative stress in goldfish brain. Chemosphere 75: 56–62.
Lusher, A., P. Hollman and J. Mendoza-Hill. 2017. Microplastics in fisheries and aquaculture. Status of knowledge on their occurrence and implications for aquatic organisms and food safety. Fao Fisheries And Aquaculture Technical Paper 615, Rome, Italy.
Magarlamov, T.Y., D.I. Melnikova, A.V. Chernyshev. 2017. Tetrodotoxin-producing bacteria: detection, distribution and migration of the toxin in aquatic systems. Toxins 9: 166.

Maia, M.M.D., A. Heasley, M.M. Camargo De Morais, E.H.M. Melo, M.A. Morais, Jr., W.M. Ledingham et al. 2001. Effect of culture conditions on lipase production by *Fusarium solani* in batch fermentation. Bioresour. Technol. 76: 23–27.

Marcovecchio, J.E., M. Victor Jorge and A. Pérez. 1991. Metal Accumulation in Tissues of Sharks from the Bahía Blanca Estuary, Argentina. Mar. Environ. Res. 31: 263–274.

Mataba, G.R., V. Verhaert, R. Blust and L. Bervoets. 2016. Distribution of trace elements in the aquatic ecosystem of the Thigithe river and the fish *Labeo victorianus* in Tanzania and possible risks for human consumption. Sci. Total Environ. 547: 48–59.

Maulvault, A.L., P. Anacleto, V. Barbosa, J.J. Sloth, R.R. Rasmussen, A. Tediosi et al. 2015. Toxic elements and speciation in seafood samples from different contaminated sites in Europe. Environ. Res. 143: 72–81.

Mehta, K. 2017. Impact of temperature on contaminants toxicity in fish fauna: a review. Indian J. Sci. Technol. 10: 1–6.

Metian, M., M. Warnau, T. Chouvelon, F. Pedraza, A.M. Rodriguez y Baena and P. Bustamante. 2013. Trace element bioaccumulation in reef fish from New Caledonia: Influence of trophic groups and risk assessment for consumers. Mar. Environ. Res. 87–88: 26–36.

Moore, R.E. and P.J. Scheuer. 1971. Palytoxin: a new marine toxin from a coelenterate. Science 172: 495–498.

Mosse, P.R.L. and R.C.L. Hudson. 1977. The functional roles of different muscle fibre types identified in the myotomes of marine teleosts: a behavioural, anatomical and histochemical study. J. Fish Biol. 11: 417–430.

Nakamura, I. and K. Sato. 2014. Ontogenetic shift in foraging habit of ocean sunfish *Mola mola* from dietary and behavioral studies. Mar. Biol. 161: 1263–1273.

Nakatsubo, T. and H. Hirose. 2007. Growth of captive ocean sunfish, *Mola mola*. Aquaculture Sci. 55: 403–407.

Naranuntarat, A., L.T. Jensen, S. Pazicni, J.E. Penner-Hahn and V.C. Culotta. 2009. The Interaction of Mitochondrial Iron with Manganese Superoxide Dismutase. J. Biol. Chem. 284: 22633–22640.

Nelms, S.E., T.S. Galloway, B.J. Godley, D.S. Jarvis and P.K. Lindeque. 2018. Investigating microplastic trophic transfer in marine top predators*. Environ. Pollut. 238: 999–1007.

Neves, D., P. Sobral, J. Lia and T. Pereira. 2015. Ingestion of microplastics by commercial fish off the Portuguese coast. Mar. Pollut. Bull. 101: 119–126.

Niyogi, S. and C.M. Wood. 2004. Biotic ligand model, a flexible tool for developing site-specific water quality guidelines for metals. Environ. Sci. Technol. 38: 6177–6192.

Noguchi, T. and O. Arakawa. 2008. Tetrodotoxin—distribution and accumulation in aquatic organisms, and cases of human intoxication. Mar. Drugs 6: 220–242

Nyegaard, M., E. Sawai, N. Gemmell, J. Gillum, N.R. Loneragan, Y. Yamanoue et al. 2018. Hiding in broad daylight: molecular and morphological data reveal a new ocean sunfish species (Tetraodontiformes: Molidae) that has eluded recognition. Zool. J. Linn. Soc. 182: 631–658.

Nyegaard, M., S.G. Barcelona, N.D. Phillips and E. Sawai. 2020. Fisheries interactions, distribution modelling and conservation issues of the ocean sunfishes pp. 216–242. *In*: T.M. Thys, G.C. Hays and J.D.R. Houghton [eds.]. The Ocean Sunfishes: Evolution, Biology and Conservation, CRC Press. Boca Raton, FL, USA.

Ogino, C. and G.-Y. Yang. 1978. Requirement of Rainbow Trout for Dietary Zinc. Bulletin of the Japanese Society of Scientific Fisheries 44: 1015–1018.

Pano, I., C. Carrascosa, J.R. Jaber and A. Raposo. 2016. Puffer fish and its consumption: to eat or not to eat? Food Reviews International 32: 305–322.

Patocka, J., R.C. Gupta, Q. Wu and K. Kuka. 2015. Toxic potential of palytoxin. J. Huazhong Univ. Sci. Technol. [Med. Sci.] 35: 773–780.

Pedro, S., I. Caçador, B.R. Quintella, M.J. Lança and P.R. Almeida. 2013. Trace element accumulation in anadromous sea lamprey spawners. Ecol. Freshw. Fish 23: 193–207.

Perrault, J.R., J.P. Buchweitz and A.F. Lehner. 2014. Essential, trace and toxic element concentrations in the liver of the world's largest bony fish, the ocean sunfish (*Mola mola*). Mar. Pollut. Bull. 79: 348–353.

Phillips, N.D., L. Kubicek, N.L. Payne, C. Harrod, L.E. Eagling, C.D. Carson et al. 2018. Isometric growth in the world's largest bony fishes (genus *Mola*)? Morphological insights from fisheries bycatch data. J. Morphol. 279: 1–9.

Phillips, N.D., E.C. Pope, C. Harrod and J.D.R. Houghton. 2020. The diet and trophic role of ocean sunfishes. pp. 146–159. *In*: T.M. Thys, G.C. Hays and J.D.R. Houghton [eds.]. The Ocean Sunfishes: Evolution, Biology and Conservation, CRC Press. Boca Raton, FL, USA.

PlasticsEurope, 2018. Plastics—The Facts 2018: An Analysis of European Plastics Production, Demand and Waste Data. PlasticsEurope: Association of Plastics Manufacturers, Brussels.

Pope, E.C., G.C. Hays, T.M. Thys, T.K. Doyle, D.W. Sims, N. Queiroz et al. 2010. The biology and ecology of the ocean sunfish *Mola mola*: a review of current knowledge and future research perspectives. Rev. Fish Biol. Fish. 20: 471–487.

Protasowicki, M. and G. Morsy. 1993. Preliminary studies on heavy metal contents in aquatic organisms from the Hornsund area, with a particular reference to the Arctic charr [*Salvelinus alpinus* (L.)]. Acta Ichthyologica et Piscatoria 23(Supple): 115–132.

Ptashynski, M.D., R.M. Pedlar, R.E. Evans, C.L. Baron and J.F. Klaverkamp. 2002. Toxicology of dietary nickel in lake whitefish (*Coregonus clupeaformis*). Aquat. Toxicol. 58: 229–247.

Raimundo, J., C. Vale, M. Caetano, E. Giacomello, B. Anes and G.M. Menezes. 2013. Natural trace element enrichment in fishes from a volcanic and tectonically active region (Azores archipelago). Deep Sea Res. Part II 98: 137–147.

Raimundo, J., C. Vale, I. Martins, J. Fontes, G. Graça and M. Caetano. 2015. Elemental composition of two ecologically contrasting seamount fishes, the bluemouth (*Helicolenus dactylopterus*) and blackspot seabream (*Pagellus bogaraveo*). Mar. Pollut. Bull. 100: 112–121.

Rainbow, P.S. 1985. The biology of heavy metals in the sea. Int. J. Environ. Stud. 25: 195–211.

Rajkowska, M. and M. Protasowicki. 2013. Distribution of metals (Fe, Mn, Zn, Cu) in fish tissues in two lakes of different trophy in Northwestern Poland. Environ. Monit. Assess. 185: 3493–3502.

Ralston, N.V.C. and L.J. Raymond. 2010. Dietary selenium's protective effects against methylmercury toxicity. Toxicology 278: 112–123.

Ricketts, C.D., W.R. Bates and S.D. Reid. 2015. The Effects of Acute Waterborne Exposure to Sublethal Concentrations of Molybdenum on the Stress Response in Rainbow Trout. Oncorhynchus mykiss. PLoS ONE 10: e0115334.

Rochman, C.M., B.T. Hentschel, and S.J. The. 2014. Long-term sorption of metals is similar among plastic types: implications for plastic debris in aquatic environments. PLoS One 9(1): e85433.

Rodriguez, P., A. Alfonso, C. Vale, C. Alfonso, P. Vale, A. Tellez et al. 2008. First toxicity report of tetrodotoxin and 5,6,11-trideoxyTTX in the trumpet shell *Charonia lampas lampas* in Europe. Anal. Chem. 80: 5622–5629.

Romeo, T., B. Pietroc, C. Pedà, P. Consoli, F. Andaloro and M. Cristina. 2015. First evidence of presence of plastic debris in stomach of large pelagic fish in the Mediterranean Sea. Mar. Pollut. Bull. 95: 358–361.

Rosseland, B.O., S. Rognerud, P. Collen, J.O. Grimalt, I. Vives, J.-C. Massabuau et al. 2007. Brown trout in lochnagar: population and contamination by metals and organic micropollutants. pp. 253–285. *In*: N.L. Rose [ed.]. Lochnagar: The Natural History of a Mountain Lake. Dordrecht: Springer.

Rudneva, I. 2013. Biomarkers for Stress in Fish Embryos and Larvae, 1st ed. Boca Raton: CRC Press,Taylor and Francis Group.

Sagara, K. and T. Ozawa. 2002. Landing statistics of Molidae in four prefectures of Japan. Mem. Fac. Fish Kagoshima Univ. 51: 27–33.

Saito, T., T. Noguchi, Y. Shida, T. Abe and K. Hashimoto. 1991. Screening of tetrodotoxin and its derivatives in puffer-related species. Nippon Suisan Gakk. 57(8): 1573–1577.

Satoh, S., T. Takeuchi and T. Watanabe. 1987. Effect of deletion of several trace elements from a mineral mixture in fish meal diets on mineral composition of gonads in rainbow trout and carp. Nippon Suisan Gakk. 53: 281–286.

Shakweer, L.M. and M.M. Abbas. 2005. Effect of ecological and biological factors on the uptake and concentration of trace elements by aquatic organisms at Edku lake. Egypt. J. Aquat. Res. 31: 271–287.

Silvani, L., M. Gazo and A. Aguilar. 1999. Spanish driftnet fishing and incidental catches in the western Mediterranean. Biol. Conserv. 90: 79–85.

Sims, D.W., N. Queiroz, T.K. Doyle, J.D.R. Houghton and G.C. Hays. 2009. Satellite tracking of the World's largest bony fish, the ocean sunfish (*Mola mola* L.) in the North East Atlantic. J. Exp. Mar. Biol. Ecol. 370: 127–133.

Sloman, K.A., D.W. Baker, C.G. Ho, D.G. Mcdonald and C.M. Wood. 2003. The effects of trace metal exposure on agonistic encounters in juvenile rainbow trout, *Oncorhynchus mykiss*. Aquat. Toxicol. 63: 187–196.

Sousa, L.L., R. Xavier, V. Costa, N.E. Humphries, C. Trueman, R. Rosa et al. 2016a. DNA barcoding identifies a cosmopolitan diet in the ocean sunfish. Sci. Rep. 6.

Sousa, L.L., N. Queiroz, G. Mucientes, N.E. Humphries and D.W. Sims. 2016b. Environmental influence on the seasonal movements of satellite-tracked ocean sunfish *Mola mola* in the north-east Atlantic. Anim. Biotelemetry 4: 7.

Sousa, L.L., I. Nakamura and D.W. Sims. 2020. Movements and foraging behavior of ocean sunfish. pp. 129–145. *In*: T.M. Thys, G.C. Hays and J.D.R. Houghton [eds.]. The Ocean Sunfishes: Evolution, Biology and Conservation, CRC Press. Boca Raton, FL, USA.

Stohs, S. and D. Bagchi. 1995. Mechanisms in the Toxicity of Metal Ions. Free Radic. Biol. Med. 18: 321–336.

Syväranta, J., C. Harrod, L. Kubicek, V. Cappanera and J.D.R. Houghton. 2012. Stable isotopes challenge the perception of ocean sunfish *Mola mola* as obligate jellyfish predators. J. Fish Biol. 80: 225–231.

Szarek-Gwiazda, E., A. Amirowicz and R. Gwiazda. 2006. Trace element concentrations in fish and bottom sediments of a eutrophic dam reservoir. Oceanol. Hydrobiol. Stud. XXXV: 331–352.

Szefer, P., M. Domagała-Wieloszewska, J. Warzocha, A. Garbacik-Wesołowska and T. Ciesielski. 2003. Distribution and relationships of mercury, lead, cadmium, copper and zinc in perch (*Perca fluviatilis*) from the Pomeranian Bay and Szczecin Lagoon, southern Baltic. Food Chem. 81: 73–83.

Takeuchi, T., T. Watanabe, C. Ogino, M. Saito, K. Nishimura and T. Nose. 1981. Effects of low protein-high calorie diets and depletion of trace elements from a fish meal diet on reproduction of rainbow trout. Bull. Japan. Soc. Sci. Fish 47: 645–654.

Taniyama, S., Y. Mahmud, M. Terada, T. Takatani, O. Arakawa, T. Noguchi. 2002. Occurrence of a food poisoning incident by palytoxin from a serranid *Epinephelus* sp. in Japan. J. Nat. Toxins 11: 277–282.

Tjälve, H. and J. Henriksson. 1999. Uptake of metals in the brain via olfactory pathways. Neurotoxicology 20: 181–195.

U.S. EPA. 1998 Method 6020A: Inductively Coupled Plasma-Mass Spectrometry. US Environmental Protection Agency, Washington, DC.

Usami, M., M. Satake, S. Ishida, A. Inoue, Y. Kan and T. Yasumoto. 1995. Palytoxin analogs from the dinoflagellate Ostreopsis siamensis. J. Am. Chem. Soc. 117: 5389–5390.

Vlamis, A., P. Katikou, I. Rodriguez, V. Rey, A. Alfonso, A. Papazachariou et al. 2015. First Detection of tetrodotoxin in Greek shellfish by UPLC-MS/MS potentially linked to the presence of the dinoflagellate *Prorocentrum minimum*. Toxins 7: 1779–1807.

Walker, J.R., P.A. Novick, W.H. Parsons, M. McGregor, J. Zablocki, V.S. Pande et al. 2012. Marked difference in saxitoxin and tetrodotoxin affinity for the human nociceptive voltage-gated sodium channel (Nav1.7). Proc. Natl. Acad. Sci. USA. 109: 18102–18107.

Walker, T.I. 1976. Effects of Species, Sex, Length and Locality on the Mercury Content of School Shark *Galeorhinus australis* (Macleay) and Gummy Shark *Mustelus antarcticus* Guenther from South-eastern Australian Waters. Aust. J. Mar. Freshw. Res. 27: 603–616.

Watanabe, T., V. Kiron and S. Satoh. 1997. Trace minerals in fish nutrition. Aquaculture 151: 185–207.

Watanabe, Y. and K. Sato. 2008. Functional Dorsoventral Symmetry in Relation to Lift-Based Swimming in the Ocean Sunfish *Mola mola*. PLoS ONE 3: e3446.

Watanabe, Y.Y. and J. Davenport. 2020. Locomotory systems and biomechanics of ocean sunfish. pp. 72–86. *In*: T.M. Thys, G.C. Hays and J.D.R. Houghton [eds.]. The Ocean Sunfishes: Evolution, Biology and Conservation, CRC Press. Boca Raton, FL, USA.

Wood, C.M., A.P. Farrell and C.J. Brauner. 2012a. Homeostatis and toxicology of essential metals. Fish physiology: vol 31A. Academic Press, New York.

Wood, C.M., A.P. Farrell and C.J. Brauner. 2012b. Homeostatis and toxicology of non-essential metals. Fish physiology: vol 31B. Academic Press, New York.

Wright, S.L., R.C. Thompson and T.S. Galloway. 2013. The physical impacts of microplastics on marine organisms: A review. Environ. Pollut. 178: 483–492.

Yeltekin, A.Ç. and A.R. Oğuz. 2018. Some macro and trace elements in various tissues of Van fish variations according to sex and weight. Arq. Bras. Med. Vet. Zootec. 70: 231–237.

Yi, Y., Z. Yang and S. Zhang. 2011. Ecological risk assessment of heavy metals in sediment and human health risk assessment of heavy metals in fishes in the middle and lower reaches of the Yangtze River basin. Environ. Pollut. 159: 2575–2585.

Yi, Y.J. and S.H. Zhang. 2012. The relationships between fish heavy metal concentrations and fish size in the upper and middle reach of Yangtze River. Procedia Environ. Sci. 13: 1699 – 1707.

Yilmaz, A.B. and M. Doğan. 2008. Heavy metals in water and in tissues of himri (*Carasobarbus luteus*) from Orontes (Asi) River, Turkey. Environ. Monit. Assess. 144: 437–444.

Zaccaroni, A., M. Silvi, A. Gustinelli and D. Scaravelli. 2008. First data on heavy metals in sunfish (*Mola mola*) from northern Adriatic sea. Biol. Mar. Mediterr. 15: 446–447.

Zaki, M.S., N.E. Sharaf and M.H. Osfor. 2010. Effect of Vanadium Toxicity in *Clarias lazera*. J. Am. Sci. 6: 291–296.

Zobkov, M.B. and E.E. Esiukova. 2018. Microplastics in a marine environment: review of methods for sampling, processing, and analyzing microplastics in water, bottom sediments, and coastal deposits. Oceanology 58: 137–143.

Chapter 12

Fisheries Interactions, Distribution Modelling and Conservation Issues of the Ocean Sunfishes

Marianne Nyegaard,[1,*] *Salvador García-Barcelona*,[2] *Natasha D. Phillips*[3] and *Etsuro Sawai*[4]

Introduction

In today's changing world—the Anthropocene epoch—it seems most of our natural world is under threat. If human activities are not directly causing detrimental impacts, climate change indirectly looms as a formidable challenge to life on Earth as we know it (Hoegh-Guldberg et al. 2018). In this context, marine megafauna, such as whales, dolphins, sea turtles, sharks, etc., are vulnerable to anthropogenic activities given their typical life history characteristics of slow growth, late maturity and low fecundity (Lewison et al. 2004). While it may be tempting to group the large species of ocean sunfish (Family Molidae; genus *Mola* and *Masturus*) with such 'typical' marine megafauna, they are in fact life-strategy outliers, most notably with extremely high reproductive potentials and capacities for fast growth (Pope et al. 2010, Forsgren et al. 2020 [Chapter 6]). Combined with widespread distributions throughout tropical and temperate ocean zones (Froese and Pauly 2019) and a history lacking widespread human exploitation, this group has very different pressures compared to other marine megafauna. However, while sunfish research efforts globally have increased significantly in recent years, many aspects of sunfish biology are still unclear, and their conservation status is debated and poorly understood.

This chapter reviews current knowledge of Molidae fisheries interactions, identifies knowledge gaps for improving species-level assessments and distribution modelling, and briefly comments on other relevant anthropogenic pressures.

[1] Auckland War Memorial Museum Tamaki Paenga Hira, Natural Sciences, The Domain, Private Bag 92018, Victoria Street West, Auckland 1142, New Zealand.

[2] Mediterranean Large Pelagics Group, Spanish Institute of Oceanography, Puerto Pesquero s/n, 29640, Fuengirola, Malaga, Spain.

[3] Queen's Marine Laboratory, Queen's University Belfast, 12-13 The Strand, Portaferry, BT22 1PF, United Kingdom.

[4] Ocean Sunfishes Information Storage Museum (online), C-102 Plaisir Kazui APT, 13-6 Miho, Shimizu-ku, Shizuoka-shi 424-0901, Japan.

* Corresponding author: mnyegaard@aucklandmuseum.com

Conservation Status

Species Considerations

A complicating factor in vulnerability assessments of the Molidae is the long and confusing taxonomic history of the family dating back nearly five centuries (Nyegaard et al. 2018b, Sawai et al. 2018). Following recent taxonomic revisions in the genus *Mola*, five species in three genera are currently recognized (Fricke et al. 2019). Of the five recognized species, *Masturus lanceolatus* (Liénard 1840), *Mola mola* (Linnaeus 1758) and *Mola alexandrini* [(Ranzani 1839); senior synonym of *Mola ramsayi* (Giglioli 1883)], all attain sizes in excess of 3 m total length, with a maximum of 2.4 m recorded to date for *Mola tecta* Nyegaard et al. 2017 (see Sawai et al. 2020, Chapter 2). Superficially, these four large species exhibit limited morphological diversity, particularly so for *Mola* spp. For the past ~100 years, *M. mola* has been near-synonymous with the unique Molidae body form, resulting in an under-appreciation of other species in the family, confusion between species, and a confounded understanding of species level distributions. In this chapter, these four large species are referred to as the 'large molids' to distinguish them from their small congener, *Ranzania laevis* (Pennant 1776); the little studied morphological outlier of the family, which is discussed separately in Section "*Ranzania laevis*."

The legacy taxonomic confusion in Molidae has manifested in a global prevalence of reporting fisheries interactions as "*M. mola*", "ocean sunfish" or simply "sunfish" (Nyegaard et al. 2018a), however, interactions may comprise other molid species in addition to, or even instead of *M. mola* (e.g., Sawai et al. 2011, Chang et al. 2018, Nyegaard et al. 2018a, Mangel et al. 2019). Given the superficially similar morphology of the *Mola* species, in particular at smaller sizes (Sawai et al. 2018), collecting fisheries data to species level is challenging. Furthermore, while *Ma. lanceolatus* may be reported separately (e.g., Fulling et al. 2007, National Marine Fisheries Service 2011) and can readily be identified in market situations (Liu et al. 2009, Chang et al. 2018), robust identification cannot always be taken for granted in fisheries data (Nyegaard et al. 2018a). In this chapter, the use of species names reflects those given in the cited literature, however, the possibility of multi-species situations and mistaken identities should be kept in mind.

IUCN Assessment

In 2015, the International Union for the Conservation of Nature (IUCN) published conservation status assessments of *R. laevis*, *Ma. lanceolatus* and *M. mola* undertaken between 2007–2015, mostly resulting in 'data deficient' or 'of least concern' (Table 1). However, *M. mola* was listed as 'vulnerable to extinction' on a global scale, partly motivated by reports of rapid declines in sunfish landings in some fisheries, and high levels of sunfish bycatch or rapid declines in Catch Per Unit Effort (CPUE)

Table 1. List designation, population trend and year of assessment (in parenthesis) of three Molidae species by the International Union for the Conservation of Nature (https://www.iucnredlist.org).

Species	Assessment	Global	Europe	Mediterranean	Gulf of Mexico
Mola mola	Liu et al. 2015a	Vulnerable Decreasing (2011)	Data deficient Decreasing (2013)	Data deficient Unknown (2007)	Least concern Decreasing (2015)
Masturus lanceolatus	Leis et al. 2015	Least concern Unknown (2011)	NA	NA	Least concern Unknown (2015)
Ranzania laevis	Liu et al. 2015b	Least concern Unknown (2011)	Data deficient Unknown (2013)	NA	Data deficient Unknown (2015)

NA: not assessed

in others. The assessment concluded "... it is likely that other fisheries using these same methods are taking large, but unreported, bycatch of *M. mola* throughout the majority of its range." In light of the taxonomic uncertainties of *Mola* spp. at the time (Liu et al. 2015a), the *M. mola* IUCN assessment is here considered best regarded at genus level.

Fisheries Interactions

Definitions

Globally, few commercial markets exist for Molidae, and consequently few fisheries target them. Instead, sunfish are predominantly caught as non-target species in other fisheries, where they are either retained for use or returned to sea. Fisheries interactions may be categorized in a variety of ways, often hindering direct comparison between studies (Hall et al. 2017, Gray and Kennelly 2018). Here, 'catch' broadly refers to all that is retained irrespective of the intent to catch it, while 'bycatch' is all that is discarded dead or released alive (Fig. 1). The latter also includes post-release mortality as a consequence of the fisheries interaction, as well as other types of cryptic mortality not easily observed (see also Section "Fisheries Related Mortality").

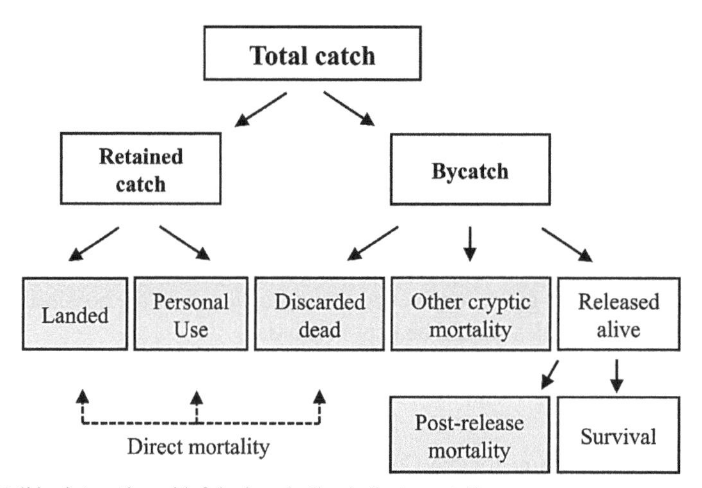

Figure 1. Fate of Molidae interacting with fisheries; shading indicate mortality.

Commercial and Artisanal Fisheries for Large Molids

The oldest known Molidae fishery dates back to the early Holocene. Ossicles and beak fragments in archaeological middens point to a substantial *M. mola* fishery for sustenance off Southern California (Porcasi and Andrews 2001). In the Mediterranean, traditional trap-net fisheries for pelagic fishes date back to the Middle ages (Cattaneo-Vietti et al. 2015). A small number of these trap-net fisheries are still in operation in Spain, Portugal, Italy and Morocco, including the *tonnarella* fishery in the Camogli Marine Protected Area in the Ligurian Sea, Italy. Here, *M. mola* (*ca.* 40–100 cm Total Length; TL) make up a large proportion of the total catches (Phillips et al. 2018; Fig. 2A,B), with reconstructed historical catch data (1950–1970) indicating that in some years sunfish catches exceeded that of the other top nine fished species combined (Relini and Vallarino 2016). High bycatches of small *M. mola* (*ca.* 30–60 cm TL) also occur in the extant trap-net fishery off Olhão in Portugal (Baptista et al. 2018), off La Azohía, Murcia, in Spain (generally < 100 cm TL; S. García-Barcelona unpublished data) and presumably in other extant Mediterranean trap-net fisheries as well (e.g., Addis et al. 2013). Traditionally, sunfish were part of the Mediterranean cuisine, but commercial use of Molidae products

Figure 2. (A) Hand pulling, and (B) *Mola mola* bycatch in the extant traditional *tonnarella* trap-net fishery in the Camogli Marine Protected Area (MPA) in Italy, managed and monitored by the MPA Authorities to ensure compliance with local marine protective measures. Images by Natasha D. Phillips (2016), reprinted with permission.

Figure 3. (A) *Masturus lanceolatus* at Hualien (2019) and (B) Shingang (2018) markets in Taiwan. Images by Ching-Tsun, reprinted with permission.

was prohibited in 2004 by the European Commission due to concerns (as yet unsubstantiated) about toxins harmful to human health, and commercial landings of sunfish were discontinued (Relini and Vallarino 2016).

Today, the largest consumers of sunfish are Taiwan, Japan and South Korea. In Taiwan, large molids are caught in targeted harpoon fisheries, and in set net, longline and driftnet fisheries targeting other species (Liu et al. 2009, Chang et al. 2018). Most molids are sold through the Hualien (> 80 percent), Shingang and Nanfangao markets. Landings at the former two markets comprise mainly *Ma. lanceolatus* (*ca.* 80 percent) (Fig. 3A,B) as well as *M. mola* and *M. alexandrini* (*ca.* 20 percent), (Chang et al. 2018; E. Sawai, personal observation). At the Nanfangao market, *Mola* spp. dominate the landings (C.-T. Chang, personal communication). Traditionally, only intestines and gonads were consumed, but a sunfish food festival in Hualien County (2003–2006) promoted eating sunfish in other ways, and other parts were subsequently utilised (Chin 2008, Thys et al. 2020 [Chapter 14]). An ensuing increase in popular demand saw annual landings increase considerably (Liu et al. 2009), and these have remained high subsequently with a peak in 2015 of nearly 1,000 tons (Fig. 4). Although other landing estimates are slightly lower (e.g., Chang et al. 2018), the data in Fig. 4 are likely reliable, as landings are recorded by each market and submitted to the central fisheries agency in Taiwan (C.-T. Chang, personal communication). Following the recommendations in the IUCN assessment of *Ma. lanceolatus* (Table 1), significant research efforts have been directed at the Taiwanese Molidae

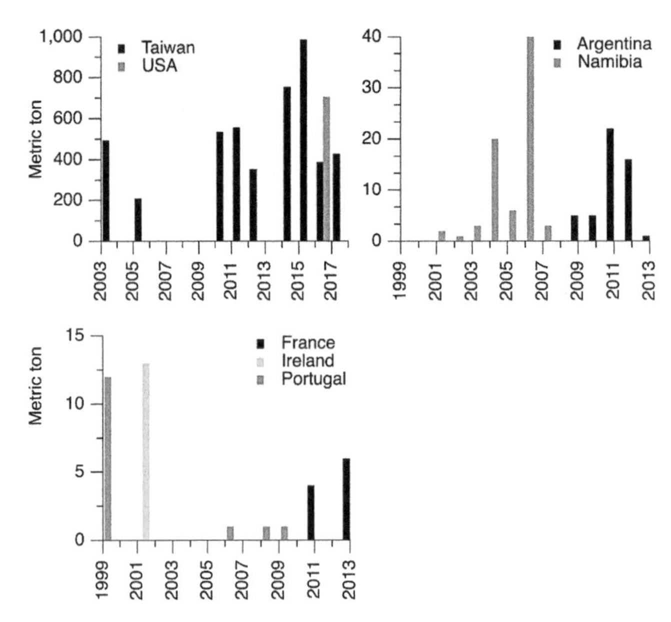

Figure 4. Landing data for "ocean sunfish", probably best interpreted as large molids. *Source*: data from FAO (2019) and Liu et al. (2009) (Taiwan 2003, 2005).

fishery (e.g., Chang et al. 2018). Across mainland China, large molids are not typically consumed, except occasionally in coastal areas (C.-T. Chang, personal communication), but have some use in traditional Asian medicine (Tang 1987, Thys et al. 2020, Chapter 14).

Sunfish consumption in Japan can be traced back nearly four centuries to a 1636 Japanese cook book (Sawai 2017). In areas of traditional sunfish cuisine, such as the Iwate, Miyagi, Kochi and Wakayama Prefectures, sunfish are today targeted in local set net and harpoon fisheries. Across wider Japan, sunfish are caught in set net, gillnet, purse seine and longline fisheries targeting other species (Sagara and Ozawa 2002a). Sunfishes are generally processed at sea and only parts for consumption, such as muscles, intestines and livers, are landed (Sawai 2017). Central sunfish statistics are not collected, and landings are instead noted by local fisheries research institutes, markets and fishing co-operations. A study by Sagara and Ozawa (2002a) of Molidae landings from 1994–2000 at 27 markets revealed substantial variation in both magnitude and seasonality between prefectures, as well as large fluctuations within markets over time. Considering the large number of markets in Japan, such heterogeneity impedes efforts to estimate the national scale of sunfish catches, which may be substantial. As an example of scale, 126.4 metric tons of sunfish were landed in 1996 across four markets in the Miyagi prefecture alone (Sagara and Ozawa 2002b). In Japan, landed sunfish mainly comprise *M. mola,* with a small proportion of *M. alexandrini* and *Ma. lanceolatus* (Sagara and Ozawa 2002a, Yoshita et al. 2009, Sawai et al. 2011, 2018), however, species-level data are limited, and the scale of the present day Molidae fishery in Japan is unknown.

In South Korea, sunfish are predominantly caught as non-target species in the large purse seine fishery off Jeju island and landed for human consumption, mainly through the port of Busan (Lee et al. 2013, Kang et al. 2015). Landings are reported as *M. mola* and between 2010–2012, these were noted as hundreds of individuals per annum. However, catch statistics and associated effort are lacking in the literature. In Thailand, sunfish are landed occasionally from foreign longline vessels (Maeroh et al. 2018), but the end use is uncertain.

Commercial markets for molids are rare outside Asia, and although small-scale landings have been reported in Peru (Mangel et al. 2019), and sporadic landings evidently occur in a number of countries (Fig. 4), the global extent is unknown. Data from the Food and Agricultural Organization (FAO) on

Molidae landings (Fig. 4) provide some insight into the global landings of large molids, although the comprehensiveness of the data is uncertain. These catch data were also reviewed in the IUCN assessment of *M. mola*, where the rapid declines in Namibia (2006 to 2007), Ireland (2001 to 2002) and Portugal (1999 to 2009), and rapid increase in Argentina (2008–2012), were noted as of concern.

The role of Molidae in artisanal fisheries is not well documented, but sunfish are undoubtedly caught and utilized locally, at least in some parts of the world. A report on marine megafauna harvesting by the 'whale hunters' of the Indonesian villages of Lamalera on Lembata Island, and Lamakera on Solor Island reported that sunfish were hunted seasonally (with spears) but were not a target species. A logbook project between May 2003 and May 2004 recorded modest catches of eight sunfish by Lamakera, and ten by Lamalera hunters (Atapada 2004). This project also documented that sunfish were not sold but were consumed locally by the villagers. Barnes (1996) recounted that the sea hunters from Lamalera considered sunfish meat a delicacy, sharing it with crew and with other boats, and consumed it along with palm wine. Another study recorded a total of four sunfish taken by large boats at Lamalera between 15 May and 9 September 2006, a minor contribution compared to the overall catch of other megafauna during the same period (Nolin 2008). Sunfish rarely end up at fish markets in Indonesia (M. Nyegaard, unpublished data) although it happens occasionally; on 1 August 2012 a large *Mola* sp. was for example seen at the Tanjung Luar fish market on Lombok (V. Jaiteh, personal communication 2012). Sunfish are also caught and consumed by local communities in the Philippines (M.D. Dimzon, personal communication) but again the extent is unknown. They appear to be consumed locally and occasionally end up at fish markets, as in the case of a *Ma. lanceolatus* specimen at the Anilao fish market, Batangas, Philippines, which is now registered at the Smithsonian National Museum of Natural History (specimen SNM 431645).

Bycatch of Large Molids in Commercial Fisheries

Fisheries interactions with large molids are widely reported from bycatch studies of longline, driftnet, purse seine and trawl fisheries, however to date, few dedicated studies have been undertaken. Catch and bycatch information appears to be available in a wide range of published and unpublished fisheries reports and databases, typically stemming from both commercial fisheries logbooks and fisheries observer programs. An exhaustive review is beyond the scope of this chapter; instead, the fisheries mentioned in the *M. mola* IUCN assessment are briefly reviewed and placed in a broader context.

Driftnet

The seasonal Californian driftnet fishery has been notorious for its high sunfish bycatch (Hahlbeck et al. 2017), with an often cited estimate of > 26,000 annual sunfish interactions during 1990–1998 (Dewar et al. 2010, H. Dewar personal communication 2019) considered of concern in the *M. mola* IUCN assessment. Large molids reportedly constituted over 33% of the total bycatch during 1990–2015 (Mason et al. 2019). However, since 1998, total fishing effort and sunfish bycatch have both decreased significantly (Fig. 5), and the fishery is set to phase out by 2023 (California Legal Information 2018, Mason et al. 2019). An estimated 95 percent of the sunfish are returned to sea alive, based on data from observers on 20 percent of the fleet (Hahlbeck et al. 2017).

Two other driftnet fisheries were mentioned as of concern in the *M. mola* IUCN assessment, both in the Mediterranean Sea. In a 1992–1994 study, small sunfish (*ca.* 0.5–7 kg, corresponding to *ca.* 21–50 cm TL; Phillips et al. 2018) made up 71–93 percent (by number) of the total catch in the illegal Spanish driftnet fishery, with 94 percent returned to sea alive (Silvani et al. 1999). The study found that in 1993 and 1994, an average of 16 and 13 *M. mola*, respectively, were caught per kilometer of net, with a simple upscaling to fleet level yielding ca. 53,400 and 25,100 annual sunfish interactions for the two years. High *M. mola* bycatches were also noted in the Moroccan driftnet fishery (Tudela et al. 2005), with a simple upscaling indicating a total of 36,450 sunfish interactions

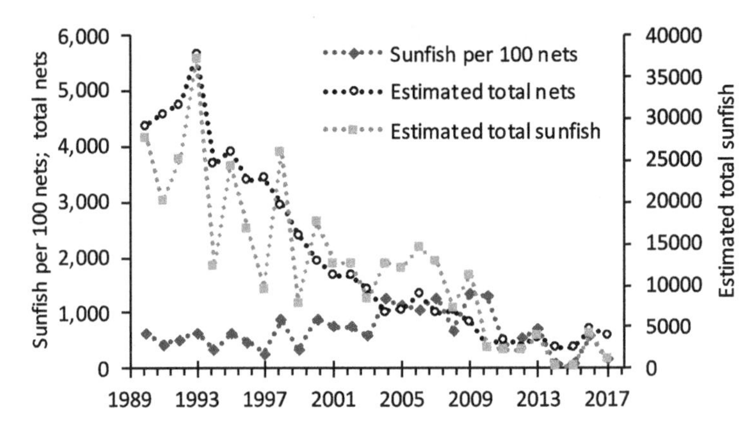

Figure 5. *Mola mola* bycatch in the California drift gillnet swordfish fishery (seasonal averages; 1 May – 31 January), set to phase out by 2023. *Source*: Estimated total nets (ETN) and sunfish per 100 nets (M) are observer data from NOAA (2019); Estimated total sunfish is ETN*M/100.

between December 2002 and September 2003 across the fishery (Pope et al. 2010, Liu et al. 2015a). As also noted by the authors, a simple upscaling to fleet level only yields rough estimates as spatio-temporal variation in bycatch rates is not taken into account. Furthermore, the total driftnet effort in the Mediterranean at that time was exceedingly difficult to estimate with accuracy (e.g., Serna et al. 1992), however, these bycatch estimates nevertheless provide an important indication of the scale of potential sunfish interactions in driftnet fisheries. A 2000–2003 study of the French driftnet fishery in the North-western Mediterranean, did not find similarly high *M. mola* bycatch levels as Silvani et al. (1999) and Tudela et al. (2005). Here, only 0.16 *M. mola* were caught per kilometer of net (and released alive), highlighting the potential high variability in bycatch levels between similar, nearby fisheries. Sunfish also interact with other driftnet fisheries in the Mediterranean (Ceyhan and Akyol 2014, Garibaldi 2015), although the scale is uncertain. Sunfish interactions have been reported from driftnet fisheries in Ireland (~186 and 2,867 individuals in 1996 and 1998, respectively; Rogan and Mackey 2007), the Tasman Sea (~2,700 individuals during the 1989–90 fishing season; Coffey and Grace 1990 in: Northridge 1991) and the North Pacific (interactions in 24–32.5 percent of sets in 1990–1991; McKinnell and Seki 2017). In the extant Peruvian small-scale gill net fishery, it is estimated that "thousands of sunfish [are] captured annually" (Mangel et al. 2019).

Longline

The large molids are widely reported as bycatch in pelagic longline fisheries for tuna and swordfish, including in the Mediterranean (Macías et al. 2004, Garibaldi 2015), tropical North East Atlantic (Fernandez-Carvalho et al. 2015), North West Atlantic (MacNeil et al. 2009), South Atlantic (Lucena-Frédou et al. 2017), West Central Pacific (SPC-OFP 2010, Clarke et al. 2014), off Hawai'i (National Marine Fisheries Service 2011), South West Pacific (Marin et al. 1998, Ward et al. 2004, Nyegaard et al. 2018a), Indian Ocean basin (Huang and Liu 2010, Ardill et al. 2012), Arabian Sea and Bay of Bengal (Varghese et al. 2013), North Gulf of Mexico (Fulling et al. 2007), and off South Africa (Petersen 2005), as well as in the small-scale shark and dolphinfish longline fishery off Peru (Mangel et al. 2019). In a study off South Africa, Petersen (2005) estimated annual CPUE of 0.08–0.29 sunfish per 1,000 hooks (2000–2003), yielding fleet-wide annual average bycatches of 340 sunfish. These estimates were later reported as 340,000 annual interactions (Sims et al. 2009), and subsequently considered of concern in the IUCN assessment of *M. mola* (Liu et al. 2015a). The concern was specifically linked to the possibility that similarly large bycatch rates occurred in longline fisheries more broadly. While comparison of sunfish interaction rates between fisheries are inherently difficult due to differences in reporting, none of the above listed fisheries reveal interactions on the scale considered of concern in Liu et al. (2015a) (Table 2). For example, the molid bycatch in the longline fishery off eastern Australia (Eastern Tuna and Billfish Fishery) was reportedly 95,438 kg during a 10 month period in 2018 (AFMA 2019), and although this could seem extensive, the actual numbers are moderate; the molid bycatch in this fishery consist of relatively large animals (generally 75–175 cm TL with some > 200 cm TL; Nyegaard et al. 2018a). Assuming an average weight of 200 kg (corresponding to *ca.* 150 cm TL; Phillips et al. 2018), a 95,438 kg bycatch equates to ca. 477 individuals, similar in scale to the molid bycatch of 273–498 individuals (annual average of 389) recorded during 2010–2012 in that same fishery (AFMA 2013). The annual molid bycatch in the New Zealand longline fishery is reportedly somewhat higher; during 2012–2016, 770–4,849 individuals were caught and released (annual average of 2,560) (MPI 2016, 2017).

Estimating total annual Molidae interactions in the global longline fisheries sector, as has been done, for example, for sharks (Worm et al. 2013), would require careful and detailed evaluation of known CPUE rates between fisheries combined with appropriate up-scaling to global fleet level. For the sunfishes, CPUE rates can be particularly tricky to standardize between fisheries given their large growth spectrum (see also Section Challenges in Fisheries Data Analysis). Importantly, any gross over- or under-estimates of global interactions would be equally unhelpful in the debate about Molidae conservation status.

Trawl

Trawl fisheries occur extensively in coastal waters throughout the world (Kroodsma et al. 2018) and comprise a highly heterogenous sector. The scale of interactions with large molids is uncertain and likely varies substantially between fisheries. The South African mid-water trawl fishery for horse mackerel (*Trachurus capensis,* Castelnau 1861) was cited as of concern in the *M. mola* IUCN assessment. Here, sunfish were found to comprise 51 percent of non-target species (by number) in 2002, followed by a sharp decrease in CPUE in 2003, which remained low until the end of the study in 2005 (Petersen and McDonell 2007). No follow-up study has since been undertaken, but Reed et al. (2017) reported consistently low *M. mola* bycatches (0.01 percent of the total catch by weight; annual average of 42 individuals) onboard a large freezer trawler between 2004–2014. However, directly comparing the two reports is problematic due to differences in fishing gear and methods. Elsewhere, Zeeberg et al. (2006) estimated that 0–273 individual *M. mola* were caught monthly in the North West African pelagic trawl fishery (seasonally, 2001–2004), and large molids were also reported as bycatch in the Australian midwater small pelagic trawl fishery (AFMA 2006), where annual bycatches in 2015–2018 ranged between 875–8,450 kg (2015–2018), with none retained (AFMA 2019).

Purse Seine

Large molids also interact with purse seine fisheries, but the global extent and scale are uncertain. Both *M. mola* and *Ma. lanceolatus* have been reported as bycatch in the purse seine fisheries in New Zealand (Paulin et al. 1982, Francis and Jones 2017), as well as in the tropical East Atlantic and Indian Ocean basins (Ardill et al. 2012, Torres-Irineo et al. 2014, Lezama-Ochoa et al. 2018, Escalle et al. 2019), and are also included in Fukofuka and Itano's (2007) guide to purse seine bycatch species from the West Central Pacific Ocean basin. Large molids are reportedly also caught to an unknown extent in the world's largest single-species fishery (purse seine for anchovies) off Peru (Mangel et al. 2019).

Ranzania laevis

The extent of fisheries interactions globally for *R. laevis* is unclear as this species generally receives little attention. Commercial exploitation is not known and although this species is undoubtedly consumed by humans occasionally, documentation is largely lacking. For example, consumption of

Table 2. Summary of literature search on bycatch of large molid (genus *Mola* and *Masturus*) in longline fisheries.

Area	Longline fishery	Data source and fishing effort	Year	Species, catch data and fate	Reference
Spain (W Mediterranean)	Swordfish	Commercial landings and onboard observers (effort: ND)	2001–2002	*M. mola* 359 individuals (6,258 kg) in 2001 (0.3% of total catch, by number); 0 in 2002 Fate: ND	Macias et al. 2004
Ligurian Sea (Mediterranean)	Swordfish (mesopelagic)	Onboard observers (32,000 hooks)	2010–2013	*M. mola* Ca. 23% of non-commercial bycatch (by number) (overall non-commercial bycatch comprise 8.6% of total catch, by number) Fate: released alive and in healthy condition	Garibaldi 2015b
Tropical NE Atlantic	Tuna and swordfish (pelagic), experimental	Onboard observers (254,520 hooks)	Aug 2008–Dec 2011	*M. mola* 0.02–0.3 sunfish per 1,000 hooks, depending on hook and bait Fate: discarded (condition not given)	Fernandez-Carvalho et al. 2015
NW Atlantic	Pelagic	Onboard observers (effort: ND)	1992–2006	*Mola* spp. 684 sunfish (< 0.3% of total catch, by number) Fate: ND	MacNeil et al. 2009
S Atlantic	Tuna	Various published and unpublished sources (effort: ND)	ND	*M. mola* In 'moderate risk' category, based on semi-quantitative Productivity and Susceptibility Analysis (see text for details) Fate: discarded (condition not given)	Lucena-Frédou et al. 2017
W and Central Pacific	Tuna and billfish (various fisheries)	Onboard observers (3,198 trips)	1994–2009	*M. mola* 499 tonnes: 1.1% of total catch and 15th most common species caught (by weight) across all fisheries; hook rate (0–0.75 kg sunfish per 1,000 hooks^{-1}) varies across years and fisheries. Highest ranking bycatch species in the shallow set longline fishery. Fate: 48–98% discarded (0–15% dead), varies between fisheries	SPC-OFP 2010, Clarke et al. 2014
Hawaii	Tuna (pelagic: deep- and shallow set)	Logbook-, observer- and state dealer data (20–26.1% cover on deep-set (2005–2008); 100% cover on shallow set (2005–2008)	2005	*M. mola* 85 individuals (deep and shallow set combined): 37,968 pounds [17,222 kg] (deep-set) 5,767 pounds [2,617 kg] (shallow-set) Fate: 123 lb landed [56 kg], but mostly discarded (condition not given) *Ma. lanceolatus* 6,217 pounds [2820 kg] (deep-set) Fate: rarely landed but fate unknown	National Marine Fisheries Service 2011
SW Atlantic (off Uruguay)	Swordfish	Observer data (21 trips, 202 sets, 155,040 hooks)	1993–1996	*M. mola* 0.11–2.2% of total catch (by number) *Ma. lanceolatus* 0–0.34% of total catch (by number) Fate: discarded (condition not given)	Marin et al. 1998

South Pacific (E Australia)	Yellowfin tuna (NE Australia) Bluefin tuna (SE Australia)	Observer data (4,441,470 and 2,055,276 hooks, respectively)	1992–1997	*M. ramsayi* and *M. alexandrini* 0.48 % and 0.15% of total catch (by number), respectively Fate: ND	Ward et al. 2004
E Australia	Tuna and billfish	Observer data	2001–2013	*Mola* spp. and *Ma. lanceolatus* combined Annual mean catch rate of 0–4.5 sunfish per 1,000 hooks^{-1}, depending on area Fate: live release	Nyegaard et al. 2018a
New Zealand	Tuna and billfish	Observer data	2001–2013	*Mola* spp. (live release) Annual mean catch rate of 0–6.5 sunfish per 1,000 hooks^{-1}, depending on area Fate: live release	Nyegaard et al. 2018a
Indian Ocean basin	Bigeye tuna , albacore , yellowfin tuna, southern bluefin tuna	Observer data (14,121 hooks across the four fisheries	2004–2008	*M. mola* 0–0.4% of total catch (by number) across the four fisheries Discard rate of < 0.01–100 % across the four fisheries (at least some of these were likely live releases)	Huang and Liu 2010
Indian Ocean basin	Experimental, Spanish fleet	539 sets, 531,916 hooks	Unclear	*M. mola* No catch estimate, but noted as among 86 ton of discard (of a total catch of 1,162 ton)	Ardill et al. 2012*
Arabian Sea, Bay of Bengal, Andaman & Nicobar waters	Research	Total of 11,990,000 hooks across the three areas	2004–2010	*M. mola* Average across the three areas: 0.01 sunfish per 1,000 hooks^{-1}, 0.319 kg per 1,000 hooks^{-1}, 0.104% of total catch (by number) Fate: ND	Varghese et al 2013
Northern Gulf of Mexico	Swordfish/tuna	Observer data (effort not given)	1992–2005	*M. mola* (n = 11 individuals); *Ma. lanceolatus* (n = 45), unidentified Molidae (n = 265) Fate: ND	Fulling et al. 2007
South Africa	Tuna and swordfish	Observer data (ca. 231,000 hooks)	2000–2003	*M. mola* Mean annual hook rates 0.08–0.29 sunfish per 1,000 hooks^{-1}, varied between years and areas Fate: ND	Petersen 2005
Peru	Small-scale	Observer data (opportunistic monitoring, effort not given)	2005–2017	*Mola* spp. and *Ma. lanceolatus* Total of 14 bycatch observations; "…*sunfish by-catch events in longlines were scarce*…[but due to the overall effort in the fishery] *sunfish by-catch could still be considerable*…" Fate: released (generally not hauled)	Mangel et al. 2019
E Australia	Tuna and billfish	Total catch (commercial log books)	2010–2012, 2018 (10 months)	*M. mola, M. ramsayi, Ma. lanceolatus* 2010–2012: 273–498 individuals annually (average of 389) 2018: 95,438 kg (see text for details) Fate: live release	AFMA 2013, 2019
New Zealand	Tuna	Total catch (commercial log books)	2012–2016	*M. mola* 770–4,849 individuals annually (average of 2,560) Fate: live release	MPI 2016, 2017

*Original source of data unclear.

R. laevis by the Malgache crew on Reunion Island longliners only came to light when specimens for scientific studies became scarce and the reason was investigated; consumption, however, appears to be a recent trend (since 2015), with this species released and/or discarded by fishing crew in the past (E. Romanov, personal communication). *Ranzania laevis* is rare among Molidae landings in both Japan (Sagara and Ozawa 2002a), and Taiwan (Liu et al. 2009, C.-T. Chang, personal communication).

Reports of *R. laevis* fisheries bycatch are infrequent in the literature, but include the deep-set Hawai'i-based longline fishery for tuna (Curran and Bigelow 2011), where 15,675 kg were caught in 2005 (National Marine Fisheries Service 2011). Assuming an average weight of 5 kg (Smith et al. 2010), this would correspond to *ca.* 3,000 individuals. *Ranzania laevis* has also been reported in the American Samoan longline fishery, where a 2007–2011 study based on fisheries observer data found it to be the sixth most common species caught after Albacore (*Thunnus alalunga*), skipjack tuna (*Katsuwonus pelamis*), yellow fin tuna (*T. albacares*), Wahoo (*Acanthocybium solandri*) and bigeye tuna (*T. obesus*) (Watson and Bigelow 2014). A simple upscaling based on the average CPUE of 1.05 *R. laevis* per 1,000 hooks, based on the 2011 total effort (PIFSC 2012) yields *ca.* 11,300 *R. laevis* interactions across the fleet. The mortality is unknown; 99.6 percent was reportedly discarded but the state of the fish was not specified. *Ranzania laevis* is reportedly also caught in the tuna longline fisheries operating in the South Atlantic and Indian Ocean basins, where it is likewise a discard species (Lucena-Frédou et al. 2017).

Not all longline fisheries report high *R. laevis* bycatches. The annual bycatch of *R. laevis* apparently does not exceeds 100 individuals in the longline fishery off Reunion Island (E. Romanov personal communication), and was negligible in the shallow-set Hawai'i-based longline fishery in 2005 (National Marine Fisheries Service 2011). Furthermore, *R. laevis* was not reported by fisheries observers during observation of > 17 million longline hooks off Eastern Australia and New Zealand (2009–2013) (Nyegaard et al. 2018a), and was not noted during a study of Molidae bycatch in the Peruvian small-scale fisheries (gillnet and longline) (Mangel et al. 2019). *Ranzania laevis* was briefly mentioned as bycatch in the Mediterranean drift net fishery by Northridge (1991), and have also been reported from the tropical purse seine fisheries in the Atlantic and Indian Ocean basins (Escalle et al. 2019), but is not included in Fukofuka and Itano's (2007) guide to bycatch species in purse seine fisheries in the West Central Pacific.

Using Fisheries Data for Abundance Trends

Challenges in Fisheries Data Analysis

Estimates of global sunfish abundance have not been attempted, and studies of local abundance using aerial or ship-based surveys lack long-term trends (e.g., Thys and Williams 2013, Breen et al. 2017, Grémillet et al. 2017), hampering assessments of population trends both locally and globally. Fisheries data potentially offer valuable insights of relative abundance over time, but as for marine megafauna, collecting reliable bycatch information for large molids is generally challenging. Firstly, logbook data by fishers may not accurately reflect megafauna bycatch (Lewison et al. 2004), and while fisheries observer data are more reliable, they may lack broad coverage (Hall et al. 2017). Secondly, the assigned Molidae species identity/identities may be unreliable and the data may potentially include more than one species. Thirdly, fisheries data typically comprise total catch as either total weight or total numbers (of individuals), but rarely both, and rarely with information on fish sizes. Either of these formats present challenges given the impressive growth spectrum of the large molids. For example, a catch of 10,000 metric tons may constitute as little as four to five large individuals of > 2 tons each, or as many as several thousand small fish of < 1 kg. Conversely, a catch of 10,000 individuals would be of very different concern in a conservation context if comprising large, mature adults versus small immature fish (e.g., Hall et al. 2017). Furthermore, observer data seldom include sex and maturity for the large molids, as this would require killing the fish, countering intentions to release them alive.

For robust interpretation of fisheries data, dedicated studies may be needed to ascertain which species interact with the fishery, preferably including genetic analysis to verify species identity/identities (e.g., Mangel et al. 2019, Nyegaard et al. 2018a). Information on size compositions (e.g., Petersen and McDonell 2007, Baptista et al. 2018), as well as sex and maturity (Liu et al. 2009, Chang et al. 2018) are invaluable, although the latter is challenging to obtain outside commercial fish markets. Sex can at present not be determined based on external morphology, as sexual dimorphism is uncertain (Nakatsubo 2008, Sawai et al. 2018), and an assay for genetically or biochemically determining sex has yet to be developed for Molidae.

Catchability

CPUE time series are widely used as proxies for abundance trends of marine organisms, based on the assumption that the organism catchability remains constant over time (Forrestal et al. 2019). However, fisheries interactions and catchability are influenced by numerous factors and are exceedingly complex.

Fisheries interactions depend on the spatio-temporal behavior and movement of fish and fishing fleets in relation to one another. Several studies of large molids have confirmed affinities to fronts and upwelling systems (Thys et al. 2015, Sousa et al. 2016a, Breen et al. 2017, Hahlbeck et al. 2017), as well as seasonal movements between high and low latitudes (Dewar et al. 2010, Potter et al. 2011, Sousa et al. 2016, Breen et al. 2017), likely linked to thermal envelopes (Sousa et al. 2016a, Phillips et al. 2017). Within a fishery, the average annual sunfish CPUE may consequently be influenced by sunfish movements and aggregations, in relation to the spatio-temporal distribution of fishing effort.

The likelihood of interactions between sunfish and fishing gear further depends on the fishing depth in relation to sunfish water column occupancy. In longline fisheries, interactions also depend on fish feeding behavior to detect, locate and prey upon baited hooks (Hall et al. 2017). Sunfish water column occupancy is complex, but several studies have shown that *M. mola* predominantly forage below the thermocline during the day, interspersed by periods at or near the surface for thermoregulation purposes (Dewar et al. 2010, Nakamura et al. 2015, Thys et al. 2015, Sousa et al. 2016a). *M. mola* night-time depths are typically shallower than during the day with relative inactivity revealed by accelerometers in one study off Japan (Nakamura et al. 2015), while diurnal foraging was suggested for *M. mola* in a study in the North East Atlantic (Sims et al. 2009). Diurnal patterns in foraging are presumably linked to both local prey availability as well as sunfish size, with evidence pointing to ontogenetic changes in prey preferences with sunfish size (Nakamura and Sato 2014, Nakamura et al. 2015, Sousa et al. 2016b). While research on catchability of large molids is scarce, indications of diurnal differences in catches were noted by Gamblin et al. (2007) where *M. mola* were only caught at night on longlines off the Seychelles (n = 3; 25–75 m depth). Similarly, sunfish were only caught at night and early mornings in trawls off Tasmania (n = 5; 100–150 m depth; AFMA 2006). Further, depth dependent catchability of sunfish was demonstrated in the Ligurian swordfish fishery, where a sharp increase in sunfish bycatch (numerical percentage of the target catch) occurred after transitioning from surface to mesopelagic longlines (Garibaldi 2015). Furthermore, Cartamil and Lowe (2004) estimated a potential 20–25 percent reduction in sunfish bycatch could be achieved by increasing the minimum target depth to 20 m, thereby decreasing interactions in the surface waters.

Other studies have shown that factors such as longline soak time (Ward et al. 2004) and hook type (Fernandez-Carvalho et al. 2015) may affect sunfish CPUE. Furthermore, the potential for scavenging of fisheries discards by sunfish have been suggested by Phillips et al. (2020) as a way of explaining the occurrence of fast-moving prey in the diet of *M. mola*. Any such scavenging could potentially influence the overall catchability of sunfish within a fishery, if this feeding mode resulted in increased sunfish proximity to fishing gear and/or increased presence on fishing grounds.

In summary, within a fishery, molid CPUE can be expected to potentially vary according to factors such as (but not limited to) fishing depth, fishing area, time of day and time of year, according to fishing gear and fishing fleet behavior, as well as sunfish behavior on the fishing ground. An increased understanding of the catchability of large molids, including their ability (or inability) to escape fishing gear via exclusion devices, may offer avenues for developing bycatch mitigation strategies. Furthermore, knowledge of Molidae sensory biology, feeding ecology and behaviour could potentially form the basis for deterrent-based bycatch mitigation practices, as has been explored for other marine megafauna, for example through targeted use of light and sound (e.g., Jordan et al. 2013).

CPUE Heterogeneity—a *Mola mola* Case Study from Spain

Large molids (recorded as *M. mola*) are caught in the Spanish longline fishery for large pelagics in the Western Mediterranean Sea and are recorded as part of the fisheries observer program run by the Spanish Institute of Oceanography. Across the fishing ground, *M. mola* bycatch data (2009–2018; *ca.* 8.25 million hooks) reveal both temporal and spatial heterogeneity in average sunfish CPUE. For example, a CPUE 'hot spot' is apparent in the Tortosa Canyon (Fig. 6) where the average monthly CPUE is elevated during March–October with a distinct peak in April; a temporal pattern not evident across the wider fishing ground (Fig. 7A).

The Spanish longline fishery consists of three main métiers. Métiers are a group of fishing operations targeting a specific assemblage of species, using a specific gear, during a precise period of the year and/or within the specific area (Deporte et al. 2012). The three Spanish longline métiers differ in regard to target depth, namely surface (15–40 m), mesopelagic (150–700 m) and bottom longlining, with the latter typically targeting the edge of the continental shelf at 150–200 m. Comparatively little observer data exist for the latter métier, but a simple comparison reveals that the average *M. mola* bottom longlining CPUE (2009–2018) is higher than in the other métiers, both in the Tortosa Canyon and across the wider fishing ground (Fig. 7B).

In the Tortosa Canyon the surface longline fishery comprises several sub-métiers according to the target species, which include bluefin tuna (*T. thynnus*) and swordfish (*Xiphias gladius*). While the target depth and gear are superficially similar between these two sub-métiers, the trend in annual average *M. mola* CPUE (2009–2018) is substantially different (Fig. 7C). The decreasing sunfish CPUE in the swordfish sub-métier is, for example, not replicated in the bluefin tuna sub-métier. Overall, the Spanish longline fishery in the Western Mediterranean is highly heterogenous (Garcia-Barcelona et al. 2010) and exemplifies the importance of considering such heterogeneity when evaluating temporal CPUE trends for conservation or management purposes. If analyzed carefully, detailed observer bycatch data sets may help shed light on various aspects of Molidae ecology such as habitat preferences, diel migrations, and potential oceanographic influences on seasonality (e.g.,

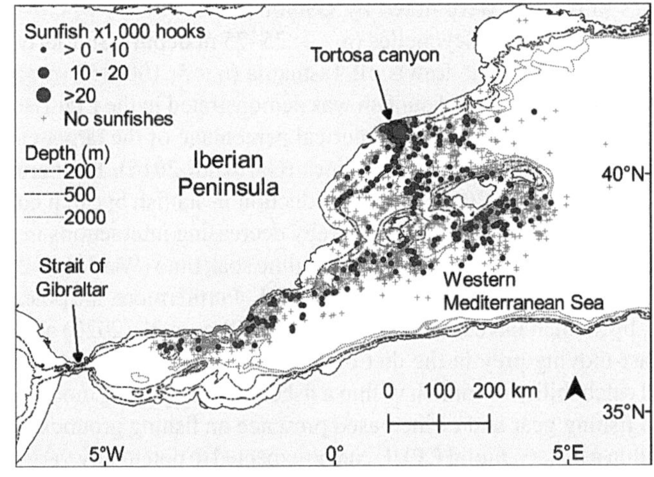

Figure 6. (A) *Mola mola* Catch per Unit Effort (CPUE) (sunfish per 1,000 hooks) in individual sets in the extant Spanish longline fishery for large pelagics in the Western Mediterranean Sea (2009 – 2018). *Source*: Fisheries observer data (~ 8.25 million hooks) from the Spanish Institute of Oceanography.

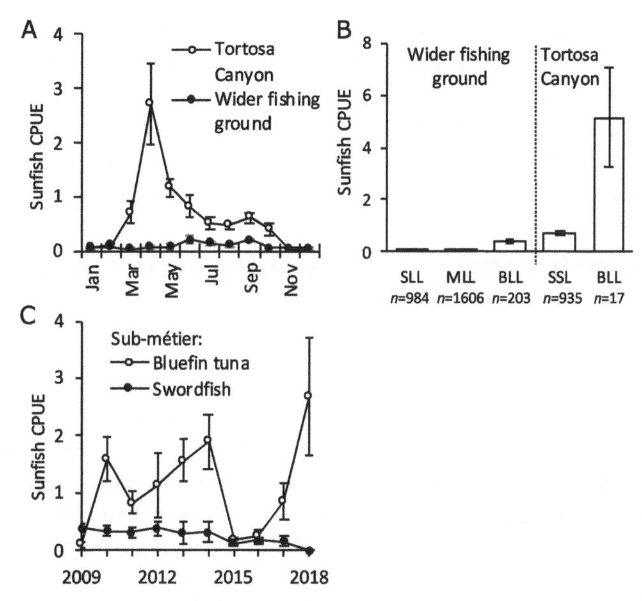

Figure 7. *Mola mola* bycatch in the extant Spanish longline fishery for large pelagics in the Western Mediterranean Sea (2009 – 2018), shown as A) Average monthly Catch per Unit Effort (CPUE; sunfish per 1,000 hooks ±SE) in the Tortosa Canyon and across the wider fishing ground; B) Average sunfish CPUE (±SE) in the surface- (SLL), bottom- (BLL) and mesopelagic- (MLL) longline métiers in the Tortosa Canyon and across the wider fishing ground; C) Average annual sunfish CPUE (±SE) in the SLL sub-métiers targeting bluefin tuna and swordfish, respectively, in the Tortosa Canyon. Source: Fisheries observer data from the Spanish Institute of Oceanography.

Hahlbeck et al. 2018). Coupled with dedicated surveys to examine sunfish species and demography, existing fisheries observer data could provide a powerful avenue for advancing our understanding of Molidae ecology.

Trends in Molidae Fisheries Interactions

The paucity of long-term sunfish-fisheries interaction data in the literature renders assessment of sunfish vulnerability to fishing pressure challenging. In lieu of catch data, the *M. mola* IUCN assessments highlighted abrupt declines in Molidae landings as a cause for concern (Fig. 4; Liu et al. 2015a). Such abrupt declines could potentially point to stock collapse; however, other factors may also be at play. For example, market dynamics, target species availability, fishing vessel storage capacity and distance to port may impact on the profitability of landing sunfish (Huang and Liu 2010). Indeed, sunfish are not always landed when caught in the Taiwanese tuna longline fishery in the Indian Ocean basin (op. cit.) despite the existence of a domestic market. Furthermore, regulatory changes may impact landings, as exemplified in the Camogli trap-net fishery in the Mediterranean, where European Commission regulation in 2004 resulted in discontinuation of commercial Molidae landings.

Of further concern in the IUCN assessment was the abrupt decline in sunfish CPUE in the South African trawl fishery for horse mackerel observed by Petersen and McDonell (2007). Here, longer time series would be invaluable to gain an understanding of inter-annual variability and long-term trends. For example, both short-term decreases and increases in sunfish CPUE were seen in the Australian (2001–2013) and New Zealand (2001–2015) longline fisheries, with no significant increasing or decreasing trends over time (Nyegaard et al. 2018a). Furthermore, the historical *M. mola* catch data from the Camogli trap-net fishery (1950–1974) (which may be considered CPUE; Relini et al. 2010), likewise varied substantially between 5–36 ton annually but with no long-term decline (Relini and Vallarino 2016). The seasonal CPUE in the Californian driftnet fishery also fluctuated widely over time

(94–1,354 sunfish per 100 nets; Fig. 5). Here, further analysis is required to ascertain if the decrease in recent years may be an artefact of the decrease in overall effort (and thereby observer effort; NOAA 2019), combined with the spatio-temporal heterogeneity in sunfish bycatches (Hahlbeck et al. 2017). Long term data sets are also invaluable for examining the complex interconnection between population trends and broad scale climatic phenomena such as for example the North Atlantic Oscillation (e.g., Condon et al. 2013) and the El Nino Southern Oscillation (e.g., Castillo et al. 2018).

Potential Impacts of Fisheries Interactions

Resilience

Traditional stock assessment methods are often difficult to apply to marine megafauna due to data deficiency (Lewison et al. 2004), but alternative approaches may be applied to gauge resilience. For example, applying the fuzzy logic expert system (Cheung et al. 2005),'extinction vulnerabilities to fisheries' were considered "high–very high" for the large molids and "moderate" for *R. laevis* by Froese and Pauly (2019). Application of another assessment method for data poor stocks, the Productivity and Susceptibility Analysis (Hobday et al. 2011), found *R. laevis* to rank among the top ten stocks at risk (of 60 evaluated stocks) caught in the tuna longline fisheries in the South Atlantic (Lucena-Frédou et al. 2017). The same study found *M. mola* to be 'medium risk' and *R. laevis* to be 'low risk' in the longline fisheries in the South Atlantic and Indian Ocean, respectively. However, such approaches typically require knowledge of key life history parameters. For Molidae, many life history characteristics are uncertain, including early life history, growth rates (outside captivity), age at maturity, spawning aggregation behaviour and mortality rates, particularly at species-level, as also noted by Lucena-Frédou et al. (2017). However, the large molids clearly do not fit the typical "k-bycatch" profile of other marine megafauna (Hall et al. 2017), as the combination of extremely high fecundity and large adult size (Pope et al. 2010), are typically associated with high and low resilience to fishing pressures, respectively (Lewison et al. 2004, Cheung et al. 2005).

To aid in gauging vulnerability to fishing pressures, the IUCN assessment for *M. mola* estimated adult mortality (M) (Then et al. 2015, 2018) and generation length (= 1/M + age at first maturity) based on inferences from data available at the time (longevity of 20–23 years; age at first maturity of 5–7 years), yielding a generation time of 8–10 years (Liu et al. 2015a). The accuracy of this estimate is unknown and should be viewed with caution. For example, using data for female *Ma. lanceolatus* (longevity: 105 years; age at first maturity: 20–21 years; Chang et al. 2018, Liu et al. 2009) yields a generation time of 35–36 years. Aging large molids is challenging due to the otolith size and structure (Nolf and Tyler 2006), and while vertebrae growth zone chronology and growth of captive specimens over time have been used to age *Ma. lanceolatus* (Liu et al. 2009) and *M. mola* (Nakatsubo 2008), verification studies using other techniques based on non-captive sunfish would be highly valuable.

With important new insights of Molidae biology and ecology continuing to emerge from research groups across the globe, including determination of important life history traits, a robust re-evaluation of Molidae vulnerability to anthropogenic pressures is increasingly achievable.

Fisheries Related Mortality

A critical aspect of gauging the resilience of Molidae to fisheries interactions is understanding fisheries related mortality (Fig. 1). While direct (observable) mortality can be quantified onboard each fishing vessel, fisheries interactions also include, 'cryptic' deaths which are not easily observed, such as post-release mortality resulting from stress and injury sustained during fishing gear interactions (Cartamil and Lowe 2004, Gilman et al. 2013). Other causes of cryptic mortality include depredation by predators such as sharks and marine mammals (MacNeil et al. 2009), and ghost fishing from interactions with abandoned, discarded or lost fishing gear (Gilman et al. 2013). Cryptic mortalities are complex and challenging to quantify, and the focus here is on direct and post-release mortality, with brief comments on depredation.

Direct and Post-Release Mortality

Fisheries related mortality in Molidae bycatch is largely unknown. In longline fisheries, large molids may be released alive in-water with little handling, by cutting off the branch line and leaving the hook in place (S. Hall, personal communication; AFMA 2011), potentially resulting in high post-release survival. Large molids are indeed released alive (Nyegaard et al. 2018a, Mangel et al. 2019) and in good condition (Garibaldi 2015) in some longline fisheries (e.g., in Australia, New Zealand, Peru, Western Mediterranean), although in many studies such information is not available (e.g., Marin et al. 1998, Petersen 2005, Huang and Liu 2010, National Marine Fisheries Service 2011). Satellite tagging of four *M. mola* (Thys et al. 2015) and one *Ma. lanceolatus* (Seitz et al. 2020) caught on longlines (the former during scientific cruises) indicate post-release survival is possible, as evidenced by tag deployments between 90–241 days and 61 days respectively, however, bycatch handling can vary dramatically from ship to ship. Consequently, assuming universally high survival rates is risky considering longline fisheries cover much of the world's ocean and combined, represent enormous fishing effort (Kroodsma et al. 2018).

In driftnet fisheries, physical trauma from gear interaction and air exposure is likely more prevalent than during longlining. Nevertheless, direct sunfish mortality appears to be low, with sunfish bycatch reportedly released alive in some fisheries (Silvani et al. 1999, Mangel et al. 2019), although many reports are unclear in regard to the fate and state of sunfish bycatch (e.g., Tudela et al. 2005, McKinnell and Seki 2017, Alfaro-Shigueto et al. 2018). In the Californian driftnet fishery, most sunfish are released alive, but *ca.* 40 percent show obvious signs of fishery-induced trauma such as loss of protective mucus coating, abrasions and bleeding, with unknown post-release survival (Cartamil and Lowe 2004). Other factors than physical interaction with the fishing gear may influence survival; in the Spanish driftnet fisheries in the late 1980s it was not uncommon to leave sunfish bycatch on deck from the first set of the day, to avoid catching them again in the second set. This practice resulted in dead discards from the first set and live releases from the second set (S. García-Barcelona unpublished data). Combined with the high bycatch rates reported from some drift net fisheries, the potential for relatively high post-release mortality render this fishing sector of some concern in the context of anthropogenic impacts on the Molidae. While a moratorium was introduced on high seas large-scale pelagic driftnetting in 1992 (McKinnell and Seki 2017), medium- and small-scale driftnet effort (combined with illegal driftnetting) remain high in some places, including the Mediterranean and off Peru (Lucchetti et al. 2017, Alfaro-Shigueto et al. 2018, Mangel et al. 2019).

Survival of large molids in trawl fisheries may be relatively low. Video from trawl nets with grid-like exclusion devices reveal that large molids remain trapped against these during the fishing operation and that mortality is potentially high (AFMA 2006, Zeeberg et al. 2006). In nets without exclusion devices, sunfish are presumably hauled onboard with the catch, further increasing the risk of mortality. For example, in the South African horse mackerel (*Trachurus capensis*) fishery, only 11.8 percent of large bycatch species (including large molids) were released alive, with 6.1 percent of these deemed unlikely to survive due to injuries (Petersen and McDonell 2007).

In purse seine fisheries, the release of megafauna typically depends on the actions of the fishers (Hall et al. 2017), however, it is uncertain if the large molids count among the megafauna routinely released, and if so, if the release methods are successful. If hauled onboard with the catch, sunfish survival presumably further depends on handling and the duration out of the water, as seen in silky sharks (*Carcharhinus falciformis*), and suggested for devil rays (*Mobula japonica*) in the New Zealand purse seine fishery (Francis and Jones 2017). For the latter, post-release mortality occurred in four of nine devil rays despite all swimming away vigorously.

In the Camogli trap-net fishery, sunfish are released alive as part of the operational permit for fishing in the Marine Protected Area. Accelerometer data (n = 3) have demonstrated survival

Figure 8. Sunfish bycatch in the extant trap-net fishery off La Azohía, Murcia, in Spain, released alive using boathooks (2012). Image by Salvador G. Barcelona, reprinted with permission.

for at least 48 hr post-release, at which time the data loggers were set to release from the fish (N. Phillips unpublished data). In other trap-net fisheries in the Mediterranean, live release of *M. mola* is presumably common (Storai et al. 2011), but the extent of direct mortality and degree of physical injury from entanglement (Addis et al. 2013) and handling is uncertain, as is post-release mortality. For example, sunfish bycatch is commonly released alive from Spanish trap-net fisheries with the use of boathooks (S. Garcia-Barcelona, personal observation, Fig. 8), with unknown consequences for survival.

In addition to using acceleration data loggers (Whitney et al. 2016), post-release mortality of large molids may be investigated through the use of satellite tags applied prior to release, as has successfully been done for other marine megafauna (e.g., Francis and Jones 2017). In both cases, an important consideration for robust results is ensuring minimal changes to the routine bycatch handling and release process during tagging.

Depredation

Depredation in a fisheries context refer to consumption (partial or whole) of marine organisms caught in fishing gear for example, by predators such as sharks and toothed cetaceans (Mitchell et al. 2018). Limited information is available on depredation on Molidae, but in a study in the NW Atlantic pelagic longline fishery, depredation on *M. mola* was seen only once in 684 sunfish observations, yielding similarly low depredation levels to those of hooked sandbar sharks (*Carcharhinus plumbeus*) and tiger sharks (*Galeocerdo cuvier*) (MacNeil et al. 2009). Ward et al. (2004) found a positive correlation between soak time and sunfish bycatch in the longline fisheries off Eastern Australia, suggesting that capture rates outweigh loss rates. As nearly all sunfish are alive when retrieved in this fishery (Nyegaard et al. 2018a), they are presumably able to avoid predators while hooked (Ward et al. 2004). Alternatively, large molids may comprise less desirable prey compared with for example, tuna (*Thunnus obesus*, *T. albacares*, *T. atlanticus*, *T. thynnus*), for which depredation levels ranged between 4.1–7.4 percent in the same study (MacNeil et al. 2009).

Species Distribution

Species Distribution Models

Effective marine conservation, encompassing the potential impacts of fisheries interactions (Moore et al. 2013), is often limited by a lack of data on the spatial distribution and biogeography of aquatic taxa (Ricklefs 2004, Rushton et al. 2004). This challenge is often exacerbated by the difficulties associated with studying widespread or cryptic species (Pearson et al. 2007). In such cases, species distribution models (SDM) can provide insights into species ecology, even from limited occurrence records, and aid in assessing broad scale temporal and spatial patterns, predicting habitat suitability and identifying environmental drivers of distribution (e.g., Elith et al. 2006, Elith and Leathwick 2009, Franklin and Miller 2010). For large molids, SDMs have been used to assess distribution from regional to global scales where data are scarce (Breen et al. 2017, Phillips et al. 2017). These models have indicated areas of favourable habitat and annual distribution shifts across varying spatio-temporal scales, which can be linked to highly valuable predictors including (amongst others); thermal envelopes, mixed layer depths and seasonality. However, there are potential limitations of such model predictions which must be considered, and future empirical validation is highly recommended as more data become available, for example through expanded species-specific sightings schemes, satellite telemetry or capture and release studies.

Applying SDM to sightings data presents challenges, requiring careful selection of relevant variables and appropriate methods (Elith and Leathwick 2009). Despite the potential limitations, if sightings data are recorded accurately with suitable metadata, in models that account for sampling bias, this approach can provide broad scale insights beyond the scope of most alternative methods, such as dedicated surveys or satellite tagging programs (Elith et al. 2006, Elith and Leathwick 2009, Danielsen et al. 2014). In this way, SDMs are an important tool for ecologists to predict species distribution, inform conservation management strategies and predict overlaps with fishing operations (e.g., Moore et al. 2013). Future developments in this field would greatly benefit from additional data however, realistically, the dedicated collection of detailed, broad scale datasets on non-commercial species is unlikely. To further developments, it would instead be beneficial to focus on boosting collaborative efforts with statutory monitoring surveys or by highlighting specific areas where researchers may be able to gather smaller, but highly detailed datasets.

Advancing Species Distribution Models

Here, we set out three main recommendations to advance data collection in support of species distribution models:

Firstly, collection of species-specific data for validation of models, and to compare and contrast Molidae distribution patterns is highly recommended. To date, species-specific studies have been lacking due to the limited morphological diversity between *Mola* species, particularly at smaller sizes, which results in most large molids being recorded at genus level. This makes species level research highly challenging, particularly when undertaking analysis of historical sightings reports. Opportunistic genetic sampling of sunfishes during statutory monitoring programs may offer the opportunity to clearly define species encounters and validate modelling efforts, however large-scale sampling programs can be logistically challenging, time consuming and cost prohibitive. By developing a standardized sample collection protocol to be undertaken during routine monitoring surveys, it would be possible to build a database of samples (referenced by species and location) with which to improve our understanding of species-specific distribution patterns.

Secondly, additional surveys or sightings programs are strongly recommended across data deficient regions, as it has been noted that data are mostly limited to coastal regions in the North Atlantic and the North East Pacific. There are significant challenges associated with expanding such data collection, as research involving offshore surveys is often prohibitively expensive, but again, it may be possible to incorporate collection of sunfish sightings opportunistically. Such collaborative efforts may provide the opportunity to obtain information from data deficient regions through statutory monitoring and commercial surveys (e.g., Breen et al. 2017).

Thirdly, a key recommendation here is to assess sunfish distribution at depth. This is a highly challenging task, however data on sunfish distribution to date is almost entirely based on surface sightings which only represent a small proportion of the sunfish's daily use of the water column. In light of this, even small-scale studies at depth would be illuminating for modelling efforts. In order to address this concern, smaller dedicated research teams could use satellite tracking and biologging devices to provide detailed data of fish movement patterns at depth that can feed back with ocean layer mapping into broad scale modelling.

Taken together, these recommendations have the potential to facilitate data collection and greatly improve our understanding of sunfish distribution patterns. If collaborations can be created with existing monitoring programs that routinely survey for commercial species, a database can be built for understudied species across broad spatial and temporal scales. By combining such data with outputs from smaller, more detailed telemetry and biologging studies, there is potential to build highly accurate models and offer a better understanding of sunfish spatial ecology.

Climate Change and Range Shifts

Anthropogenically induced climate change is one of the greatest challenges currently affecting global ecosystems (IPCC 2007, Hoegh-Guldberg et al. 2018), however, predicting the impacts of such rapid, broad scale changes is highly complicated and assessing the potential effects on data-poor pelagic species presents a significant hurdle (e.g., McMahon and Hays 2006, Newson et al. 2009). Broadly speaking, the effects of climate change in open ocean areas are predicted to include rising sea temperatures, lower oxygen levels and shoaling of oxygen minimum zones, changing ocean currents and ocean acidification (IPCC 2007, Doney et al. 2012, McCauley et al. 2015). Under such climate change scenarios, significant changes for marine life will occur, both positive and negative, however it is difficult to predict species-specific effects and so here we consider the overall robustness of the Molidae in light of what is known about their thermal capacity and tolerance for low oxygen environments.

The thermal tolerance of the large molids appears relatively broad, suggested to range from 1.8–27.5°C (Dewar et al. 2010, Nakamura et al. 2015, Thys et al. 2015, Sousa et al. 2016a), however within this range, it appears the *Mola* genus may occupy an ideal thermal envelope of 16–23°C (Phillips et al. 2017), with suggested slight inter-species differences (Sawai et al. 2011, Nyegaard et al. 2018a). Whilst a thorough empirical assessment of thermal optima within the Molidae family is required to fully investigate potential climate driven changes, we suggest that a global rise in ocean temperatures would be highly likely to affect the distribution and habitat use of sunfishes. An increase in temperature may encourage population movements, driving fishes to higher latitudes or greater depths to maintain their preferred thermal range. The predicted rise in water temperature may alter broad scale ocean currents and reduce the oxygen carrying capacity of water, particularly below the thermocline (e.g., IPCC 2007, Bollmann et al. 2010). Such significant changes are likely to alter sunfish distribution patterns further as noted with other fishes, such as restricting foraging dives to shallower waters to avoid deeper hypoxic conditions (Coffey et al. 2017). Although *M. mola* have been noted to briefly venture into hypoxic zones at 60 m (Thys et al. 2015), modelling of the *Mola* genus suggests that sunfishes occupy dissolved oxygen levels between 5–7 ml/L (Phillips et al. 2017). Recovery in well-oxygenated waters is required following exposure to hypoxic conditions, rendering access to high levels of oxygen critical for this to remain a possibility (Cartamil and Lowe 2004). If low oxygen levels occur for sustained periods, this may also extend 'ocean dead zones' with significant impacts for benthic feeding species in such areas (Diaz and Rosenberg 2008).

While it is difficult to predict how climate change will affect sunfishes directly, such rapid changes in the marine environment are of serious concern for all marine life. It should also be considered that climate change is likely to affect the distribution of fishing efforts, putative prey species and many other factors that may have knock-on effects for sunfish abundance and distribution patterns. Although marine life is fundamentally tolerant to variable environmental conditions and can adapt to brief extreme situations, it must be kept in mind that it is the cumulative effects of climate change that may push species beyond their limits (Doney et al. 2012).

Other Conservation Issues

Information relating to non-fisheries conservation issues among the Molidae is scarce. Fishing related mortality was the only threat listed in the IUCN assessment for *M. mola* (Liu et al. 2015a), while no (known) threats were listed for *Ma. lanceolatus* (Leis et al. 2015) and *R. laevis* (Liu et al. 2015b).

Plastic debris is a rapidly emerging issue of serious and widespread concern for marine species and ecosystems (Markic et al. 2018). For the large molids, the superficial similarity between soft, translucent plastics and gelatinous zooplankton plausibly exposes them to risks of plastic ingestion, however, limited gut content studies in Molidae leaves this unverified (e.g., Bakenhaster and Knight-Gray 2016). Furthermore, the rapid increase of microplastics in the world's oceans could potentially pose a risk to large molids through bioaccumulation of toxins (see Baptista et al. 2020 [Chapter 11]). Other potentially large-scale threats include increased shipping traffic globally, which exposes the large molids to risks of ship strikes during sea surface occupancy. Strikes may occur by small boats and large commercial vessels alike (e.g., Nyegaard and Sawai 2018), as well as by yachts, as exemplified by the sunfish collisions that occur nearly every year during the Sydney to Hobart race (e.g., The Daily Telegraph 2019). However, while strikes certainly occur (e.g., Fig. 9), the extent is not known. Likewise, underwater noise pollution such as seismic airguns, sonar and ship engines, is a widespread concern for marine megafauna in the Anthropocene, and may negatively impact marine fish physically, physiologically and behaviorally (e.g., Carroll et al. 2017). However, noise pollution is complex, and the effects have not yet been investigated for Molidae.

Nature-based tourism targeting charismatic megafauna is an increasingly popular activity globally, but may potentially cause detrimental impacts to the targeted populations if left unmanaged (e.g., Bejder et al. 2006, Higham et al. 2015). Sunfish tourism is currently limited on a global scale; instead sunfish generally make for welcome sightings, for example during whale watching tours. However, sunfish-specific tourism does exist, as for example off Bali in Indonesia (Thys et al. 2016), where sunfish make up an important drawcard for the local SCUBA diving tourism (Nyegaard 2018). The sunfish appear on a seasonal basis on the reefs off the small island group of Nusa Penida near Bali (*Op. Cit.*), where they seek cleaner fish interactions for removal of skin parasites (Konow et al. 2006). While little is known of the importance of these cleaning events to the individual sunfish, it is plausible that repeated disturbances by SCUBA divers may lead to population level impacts through overall decreased fitness associated with increased parasite loads. With little information on the relative importance of the tourism targeted reefs for sunfish-cleaner fish interactions, and the potential existence of alternative, undisturbed cleaning areas, it is at present difficult to gauge the level of tourism related impacts (e.g., Nyegaard 2018), but the potential is certainly there.

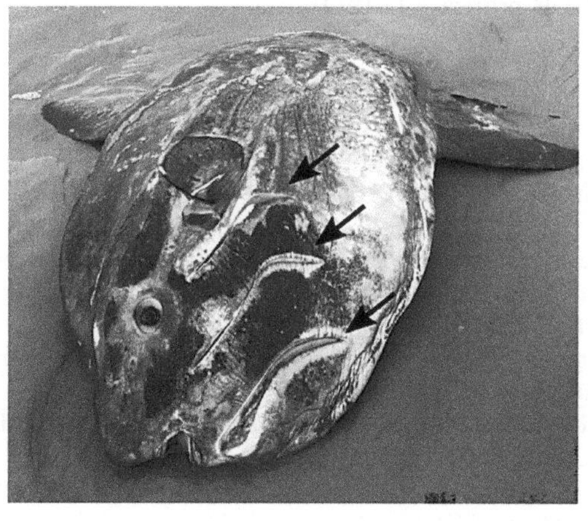

Figure 9. *Mola mola* with propeller injuries (black arrows), found dead in November 2018 in North Carolina. Image by Anita McLeod, reprinted with permission.

Remaining Questions

The global extent and scale of Molidae fisheries interactions are poorly understood. While generally not retained, post-release mortality is unknown. Coupled with uncertainties regarding life history traits and ecological constraints, a robust risk-based assessment of the large Molidae is difficult, in particular at species level. Furthermore, with limited understanding of global sunfish abundance and ecological importance as the large molids grow from small larvae to enormous adults, their significance in marine food webs is uncertain, and the consequences of (potentially) reducing their abundance in coastal and oceanic ecosystems via fisheries interactions is unknown (Pope et al. 2010, Syväranta et al. 2012). In light of these uncertainties, key areas for future research to advance our understanding of the conservation status of the Molidae include (1) Estimates of post release mortality in key fisheries, including trap-net, drift net, purse seine, longline and trawl fisheries; (2) Establishment of species level life history traits, including size at maturity, fecundity, larval and juvenile survival and growth rates in the wild; (3) Species-level range and migrations, and critical spawning areas and times to enable assessments of overlap with fisheries; (4) Detailed analysis of existing fisheries bycatch data, coupled with targeted studies of species composition, sex ratios and demography to support investigations of catch trends over time. Underpinning such research efforts is the need for taxonomic clarity in the Molidae, to allow robust species-specific data collection in both fisheries and research scenarios. Furthermore, a clear and user-friendly field guide for species identification is needed to facilitate collection of species-level data from sunfish from larvae to giants. Development of genetic or biochemical assays for sex determination of live sunfish is also needed. To achieve these goals, global collaboration and knowledge sharing are crucial. While modern technology increasingly provides excellent platforms for communication, personal connections between researchers, fostered for example through sunfish workshops, symposiums and large collaborative projects, are invaluable to enable globally focused research of the cosmopolitan Molidae. A global perspective on Molidae research efforts will be particularly beneficial in continuing to increase our knowledge on a species level, to better gauge the resilience of the Molidae in the Anthropocene, as well as to understand their role in the world's changing marine ecosystems.

Acknowledgments

We are thankful to Miguel Baptista, Ching-Tsun Chang, Daniel Cartamil, Heidi Dewar, and Evgeny V. Romanov for sharing valuable insights and knowledge, and to several fisheries observers at the Spanish Institute of Oceanography (IEO). The collection of IEO data was supported by two projects carried out by the IEO centre in Malaga, GPM16 (IEO) and the National Basic Data Program of the Spanish fishing sector (PNDB project, EU-IEO).

References

Addis P., M. Secci and Cau, A. 2013. The effect of Mistral (a strong NW wind) episodes on the occurrence and abundance of Atlantic bluefin tuna (*Thunnus thynnus*) in the trap fishery of Sardinia (W Mediterranean). Sci. Mar. 77: 419–427.

AFMA. 2006. Dolphin and seal interactions with mid-water trawling in the Commonwealth small pelagic fishery, including an assessment of bycatch mitigation. Final report R05/0996. Australian Fisheries Management Authority, Canberra, ACT.

AFMA. 2011. Australian Tuna and Billfish Fisheries bycatch and discarding workplan 2011–2013. Australian Fisheries Management Authority, Canberra, ACT.

AFMA. 2013. AFMA submission for Reassessment of the Eastern Tuna and Billfish Fishery 2014. Australian Fisheries Management Authority, Canberra, ACT.

AFMA. 2019. AFMS submission for reassessment of the Eastern Tuna and Billfish Fishery 2019. Australian Fisheries Management Authority, Canberra, ACT.

AFMA. 2019. SPF catch data. https://www.afma.gov.au/fisheries/small-pelagic-fishery/spf-catch-data. Accessed 20 Oct 2019.

Alfaro-Shigueto, J., J.C. Mangel, J. Darquea, M. Donoso, A. Baquero, P.D. Doherty et al. 2018. Untangling the impacts of nets in the southeastern Pacific: Rapid assessment of marine turtle bycatch to set conservation priorities in small-scale fisheries. Fish Res. 206: 185–192.

Ardill, D., D. Itano and R. Gillett. 2012. A review of bycatch and discard issues in Indian Ocean tuna fisheries. 7th Session of the Working Party on Environment and Bycatch, Indian Ocean Tuna Commission, Maldives.

Atapada, Z., W. Prayitno, M. Hilmi, C. Holeng, P.L.K. Mustika and P. Pet-Soede. 2004. Manta harvesting in the Alor and Solor Waters in Eastern Indonesia – a report of monitoring activities and recommendations for follow-up. World Wide Foundation Indonesia.

Bakenhaster, M.D. and J.S. Knight-Gray. 2016. New diet data for *Mola mola* and *Masturus lanceolatus* (Tetraodontiformes: Molidae) of Florida's Atlantic coast with discussion of historical context. Bull. Mar. Sci. 92: 497–511.

Baptista, M., A. Couto, J.R. Paula, J. Raimundo, N. Queiroz, R. Rosa et al. 2018. Seasonal variations in the abundance and body size distribution of the ocean sunfish *Mola mola* in coastal waters off southern Portugal. J. Mar. Biol. Assoc. United Kingdom 1–7.

Baptista, M., C. Figueiredo, C. Lopes, P.R. Costa, J. Dutton, D.H. Adams et al. 2020. Biotoxins, trace elements and microplastics in the ocean sunfishes (Molidae). pp. 186–215. *In*: T.M. Thys, G.C. Hays and J.D.R. Houghton [eds.]. The Ocean Sunfish: Evolution, Biology and Conservation, CRC Press. Boca Raton, FL, USA.

Barnes, R.H. 1996. Sea Hunters of Indonesia. Fishers and Weavers of Lamalera. Clarendon Press, Oxford.

Bejder, L., A. Samuels, H. Whitehead, N. Gales, J. Mann, R. Connor et al. 2006. Decline in relative abundance of bottlenose dolphins exposed to long-term disturbance. Conserv. Biol. 20: 1791–1798.

Bollmann, M., T. Bosch, F. Colijn, R. Ebinghaus, R. Froese, K. Gussow et al. 2010. The world oceans, global climate drivers. *In*: World ocean review: Living with the oceans. Maribus GmbH, Hamburg, pp. 1–234.

Breen, P., A. Cañadas, O.Ó. Cadhla, M. Mackey, M. Scheidat, S.C.V. Geelhoed et al. 2017. New insights into ocean sunfish (*Mola mola*) abundance and seasonal distribution in the northeast Atlantic. Sci. Rep. 7: 2025.

California Legal Information. 2018. Senate Bill No. 1017, Chapter 844. https://leginfo.legislature.ca.gov/. Accessed 15 Jun 2019.

Carroll, A.G., R. Przeslawski, A. Duncan, M. Gunning and B. Bruce. 2017. A critical review of the potential impacts of marine seismic surveys on fish & invertebrates. Mar. Pollut. Bull. 114: 9–24.

Cartamil, D.P. and C.G. Lowe. 2004. Diel movement patterns of ocean sunfish *Mola mola* off southern California. Mar. Ecol. Prog. Ser. 266: 245–253.

Castillo, R., L. Dalla Rosa, W. García Diaz, L. Madureira, M. Gutierrez, L. Vásquez et al. 2018. Anchovy distribution off Peru in relation to abiotic parameters: A 32-year time series from 1985 to 2017. Fish Oceanogr. 28(4): 389–401.

Cattaneo-Vietti, R., V. Cappanera, M. Castellano and P. Povero. 2015. Yield and catch changes in a Mediterranean small tuna trap: A warming change effect? Mar. Ecol. 36: 155–166.

Ceyhan, T. and O. Akyol. 2014. On the Turkish surface longline fishery targeting swordfish in the Eastern Mediterranean Sea. Turkish J. Fish. Aquat. Sci. 14: 825–830.

Chang, C.-T., C.-H. Chih, Y.-C. Chang, W.-C. Chiang, F.Y. Tsai, H.-H. Hsu et al. 2018. Seasonal variations of species, abundance and size composition of molidae in Eastern Taiwan [臺灣東部海域翻車魨科種類、數量與體長的季節性變動研究]. J. Taiwan Fish. Res. 26: 27–42. Chinese with English abstract.

Cheung, W.W.L., T.J. Pitcher and D. Pauly. 2005. A fuzzy logic expert system to estimate intrinsic extinction vulnerabilities of marine fishes to fishing. Biol. Conserv. 124: 97–111.

Chin, C.-T. 2008. Observation of Ocean Sunfish Festival Fever at Hualien and Its Conflict Issues [花蓮曼波魚季熱潮觀察及引發之衝突議題研究]. Master thesis, National Donghua University, China. Chinese with English abstract.

Clarke, S., M. Sato, C. Small, B. Sullivan, Y. Inoue and D. Ochi. 2014. Bycatch in longline fisheries for tuna and tuna-like species. A global review of status and mitigation measures. Food and Agriculture Organisation of the United States (FAO), Rome.

Coffey, D.M., A.B. Carlisle, E.L. Hazen and B.A. Block. 2017. Oceanographic drivers of the vertical distribution of a highly migratory, endothermic shark. Sci. Rep. 7: 1–14.

Condon, R.H., C.M. Duarte, K.A. Pitt, K.L. Robinson, C.H. Lucas and S.H.D. Haddock. 2013. Recurrent jellyfish blooms are a consequence of global oscillations. Proc. Natl. Acad. Sci. USA 110: 1000–1005.

Curran, D. and K. Bigelow. 2011. Effects of circle hooks on pelagic catches in the Hawai'i-based tuna longline fishery. Fish. Res. 109: 265–275.

Danielsen, F., K. Pirhofer-Walzl, T.P. Adrian, D.R. Kapijimpanga, N.D. Burgess, P.M. Jensen et al. 2014. Linking public participation in scientific research to the indicators and needs of international environmental agreements. Conserv. Lett. 7: 12–24.

Deporte, N., C. Ulrich, S. Mahévas, S. Demanèche and S. Bastardie. 2012. Regional métier definition: a comparative investigation of statistical methods using a workflow applied to international otter trawl fisheries in the North Sea. ICES Journal of Marine Science 69(2): 331–342,

Dewar, H., T. Thys, S.L.H. Teo, C. Farwell, J. O'Sullivan, T. Tobayama et al. 2010. Satellite tracking the world's largest jelly predator, the ocean sunfish, *Mola mola*, in the Western Pacific. J. Exp. Mar. Bio. Ecol. 393: 32–42.

Diaz, R.J. and R. Rosenberg. 2008. Spreading dead zones and consequences for marine ecosystems. Science 321: 926–929.

Doney, S.C., J.E. Duffy, S.C. Doney, M. Ruckelshaus, J.E. Duffy, J.P. Barry et al. 2012. Climate change impacts on marine ecosystems. Ann. Rev. Mar. Sci. 4: 11–37.

Elith, J., C.H. Graham, R.P. Anderson, M. Dudik, S. Ferrier, A. Guisan et al. 2006. Novel methods improve prediction of species' distributions from occurrence data. Ecography 29: 129–151.

Elith, J. and J.R. Leathwick. 2009. Species distribution models: ecological explanation and prediction across space and time. Annu. Rev. Ecol. Evol. Syst. 40: 677–697.

Escalle, L., D. Gaertner, P. Chavance, H. Murua, M. Simier, P.J. Pascual-Alayón et al. 2019. Catch and bycatch captured by tropical tuna purse-seine fishery in whale and whale shark associated sets: comparison with free school and FAD sets. Biodivers. Conserv. 28: 467–499.

FAO. 2019. Global Production Statistics 1950–2017. Version March 2019. http://www.fao.org/fishery/statistics/global-production/query/en. Accessed 15 Jun 2019.

Fernandez-Carvalho, J., R. Coelho, M.N. Santos and S. Amorim. 2015. Effects of hook and bait in a tropical northeast Atlantic pelagic longline fishery: Part II-Target, bycatch and discard fishes. Fish. Res. 164: 312–321.

Forsgren, K., R. McBride, T. Nakatsubo, T. Thys, C. Carson, E. Tholke et al. 2020. Reproductive biology of the ocean sunfishes. pp. 87–104. *In*: T.M. Thys, G.C. Hays and J.D.R. Houghton [eds.]. The Ocean Sunfishes: Evolution, Biology and Conservation, CRC Press. Boca Raton, FL, USA.

Forrestal, F.C., M. Schirripa, C.P. Goodyear, H. Arrizabalaga, E.A. Babcock, R. Coelho et al. 2019. Testing robustness of CPUE standardization and inclusion of environmental variables with simulated longline catch datasets. Fish. Res. 210: 1–13.

Francis, M.P. and E.G. Jones. 2017. Movement, depth distribution and survival of spinetail devil rays (*Mobula japanica*) tagged and released from purse-seine catches in New Zealand. Aquat. Conserv. Mar. Freshw. Ecosyst. 27: 219–236.

Franklin, J. and J.A. Miller. 2010. Mapping species distributions: Spatial inference and prediction. Cambridge University Press, Cambridge.

Fricke, R., W.N. Eschmeyer and R. Van Der Laan. 2019. Eschmeyer's catalog of fishes. https://www.calacademy.org/scientists/projects/eschmeyers-catalog-of-fishes. Accessed 15 Jun 2019.

Froese, R. and D. Pauly. 2019. Fishbase. https://www.fishbase.org. Accessed 15 Jun 2019.

Fukofuka, S. and D.G. Itano. 2007. Photographic identification guide for non-target fish species taken in WCPO purse seine fisheries. Fishing Technology Specialist Working Group, Western and Central Pacific Fisheries Commission, Honolulu.

Fulling, G.L., D. Fertl, K. Knight and W. Hoggard. 2007. Distribution of molidae in the Northern Gulf of Mexico. Gulf Caribb. Res. 19: 53–67.

Gamblin, C., B. Pascal and V. Lucas. 2007. Comparison of bycatch species captured during daytime and nightime: preliminary results of longline experiments carried out in Seychelles waters. 3rd Working Party on Ecosystems and Bycatch, Indian Ocean Tuna Commission, Seychelles.

Garcia-Barcelona, S., J.M. Ortiz de Urbina, J.M. de la Serna, E. Alot and D. Macías. 2010. Seabird bycatch in Spanish Mediterranean large pelagic longline fisheries, 2000–2008. Aquat. Living Resour. 23: 363–371.

Garibaldi, F. 2015. By-Catch in the mesopelagic swordfish longline fishery in the Ligurian Sea (Western Mediterranean). Collect. Vol. Sci. Pap. ICCAT 71: 1495–1498.

Gilman, E., P. Suuronen, M. Hall and S. Kennelly. 2013. Causes and methods to estimate cryptic sources of fishing mortality. J. Fish. Biol. 83: 766–803.

Gray, C.A. and S.J. Kennelly. 2018. Bycatches of endangered, threatened and protected species in marine fisheries. Rev. Fish. Biol. Fish. 28: 521–541.

Grémillet, D., C.R. White, M. Authier, G. Dorémus, V. Ridoux, E. Pettex et al. 2017. Ocean sunfish as indicators for the 'rise of slime'. Curr. Biol. 27: 1263–1264.

Hahlbeck, N., K.L. Scales, H. Dewar, S.M. Maxwell, S.J. Bograd and E.L. Hazen. 2017. Oceanographic determinants of ocean sunfish (*Mola mola*) and bluefin tuna (*Thunnus orientalis*) bycatch patterns in the California large mesh drift gillnet fishery. Fish. Res. 191: 154–163

Hall, M., E. Gilman, H. Minami, T. Mituhasi and E. Carruthers. 2017. Mitigating bycatch in tuna fisheries. Rev. Fish. Biol. Fish. 27: 881–908.

Higham, J.E.S., L. Bejder, S.J. Allen, J. Peter and D. Lusseau. 2015. Managing whale-watching as a non-lethal consumptive activity. J. Sustain. Tour. 24: 73–90.

Hobday, A.J., A.D.M. Smith, I.C. Stobutzki, C. Bulman, R. Daley, J.M. Dambacher et al. 2011. Ecological risk assessment for the effects of fishing. Fish. Res. 108: 372–384.

Hoegh-Guldberg, O., D. Jacob, M. Taylor, M. Bindi, S. Brown, I. Camilloni et al. 2018. Impacts of 1.5°C Global warming on natural and human systems. *In*: V. Masson-Delmotte, P. Zhai, H.-O. Pörtner, D. Roberts, J. Skea and P.R. Shukla [eds.]. IPCC special report: Global Warming of 1.5°C. The Intergovernmental Panel on Climate Change.

Huang, H.W. and K.M. Liu. 2010. Bycatch and discards by Taiwanese large-scale tuna longline fleets in the Indian Ocean. Fish. Res. 106: 261–270.

IPCC. 2007. Climate Change 2007: The physical science basis. *In*: S. Solomon, D. Qin, M. Manning and Z. Chen [eds.]. Contribution of Working Group I to the Fourth Assessment Report of the Intergovernmental Panel on Climate Change. Cambridge University Press, Cambridge.

Jordan, L.K., J.W. Mandelman, D.M. McComb, S.V. Fordham, J.K. Carlson and T.B. Werner. 2013. Linking sensory biology and fisheries bycatch reduction in elasmobranch fishes: a review with new directions for research. Conserv. Physiol. 1: cot002.

Kang, M.J., H.J. Baek, D.W. Lee and J.H. Choi. 2015. Sexual maturity and spawning of ocean sunfish *Mola mola* in Korean waters. Korean J. Fish. Aquat. Sci. 48: 739–744.

Konow, N., R. Fitzpatrick and A. Barnett. 2006. Adult emperor angelfish (*Pomacanthus imperator*) clean giant sunfishes (*Mola mola*) at Nusa Lembongan, Indonesia. Coral Reefs 25: 208.

Kroodsma, D.A., J. Mayorga, T. Hochberg, N.A. Miller, K. Boerder, F. Ferretti et al. 2018. Tracking the global footprint of fisheries. Science 359: 904–908.

Lee, D., J.-H. Choi and K.-H. Choi. 2013. Catch distribution of ocean sunfish *Mola mola* off Korean waters. Korean J. Fish. Aquat. Sci. 46: 851–855.

Leis, J.L., K. Matsuura, K.-T. Shao, G. Hardy, G. Zapfe, M. Liu et al. 2015. *Masturus lanceolatus* (errata version published in 2017). The IUCN Red List of Threatened Species 2015: e.T193634A115330232.

Lewison, R.L., L.B. Crowder, A.J. Read and S.A. Freeman. 2004. Understanding impacts of fisheries bycatch on marine megafauna. Trends Ecol. Evol. 19: 598–604.

Lezama-Ochoa, N., H. Murua, J. Ruiz, P. Chavance, A. Delgado de Molina, A. Caballero et al. 2018. Biodiversity and environmental characteristics of the bycatch assemblages from the tropical tuna purse seine fisheries in the eastern Atlantic Ocean. Mar. Ecol. 39: e12504.

Liu, J., G. Zapfe, K.-T. Shao, J.L. Leis, K. Matsuura, G. Hardy et al. 2015a. *Mola mola* (errata version published in 2016). The IUCN Red List of Threatened Species 2015: e.T190422A97667070.

Liu, J., G. Zapfe, K.-T. Shao, J.L. Leis, K. Matsuura, G. Hardy, et al. 2015b. *Ranzania laevis* (errata version published in 2016). The IUCN Red List of Threatened Species 2015: e.T193615A97668925.

Liu, K.M., M.L. Lee, S.J. Joung and Y.C. Chang. 2009. Age and growth estimates of the sharptail mola, *Masturus lanceolatus*, in waters of eastern Taiwan. Fish. Res. 95: 154–160.

Lucchetti, A., P. Carbonara, F. Colloca, L. Lanteri, M.T. Spedicato and P. Sartor. 2017. Small-scale driftnets in the Mediterranean: Technical features, legal constraints and management options for the reduction of protected species bycatch. Ocean Coast. Manag. 135: 43–55.

Lucean-Frédou, F., L. Kell, T. Frédou, D. Gaertner, M. Potier, P. Bach et al. 2017. Vulnerability of teleosts caught by the pelagic tuna longline fleets in South Atlantic and Western Indian Oceans. Deep-Sea Research II. 140: 230–241.

Macías, D., M.J. Gómez-Vives and J.M. de la Serna. 2004. Desembarcos de especies asociadas a la pesquería de palangre de superficie dirigido al pez espada (*Xiphias gladius*) en el Mediterráneo durante 2001 y 2002. Collection ICCAT Scientific Papers 56(4): 981–986.

MacNeil, M.A., J.K. Carlson and L.R. Beerkircher. 2009. Shark depredation rates in pelagic longline fisheries: A case study from the northwest Atlantic. ICES J. Mar. Sci. 66: 708–719.

Maeroh, K., S. Hoimuk and N. Somkliang. 2018. Bycatch landings in Phuket Ports by foreign vessel, 2017. 14th Working Party on Ecosystems and Bycatch, Indian Ocean Tuna Commission, Cape Town.

Mangel, J.C., M. Pajuelo, A. Pasara-Polack, G. Vela, E. Segura-Cobeña and J. Alfaro-Shigueto. 2019. The effect of Peruvian small-scale fisheries on sunfishes (Molidae). J. Fish. Biol. 94: 77–85.

Marin, Y.H., F. Brum, L.C. Barea and J.F. Chocca. 1998. Incidental catch associated with swordfish longline fisheries in the south-west Atlantic Ocean. Mar. Freshw. Res. 49: 633–639.

Markic, A., C. Niemand, J.H. Bridson, N. Mazouni-Gaertner, J.C. Gaertner, M. Eriksen et al. 2018. Double trouble in the South Pacific subtropical gyre: Increased plastic ingestion by fish in the oceanic accumulation zone. Mar. Pollut. Bull. 136: 547–564.

Mason, J.G., E.L. Hazen, S.J. Bograd, H. Dewar and L.B Crowder. 2019. Community-level effects of spatial management in the California drift gillnet Fishery. Fish. Res. 214: 175–182.

McCauley, D., M. Pinsky, S. Palumbi, J.A. Estes, F.H. Joyce and R.R. Warner. 2015. Marine defaunation: Animal loss in the global ocean. Science 347: 1255641–1255641.

McKinnell, S. and M.P. Seki. 2017. Arcane epipelagic fishes of the subtropical North Pacific and factors associated with their distribution. Prog. Oceanogr. 150: 48–61.

McMahon, C.R. and G.C. Hays. 2006. Thermal niche, large-scale movements and implications of climate change for a critically endangered marine vertebrate. Glob. Chang. Biol. 12: 1330–1338.

Mitchell, J.D., D.L. McLean, S.P. Collin and T.J. Langlois. 2018. Shark depredation in commercial and recreational fisheries. Rev. Fish. Biol. Fish. 28: 715–748.

Moore, J.E., K.A. Curtis, R.L. Lewison, P.W. Dillingham, J.M. Cope, S.V. Fordham et al. 2013. Evaluating sustainability of fisheries bycatch mortality for marine megafauna: a review of conservation reference points for data-limited populations. Environ. Conserv. 40: 329–344.

MPI. 2016. Fisheries Assessment Plenary November 2016: stock assessments and stock status. Compiled by the Fisheries Science Group, Ministry for Primary Industries, Wellington, New Zealand.

MPI. 2017. Fisheries Assessment Plenary November 2016: Stock Assessments and Stock Status. Compiled by the Fisheries Science Group, Ministry for Primary Industries, Wellington, New Zealand.

Nakamura, I. and K. Sato. 2014. Ontogenetic shift in foraging habit of ocean sunfish *Mola mola* from dietary and behavioral studies. Mar. Biol. 161: 1263–1273.

Nakamura, I., Y. Goto and K. Sato. 2015. Ocean sunfish rewarm at the surface after deep excursions to forage for siphonophores. J. Anim. Ecol. 84: 590–603.

Nakatsubo, T. 2008. A study on the reproductive biology of ocean sunfish *Mola mola*. PhD Thesis, Kamagora Sea World, Japan.

National Marine Fisheries Service. 2011. U.S. National Bycatch Report. Karp, W.A., Desfosse, L.L. and S.G. Brooke. [eds.]. NOAA Tech. Memo. Nmfs-F-/Spo-117C and Nmfs-F-/Spo-117E, U.S. Department of Commerce, National Oceanic and Atmospheric Administration, National Marine Fisheries Service.

Newson, S.E., S. Mendes, H.Q.P. Crick, N.K. Dulvy, J.D.R. Houghton, G.C. Hays et al. 2009. Indicators of the impact of climate change on migratory species. Endanger. Species Res. 7: 101–113.

NOAA. 2019. West Coast Region Observer Program data summaries and reports. https://www.westcoast.fisheries.noaa.gov/fisheries/wc_observer_programs/sw_observer_program_info/data_summ_report_sw_observer_fish.html. Accessed 15 May 2019.

Nolf, D. and J.C. Tyler. 2006. Otolith evidence concerning interrelationships of caproid, zeiform and tetraodontiform fishes. Bull. L'Inst. Royal Sci. Nat. Belgique Biol. 76: 147–189.

Nolin, D.A. 2008. Doctoral dissertation. Seattle: University of Washington. Food-sharing networks in Lamalera, Indonesia: Tests of adaptive hypotheses.

Northridge, S.P. 1991. Driftnet fisheries and their impacts on non-target species: a worldwide review. FAO Fisheries Technical Paper, Food and Agriculture Organisation of the United Nations Rome.

Nyegaard, M. 2018. There be giants! The importance of taxonomic clarity of the large ocean sunfishes (genus *Mola*, Family Molidae) for assessing sunfish vulnerability to anthropogenic pressures. PhD Thesis, Murdoch University, Murdoch, Australia.

Nyegaard, M. and E. Sawai. 2018. Species identification of sunfish specimens (Genera Mola and Masturus, Family Molidae) from Australian and New Zealand natural history museum collections and other local sources. Data Br. 19: 2404–2415.

Nyegaard, M., N. Loneragan, S. Hall, J. Andrew, E. Sawai and M. Nyegaard. 2018a. Giant jelly eaters on the line: Species distribution and bycatch of three dominant sunfishes in the Southwest Pacific. Estuar. Coast Shelf Sci. 207: 1–15.

Nyegaard, M., E. Sawai, N. Gemmell, J. Gillum, N.R. Loneragan, Y. Yamanoue et al. 2018b. Hiding in broad daylight: molecular and morphological data reveal a new ocean sunfish species (Tetraodontiformes: Molidae) that has eluded recognition. Zool. J. Linn. Soc. 182: 631–658.

Paulin, C.D., G. Habib, C.L. Carey, P.M. Swanson and G. Vos. 1982. New records of Mobula japanica and *Masturus lanceolatus*, and further records of *Luvaris imperialis* (Pisces: Mobulidae, Molidae, Louvaridae) from New Zealand. New Zeal. J. Mar. Freshw. Res. 16: 11–17.

Pearson, R.G., C.J. Raxworthy, M. Nakamura and A. Townsend Peterson. 2007. Predicting species distributions from small numbers of occurrence records: A test case using cryptic geckos in Madagascar. J. Biogeogr. 34: 102–117.

Petersen, S. 2005. Initial bycatch assessment: South Africa's domestic pelagic longline fishery, 2000–2003. BirdLife South Africa, Cape Town.

Petersen, S. and Z. McDonell. 2007. A bycatch assessment of the cape horse mackerel *Trachurus trachurus capensis* mid-water trawl fishery off South Africa. BirdLife/WWF Responsible Fisheries Programme, Johannesburg.

Phillips, N.D., N. Reid, T. Thys, C. Harrod, N.L. Payne, C.A. Morgan et al. 2017. Applying species distribution modelling to a data poor, pelagic fish complex: The ocean sunfishes. J. Biogeogr. 44: 2176–2187.

Phillips, N.D., L. Kubicek, N.L. Payne, C. Harrod, L.E. Eagling, C.D. Carson et al. 2018. Isometric growth in the world's largest bony fishes (genus *Mola*)? Morphological insights from fisheries bycatch data. J. Morphol. 279: 1312–1320.

Phillips, N.D., E.A. Smith, S.D. Newsome, J.D.R. Houghton, C.D. Carson, J. Alfaro-Shifueto et al. 2020. Bulk tissue and amino acid stable isotope analyses reveal global ontogentic patterns in ocean sunfish trophic ecology and habitat use. Mar. Ecol. Prog. Ser. 633: 127–140.

PIFSC. 2012. PIFSC report on the American Samoa longline fishery year 2011. Fisheries Research and Monitoring Division, Pacific Islands Fisheries Science Center. https://repository.library.noaa.gov/view/noaa/5245. Accessed 11 November 2019.

Pope, E.C., G.C. Hays, T.M. Thys, T.K. Doyle, D.W. Sims, N. Queiroz et al. 2010. The biology and ecology of the ocean sunfish *Mola mola*: A review of current knowledge and future research perspectives. Rev. Fish. Biol. Fish. 20: 471–487.

Porcasi, J. and S. Andrews. 2001. Evidence for a prehistoric *Mola mola* fishery on the Southern California Coast. J. Calif. Gt. Basin Anthropol. 23: 51–66.

Potter, I.F., B. Galuardi and W.H. Howell. 2011. Horizontal movement of ocean sunfish, *Mola mola*, in the northwest Atlantic. Mar. Biol. 158: 531–540.

Reed, J.R., S.E. Kerwath and C.G. Attwood. 2017. Analysis of bycatch in the South African midwater trawl fishery for horse mackerel *Trachurus capensis* based on observer data. African J. Mar. Sci. 39: 279–291.

Relini, L.O., L. Lanteri and F. Garibaldi. 2010. Medusivorous fishes of the Mediterranean. A coastal safety system against jellyfish blooms. Biol. Mar. Mediterr. 17: 348–349.

Relini, L.O. and G. Vallarino. 2016. An artisanal fishery, a centuries-long experience for the future and intriguing relations between ocean sunfish and sharks. Biol. Mar. Mediterr. 23: 17–20.

Ricklefs, R.E. 2004. A comprehensive framework for global patterns in biodiversity. Ecol. Lett. 7: 1–15.

Rogan, E. and M. Mackey. 2007. Megafauna bycatch in drift nets for albacore tuna (*Thunnus alalunga*) in the NE Atlantic. Fish. Res. 86: 6–14.

Rushton, S.P., S.J. Ormerod and G. Kerby. 2004. New paradigms for modelling species distributions? J. Appl. Ecol. 41: 193–200.

Sagara, K. and T. Ozawa. 2002a. Report on the questionnaire about molids in Japan. Bull. Japanese Soc. Fish. Oceanogr. 66: 164–167.

Sagara, K. and T. Ozawa. 2002b. Landing statistics of molids in four prefectures in Japan. Mem. Fac. Fish. Kagoshima Univ. 51: 27–33.

Sawai, E., Y. Yamanoue, Y. Yoshita, Y. Sakai and H. Hashimoto. 2011. Seasonal occurrence patterns of *Mola* sunfishes (*Mola* spp. A and B; Molidae) in waters off the Sanriku region, eastern Japan. Japanese J. Ichthyol. 58: 181–187.

Sawai, E. 2017. The mystery of ocean sunfishes [Manbou no Himitsu]. Iwanami Shoten Publishers, Tokyo (In Japanese).

Sawai, E., Y. Yamanoue, M. Nyegaard and Y. Sakai. 2018. Redescription of the bump-head sunfish *Mola alexandrini* (Ranzani 1839), senior synonym of *Mola ramsayi* (Giglioli 1883), with designation of a neotype for *Mola mola* (Linnaeus 1758) (Tetraodontiformes: Molidae). Ichthyol. Res. 65: 142–160.

Sawai, E., M. Nyegaard and Y. Yamanoue. 2020. Phylogeny, taxonomy and size records of the ocean sunfishes. pp. 18–36. *In*: T.M. Thys, G.C. Hays and J.D.R. Houghton [eds.]. The Ocean Sunfishes: Evolution, Biology and Conservation, CRC Press. Boca Raton, FL, USA.

Secretariat of the Pacific Community-Oceanic Fisheries Programme (SPC-OFP). 2010. Non-target species interactions with the tuna fisheries of the western and central Pacific Ocean. Paper prepared for the Joint Tuna RFMOs International Workshop on Tuna RFMO Management Issues relating to Bycatch. Brisbane, Australia, 23–25 June 2010.

Seitz, A.C., K. Weng, A.M. Boustany and B.A. Block. 2002. Behaviour of a sharptail mola in the Gulf of Mexico. J. Fish. Biol. 60: 1597–1602.

Serna, M.J. de la, E. Alot and E. Rivera. 1992. Análisis de las CPUEs por grupos de tallas del pez espada (*Xiphias gladius*) capturado con artes de superficie y enmalle a la deriva en el área del Estrecho de Gibraltar, durante los años 1989 y 1990. Relación con la fase lunar y otros factores ambientales. Collect Vol Sci Pap - ICCAT 39:626–634.

Silvani, L., M. Gazo and A. Aguilar. 1999. Spanish driftnet fishing and incidental catches in the western Mediterranean. Biol. Conserv. 90: 79–85.

Sims, D.W., N. Queiroz, T.K. Doyle, J.D.R. Houghton and G.C. Hays. 2009. Satellite tracking of the world's largest bony fish, the ocean sunfish (*Mola mola* L.) in the North East Atlantic. J. Exp. Mar. Bio. Ecol. 370: 127–133.

Smith, K.A., M. Hammond and P.G. Close. 2010. Aggregation and stranding of elongate sunfish (*Ranzania laevis*) (Pisces: Molidae) (Pennant, 1776) on the southern coast of Western Australia. J. R. Soc. West. Aust. 93: 181–188.
Sousa, L.L., N. Queiroz, G. Mucientes, N.E. Humphries and D.W. Sims. 2016. Environmental influence on the seasonal movements of satellite-tracked ocean sunfish *Mola mola* in the north-east Atlantic. Anim. Biotelem. 4: 7.
Sousa, L.L., R. Xavier, V. Costa, N.E. Humphries, C. Trueman, R. Rosa et al. 2016a. DNA barcoding identifies a cosmopolitan diet in the ocean sunfish. Sci. Rep. 6: 28762.
Storai, T., L. Zinzula, S. Repetto, M. Zuffa, A. Morgan, J. Mandelman et al. 2011. Bycatch of large elasmobranchs in the traditional tuna traps (tonnare) of Sardinia from 1990 to 2009. Fish. Res. 109: 74–79.
Syväranta, J., C. Harrod, L. Kubicek, V. Cappanera and J.D.R. Houghton. 2012. Stable isotopes challenge the perception of ocean sunfish *Mola mola* as obligate jellyfish predators. J. Fish. Biol. 80: 225–231.
Tang, W.-C. 1987. Chinese medical materials from the sea. Abstr. Chin. Med. 1: 571–600.
The Daily Telegraph. 2019. Sydney to Hobart delivers big-time as Wild Oats XI holds on in supermaxi duel to win for a ninth time. Https://www.dailytelegraph.com.au/sport/more-sports/sydney-to-hobart-delivers-bigtime-as-wild-oats-xi-holds-on-in-supermaxi-duel-to-win-for-a-ninth-time/news-story/cab059ba568c8c1b8c9efef4da4f02fb. Accessed 30 Oct 2019.
Then, A.Y., J.M. Hoenig, N.G. Hall and D.A. Hewitt. 2018. Corrigendum, Evaluating the predictive performance of empirical estimators of natural mortality rate using information on over 200 fish species. ICES J. Mar. Sci. 75: 1509.
Then, A.Y., J.M. Hoenig, N.G. Hall and D.A. Hewitt. 2015. Evaluating the predictive performance of empirical estimators of natural mortality rate using information on over 200 fish species. ICES J. Mar. Sci. 72: 82–92.
Thys, T. and R. Williams. 2013. Ocean sunfish in Canadian Pacific waters: Summer hotspot for a jelly-eating giant? *In*: 2013 OCEANS - San Diego. MTS, pp. 1–5.
Thys, T.M., J.P. Ryan, H. Dewar, C.R. Perle, K. Lyons, J. O'Sullivan et al. 2015. Ecology of the ocean sunfish, *Mola mola*, in the southern California current system. J. Exp. Mar. Bio. Ecol. 471: 64–76.
Thys, T., J.P. Ryan, K.C. Weng, M. Erdmann and J. Tresnati. 2016. Tracking a marine ecotourism star: movements of the short ocean sunfish *Mola ramsayi* in Nusa Penida, Bali, Indonesia. J. Mar. Biol. 2016: Article ID 8750193.
Thys, T., M. Nyegaard, and L. Kubicek. 2020. Ocean sunfishes and society. pp. 263–279. *In*: T.M. Thys, G.C. Hays and J.D.R. Houghton [eds.]. The Ocean Sunfish: Evolution, Biology and Conservation, CRC Press. Boca Raton, FL, USA.
Torres-Irineo, E., M.J. Amandè, D. Gaertner, A.D. Molina H. de Murua, P. Chavance et al. 2014. Bycatch species composition over time by tuna purse-seine fishery in the eastern tropical Atlantic Ocean. Biodivers. Conserv. 23: 1157–1173.
Tudela, S., A. Kai Kai, F. Maynou, M. Andalossi and P. El Guglielmi. 2005. Driftnet fishing and biodiversity conservation: the case study of the large-scale Moroccan driftnet fleet operating in the Alboran Sea (SW Mediterranean). Biol. Conserv. 121: 65–78.
Varghese, S.P., K. Vijayakumaran and D.K. Gulati. 2013. Pelagic megafauna bycatch in the tuna longline fisheries off India. 9th Working Party on Ecosystems and Bycatch, Indian Ocean Tuna Commission, La Réunion.
Ward, P., R.A. Myers and W. Blanchard. 2004. Fish lost at sea: The effect of soak time on pelagic longline catches. Fish. Bull. 102: 179–195.
Watson, J.T. and K.A. Bigelow. 2014. Trade-offs among catch, bycatch, and landed value in the American Samoa longline fishery. Conserv. Biol. 28: 1012–1022.
Whitney, N.M., C.F. White, A.C. Gleiss, G.D. Schwieterman, P. Anderson, R.E. Hueter et al. 2016. A novel method for determining post-release mortality, behavior, and recovery period using acceleration data loggers. Fish. Res. 183: 210-221.
Worm, B., B. Davis, L. Kettemer, C.A. Ward-Paige, D. Chapman, M.R. Heithaus et al. 2013. Global catches, exploitation rates, and rebuilding options for sharks. Mar. Policy 40: 194–204.
Yoshita, Y., Y. Yamanoue, K. Sagara, M. Nishibori, H. Kuniyoshi, T. Umino et al. 2009. Phylogenetic relationship of two *Mola* sunfishes (Tetraodontiformes: Molidae) occurring around the coast of Japan, with notes on their geographical distribution and morphological characteristics. Ichthyol. Res. 56: 232–244.
Zeeberg, J., A. Corten and E. de Graaf. 2006. Bycatch and release of pelagic megafauna in industrial trawler fisheries off Northwest Africa. Fish. Res. 78: 186–195.

Chapter 13

Sunfish on Display

Husbandry of the Ocean Sunfish *Mola mola*

Michael J. Howard,[1,*] *Toshiyuki Nakatsubo*,[2] *João Pedro Correia*,[3] *Hugo Batista*,[4] *Núria Baylina*,[4] *Carlos Taura*,[5] *Kristina Skands Ydesen*[6] and *Martin Riis*[6]

Introduction

The ocean sunfish, *Mola mola* (Linnaeus 1758), is an odd-looking creature requiring constant vigilance for successful display in public aquariums. While much progress has been made in captive management strategies over recent decades, displaying ocean sunfishes publicly still presents many challenges due to their numerous atypical life history traits. For example, relative to their body size, ocean sunfishes, require an enormous amount of space (ideally hundreds of thousands of liters per animal). Their unusually shaped bodies require careful handling. Despite being encased in a notably thick collagenous underlayer of hypodermis (Bemis et al. 2020 [Chapter 4], Watanabe and Davenport 2020 [Chapter 5]), their external dermis is highly sensitive and wears off easily through excessive contact with enclosure surfaces. Such abrasions can lead to secondary infections. Additionally, their large and rapid growth rates must be kept in check by closely monitoring ontogenetic shifts in dietary needs and adjusting the volume, calories and food composition of the diet. This chapter highlights best practices and guidelines from key institutions that have achieved success in the long-term display of ocean sunfish. It also acknowledges that each sunfish specimen is an individual who will experience unique challenges in its journey through any aquarium program. However, the rewards of presenting such an ocean oddity to the public far outweigh the challenges of captive management.

[1] Monterey Bay Aquarium, 886 Cannery Row, Monterey, CA 93940, U.S.A.
[2] ORIX Aquarium Corporation, Nippon Life Hamamatsucho Crea Tower 14F, 2-3-1 Hamamatsu-cho, Minato-ku, Tokyo 105-0013, Japan; Email: t.nakatsubo@aqua.email.ne.jp
[3] Flying Sharks, Rua Farrobim do Sul 116, 9900-361, Horta, Portugal; MARE – Marine and Environmental Sciences Centre, ESTM, Instituto Politécnico de Leiria, 2520-641 Peniche, Portugal; Email: info@flyingsharks.eu
[4] Oceanário de Lisboa, Esplanada D. Carlos I, 1990-005 Lisboa, Portugal; Email: hbatista@oceanario.pt; nbaylina@oceanario.pt
[5] Oceanogràfic de Valéncia, C/Eduardo Primo Yufera 1-B, 46013, Valéncia, Spain; Email: ctaura@oceanografic.org
[6] North Sea Oceanarium, Willemoesvej 2, 9850 Hirtshals, Denmark; Email: ky@nordsoemail.dk; mr@nordsoemail.dk
* Corresponding author: mhoward@mbayaq.org

The History of Sunfish in Aquariums

The history of ocean sunfish, *Mola mola*, in public aquariums dates to the early 1900s. Suyehiro and Tsutsumi (1973) reported that the New York Aquarium exhibited a 75 kg ocean sunfish as early as 1919 while the Steinhart Aquarium in San Francisco displayed an individual for seven days in 1961, noting the successful delivery of 'clam' as a food item. Arakawa and Masuda (1961) make reference to twenty-one days of rearing data from captive mola at the Miyajima Aquarium, Hiroshima, Japan in 1960. This paper also provides an account of sunfish swimming motion and speed, buoyancy control by way of the hypodermis and eating habits.

Progress towards successful public display of ocean sunfish (here referring primarily to *Mola mola*) steadily increased in Japan in the 1970s (Nishimura et al. 1971, Araga et al. 1973, Tatsuki et al. 1973, Suyehiro and Tsutsumi 1973, Shimoyama and Kawamura 1978). Taking stock of these advancements, Suyehiro and Tsutsumi (1973) reviewed the collection, transport, handling, size and rearing enclosures for ocean sunfish. Based on 47 days of data (the longest recorded tenure at that time) Tatsuki et al. (1973) concurrently examined rearing conditions and captive behavior with reference to swimming speed, suitable water temperature, tank depth and overall utilization and diet. This was one of the earliest accounts of target feeding, with recommendations on the minimum essential feeding rate. Common to all these studies was an effort to understand the species, *Mola mola*, through captive observations and necropsy results. While Shimoyama and Kawamura (1978) reported basic biometric information on the relationship between total length and total height, their main goal was to improve the rearing environment by focusing on water quality which was a unique approach at the time.

In the 1980s, rearing practices for ocean sunfish continued to develop rapidly in Japan. Tsuzaki (1986) and Kondo (1986) described rearing efforts at Kamogawa Sea World, including the first documented use of a transparent polyester film fence to minimize collisions with the enclosure walls. This approach led to significant progress in reared sunfish longevity, with Kamogawa Sea World achieving the world's first year-long tenancy for an individual sunfish in 1979. Japanese aquariums established modern husbandry management practice for the species, providing clear guidance on training and hand feeding by means of a feeding rod (target) and adjustments of feeding rations based on daily observations. Furthermore, they standardized the routine collection of biometric data on reared specimens by recording total length (TL) over time and body mass (BM) based on TL-BM relationships.

This groundbreaking work paved the way for the worldwide aquarium community. In the mid-1980s, public aquariums in the United States including the New Jersey Aquarium on the east coast and Sea World San Diego and Monterey Bay Aquarium (MBA) on the west coast, began working with locally sourced specimens. However, these early rearing efforts outside of Japan resulted only in modest success. For example, MBA's semi-open enclosure system had seasonal drops in water temperature below 13°C which could not be tolerated long-term by individuals of 40–70 cm TL (Sommer et al. 1989; F. Sommer, personal observations). Nonetheless, this learning curve motivated MBA to consider the needs of ocean sunfish specifically while designing a new exhibit that opened in 1996. Since then, MBA has reared several large specimens (100–200 cm TL), thus far returning more than 20 ocean sunfish back to the wild. Numerous European facilities have also embarked on their own efforts with sunfish in recent decades. In 2000, the North Sea Oceanarium in Denmark landed a sunfish in local waters (an unusual occurrence) and reared it to nearly 300 kg. Since 2002, many other facilities such as the Oceanogràfic de Valéncia in Spain and the Oceanário de Lisboa in Portugal have kept sunfish consistently. Presently, 12 public aquariums worldwide have active ocean sunfish husbandry programs, displaying them to the amazement and delight of the general public (Table 1).

Table 1. List of public aquariums that support active ocean sunfish husbandry programs and commonly display the species, *Mola mola*.

Continent	Country	Aquarium
Asia	Japan	Aqua World Ibaraki Prefectural Oarai Aquarium (Oaraimachi, Ibaraki)
		Ashizuri Kaiyoukan Aquarium (Tosashimizu, Kochi)
		Echizen Matsushima Aquarium (Sakai, Fukui)
		Kamogawa Sea World (Kamogawa, Chiba)
		Osaka Aquarium Kaiyukan (Osaka, Osaka)
		Shima Marineland (Shima, Mei)
		Sunshine Aquarium (Toshima City, Tokyo)
Europe	Denmark	Nordsøen Oceanarium (Hirtshals, Nordjylland)
	Portugal	Oceanário de Lisboa (Lisboa, Estramadura)
	Spain	L'Aquarium de Barcelona (Barcelona, Barcelona)
		Oceanogràfic de Valéncia (Valéncia, Valéncia)
North America	USA	Monterey Bay Aquarium (Monterey, California)

Collection/Acquisition

Ocean sunfishes occur in sub-tropical and temperate zones of the world ocean (Pope et al. 2010, Phillips et al. 2017). The record holding ocean sunfish, caught in Kamogawa, Japan in 1996, measured 2.72 m long and weighed 2,300 kg (see Pope et al. 2010). In 2004, fishers recorded a 3.32 m *M. alexandrini* (Ranzani 1839), near Aji Island, Japan (Sawai et al. 2018), but unfortunately it was not weighed so cannot be confirmed as a record holder. Nevertheless, given their unusual appearance, ability to grow to immense sizes, typically slow, deliberate movements and charismatic nature, ocean sunfishes are highly sought after by large public aquariums.

To that aim, the best way to capture ocean sunfishes depends on location and local regulations. Successful captures and transportation minimizes handling and reduces the impacts of physical contact through the use of vinyl stretchers or hoops, rubberized dip nets, and latex/vinyl gloves. The most common methods of capture include: small scale set nets (Japan), small scale set nets (Mediterranean: Almadraba, Armação, and Tonnarella), purse seines, and targeted dip netting. Permanently anchored set nets off the coast of Japan funnel all incoming fauna into smaller and smaller nets (leader, impounding and bag). The nets are checked routinely (every day or every few days) and the final bag net is pursed by means of several boats working in synchrony. While there is some bycatch using this method, Ishidoya and Ishizaki (1995) report a very low discard ratio. Public aquariums often broker deals with the local set net owners to procure sunfish via this fishery. Sunfish are easily visible as the net is raised and, once restrained, can be hoisted out of the water and placed into a live well (Fig. 1).

In the Mediterranean Sea, set nets of a slightly different design are employed. Developed to capture tuna during their spawning migrations, they function by running a barrier net from shore to a large catch net with several opposing, angled box barrier nets that lead fish into the final catch net. These are known as Almadrabas in Spain (García Vargas and Florido del Corral 2007), Armação in Portugal (Batista and Gonçalves 2017) and Tonnarella in Italy (Di Natale 2014).

Ocean sunfish often constitute substantial bycatch from these nets, which is either released (i.e., the Italian Tonarella) or sold at local markets. As in Japan, local aquariums, researchers (e.g., Phillips et al. 2018) and collection firms arrange with fishers to collect healthy sunfish. Likewise, Peniche on the west coast of Portugal, hosts multiple commercial seine fishing companies that target *Sardina pilchardus*, *Scomber* spp. and *Trachurus trachurus*. Local fishermen report that *Mola mola* never

Figure 1. Large ocean sunfish being lifted from set net for satellite tagging. Chiba, Japan. Photo taken and permission granted by: Michael J. Howard.

occur during *Sardina pilchardus* hauls but are common when *Scomber* spp. are prevalent (P. Leitão, personal communication). This purse seining fleet has served as another source of sunfish since 2017.

In the United States, where set nets are banned, sunfish are targeted individually. This method requires calm waters with little to no breeze so that collectors can spot sunfish fins when they break the surface. The presence of cetaceans or sea lions typically frighten sunfish away (M. Howard, personal observations). Gulls sitting on the sea surface and/or pecking into the water (e.g., Western gulls, *Larus occidentalis*, and Heermann's gulls, *L. heermanni*) can signal a sunfish's presence as they are known to remove ectoparasites from sunfish at the surface (Tibby 1936, King 1978). Similar observations are reported for Laysan albatrosses, *Phoebastria immutabilis*, from coastal waters off Japan (Abe et al. 2012). When a sunfish is engaged with a gull that is picking parasites, it is focused on that activity and usually much easier to capture than free-swimming sunfish. A spotter plane may also be used to help locate individuals, but this greatly increases the costs of capture. As drone technology continues to improve, it is likely that this technology will come into play increasingly.

Swimming individuals are difficult to net and require a fast ambush approach. Alternatively, animals that are basking or being cleaned are better approached slowly and captured with a rapid thrust of the net into the water to secure the front of the fish. In all cases, a rubber net is employed, and the fish is lifted out of the water immediately to restrain it under its own weight. This approach prevents fin tips from entangling in the net by minimizing fin movement. The fish can then be placed immediately into a live well on the vessel, with fin tips protected at all times during capture and transport.

Transportation

Short Distance, By Sea or Land

Like the targeted sunfish, those caught in the set-nets and purse-seine nets should be removed carefully from the water using non-abrasive vinyl stretchers or rubber nets, operated by aquarists wearing latex gloves to minimize the removal of mucous and damage to the epidermis. Once on a fishing vessel, there are several transportation methods available to reduce fish movement and minimize contact with tank walls. One option maintains sunfish in a tank with a sealed combing lid top which eliminates the sloshing of water; this, combined with dissolved oxygen (DO) levels maintained at 120–150 percent, calms the fish and reduces overall movement during transportation. In this case, sunfish usually stay at the bottom of the tank and move very little. Another option carries sunfish in a free standing, tall, circular tank (fiberglass or polyethylene) with a significant (40–50 cm) air gap between water line and tank lid. As sloshing occurs during transportation, causing some fish movement, the air gap should be great enough to prevent dorsal fin contact with the top of the tank. Transportation in Portugal and Spain is based on round polyethylene tanks (1.4–1.6 m diameter) filled to a depth of 0.8–0.9 m of natural seawater. Transit times are typically one to two hours during which DO levels are maintained above 100 percent saturation to help keep the fish relaxed and still. For the Peniche collections, Betadine® (10 percent povidone-iodine) is added to the transportation water, at a concentration of 10 ml/m³ to serve as a prophylactic antiseptic treatment to address any minor damage to the dermis incurred during capture (from purse seine netting and handling).

Long Distance, By Land and Air

Long distance transportation presents additional challenges. It should not exceed 45 hours and must align with well-established welfare protocols for the physiological and operational aspects involved during such transportation (Smith 1992, Correia 2001, Young et al. 2002, Smith et al. 2004, Correia et al. 2011, Rodrigues et al. 2013, Correia and Rodrigues 2017). Correia et al. (2008) provide specific details on the transport of *Mola mola* (although the methods described by those authors have been modified since airlines have banned the inflight use of lithium batteries and compressed oxygen). Due to dimensional limits to cargo transportation tanks, sunfish designated for shipping should be small (< 50 cm TL). Other concerns for shipments lasting more than two hours include: (1) a gradual decrease in pH, (2) elevated levels of ammonia, and (3) a steady decline in DO which occurs from the build-up of carbon dioxide, nitrogenous waste and stress-related metabolites, and through consumption of DO by means of respiration. These three issues can be addressed through water filtration and the addition of chemical supplements and oxygen. Likewise, the control of pH can be achieved via buffering agents such as common baking soda (sodium bicarbonate, $NaHCO_3$) or soda ash (sodium carbonate, Na_2CO_3). Ammonia (NH_3 and NH_4^+) can be removed with the assistance of quenching agents such as AmQuel® ($HOCH_2SO_3$) (Novalek Inc., U.S.A.), which binds to ammonia and transforms it into non-toxic aminomethanesulfonate ($H_2NCH_2SO_3^-$) and water. This substance has been used successfully in the transportation of marine species for many years (Visser 1996, Young et al. 2002, Smith et al. 2004, Correia et al. 2008, 2011, Rodrigues et al. 2013, Correia and Rodrigues 2017). The decrease in oxygen saturation rate, a direct result of respiration, may be counteracted by supplying oxygen through the use of an air-stone connected to a cylinder of compressed medical grade oxygen.

Post-capture, all candidates for long distance transportation should first receive a gross physical examination to assess condition. When deemed 'healthy' and in an unstressed state (i.e., orienting well within the enclosure and displaying appropriate targeting and feeding responses) all animals should be fasted for two days before transportation to limit the build-up of nitrogenous waste during transit. Transportation details for several successful efforts are listed in Table 2, and shown in Figs. 2, 3 and 4.

Figure 3 shows transportation tanks ready to be loaded onto a plane. They are 1.4 m diameter × 0.9 m high, filled with 0.7 m of seawater, creating a usable volume of 1.1 m³ to accommodate one or two four kg sunfish, which equates to a bioload (i.e., the amount of living matter in a tank) of either 3.6 or 7.3 kg/m³ (Table 2). Table 2 illustrates how various water volumes and bioloads can achieve successful transportation as long as the water is buffered properly, and adequate nitrogen quenching agents are employed. To achieve the most stable water quality, 50 g of sodium bicarbonate and 50 g of sodium carbonate are added to the oxygenated water along with 10 g of AmQuel® which lowers pH. Like other teleosts, ocean sunfish remain practically motionless in this type of transit. The addition of AmQuel® coupled with pH buffering agents contributes significantly to the successful delivery of these fish. In the absence of buffering agents, the water chemistry degrades to lethal levels.

Tanks shipped via air cargo are subject to strict rules and inspected thoroughly by airport officials. It is important to use a fiberglass reinforced lid bolted to the polyethylene tank to ensure a leak-proof seal. A Plexiglas® hatch allows for visual inspections of the animals, including their positioning within

Table 2. Operational specifications for *Mola mola* transport using tanks fitted with filtration (up to 2010) and sealed, with no filtration (2014–present) to multiple international destinations. Transportation with O_2 fed continuously targets 150 percent saturation. All other transports involve tanks fed with oxygen until delivery to the airport, approximately four hours before each flight, at which point the tank is sealed with 300 percent saturation and oxygen supply is discontinued.

Date	No. Fish	Indiv. Weight (kg)	Destination	No. Fish/Tank	Tank Dimensions (Diam. x Height, m)	Volume (L)	Bioload (kg/m^3)	Duration (hr.)	Transport	O_2 on arrival (%)	Survivorship in Transit	Introduction	Survivorship after Delivery (days)	Necropsy
05/04/2007	4	4	Atlanta, USA	1	1.4 x 1.1	1100	3.6	43	Road & Air	O_2 continuously	100%	Quarantine	Approx. 14	
11/09/2008	4	4	Dubai, UAE	2	1.4 x 1.1	1100	7.3	44	Road & Air	O_2 continuously	100%	Quarantine	Approx. 14	
05/06/2010	2	4	Stralsund, Germany	1	1.6 x 1.06	1400	2.9	24	Road & Air	O_2 continuously	100%	Main exhibit		Liver damage (parasites); Bacterial infection; Construction may have disturbed
13/05/2014	4	4	Singapore, Singapore	2	1.4 x 1.1	1100	7.3	50	Road & Air	~80	75%	Quarantine	Approx. 14	
01/07/2014	2	3	Hirtshaals, Denmark	2	1.35 x 0.9	540	11.1	17	Road & Air	~60	100%	Main exhibit	5 years + est. 500 kg & alive at time of publishing); 2.1 years (euthanized at 168 kg due to severe lesions)	
11/07/2014	1	3	Copenhagen, Denmark	1	1.35 x 0.9	617	4.9	12	Road & Air	~80	100%	Quarantine	a few days; approx. 30 days	Heavy parasite load in gills and liver
13/08/2014	1	5	Copenhagen, Denmark	1	1.35 x 0.9	500	10.0	12	Road & Air	~80	100%	Quarantine		
11/05/2017	1	4	Moscow, Russia	1	1.4 x 0.9	700	5.7	16	Road & Air	~80	100%	Quarantine	11 days	Internal parasites; Multiple organ failure
16/10/2017	2	4	Boulogne-Sur-Mer	2	2.4 x 1.0	3600	2.2	30	Road & Air	O_2 continuously	100%	Main exhibit	3 days; 5 days	
20/10/2017	2	4	Dubai, UAE	1	1.2 x 1.2	700	5.7	16	Road & Air	~150	100%	Quarantine	1 day; 2 days	
30/05/2018	1	4	Hirtshaals, Denmark	1	1.4 x 0.9	750	5.3	17	Road & Air	~100	100%	Quarantine	1+ years (alive at time of publishing)	

Figure 2. Polyethylene transport tank used for *Mola mola*: (1.4 m diam. × 0.92 m high) with 750 L capacity. The tank flies completely sealed, (IATA LAR 60) with a small aeration unit mounted under the lid. The unit dissolves oxygen that was added previously into the water. The water level is approximately half of the height. Photo taken and permission granted by: João Correia, Flying Sharks.

Figure 3. Four tanks loaded with one *Mola mola* each at JFK airport on the 5th of April 2007, en route to Georgia Aquarium, in Atlanta, USA. Each tank is loaded with 1100 L of seawater and one four kg animal, yielding a bioload of 3.6 kg/m^3. Total transit time was 43 hours and all animals arrived alive to their destination. Each tank was equipped with a 12 V bilge pump and was fed continuously with oxygen, for a target saturation of 200 percent. Photo taken and permission granted by: João Correia, Flying Sharks.

Figure 4. SCUBA diver hand feeding ocean sunfish, *Mola mola*, at the North Sea Oceanarium, Denmark, taken in the spring of 2019, with an estimated mass of over 500 kg. The specimen was originally shipped via air cargo in July, 2014 (see Table 2). Photo taken and granted permission by: North Sea Oceanarium.

the tank. As a rule, whenever aquarists have access to the transportation tanks, they should check the system and animals at regular intervals. Checks should include the animal's behavior and respiration rate, equipment functionality and water quality parameters, such as temperature and DO (using, for example, a hand held OxyGuard® Handy Oxygen probe®—OxyGuard Intl., Denmark), pH (using, for example, a hand held OxyGuard® Handy pH® probe) and ammonia (using, for example, Tetra® Ammonia test kits—Tetra Werke, Germany).

Current aeronautical regulations require the use of completely self-contained, sealed tanks. The sealed tanks contain only seawater, the animal, and a small, three V aeration unit (mounted on the lid underside) to dissolve pure oxygen into the water. Water is exchanged by 70–80 percent approximately 30 minutes after the animals are placed in the tanks before driving to the airport. Oxygen should be administered continuously while travelling. Both pH and ammonia buffering agents are used to keep parameters stable and, in the case of pH, above normal before sealing the lid (to account for gradual decline during shipment). Arrangements should be made with ground handling agents at the airport to ensure access to the shipping containers until the last possible moment before the actual loading of the aircraft, at which point final measurements of oxygen, pH and ammonia are made, and buffering agents can be applied as needed. DO should be raised to 300 percent to supply enough DO for the sunfish on the flight, during which supplemental oxygen cannot be administered after sealing the lid with silicone. When transportation containers arrive at the destination, all animals need to be acclimated to the system water of their new enclosure. All sunfish (previously target trained and having fasted for two days prior to shipment) should be offered food as soon as possible using their targets as a cue in order to resume the target training process. Figure 4 shows a sunfish on exhibit at North Sea Oceanarium five years after being shipped using these methods.

Despite such precautions, mortalities during transportation do occur. Necropsy data from sunfish that do not survive long term post-delivery usually reveal moderate to heavy internal parasite loads, particularly in the liver and/or systemic bacterial infection. It is difficult to know if such infections were present prior to transportation or as a result of sustained stress once at the receiving institution. Blood sample analyses pre and post transportation may provide greater insight during future efforts. Marked shifts in water temperature between capture sites and quarantine/exhibition tanks should be avoided where possible.

Accession/Training/Quarantine

Even for small sunfish (< 50 cm TL), water volume and depth in particular (> 150 cm) are extremely important for individuals to thrive during quarantine. It is also good practice to include a soft vinyl curtain that hangs loosely, a set distance away from the hard sides of the enclosure to reduce abrasion while the individual progresses through quarantine and training. The quarantine enclosure should be as large as possible, ranging from 3.5–10 m in diameter and 0.8–2 m in depth. To avoid the need for lengthy temperature acclimations, the holding enclosure should match (within 1–2°C) the sea surface temperature at the point of capture. Throughout the quarantine and training period, the temperature can be increased up to 22°C (or set to match the destination enclosure's temperature) over the course of several days. All closed systems should employ some form of mechanical and biological filtration components in addition to temperature control. When a semi-enclosed system is used, flow rates must not exceed the chilling or heating capacity to maintain stable temperatures. DO (90–100 percent) and pH (7.7–8.2) levels should be maintained close to normal ocean surface levels.

Successful sunfish display programs do not follow a traditional marine fish quarantine process (which typically involves treatment periods of weeks) as the risk of mortality scales with time. Subsequently, any prophylactic treatments for sunfish should be minimal (oral medications are preferable to injections), usually of short duration (immersion baths in the minutes as opposed to days or weeks) and be as unobtrusive as possible when physical handling is required. Sunfish quarantine is largely an observational period while the fish is target trained. Aquarists need to assess and monitor enclosure utilization, swimming behaviors, respiration rate, gross physical condition (especially the eyes and dermis) and inter and intra-species interactions.

All sunfish should receive an initial health assessment while being accessioned into an aquarium collection or placed into a quarantine enclosure, with all ectoparasites removed at this stage (e.g., via a five-minute freshwater immersion bath, where pH and alkalinity match the transportation water). If the sunfish is visibly distressed in the freshwater bath, ectoparasites can be removed manually by: (1) supporting the sunfish on a round vinyl stretcher just below the water surface of the transportation tank to allow normal respiration; (2) increasing the DO level to 120–150 percent; (3) placing a soft chamois cloth over the upwards facing eye to keep the fish in a 'relaxed' state; (4) using hard plastic forceps, remove all external parasites as quickly as possible. A steady hand is required to avoid scratching the dermis or disrupting the mucous layer. While the sunfish is in this position, basic morphometric information should be recorded using a soft vinyl measuring tape. After the ectoparasites have been removed from both sides of the fish, attach the vinyl stretcher to a digital scale. Briefly lift the fish and stretcher from the water and record the mass once all of the water has drained out. Be sure to record the wet mass of the stretcher, bridle and chamois separately and deduct from gross weight to obtain the sunfish's actual mass. This value can be used to determine an initial dose of oral Praziquantel, an antihelmintic (dosage—12 mg/kg at 56.8 mg/ml). Once all notes on gross physical condition are recorded, introduce the sunfish to the holding enclosure and offer food.

Once in the holding enclosure, the sunfish can be offered 3–10 percent body weight (BW) per day of solid foods (mollusc or crustacean) cut into appropriately sized pieces. Gelatin capsules filled with the appropriate dosage of liquid oral Praziquantel can be implanted into food items and administered at this time. If rejected, food items can be recaptured and offered again until consumed. This process of initially associating food with a visual 'target' is sufficient for the first day, after which the individual should be left to acclimate to its new surroundings until the following day, when the training process can begin in earnest (Fig. 5).

Target training is best performed when sunfish are placed into a holding enclosure by themselves to avoid inter-specific competition. However, this is not always possible, and it is helpful that they eventually have 'tank-mates' prior to any introduction to a display enclosure (discussed later). For simplicity, we describe the process of training one fish at a time. However, because small sunfish (< 60 cm TL) often travel in small schools in the ocean, they should be kept in small groups within aquarium enclosures (whenever possible), from an animal welfare perspective. Starting with feeding, it can be difficult to deliver food to an untrained sunfish in large initial holding enclosures (up to 10 m diameter). Therefore, a visual target (occasionally coupled with a specific audio cue) and hand feeding at stations are critical components to successful sunfish husbandry. Feeding at stations via target training serves many purposes. Primarily, it allows aquarists to deliver a set amount (percent BW per day) of specific food items (Kcal/kg * day) and minimizes 'free-feeding' (which can lead to obesity) or stealing of other food items designated for other animals. It also provides multiple

Figure 5. Target feeding at Monterey Bay Aquarium, 7 September 2016 using a green ball. Photo taken and permission granted by Lawrence Eagling.

daily opportunities for close inspection of body condition and behavior and assists with medication delivery. Finally, target feeding (often delivered at a set 'station') serves as a great enrichment tool.

After accession, the training process usually begins with the target deployed at the start of each training session. As long as the target is distinctly different from any other species-specific targets used within the same enclosure it should be effective. If the sunfish does not respond by swimming towards the target, food can be delivered by means of attaching a piece to the end of a long pole. Since sunfish are surprisingly agile and can swim backwards away from unknown foreign objects, it is important to present food initially from a known blind spot along the dorsal ridge and above its eyes (see Kino et al. 2009) so that the piece arrives smoothly at its mouth and drawn in before it can retreat. If the first few tries are unsuccessful and the fish reacts negatively by swimming away from the food/pole, attempts should cease, and the session should end by removing the target. However, after a few successful deliveries, sunfish swim directly to the food. As this happens, each offering should occur closer and closer to the target so that the association is made. Early in the process, there should be several sessions provided per day (up to six, as time allows) to reinforce the behavior, but the total amount of food offered daily should not exceed its prescribed daily ration by a significant amount. Once a sunfish is routinely targeting and stationing for food, sessions per day can be reduced. Two to three is ideal in order to continue the reinforcement of the training process while dividing its daily rations into more manageable, smaller meals. If at this point the fish appears well acclimatized to its surroundings, it is time to introduce other species as 'tank-mates' into the quarantine enclosure. Allowing an individual to become accustomed to other fish is important, especially in the case of fast-swimming, schooling species that will occupy the same areas within the display enclosure.

A healthy sunfish that has been impacted minimally by the processes of capture, transport and accession should pick up target training in as little as a one day, but for reasons that are unclear, it can take as many as 14 days or more (Monterey Bay Aquarium unpublished data). Force feeding of the sunfish is required in this case. While the training time for some sunfish may be relatively lengthy, their ability to learn target training is quite remarkable, particularly considering its reduced brain size (Chanet et al. 2012). Once a sunfish is well adapted and target trained, it should be transferred to the display enclosure. The process may cause some stress to the individual resulting in a temporary cessation of targeting behaviors (temporarily increasing its daily rations up to a week in advance of a transfer can mitigate the stress from a short-term disruption in food intake). Likewise, in a new and larger display enclosure, a sunfish may simply not know where to find its target. In this case, it may be necessary to deliver food via SCUBA divers. While approaching the sunfish, the dive team should deploy its target while offering food to the sunfish. Over time these targets will facilitate a response without the dive team, usually over a period of two to eight days.

Display Enclosure Styles/Concepts

There are three main styles of display enclosures that have been used successfully for exhibiting the ocean sunfish at public aquariums. Each approach has advantages and challenges, but all follow a similar regime to maintain water quality parameters. The first style, prevalent in Japan, houses sunfish singularly or in small groups within a large, single species display enclosure. These typically are devoid of any reef structure or elements of aquarium décor (e.g., pier pilings, shipwrecks) and are lined with a soft, clear protective vinyl curtain to minimize abrasions. However, curtains also create additional surfaces for aquarists to keep clean and free from diatoms and other algae and their presence is often troublesome to other fishes. However, there are a handful of species, primarily smaller schooling types (e.g., Japanese butterfish, *Psenopsis anomala*) or small juvenile fish that associate with floating material (e.g., blacksmith, *Chromis punctipinnis*, rockfish, *Sebastes* spp.) that can be displayed safely with sunfish in this manner. The second style of display enclosure is a multi-species, sub-tropical, open ocean display which has been used successfully in aquariums in the United States. Notably, there are no structures (natural or foreign) within the display. In the absence of a clear curtain and with no discernable reef structures for orientation, it is essential to monitor enclosure utilization (wall avoidance, especially) and negative interactions with other species (e.g., sea turtles, elasmobranchs). The final approach, used widely in European aquariums, is a multi-species, sub-tropical outer reef habitat. In this type of display, the species list usually includes several different teleosts and large elasmobranchs and occasionally sea turtles. These tanks typically comprise a combination of open spaces along with patches of low-lying reef or habitat structures. This set-up provides an excellent mixture of large open water space for midwater and near surface swimming while also providing spatial context for the sunfish to navigate within the water column and during time spent closer to the bottom and near structures. However, an agitated sunfish may make incidental contact with reef structures incurring injury to fin tips, flanks, eyes, or mouth.

Whatever the enclosure type, sunfish should be maintained in aquariums with water temperatures ranging between 16.5–22°C. They can be exposed to natural or artificial light with different photoperiods (from 7 to 17 hours of light) with salinity ranging from 32.8–34.5 ppt. DO levels should be maintained between 6.9–7.2 mg/l or 96–100 percent. Sunfish have been known to react poorly when DO levels drop suddenly or are consistently at levels below 90 percent for long durations, suggesting a lack of efficiency in gas exchange through the gills which are notably pale in healthy, living specimens (M. Howard, personal observations). Likewise, sunfish should not be held in aquarium systems that have poorly functioning nitrogen fixing bacteria or incompletely cycled systems. Other dissolved gases (i.e., $NH^3/NO^2/NO^3$) should be maintained at typically acceptable levels for well-run marine systems (0/0–0.005/< 80 mg/l respectively).

Feeding Strategies

New insights from wild sunfish diet research are helping to guide captive practices as well. Recent studies show that wild juveniles consume a rich, primarily benthic diet and, as they grow through sub-adult to adulthood, transition to a diet composed increasingly of gelatinous taxa (Syväranta et al. 2012, Nakamura and Sato 2014, Sousa et al. 2016, Phillips et al. in press). It has been hypothesized (Sousa et al. 2020 [Chapter 8]) that small individuals are more vulnerable to predation and may require food that is richer in order to grow more rapidly and reduce the number of potential predators. However, a closer look at the bioenergetics of different size classes of sunfish through experimental trials within a mega-flume might shed light on the drivers of this dietary shift further (Payne et al. 2015). Replicating this mixed diet under captive conditions is achieved through gelatin-based foods (e.g., agar or low calorie #5BOQ gelatin—Mazuri Exotic Animal Nutrition, Land O' Lakes, Inc., USA) with high water content (60–96 percent) prepared on site. These designed gelatins often include a small portion of molluscan, crustacean, or fish-based material to improve palatability. In turn, such feeds can be supplemented by multivitamins, or additional amino acids and vitamin C to enhance nutritional value. Foods with high water content also help mitigate the risk of dehydration (evidenced by the appearance of wrinkles and/or a shift from moist, slippery mucous to dense, sticky mucous) which can occur quickly (i.e., in less than one day for sunfish < 50 cm TL). Such problems can occur if an individual misses a scheduled feeding event, but can be rectified quickly by providing additional feeding sessions (either at station or via SCUBA) with increased amounts of gelatinous items, liquidized mollusc tissue, or in severe cases, deionized freshwater poured directly into the fish's mouth (Fig. 6).

Further studies are needed to better understand appropriate consumption rates. Living in a stable aquarium environment with fewer metabolic challenges heightens differences between wild and captive food consumption as well as growth rates. Along with daily volumetric intake in percent BW per day, Monterey Bay Aquarium staff also track consumption in terms of Kcal/kg * day. The caloric needs of sunfish decrease with growth and age, making it important to adjust this value over time to prevent obesity. Figure 7 highlights the variability of consumption in Kcal/kg * day for one specimen at MBA, when it has the opportunity to feed freely as compared to days it was fed via targeted session.

Figure 6. Pouring bottled freshwater into a *Mola mola's* mouth to correct dehydration. Photo taken and permission granted by: João Correia, Flying Sharks.

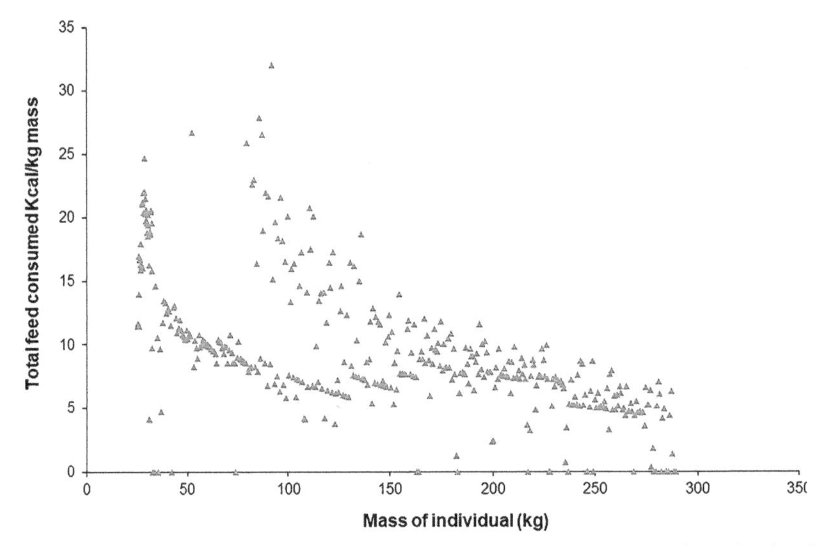

Figure 7. Kcal/kg * day consumption of one ocean sunfish over time at Monterey Bay Aquarium, showing the difference between targeted consumption (daily ration) versus estimated total daily consumption through observed free-feeding of market squids during other animals' feeding sessions. To estimate the additional Kcals consumed, individual squids were weighed, and a standard error curve was applied to ensure the sample size was adequate in relation to the deviation around the calculated mean.

Shortly after it was introduced to the exhibit enclosure, it began to free feed on market squids (*Doryteuthis opalescens*) during the broadcasted tuna feedings. At times, the sunfish was consuming three to four times the amount intended on a daily basis, potentially causing gastrointestinal stress and an obesity issue. As a result of these data, the management strategy was changed to offer two target feeding sessions per day, one of which corresponded with the tuna feedings. This allowed aquarists to keep the sunfish at station to eat its own food (prescribed daily ration) and not the tunas' food.

Understanding natural growth rates of a species in the wild is vital when designing an appropriate captive feeding regime. For ocean sunfish, length/weight data is available from Nakatsubo et al. (2007) and Kamogawa Sea World (Japan) and North Sea Oceanarium (Denmark) (previously unpublished data from North Sea bycatch) (Table 3; Fig. 8). With sufficient singular data points from wild specimens, a general guideline for captive weight management may materialize (Table 4). Nonetheless, more work is required across regions so that generalizable relationships with representative confidence intervals can be generated. A mark-recapture study underway by MBA will help towards this goal, with identifiable individuals measured and weighed over time allowing natural growth rates to be estimated.

Table 3. Average length-mass ratios for wild ocean sunfish from the western Pacific used by permission from Hiroshi Katsumata, Kamogawa Sea World.

Length (cm)	Weight (kg)
40	3.5
45	5
50	7
55	9.5
60	12
65	15
70	18
75	21
80	26
85	32
90	38
95	45
100	52
105	60
110	70
115	80
120	90
125	100
130	120
135	130
140	150
145	160
150	170
155	190
160	210
165	230
170	250
175	270
180	290
185	310
190	340
195	360
200	390

Clearly, data are required from a number of individuals (Cailliet et al. 1992) before relationships can be derived but the effort is far superseded by the potential benefits for refining sunfish husbandry.

Medical Procedures

Medical attention is required for external injuries and behavioral issues. Typical external injuries include dermal abrasions, damage to fin tips, corneal edemas/ulcers, and inflammation or damage to the mouth or jaws. Behavioral indicators of poor health include changes in appetite, stereotypic behavior, disorientation and reduced enclosure usage. When medical procedures are necessary, nitrile or latex gloves should always be worn, handling should be kept to a minimum, and when possible, the fish should remain in seawater to reduce stress and physical abrasion. For minimally invasive treatments, it is helpful to coax sunfish into a 'basking' position at the surface (Abe et al. 2012, Nakamura et al. 2015) prior to treatment to reduce handling stress. However, some procedures require physical restraint. Resting the fish on a vinyl stretcher just above the water, while providing

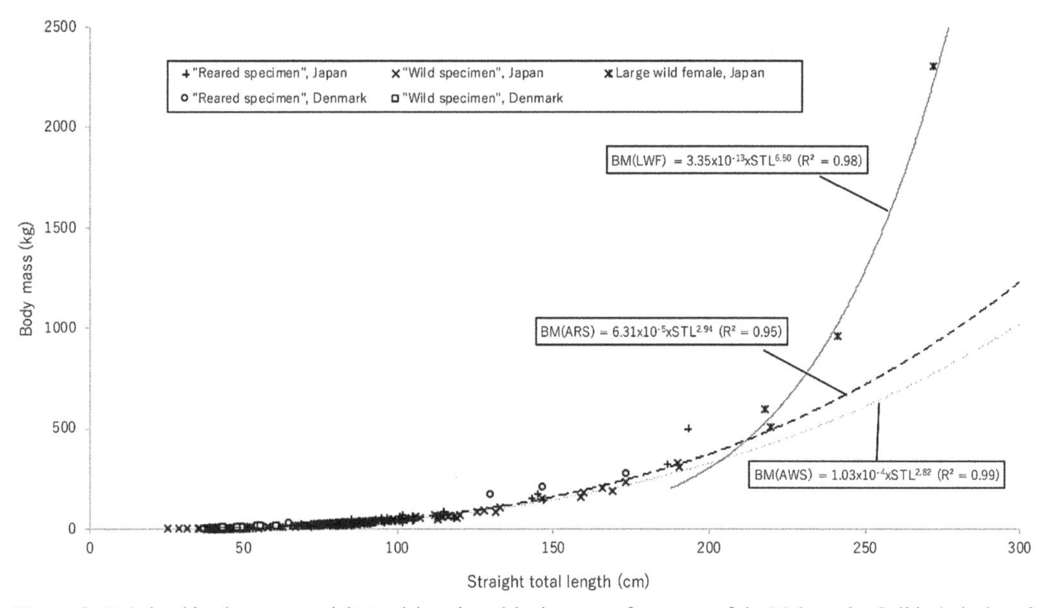

Figure 8. Relationships between straight total length and body mass of ocean sunfish, Mola mola. Solid, dashed, and dotted lines indicate regressions of large wild females (LWF), all reared specimens (ARS), and all wild specimens (AWS), respectively.

Table 4. Conversion table total length to body mass for the ocean sunfish, *Mola mola*, used by permission from Hiroshi Katsumata, Kamogawa Sea World.

Conversion table total length to body weight of ocean sunfish										
Total length rounded off the first digit (cm)	First digit of total length (cm)									
	0	**1**	**2**	**3**	**4**	**5**	**6**	**7**	**8**	**9**
0	0.00	0.00	0.00	0.00	0.00	0.00	0.01	0.01	0.02	0.03
10	0.04	0.05	0.07	0.09	0.11	0.14	0.17	0.20	0.24	0.28
20	0.33	0.39	0.45	0.52	0.59	0.67	0.75	0.85	0.95	1.06
30	1.18	1.30	1.44	1.58	1.74	1.90	2.08	2.26	2.46	2.66
40	2.88	3.11	3.35	3.61	3.87	4.16	4.45	4.76	5.08	5.42
50	5.77	6.13	6.51	6.91	7.33	7.76	8.20	8.67	9.15	9.65
60	10.2	10.7	11.3	11.8	12.4	13.0	13.7	14.3	15.0	15.7
70	16.4	17.2	17.9	18.7	19.5	20.3	21.2	22.1	23.0	23.9
80	24.9	25.9	26.9	27.9	28.9	30.0	31.1	32.3	33.5	34.7
90	35.9	37.1	38.4	39.7	41.1	42.4	43.9	45.3	46.8	48.3
100	49.8	51.4	53.0	54.6	56.2	57.9	59.7	61.4	63.3	65.1
110	67.0	68.9	70.8	72.8	74.8	76.9	79.0	81.1	83.3	85.5
120	87.8	90.1	92.4	94.8	97.2	99.7	102.2	104.7	107.3	109.9
130	112.6	115.3	118.1	120.9	123.7	126.6	129.5	132.5	135.6	138.6
140	141.8	144.9	148.2	151.4	154.7	158.1	161.5	165.0	168.5	172.1
150	175.7	179.4	183.1	186.9	190.7	194.6	198.5	202.5	206.5	210.6
160	214.8	219.0	223.2	227.5	231.9	236.3	240.8	245.3	249.9	254.6
170	259.3	264.1	268.9	273.8	278.8	283.8	288.8	294.0	299.2	304.4
180	309.8	315.1	320.6	326.1	331.7	337.3	343.0	348.8	354.6	360.5
190	366.5	372.5	378.6	384.8	391.0	397.3	403.7	410.1	416.6	423.2
	Converted body weight (kg)									
	Converted by the TL-BW formula: $BW=3\times10^{-5}\times TL^{3.11}$									

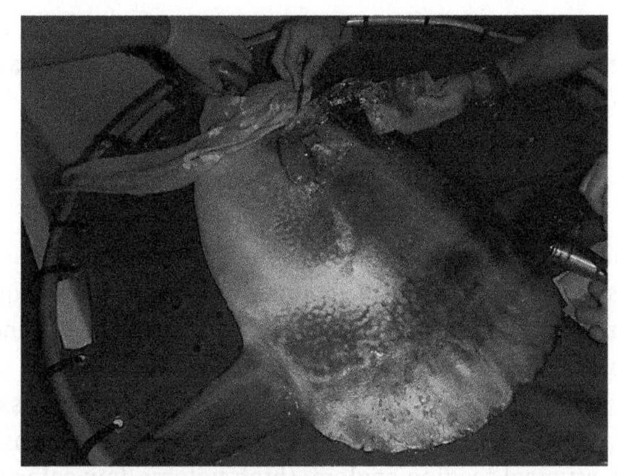

Figure 9. Using a vinyl stretcher with eye covering and water flowing into mouth and across gills in order to provide an exam. Photo taken and used with permission by: Michael J. Howard.

a flow of seawater into its mouth and through the gill areas can be an effective means of restraint (Fig. 9). Its eye is covered, and water is directed into its mouth (visible gill pumping can be seen as splashes of water coming from the gill opening). In such cases, it is important to cover the eye with a soft chamois and to saturate the respiration water to oxygen levels of 120–150 percent to subdue the fish, minimizing fin movements and reducing stress. This type of restraint provides excellent access to all caregivers and can be employed safely for up to ten minutes on sunfish < 100 cm TL.

Treatments

All treatments should be prescribed by a professional, licensed veterinarian. Superficial abrasions along the flanks or other parts of the body usually heal without treatment. However, when these occur during capture, initial handling, or transport and the sunfish is small, an immersion bath containing 0.1 mg/L of povidone-iodine ten percent (one percent available iodine, commercial name Betadine™) may prove effective in reducing infection. An entire system may be treated for 48 hours, then flushed to allow a sunfish to rest in untreated water for 24 hours. A second 24 h bath of Betadine™, using the same concentration, following the 24 h rest period may be necessary. This course of treatment can be administered once per week and repeated two or three times until all abrasions are healed. While treating an entire system imposes less stress on a sunfish (no handling, provides ample space) the amount, cost, and proper disposal of any therapeutic must be considered prior to implementation. Larger or more significant abrasions can be treated with topical ointments. Gentocil™ (based on the antibiotic, gentomicin) can be effective when combined with Betadine™ for wounds along the body and eyes and can heal lesions within three to four days. Regranex™ is also effective in this context. Most ointments are water soluble, however, which means they may need to be reapplied frequently, especially on high motion surfaces such as fin tips.

Antibiotics can help manage wounds, subsequent infections, abscesses and other growths that may occur along the fin tips due to excessive or repetitive contact with enclosure surfaces. They can also help healing after debridings, amputations and wound closure. These medicines are best administered orally since no handling is required, but this tactic requires the animal to be regularly feeding at station. Enroflaxacin is a helpful oral antibiotic that does not affect appetite. Antibiotics can also be administered intra-muscularly, although this requires additional handling and can add unwanted stress. This technique is typically used with Cefazolin, dosed at 30 mg/kg, and administered in three-day intervals for a total of five injections.

Lastly, when a sunfish's appetite is suppressed, assisted feeding is necessary while the root cause is determined. If there is no obvious cause such as a physical injury or environmental stressors relating to poor water quality, the possibility of inter- or intra-species interactions should be evaluated. If a cause is still not immediately evident, oxolinic acid, a quinolone antibiotic, may be used to stimulate appetite. However, since it is administered orally (12–20 mg/kg for 5–10 days) challenges may arise in delivery, and additional handling may be necessary.

Deaccession and Release

Public aquariums and animal research facilities should have deaccession plans (a disposition decision tree) in place for their specimens prior to the implementation of any acquisition plan. For many animals held in aquariums, husbandry management plans have evolved and advanced to the point at which an acquired specimen will live out a full and enriched life as a species ambassador. For others, including the ocean sunfish, *Mola mola*, this sometimes is not a possibility. This may be due to its growth rate and maximum size (enclosure/transport size limitations), species compatibility (inter and intra-specific), atypical feeding ecology (high volume/low caloric needs), and occasional behavioral issues. At this point, geography can pose a significant challenge as the display facility may be land-locked or located beyond the natural range of the species. In these cases, when release is not possible and quality of life has declined, euthanasia may be the only option as long as it is coupled with a thorough necropsy, sampling and data collection plan. Euthanasia is justified following several considerations. First, this species is an iconic, charismatic animal that holds intrinsic value as an ambassador that will fascinate and educate the public. Throughout a specimen's tenure, the public can learn a multitude of important topics relating both specifically and generally to the sunfish. These include evolution, biomechanics, ecology, bio-toxins, parasites, fisheries and conservation issues (as discussed throughout this book).

With recent advances in animal transportation, it is also possible to de-access animals via institutional trading, with size limitations. This is a good option when access to wild specimens is disrupted through natural disasters or other institutional limitations. The best option, although the most challenging, is a well-designed release plan. However, the path from display enclosure to ocean is a multi-step procedure (Monterey Bay Aquarium, unpublished data) and may require special permission from local authorities. Moreover, the process to move much larger specimens requires the use of SCUBA divers, industrial hoists, specialized restraining devices, and larger transportation tanks on bigger vessels. At no point during the process should the sunfish be without water flowing across its gills for more than 90 seconds (either via water pump directing seawater into its mouth or through full submersion within a transportation enclosure). Just prior to release, all sunfish should be marked, either with visual ID tags or electronic transmitters (i.e., satellite or acoustic). Indeed, best practices for attaching pop-off archival satellite tags were determined through aquarium and research scientist collaboration in the late 1990s using captive specimens and subsequently used for sunfish tagging studies (e.g., Dewar et al. 2010, Thys et al. 2015; Fig. 10).

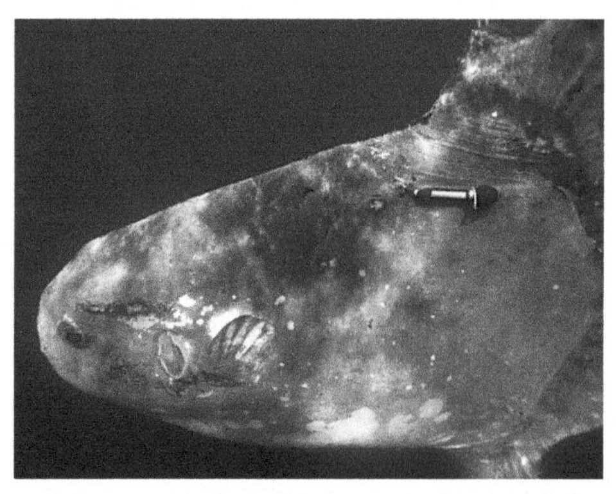

Figure 10. Ocean sunfish with pop off archival tag inserted at the base of the dorsal fin. Photo taken and permission granted by: Wyatt Patry.

Remaining Questions

Significant advances in husbandry techniques over the past several decades have made it easier to exhibit and rear ocean sunfishes. As with any discipline, there is scope for continual refinement. As field studies provide further insights into the biology of sunfishes, information on dietary composition and shifts, biomechanics and morphometrics will all help towards this goal. However, further information is required on the bioenergetics of sunfish so that scaling factors such as metabolic rate and cost of transport can be balanced against daily caloric intake. A deeper understanding of the high parasitic burden that characterize many wild sunfishes (Ahuir-Baraja 2020 [Chapter 10]) may help with the development of targeted medical procedures, and improved protocols during the early stages of transportation and quarantine.

The reciprocal is also true with information from sunfish in aquariums helping to address otherwise intractable questions arising from the field. For example, understanding the rate at which particular prey items are assimilated into different body tissues, or mucus is a fundamental challenge when seeking to reconstruct diets using biochemical approaches such as stable isotopes (Phillips et al. 2020). Under captive conditions, using isotopically labelled foods, such challenges could be overcome. Likewise, the reconciliation of electronic traces from multi-channel data loggers (e.g., tri-axial accelerometers) against known behaviors could be conducted in large display tanks with minimal stress to the study animal, collected mucus and blood samples can also provide information on nutritional status, immunological responses to stress or injury and the level of stress conditions through the presence of certain hormones (study in progress, Monterey Bay Aquarium). Maturity status may also be determined through the presence and quantities of sexual hormones (Du et al. 2017). In addition, if ultrasonography or X-ray radiography could be performed, it may be possible to determine non-invasively, not only visceral diseases but also gender by visualizing gonad appearance and shape (Martin-Robichaud and Rommens 2001, Colombo et al. 2004). Continuing in this vein, captive reproduction and larval culture of the ocean sunfish is another potentially rich research avenue. Knowledge of sunfish reproduction is limited (although see Forsgren et al. 2020 [Chapter 6]), and to date there are no known records of observed courtship rituals nor spawning in either aquarium or natural settings. Public aquariums increasingly have the ability to house multiple mature specimens. The concepts of providing suitable habitat for courtship and spawning (including seasonal adjustments to photoperiod and temperature) along with a means to collect fertilized eggs are well ingrained into the missions of public aquariums. It may be only a matter of time until the opportunity to culture larval ocean sunfish becomes a reality. Lastly, it is evident that field biologists benefit greatly from sunfish on display by boosting public interest and political awareness of this often-overlooked species.

Acknowledgements

The staff at Kamogawa Sea World deserves special recognition for their pioneering work with *Mola mola*. In particular, we thank Dr. Teruo Tobayama, Dr. Kazutoshi Arai, Hiroshi Katsumata, and Akihisa Osawa. We are grateful for the support from members of the ORIX Aquarium Corporation, including Takaaki Nitanai, Takashi Iwamaru, and Yumiko Tabata. We would like to acknowledge the kind staff of Tunipex and Mestre Comboio, Paulo Leitão, Morikawa Hirofumi, Carlos Poço, and Alfredo Poço, for their valuable contributions towards developing successful long-distance transportations. We also wish to extend our gratitude towards TAP—Air Portugal and Groundforce, for providing the necessary means for these animals to travel with minimum disturbance and ground handling time. Many thanks to staff veterinarians from contributing facilities including, Dr. Nuno Pereira, Dr. Daniel García, Dr. Theresa Álvaro, Dr. Mónica Valls, and Dr. Mike Murray. Deepest gratitude to all the primary caretakers at contributing facilities including Marga Ardao, Mario Roche and the Oceans Team at Oceanogràfic de Valéncia in Spain; Gonçalo David Nunes, Patricia Napier, Simão Santos, Tiago Martins, and Ana Ferreira at Oceanário de Lisboa in Portugal; Henrik Flintegaard at the North Sea Oceanarium; Sarah

Halbrend, Raymond Direen, Chuck Farwell and the Collections and Mola teams at Monterey Bay Aquarium in USA. The authors appreciate the support from staff and leadership at Oceanário de Lisboa, Oceanogràfic de Valéncia, North Sea Oceanarium, and Monterey Bay Aquarium. We would like to acknowledge and thank Juan Manuel Ezcurra of Monterey Bay Aquarium and an anonymous reviewer for providing helpful suggestions on the manuscript. Additionally, we gratefully thank the editors for inviting us to participate, their guidance, encouragement and immense efforts towards realizing this project.

References

Abe, T., K. Sekiguchi, H. Onishi, K. Muramatsu and T. Kamito. 2012. Observations on a school of ocean sunfish and evidence for a symbiotic cleaning association with albatrosses. Mar. Biol. 159: 1173–1176.

Ahuir-Baraja, A.E. 2020. Parasites of the ocean sunfishes. pp. 160–185. *In*: T.M. Thys, G.C. Hays and J.D.R. Houghton [eds.]. The Ocean Sunfishes: Evolution, Biology and Conservation, CRC Press. Boca Raton, FL, USA.

Araga, C., H. Tanase, S. Moriyama, M. Ota and Y. Kashiyama. 1973. Keeping the sunfish *Mola mola* (Linnaeus), in the aquarium with references to its habit. J. Jap. Assoc. Zoos Aquariums 15: 27–32.

Arakawa, K.Y. and S. Masuda. 1961. Some observations on ocean sunfish *Mola mola* (Linnaeus) with reference to its swimming behaviour and feeding habits. J. Jap. Assoc. Zoos Aquariums 3: 95–97.

Batista, N. and M. Gonçalves. 2017. Ritual of blessing the tuna fishing nets in Algarve (Portugal), between the 30's and 60's. Sharing Cultures 2017: 47–54.

Bemis, K.E., J.C. Tyler, E.J. Hilton and W.E. Bemis. 2020. Overview of the anatomy of the ocean sunfishes (Molidae Tetraodontiformes). pp. 55–71. *In*: T.M. Thys, G.C. Hays and J.D.R. Houghton [eds.]. The Ocean Sunfishes: Evolution, Biology and Conservation, CRC Press. Boca Raton, FL, USA.

Cailliet, G.M., H.F. Mollet, G.G. Pittenger, D. Bedford and L.J. Natanson. 1992. Growth and demography of the Pacific angle shark *Squatina californica*, based upon tag returns off California. Mar. Freshwater Res. 43(5): 1313–1330.

Chanet, B., C. Guintard, T. Boisgard, M. Fusellier, C. Tavernier, E. Betti et al. 2012. Visceral anatomy of ocean sunfish (*Mola mola* (L., 1758), Molidae, Tetraodontiformes) and angler (*Lophius piscatorius* (L., 1758), Lophiidae, Lophiiformes) investigated by non-invasive imaging techniques. C. R. Biol. 335(12): 744–752.

Colombo, R.E., P.S. Wills and J.E. Garvey. 2004. Use of ultrasound imaging to determine sex of shovelnose sturgeon. N. Amer. J. Fish. Man. 24(1): 322–326.

Correia, J.P.S. 2001. Long-term transportation of ratfish *Hydrolagus colliei* and tiger rockfish *Sebastes nigrocinctus*. Zoo Biol. 20: 435–441.

Correia, J.P., J. Graça and M. Hirofumi. 2008. Long-term transportation, by road and air, of Devil-ray Mobula mobular, Meagre Argyrosomus regius and Ocean Sunfish *Mola mola*. Zoo Biol. 27(3): 234–250.

Correia, J.P., J. Graça, M. Hirofumi and N. Kube. 2011. Long term transport of *Scomber japonicus* and *Sarda sarda*. Zoo Biol. 30: 459–472.

Correia, J.P. and N.V. Rodrigues. 2017. Packing and shipping. pp. 597–607. *In*: R. Calado, I. Olivotto, M.P. Oliver andG.J. Holt [eds.]. Marine Ornamental Species Aquaculture. Wiley Blackwell, West Sussex, England.

Dewar, H., T. Thys, S.L.H. Teo, C. Farwell, J. O'Sullivan, T. Tobayama et al. 2010. Satellite tracking the world's largest jelly predator, the ocean sunfish, *Mola mola*, in the Western Pacific. J. Exp. Mar. Biol. Ecol. 393(1-2): 32–42. doi:10.1016/j.jembe.2010.06.023.

Di Natale, A. 2014. Iconography of tuna traps: the discovery of the oldest recorded printed image of a tuna trap. Collect. Vol. Sci. Pap. ICCAT. 70(6): 2820–2827.

Du, H., X. Zhang, X. Leng, S. Zhang, J. Luo, Z. Liu et al. 2017. Gender and gonadal maturity stage identification of captive Chinese sturgeon, *Acipenser sinensis*, using ultrasound imagery and sex steroids. Gen. Comp. Endocrin. 245: 36–43.

Forsgren, K., R.S. McBride, T. Nakatsubo, T.M. Thys, C.D. Carson, E.K. Tholke et al. 2020. Reproductive biology of the ocean sunfishes. pp. 87–104. *In*: T.M. Thys, G.C. Hays and J.D.R. Houghton [eds.]. The Ocean Sunfishes: Evolution, Biology and Conservation, CRC Press. Boca Raton, FL, USA.

García Vargas, E.A. and D. Florido del Corral. 2007. The origin and development of tuna fishing nets (Almadrabas). *In*: International Workshop On Nets And Fishing Gears In Classical Antiquity: A First Approach, 1st, 2007, Cádiz, España.

Ishidoya, H. and H. Ishizaki. 1995. Set net fisheries for horse mackerel, mackerel and sardine. pp. 88–95. *In*: K. Matsuda [ed.]. By-catch in Japanese Fisheries. Koseishakoseikaku, Tokyo.

King, B. 1978. Feeding behavior of gulls in association with seal and sun fish. Bristol Ornith. 11: 33.

Kino, M., T. Miayzaki, T. Iwami and J. Kohbara. 2009. Retinal topography of ganglion cells in immature ocean sunfish, *Mola mola*. Environ. Biol. Fishes 85(1): 33–38.

Kondo, Y. 1986. Rearing and exhibit of ocean sunfish (*Mola Mola*). Museam Chiba: Bull. Chiba Mus. Assoc. 17: 253–260.

Martin-Robichaud, D.J. and M. Rommens. 2001. Assessment of sex and evaluation of ovarian maturation of fish using ultrasonography. Aqua. Res. 32(2): 113–120.

Nakamura, I. and K. Sato. 2014. Ontogenetic shift in foraging habit of ocean sunfish, *Mola mola* from dietary and behavioral studies. Mar. Biol. 161: 1263–1268.

Nakamura, I., G. Yusuke and K. Sato. 2015. Ocean sunfish rewarm at the surface after deep excursions to forage for siphonophores. J. Anim. Ecol. 84(3): 590–603.

Nakatsubo, T. and H. Hirose. 2007. Growth of captive ocean sunfish *Mola mola*. Aq. Sci. 55: 403–407.

Nakatsubo, T., M. Kawachi, N. Mano and H. Hirose. 2007. Estimation of maturation in wild and captive ocean sunfish *Mola mola*. Aq. Sci. 55: 259–264.

Nishimura, Y., T. Isogai, T. Takeuchi, H. Kabasawa, S. Mikami, S. Ooi et al. 1971. Observation on keeping *Mola mola*. Annu. Rep. Keikyu Aburatsubo Mar. Park Aquarium 3: 66–69.

Payne, N.L., E.P. Snelling, R. Fitzpatrick, J. Seymour, R. Courtney, A. Barnett et al. 2015. A new method for resolving uncertainty of energy requirements in large water breathers: the 'mega-flume' seagoing swim-tunnel respirometer. Methods Ecol. Evol. 6(6): 668–677.

Phillips, N.D., N. Reid, T. Thys, C. Harrod, N.L. Payne, C.A. Morgan et al. 2017. Applying species distribution modelling to a data poor, pelagic fish complex: The ocean sunfishes. J. Biogeogr. 44(10): 2176–2187. https://doi.org/10.1111/jbi.13033.

Phillips, N.D., L. Kubicek, N.L. Payne, C. Harrod, L.E. Eagling, C.D. Carson et al. 2018. Isometric growth in the world's largest bony fishes (genus *Mola*)? Morphological insights from fisheries bycatch data. J. Morph. 279(9): 1312–1320.

Phillips, N.D., E.A.E. Smith, S.D. Newsome, J.D.R. Houghton, C.D. Carson, J.C. Mangel et al. 2020. Bulk tissue and amino acid stable isotope analyses reveal global ontogenetic patterns in ocean sunfish trophic ecology and habitat use. Mar. Ecol. Prog. Ser. 633: 127–140.

Pope, E.C., G.C. Hays, T.M. Thys, T.K. Doyle, D.W. Sims, N. Queiroz et al. 2010. The biology and ecology of the ocean sunfish *Mola mola*: A review of current knowledge and future research perspectives. Rev. Fish Biol. Fish. 20: 471–487.

Rodrigues, N.V., J.P.S. Correia, R. Pinho, J.T.C. Graça, F. Rodrigues and M. Hirofumi. 2013. Notes on the husbandry and long-term transportation of Bull Ray *Pteromylaeus bovinus* and dolphinfish *Coryphaena hippurus* and *Coryphaena equiselis*. Zoo Biol. 32: 222–229. DOI 10.1002/zoo.21048.

Sawai, E., Y. Yamanoue, M. Nyegaard and Y. Sakai. 2018. Redescription of the bump-head sunfish *Mola alexandrini* (Ranzani 1839), senior synonym of *Mola ramsayi* (Giglioli 1883), with designation of a neotype for *Mola mola* (Linnaeus 1758) (Tetraodontiformes: Molidae). Ichthyol. Res. 65(1): 142–160.

Shimoyama, T. and T. Kawamura. 1978. A few knowledge on keeping ocean sunfish *Mola mola*. J. Jpn. Assoc. Zoos Aquariums 20: 73–77.

Smith, M.F.L. 1992. Capture and transportation of elasmobranchs, with emphasis on the grey nurse shark *Carcharias taurus*. Aust. J. Mar. Fresh. Res. 43: 325–43.

Smith, M.F.L., A. Marshall, J.P. Correia and J. Rupp. 2004. Elasmobranch transport techniques and equipment. pp. 105–132. *In*: M. Smith, D. Warmolts, D. Thonney and R. Hueter [eds.]. Elasmobranch Husbandry Manual: Captive Care of Sharks, Rays, and Their Relatives. The Ohio Biological Survey, Columbus, OH.

Sommer, F., J. Christiansen, P. Ferrante, R. Gary, B. Grey, C. Farwell et al. 1989. Husbandry of the ocean sunfish, *Mola mola*. Am. Assoc. Zool. Parks. Aquar. Reg. Conf. Proc. USA: 410–417.

Sousa, L.L., R. Xavier, V. Costa, N.E. Humphries, C. Trueman, R. Rosa et al. 2016. DNA barcoding identifies a cosmopolitan diet in the ocean sunfish. Sci. Rep. 6: 1–9. doi.org/10.1038/rep28762.

Sousa, L.L., I. Nakamura and D.W. Sims. 2020. Movements and foraging behavior of ocean sunfish. pp. 129–145. *In*: T.M. Thys, G.C. Hays and J.D.R. Houghton [eds.]. The Ocean Sunfishes: Evolution, Biology and Conservation, CRC Press. Boca Raton, FL, USA.

Suyehiro, Y. and T. Tsutsumi. 1973. On the rearing of ocean sunfish in an aquarium. Sci. Rep. Keikyu Aburatsubo Mar. Park Aquarium 5(6): 12–15.

Syväranta, J., C. Harrod, L. Kubicek, V. Cappanera and J.D.R. Houghton. 2012. Stable isotopes challenge the perception of ocean sunfish *Mola mola* as obligate jellyfish predators. J. Fish. Biol. 80: 225–231.

Tatsuki, T., H. Misaki and I. Miyawaki. 1973. On short rearing of the sunfish, *Mola mola*, in the Aquarium. J. Jpn. Assoc. Zoos Aquariums 15: 33–36.

Thys, T.M., J.P. Ryan, H. Dewar, C.R. Perle, K. Lyons, J. O'Sullivan et al. 2015. Ecology of the ocean sunfish, *Mola mola*, in the southern California current system. J. Exp. Mar. Biol. Ecol. 471: 64–76.

Tibby, R.B. 1936. Notes on the ocean sunfish, *Mola mola*. Calif. Fish Game 22(1): 49–50.
Tsuzaki, J. 1986. Record time in captivity for the ocean sunfish, *Mola mola*. Anim. Zoos 224–227.
Visser J. 1996. The capture and transport of large sharks for Lisbon Zoo. Int. Zoo News 43(3): 147–151.
Watanabe, Y.Y. and J. Davenport. 2020. Locomotory systems and biomechanics of ocean sunfish. pp. 72–86. *In*: T.M. Thys, G.C. Hays and J.D.R. Houghton [eds.]. The Ocean Sunfishes: Evolution, Biology and Conservation, CRC Press. Boca Raton, FL, USA.
Young, F.A., S.M. Kajiura, G.J. Visser, J.P.S. Correia and M.F.L. Smith. 2002. Notes on the long-term transport of the scalloped hammerhead shark *Sphyrna lewini*. Zoo Biol. 21: 243–251.

Chapter 14

Ocean Sunfishes and Society

Tierney M. Thys,[a,*] *Marianne Nyegaard*[b] and *Lukas Kubicek*[c]

Introduction and Records

In the piscine pantheon of noteworthy fishes, the ocean sunfishes hold a position of honor, possessing one of the oddest morphologies. They are true superlatives with an extensive number of records to their name (specifically the genera *Mola* and *Masturus*). Their extraordinary attributes include being the world's heaviest extant teleost (bony) fish—an adult *Mola* spp. can weigh at least 2.3 metric tons (Sawai et al. 2020 [Chapter 2], Guinness World Records 2019)—and having the greatest number of ova ever recorded in a single vertebrate female [847 million in a 2.2 m TL individual (Forsgren et al. 2020 [Chapter 6]), 300 million ova in a 1.5 m TL female (Schmidt 1921)]. While growth rates in the wild are unknown, in captivity *Mola mola* can grow from 26 to 399 kg (57 to 880 lbs) in 15 months, averaging 0.82 kg/day (Powell 2001). More typical growth rates are 0.02 to 0.49 kg/day (Nakatsubo and Hirose 2007, Pan et al. 2016). Gudger (1936) crowned the genus *Masturus* as the vertebrate growth champion of the world, reporting that it increased in weight 60 million times from hatching size to the full adult size of 545 kg . With even larger individuals on record, the genus *Mola,* however, eclipses the Gudger record—with a growth trajectory of an astounding 600 million times.[1]

The most prominent feature of the *Mola* and *Masturus* genera, however, is their strikingly abridged shape (Fig. 1): massive head-like bodies with no tail, adorned with long, twistable, lift-generating, vertical fins (Watanabe and Davenport 2020 [Chapter 5]). Notably, the molids are the only vertebrates known to possess such a vertically-oriented, lift-generating system (Watanabe and Sato 2008). Another anatomical oddity is their spinal cord which lies entirely within their cranium making it the shortest in relation of brain to body weight of any vertebrate (Chanet et al. 2013, Uehara et al. 2015).

Despite these strange anatomical features and their seeming ungainly appearance, molids can swim with surprising grace and competence against moderate ocean currents. *Mola* spp. also have a surprisingly extensive distribution, ranging as far north as the Arctic Circle in Norway (70°N) to as far south as Argentina's Beagle Channel (El Penguino 2013, See Fig. 2A in Hays et al. 2020 [Chapter 15]). *Mola mola* have been recorded sculling 20–30 km/day (e.g., Potter et al. 2010) and diving as deep as 1000 m (see Sousa et al. 2020 [Chapter 8] for review of movements). Being

[a] California Academy of Sciences, San Francisco, CA 94118, USA.
[b] Memorial Museum Tamaki Paenga Hira, Natural Sciences, The Domain, Private Bag 92018, Victoria Street West, Auckland 1142, New Zealand; Email: mnyegaard@auklandmuseum.com
[c] Aussermattweg 22, 3532, Zäziwil, Switzerland; Email: molamondfisch@gmail.com
* Corresponding author: tierneythys@gmail.com

[1] Based on a larval weight of 3.7 mg and an adult weight of 2,300 kg (Sawai et al. [Chapter 2]). The larval weight was obtained from two Molidae specimens stored in ethanol, both 2.2 mm TL, rehydrated in water prior to weighing (T. Nakatsubo unpublished data).

Figure 1. Contemplating a sunfish captured at Catalan Bay, Gibraltar, The Illustrated London News, volume LX, June 22, 1872. Rights purchased from Getty Images.

capable of impressive burst speeds [6 m/s recorded for *Mola mola* (Thys et al. 2017)], both *M. mola* and *Mola alexandrini* are known to breach, a behavior presumably employed to escape predators (Fig. 2 left) and overly zealous cleaner fish (Konow et al. 2006 [*M. alexandrini*]). One 1883 report from *Harper's Weekly* records breaching off the west coast of Ireland in response to repeated shooting by rifle-wielding fishermen (Fig. 2 right)—an apparently not uncommon, and rather unfortunate, pastime (Sweeney 1972).

While ocean sunfishes remain elusive, there is a long history of human interest in this group dating back thousands of years (Porcasi and Andrews 2001). In the public eye, the odd shape of *Mola* spp. has endeared them to aquarium viewers at sites around the world (Howard et al. 2020 [Chapter 13]). *Mola mola* is, in fact, the favorite fish of Julie Packard, long-time director of the world-renowned Monterey Bay Aquarium in California, USA where this species is often displayed in the Open Sea tank (Packard 2020 [Foreword]).

First Written Impressions

The molids captured the attention of the ancient Roman philosopher and naturalist, Pliny the Elder, aka Gaius Plinius Caecilius Secundus (63–*ca.* 113). Pliny called ocean sunfish "*Orthagoriscus*" (Hill 1752), a Latin name which literally translates to "sucking pig" likely in reference to observations that molids grunt when removed from the water (Andrews 1948). In 1554, the ocean sunfish (also known as mole) was illustrated for the first time (Fig. 3) by professor and naturalist Guillaume Rondelet (1507–1566), author of *De Piscibus Marinis in Quibus Verae Piscium Effigies Expressae Sunt* which translates to "the marine fishes in which true images of fish are portrayed". Rondelet, considered one of the fathers of Renaissance zoology, dedicated countless hours to observing the natural world, making detailed drawings (Rondelet 1558).The Italian papal physician and Renaissance naturalist Ippolito Salviani (1514–1572) also depicted sunfish in his seminal work, *De piscibus tomi duo, cum eorumdem figuris aere incisis*, which was later reprinted as *De aquatilium animalium curandorum formis* (Salviani 1558). In 1609, the polymath, writer and scholar Johan Jakob Grasser noted that sunfish shone in the night sea, frightening people because they thought it was a ghost (Grasser 1609). Whether sucking pigs or shining ghosts, the ocean sunfishes have received their fair share of strange sobriquets.

What's in a name?

The name *Mola* is Latin for millstone and describes one of three genera of ocean sunfishes, the other two being *Masturus* and *Ranzania.* When written in lowercase, 'mola' serves as a colloquial, catch-all term to describe any member of the family Molidae. The genus name, *Masturus*, is Greek and

Figure 2. (Left) Ocean sunfish *Mola* spp. breaching in Monterey Bay, CA USA. Photo used with permission by Kate Cummings, Blue Ocean Whale Watch; (Right) Sunfish Shooting Off West Coast Of Ireland 1883. *Harper's Weekly*, New York. Illustration used under public domain.

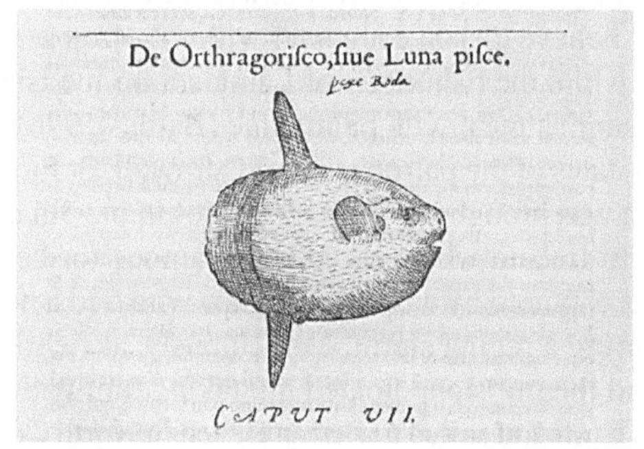

Figure 3. The first published illustration an ocean sunfish drawn by G. Rondelet 1554 De Piscibus Marinis, Rondelet, Guillaume publisher Lugduni, apud Mathiam Bonhomme. Photo reprinted courtesy of The Linda Hall Library of Science, Engineering & Technology.

means ‘nipple-tail’ in reference to the posterior protrusion from this genus's abbreviated posterior region (termed clavus: Latin for rudder; Fraser-Brunner 1951). The genus *Ranzania* is named after the Italian naturalist and professor Camillo Ranzani (1775–1841) who served as the Director of the Museum of Natural History of Bologna, Italy from 1803 to 1841 and wrote *Elementi di Zoologia* (Ranzani 1819, 1839). (See Fig. 1 in Sawai et al. 2020 [Chapter 2] for drawings of all three genera.)

Since their initial discovery, the ocean sunfishes have acquired more than 140 common names (Froese and Pauly 2020), adding to the ongoing taxonomic confusion with the group (Sawai et al. 2020 [Chapter 2], Fraser-Brunner 1951). However, this plethora of names provides insight into features deemed noteworthy by different cultures. In Taiwan, ocean sunfish are known as the *toppled wheel fish* (fān *chē yú* 翻車魚) due to their ungainly size and habit of flopping over on their side. Another more colorful common name is described later in the Taiwan section of this chapter. In Ecuador, fishermen refer to molids as *pez borracha* (Spanish for drunken fish) due to their swerving, sometimes topsy-turvy manner of swimming near the sea surface (A. Hearn personal communication). In most Spanish-speaking countries, however, as well as most European countries, sunfish are known as *pez luna* (moonfish) due to their roundish shape and grey color, e.g., *poisson lune* (France), *pesce luna* (Italian) and *månefisk* (Sweden/Norway). In Denmark, molids are called *klumpfisk* (clump fish) and in Turkey they are *pervane* which translates to moth or ship's propeller, presumably because they sometimes swim around in circles. In Japan, the common name for sunfish is *manbou* and one species, *Mola alexandrini*, is named *ushi-manbo* (ウシマンボウ) which translates to cow (ushi) mola (manbou) since this species can develop a large bulbous, bovine-like appearance upon reaching adulthood (Yamanoue et al. 2010, Sawai et al. 2020 [Chapter 2]). The most frequent common name in the English-speaking world is *sunfish*, alluding to the fish's habit of basking on the sea surface (Dewar et al. 2010, Nakamura et al. 2015). Their "shining skin" was also likened to the sun by Yarrell (1836). A note of caution is due, however, with common names since numerous other fishes sport identical ones which can lead to confusion (e.g., opah, *Lampris* spp. are commonly called moonfish (Wegner et al. 2015) and freshwater centrarchids (e.g., *Lepomis* sp.) are also known as sunfish.

Folklore

Little is known of the cultural folkloric significance of ocean sunfishes, however they are referred to as *makua* in Hawai'i which means king (Titcomb 1977). Hawai'ians are said to have revered the sunfish makua, feeling they were a type of spirit. Jenkins (1895) also recounts native [Polynesian] fishermen warning against harming *Ranzania* (slender mola) for fear that they were akua (spirits or deities). Jordan and Evermann (1898) report that native peoples referred to sunfish as King of the Mackerels and King of the Tunnies. According to Smith (1961), line fishermen off the coast of South Africa were quite superstitious of seeing sunfish at sea. If they encountered one, they would return to shore for fear of an impending storm. In contrast, Indonesian fishermen from Majene on the island of Sulawesi allegedly believe that an encounter with a mola at sea brings good fortune with an abundant catch (Nur 2019) while the fishermen of Lamalera on the island of Lembata purportedly enjoy capturing and eating sunfish meat with palm wine (Barnes 1996).

In the United Kingdom, tales of the patron saint of Cornwall, St Piran, have become intertwined with ocean sunfish. St. Piran was brought up in South Wales in the 5th century, established a church in Cardiff and then moved to Ireland where he became famous for healing the sick. According to legend, in his old age he was captured by a band of pagans who tied a millstone around his neck and flung him off a high cliff during a raging storm. Miraculously, as he sank beneath the mountainous waves, the storm calmed and the millstone floated up, taking him with it. Piran then clambered up onto the millstone, used it as a raft and sailed to Perran Beach in North Cornwall where he purportedly set up another church, discovered tin for the Cornish, and performed many further miracles. Given that *Mola* is Latin for millstone, one could whimsically speculate that when St. Piran bobbed up to the surface, instead of rafting atop a floating millstone, he simply seized hold of a basking ocean sunfish and used it as a life buoy. Today, ocean sunfish are seen swimming around the headlands in North Cornwall, most commonly in the summer months (Houghton et al. 2006). Perhaps they are checking in to see if St Piran needs their help again. (D. Herdson, Edinburgh Natural History Museum personal communication).

Sunfish on the Menu

A Note about Toxicity

While sunfishes are eaten in many regions worldwide, consumption is much more popular in eastern than western countries. In Europe, since molids are Tetraodontiformes and, by default, associated with the neurotoxin tetraodotoxin (TTX), the sale of all molid species has been banned commercially since 2004 by the European Commission (E.C.). To date, however, no TTX has ever been found in any molid species (Baptista et al. 2020 [Chapter 11]). The only recorded sunfish toxin case is a possible palytoxin or palytoxin-like compound poisoning reported after the consumption of a sharptail mola *Masturus lanceolatus* (Huang et al. 2011). Despite E.C. ban, sunfish are still consumed sporadically in Europe and Italian food blogs offer numerous recipes (e.g., RitaAmordicucina 2019).

Earliest Recorded Fishery

The oldest documented fishery for ocean sunfishes can be traced to fossil remains from the Channel Islands off the central and southern Californian coast of the USA (Porcasi and Andrews 2001). Archeological remains date back to the mid-Holocene and suggest the existence of a robust sunfish fishery between 5000–8000 years ago (Porcasi and Andrews 2001).[2] Unfortunately, molid skeletons are not heavily ossified and do not fossilize well, resulting in a depauperate fossil record, limited mostly to mouth parts and hard claval ossicles (Carnevale et al. 2020 [Chapter 1]). Hundreds of ossicles, however, have been retrieved from sites in the Channel Islands, such as sites at Little Harbor on Catalina Island. These ossicles range in size from 12 mm to an astounding 8 cm in length suggesting that the individual sunfishes may have spanned an immense size range.[3] A realistic effigy of a sunfish carved from abalone shell collected from a midden on nearby San Nicolas Island in the late 1800s further suggests that native peoples [ancestors of the Chumash (Erlandson 1994)] appreciated sunfish not only gastronomically but also aesthetically (Fig. 4).

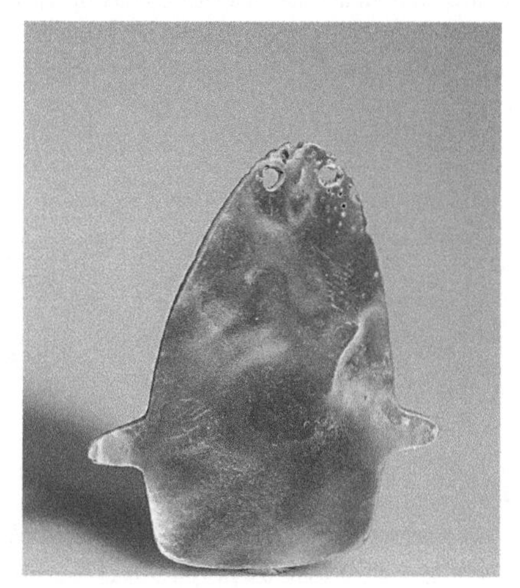

Figure 4. Polished abalone shell ornament from Native American site, Ventura County, San Nicolas Island, CA, USA. Collected by James Terry, Curator of Anthropology AMNH 1891–1894. Courtesy of the Division of Anthropology, American Museum of Natural History Catalog No. T/16762.

Consumption and Reverence in Japan

In more recent history, sunfish can be traced through writings and illustrations from Asian countries, particularly Japan, where they have been revered and consumed for centuries. Today, sunfish continue to occupy a special place in Japanese culture with more than 40 different common names—the best known name being manbou (Sawai et al. 2020 [Chapter 2]). First mentioned in a 17th century cookbook (Sawai 2017, Nyegaard et al. 2020 [Chapter 12]), sunfish were used as a form of tax payment in the 17th–18th centuries during the time of the Shōguns—military dictators who ruled in feudal Japan from 1185–1868 (Jingushicho 1910). Artist Kurimoto Tanshu (1756–1835) provides several detailed illustrations of *Mola* and *Ranzania* (Fig. 5) in a book titled *Manboko* which describes how sunfish skin was used as decoration for Japanese swords, internal organs were made into side

[2] A comical short film entitled *Fish Guys* depicts a graduate student discovering that her frustratingly mysterious and numerous roundish artifacts from Native American middens in California's Channel Islands are, in fact, hundreds of ossicles left behind from large ocean sunfishes dating back to the early Holocene. http://www.youtube.com/watch?v=X4smMyPNcA0 .

[3] For comparison, the paraxial ossicle (i.e. the 'middle' ossicle of the clavus; Fraser-Brunner 1951) on a ca. 35 cm TL *Mola mola* from California is 1 cm in length, while all the other ossicles are smaller (approximately ¼ - ½ the length) with some not yet even formed. There appears to be large variation in the ossicle size of individual molids (T. Thys unpublished data).

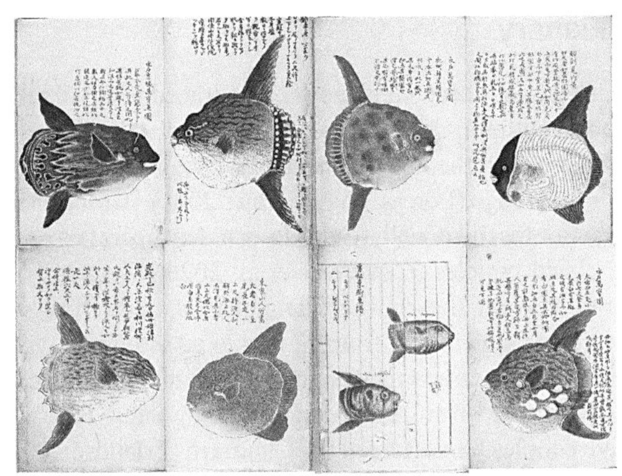

Figure 5. Detailed illustrations of ocean sunfishes (*Mola* sp. and *Ranzania* sp.) by Artist Kurimoto Tanshu (1756–1835) in Manboko [翻車考]. Reprinted with permission from the National Diet Library of Japan.

Figure 6. A. Illustration from *Manboushi* by Nanyo Hara (Hara Nan-yō) (1753–1820). Reprinted with permission from National Archives of Japan (Takebashi, Tokyo).

dishes, and skeletal bones (aside from the hard vertebrae) were pickled in plum vinegar and served as appetizers (Tanshu 1825).

Another Japanese book entitled Manboshi *[査魚志]* by Haro Nanyo contains lavish illustrations from the early 19th century (1802) depicting fishermen performing a vivisection on a large, wild sunfish, removing body parts and loading them into a shore-bound boat for presumed consumption (Fig. 6).

One measure of obsession with sunfish can be found in the number of aquariums displaying them. In Japan, seven different public aquariums display live sunfish, more displays than any other country in the world (See Table 1 in Howard et al. 2020 [Chapter 13]). In one location, Kamogawa Seaworld, where live sunfish have been on display since February 1971 (Kondo 1986), the town of Kamogawa even adopted the sunfish as their mascot between the years of 1996–2010.

Consumption and Reverence in Taiwan

Sunfish are also immensely popular in Taiwan albeit without such a long history. From 2003–2006, an annual sunfish festival was held in the town of Hualien and a statue was even erected to celebrate the sharptail sunfish (*Masturus lanceolatus*) (Chang et al. 2018) (Fig. 7). The Hualien sunfish festival at its height drew more than 120,000 visitors, succeeded in increasing the popularity of sunfish and featured a multitude of intriguing dishes (Fig. 8). Nearly every part of the

Figure 7. A *Masturus* sculpture in Hualien, Taiwan. Photo taken and printed with permission by Lukas Kubicek.

曼波魚專區 Manbo Fish

53 曼波酥豆腐	Manbo Crispy Tofu	119
116 干貝蒟蒻	Scallop Jelly	189
119 黑胡椒曼波魚排	Black Pepper Manbo Fish Fillet	459
117 酥炸曼波魚	Deep Fried Manbo Fish	489
118 麻油干貝魚	Sesme Oil Scallop And Fish	489
104 鮮炒曼波肉	Stir Fried Manbo Meat	549
105 XO醬龍腸	XO Sauce Manbo Intestine	589
49 曼波生魚片	Manbo Sashimi	499
110 曼波蛤蠣湯	Manbo And Clam Soup	279
120 曼波燕窩	Manbo And Swallow's Saliva	89
32 曼波膠原茶	Manbo Collagen Tea Jelly	29
1411 曼波冰棒	Manbo Icicle	35
12 活力曼波果汁	Energetic Manbo Icicle	119
914 曼波果汁（壺）	Energy Fruit Juice (pot)	149
969 曼波巴黎風	Manbo Paris Wind	459
916 雀巢曼波肝	Nestle Manbo a liver	179

Figure 8. Menu of mola dishes from restaurant in Hualien, Taiwan. Photo taken and printed with permission by Lukas Kubicek.

sunfish is consumed, from the fin musculature and spongy skeleton to the hypodermis and viscera. Some of the dishes made from or inspired by these fish include: *dragon intestines* (龍腸) which use the thick rope-like intestines, *energetic sunfish icicles* which are just ice spears resembling sunfish, *sunfish and swallows saliva* which uses bird nests and sunfish meat and *sunfish collagen tea jelly* which uses the skin (Fig. 8).

Sunfish flesh has several features that may detract from its gustatory appeal. For example, it possesses a characteristic odor that can be off-putting to some (T. Thys personal observation) and typically host a sizeable variety of parasites (72 species of micro and metazoan parasites; Ahuir-Bajara 2020 [Chapter 10]) which occupy nearly every body part and thus require extra preparation and cleaning. The hypodermis transforms from a white coconut meat-like consistency into a transparent, jelly-like consistency when cooked (T. Thys personal observation) which may or may not be a desirable feature depending on the chef. According to Chefs' Resources (an online culinary resource for professional chefs www.chefs-resources.com/seafood/seafood-yields/), ocean sunfishes have one of the lowest fish to fillet ratios (16% with skin off) versus other more commonly targeted fishes (e.g., salmon, sole, trout, tuna and drum) which yield between 30–80%. Relini et al. (2010) report that a five kg *Mola mola* yields only a 0.82 kg fillet. When contrasted with Tanaka's eel pout *Lycodes tanakae* and magistrate armhook squid *Berryteuthis magister*, sunfish is neither high in protein, nor high in total free amino acid content (Lee et al. 2012). The red and white fin muscle of *Mola* and *Masturus* is also quite watery (79–83% water for *M. mola* (Watanabe and Davenport 2020 [Chapter 5]) and reduces to one third its size when cooked (T. Thys personal observation).

One curious Taiwanese nickname for sunfish, 打某魚, relates to this watery property of the sunfish musculature. The name translates literally to "beating wife fish" and derives from the tale of a naive wife buying a large sunfish at the fish market and lugging it home. After she cooks it, she discovers, much to her dismay, that the meat has shrunken down to a shockingly small size. With only a miniscule dinner to offer her spouse, the husband becomes angry, thinking his wife has hidden the fish meat from him and trouble ensues. In reality, no husband would dare beat his wife since she is the one holding the knife in the story and this name is used only in a joking manner in Taiwan, not as a testament to actual domestic violence (H.Y. Young personal communication).

Consumption in the Mediterranean

Sunfish have been eaten in the Mediterranean for more than 400 years. In the Ligurian Sea near Camogli, Italy fishermen still use an ancient, two-chambered artisanal tuna trap called a tonorrella (Mouton 2004, Nyegaard et al. 2020 [Chapter 12]), that has become a hotspot for sunfish encounters. Parona (1919) recorded more than 500 individual mola in the trap in June 1912, noting that while the fish could be used commercially, its abundance was seen as hampering the fishing of more desirable species, such as bluefin tuna. Today, since all commerical trading of molids was banned in 2004 by the European Commission (Relini and Valarino 2016), sunfish are simply seen as bycatch that are thrown back alive. Their continued occurrence in these traps does however, allow for additional molid monitoring in these waters (Relini et al. 2010). For a detailed overview of ocean sunfish fisheries and global bycatch see Nyegaard et al. 2020 [Chapter 12] and Fig. 2 in Hays et al. 2020 [Chapter 15].

Waste Not Want Not: Historical Uses of Sunfish Beyond Consumption

Sunfish body parts have served a variety of non-edible uses since the 1500s (Rondelet 1558, Barlow 1740, Bennett 1840). The two most commonly mentioned body parts include oil from the liver and the hypodermis—the layer of material underlying the thin outer skin which thickens with size and can be > 7 cm thick (Bemis et al. 2020 [Chapter 4], Watanabe and Davenport 2020 [Chapter 5]). According to mariners, the hypodermis held elastic properties that could be fashioned into bouncy balls by native people for recreation (Sweeney 1972, Thys 1994). Barlow (1740) reported that the skin (presumably the hypodermis) could be made into an effective glue when boiled down. Green (1899) also found the skin (hypodermis) contained albuminoids and collagens known for their glue-like properties. In Japan, sunfish skin was used to decorate sword sheaths (Tanshu 1825) and the liver oil was touted as a folk remedy to heal sprains, contusions, rheumatic pains (Rondelet 1558, Bennett 1840, Shufeldt 1900) and human gastric ulcers (Akahora et al. 1990). The oil has also been clinically tested and found to possess potential preventative and suppressive properties for gastric lesions in rats (Akahori et al. 1990) and as a detector for anti-tumor promotors connected with the Epstein-Barr virus (Hiroyuki et al. 2001).

Ship Collisions

Beyond fisheries and bycatch, additional negative human interactions with sunfish involve ship strikes. Notable collisions include a strike with the warship HMS Leander off the Japanese coast in 1886 (Cole 2011). On September 18th 1908, 65 km from Sydney Harbour, Australia, the steamship

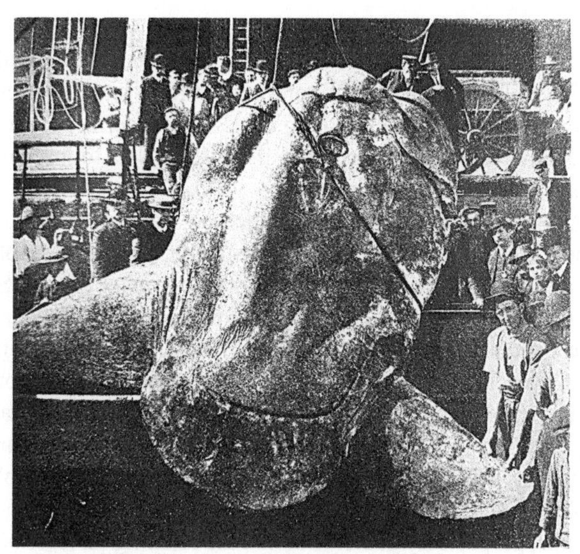

Figure 9. 3.1 m (10 ft) TL, 4.3 m (14 ft) depth, 1996 kg (4400 lb) sunfish caught on the propeller of the M/V Fiona near Sydney Australia in 1908 (Illustration from Gudger (1928) Public domain).

M/V Fiona ran into a 3.1 m TL, 1996 kg (4400 lb) ocean sunfish (Fig. 9), which jammed the ship's propeller (Gudger 1928). More recently on 13th October 1998, another large sunfish, 2.5 m TL weighing 1400 kg became stuck on the bow of the cement carrier M/V Goliath in Jervis Bay, New South Wales, Australia (Nyegaard and Sawai 2018). This fish caused the ship to slow down from 14 to 11 knots and its rough skin wore the ship's paint to bare metal.

In the case of yacht racers, ocean sunfishes pose a regular hazard with numerous reports of collisions happening during competitions such as during the annual Rolex Sydney Hobart Yacht race (e.g., The Daily Telegraph 2019, Kothe 2012). Competitors are wary of encountering a "malevolent sunfish lurking in the Bass Strait" which could bring their boats to a grinding halt and ruin their chances for a win (Clarey 2014, Montgomery 2015). In any ship collision, the softer-bodied sunfish will of course bear the brunt of the damage and, if not killed outright, carry scars as evidenced by the presence of multiple propeller gashes found regularly (e.g., see Fig. 9 in Nyegaard et al. 2020 [Chapter 12]).

Ecotourism

While boaters may seek to avoid sunfishes, snorkelers and SCUBA divers actively seek them out. Indeed, swimming with wild sunfishes is a popular bucket list item at ecotourism destinations world-wide. As with other marine megafauna, ecotourism can bring in valuable tourist dollars [e.g., diving with manta rays is reportedly worth 73 million $USD annually (O'Malley et al. 2013), while shark diving is worth at least 180 million $USD (e.g., Haas et al. 2017, Clua et al. 2013, Vianna et al. 2011–2, Jones et al. 2009)]. The economic value of ocean sunfish ecotourism is currently untallied and worthy of further research.

Two of the most reliable sites for swimming or SCUBA diving with wild sunfish include Punta Vicente Roca off Isabela Island (Galápagos Ecuador) and Nusa Penida Island, Bali, Indonesia (Jarvis 2016). In both places, *M. alexandrini* can be found with some predictability: seasonally in the case of Bali (Nyegaard 2018) and year-round in the case of Galápagos (Thys et al. 2017). Individual fish visit these reefs to solicit the cleaning services of smaller fishes like butterfly fish, angelfish and wrasses (Konow et al. 2006, Thys et al. 2016, Thys et al. 2017). Strict restrictions on the total number of tourists have been implemented in Galápagos whereas in Bali, dive operators have implemented a voluntary code of conduct (Pokja KKP NP 2012). A persistent concern in Bali is that frequent diver interactions disrupt the sunfishes' ability to solicit cleaning from smaller fishes (A. Taylor, Blue Corner Dive/Blue Corner Marine Research personal communication 2020, Thys et al. 2016, Ahuir-Baraja 2020 [Chapter 10]). Currently, the Indonesian government's Ministry of Marine Affairs and Fisheries is considering sunfish as a priority species for protection partially due to the rising concerns of tourism-related impacts in and around Bali but requires additional in-country life history data to implement formal protection measures (M. Welly personal communication).

Additional places for targeted sunfish tourism include: Palau, Sardinia and Camogli, Italy (V. Cappanera www.ziguele.it/). In Eaglehawk Neck, Tasmania, sunfish are not a targeted ecotourist feature, but constitute an added perk when seen during whale watching tours (S. Buetow, Wild Ocean Tasmania, personal communication). Similarly, while sunfish are not directly targeted during SCUBA tours, they are regularly encountered in places such as the Alboran Sea, Spain, and off the coast of Chile (D. Patricio Hombre y Territorio, R.G. Sanchez Buceando Chile, personal communication).

Japan offers two types of live sunfish interactions. The local fishery cooperative and set-net fishery near Tosashimizu-city and Tateyama on the Boso Peninsula, Chiba Prefecture, allow customers to swim with captured sunfish in floating cages from late April to August. People can swim with wild sunfish off the Shizuoka Prefecture near the Izu Peninsula (January to April) and Osezaki, Numazu (May–June) where sunfish purportedly gather to solicit the cleaning services of Japanese butterflyfish *Chaetodon nippon* from May to June (T. Nakatsubo personal communication).

Artistic Inspirations

Ocean sunfishes have also inspired the arts and artists throughout the ages, from the famous Dutch artist M.C. Escher (1898–1972), who placed a sharptail sunfish (*Ma. lanceolatus*) at the center of his masterpiece *Mosaic II* (Fig. 10), to stamp creators in the Cook Islands and Guernsey Island (Fig. 11A,B), Cuba, Monaco, Spain, and Vietnam. Sunfish have been used by street sign makers in Hualien Taiwan (Fig. 11C), by the Kuna people of Panamá in their handsewn traditional textiles [also known as molas (Fig. 11D)] and have been featured in murals by artists such as Ray Troll (Fig. 11E).

Sunfish have also served as a muse for popular video games like *Survive Mola mola*! (Fig 12C) and where the goal is to keep a digital sunfish alive by dodging often ridiculous ways of dying. Likewise, sunfish are represented in the immensely popular Japanese video game *Pokémon*, as the character Alomomola—a water-type, pink heart-shaped character with the ability to heal or hydrate. More broadly, sunfish shapes and insignias can also be found on countless trinkets and tchotchkes from hourglasses to cutting boards to ear-cleaners and as the emblem on the official dive flag of Monterey Bay, California (Fig. 12A-B).

Figure 10. M.C. Escher, Mosaic II. Reprinted with permission by Escher Foundation.

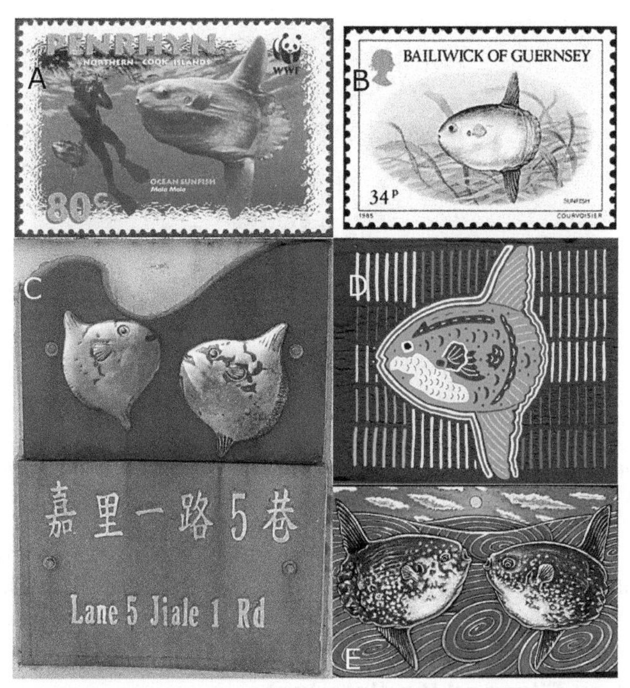

Figure 11. (A) Northern Cook Island stamp. Reproduced with permission granted by the Cook Islands Philatelic Bureau. (B) Guernsey stamp image by kind permission of Guernsey Post Ltd. (C) Street sign in Hualien, Taiwan. Photo reprinted with permission by Lukas Kubicek. (D) A Kuna mola from Panamá of a *Mola mola* compliments of the Powell family. Photo reprinted with permission by Tierney Thys. (E) Mola drawing. Reprinted with permission from artist Ray Troll.

Figure 12. (A) Sunfish tchotchkes. Photo reprinted with permission by T. Thys. (B) DiveMola™ flag designed by Patrick Anders Webster. (C) *Survive! Mola mola* game Creative Commons.

Sunfish Art Activism

Aside from aesthetic trinketry, sunfish art installments have served as useful tools for raising ocean conservation funds. In 2014, the Plymouth College of Art in the UK, as part of the *Making Waves*

Figure 13. Sunfish sculpture entitled *Smitten* painted by artist Peter Heywood as part of the *Making Waves* program. Photo used with permission by Peter Heywood.

program (e.g., Fig. 13) at the National Marine Aquarium, commissioned artists to paint fifteen ocean sunfish sculptures to be displayed across the city. The sculptures were used to increase ocean awareness and through their sales raised £12,890 for marine conservation (Fishkeeping News 2014).

Sunfish Caught on Film and the (Inter)net

The ocean sunfishes have made notable media appearances and reached millions of viewers in numerous nature documentary films (e.g., BBC's *Planet Earth* and *Blue Planet I* and *II*, Japan's NHK's *The Global Family* and National Geographic's *Strange Days on Planet Earth*). The popularity of sunfish on the internet continues to increase via Facebook and YouTube postings (e.g., a Google search brings up > 26,000 ocean sunfish videos and millions of sunfish photos are posted on sites such as Instagram and Tumblr). The TED talk by T. Thys, *Swimming with Sunfish* (www.TED.com) has amassed > 1.3 million views (as of Feb 2020). Such widespread public interest can be harnessed to promote valuable research through a variety of community science research tools.

Community Science: Sightings, Tracking and Strandings

Community science websites like iNaturalist (www.inaturalist.org), www.oceansunfish.org, and the MatchMyMola platform at www.oceansunfishresearch.org allow the public and community scientists to log photos and observations of sunfish to help scientists reveal underlying patterns in sunfish occurrences worldwide. For example, www.oceansunfish.org, established in 2002, has amassed more than 3000 sunfish observations across the globe. When carefully curated and maintained long-term, such databases can be powerful tools to increase our understanding of species-level distribution, inform species distribution models (Phillips et al. 2017) and potentially reveal range shifts over time in response to forces such as climate pollution. A recent example of the utility of globally-supported platforms involved a stranded sunfish on the central California coast, which was photographed and logged on iNaturalist. It quickly caught the eye of sunfish scientists in far flung Australia and New Zealand, and an *ad hoc* global collaboration arose to obtain further photographs and a tissue sample of the fish for robust species identification (Leachman 2019). The fish was eventually identified as *Mola tecta*—the second-ever record of this species in the northern hemisphere at the time—by the very scientist who first described the species on the other side of the world (Nyegaard et al. 2018).

Curated sunfish photo collections also hold the potential for resightings of individual sunfish over time. MatchMyMola focuses on the area of Bali, Indonesia to investigate resightings of the seasonal sunfish phenomenon targeted by the local dive tourism industry (Konow et al. 2006, Thys et al. 2016). While comparisons of individuals based on the elaborate skin patterns of the species (*M. alexandrini*) are relatively straightforward, building up a useful photo ID catalogue based on images submitted by tourist divers is challenging. Firstly, photos are rarely of sufficient quality for robust skin pattern comparison due to the low light levels at > 20 m depth where the sunfish are most commonly seen. Further, the angles at which divers take photos are highly variable, impeding comparison of all photos against one another. Likewise, while the eye area is commonly included, the skin patterns around the head are often obscured by cleaner fish. Finally, *M. alexandrini* can, like the other *Mola* species, change the contrast of its skin from uniform grey to nearly black and white (M. Nyegaard and T. Thys personal observation, Nyegaard et al. 2018, Sawai et al. 2017). If photographed when displaying its 'low contrast mode' in low light levels and/or from a distance, any existing skin patterns can be nearly impossible to detect. A final challenge is the uncertainty with regards to skin pattern stability over time. Sunfish in captivity can provide insight however, aquarium-kept *M. mola* differ in their responses. In Hirtshals, Denmark, at the North Sea Oceanarium (Nordsøen Oceanarium), aquarists have previously reported that after some time in captivity, *M. mola* tend to lose their skin patterning and instead appear uniform silvery-grey (M. Nyegaard unpublished data). Aquarists at the Monterey Aquarium in California, USA however, note that while skin patterning may fade slightly with time, age and growth, patterns displayed ultimately depend on the level of excitement and stress. During regular cruising, coloration and contrast is pale; during feed sessions however, the contrasting spots and purplish sheen become apparent. During physical examination or handling, the skin patterns show most strikingly. These observations correspond well with observations of wild sunfish off Nusa Penida, Bali, where the skin pattern of *M. alexandrini* varies from nearly uniform grey during cleanerfish interactions to stark contrast when being harassed by divers (T. Thys and M. Nyegaard personal observation). While further research is required to better understand skin pattern stability, preliminary results from the Bali area indicate that skin patterning of wild *M. alexandrini* does remain stable at least over several years (M. Nyegaard unpublished data). Despite all these caveats, gathering additional high quality photos of wild molas from the public will certainly add to our knowledge and potential for non-invasive individual tracking.

Notably, photo identification has been used successfully to identify and track re-sightings of individuals of many marine species based on skin patterns such as whale sharks (www.whaleshark.org), manta rays (www.mantamatcher.org, www.mantatrust.org/idthemanta) and teleost fish (https://spottinggiantseabass.msi.ucsb.edu). Other methods rely on the shape of and scarring on the flukes of cetaceans (www.flukebook.org). For molids, this method has proven less effective on its own yet is highly valuable when used in combination with skin patterns (M. Nyegaard unpublished data). However, pattern matching by eye can be very time-consuming and labor intensive. Possible alternatives include the development of algorithms for computerized pattern recognition (e.g., www.wildme.org). Due to the complexity of comparisons, machine learning may also prove to be a useful tool for robust matching outcomes in the future.

Fish of the Future and Ambassadors for the Ocean

In light of our human-impacted, overfished world ocean (FAO 2018), molids may possess a range of traits that enhance their adaptive capacity. These include: global distribution for each genera which can offset local extirpations, a broad thermal range (e.g., 1.8°–22°C for *Mola mola*; Dewar et al. 2010: 4.5°–23.2°C for *M. alexandrini*; Thys et al. 2017) and the ability to dive deep—e.g., 1000 m for *Mola alexandrini* (Thys et al. 2017) as a means of seeking possible thermal refuge from rising surface temps. They have an extremely high ova output of more than 800 million ova (Forsgren et al. 2020 [Chapter 6]), the ability to grow fast (Powell 2001) and can reach sizes (> 2 m) large enough to outgrow many predators. Preliminary studies suggest they may be robust enough to tolerate short stints in hypoxic ocean conditions (Thys et al. 2015). Furthermore, they are non-schooling as adults, reducing their chances of capture *en masse*. When they do happen to be captured in trap nets, they often can survive the initial capture and many can be released alive if the fishermen take the trouble to do so (Nyegaard et al. 2020 [Chapter 12]). Another attribute is that they are not gastronomically popular in the West and are banned commercially by the European Commission. As adults, they are able to capitalize on eating gelatinous zooplankton a widespread resource which is viewed in many regions as a menace (Chabin 2019), especially when numbers exceed normal background levels. Combined with their quizzical looks and ability to consume and thereby remove stinging jellies from the water, the sunfishes, as this chapter outlines, are endearing and engaging as marine megafauna. As we venture deeper in the Anthropocene, perhaps the most important role for the ocean sunfishes lies on land, serving as a charismatic ambassador. Certainly, their ability to capture the imagination, draws in members of the public to the realm of scientific discovery and ocean stewardship, and ideally, helps in a small way, to bring humanity together to restore declining ocean health before the mounting damages become irreparable.

Acknowledgements

Special thanks go to: Don Kohrs from Stanford University for invaluable and tireless research support, Toshiyuki Nakatsubo, Kamogawa Seaworld staff, Chuck Farwell from the Monterey Bay Aquarium and Etsuro Sawai for insight into the Japanese relationship with sunfish and ecotourism sites, Hong Young Yan and Stephen Huang for providing expert insight into the Taiwanese relationship with sunfish, Sylvia Earle and Eugenie Clark for additional sunfish insight and support, Natasha Phillips, Inga Potter and Lawrence Eagling for help with the figures and research, the Powell family for the Kuna mola of the *Mola mola*, Sea Studios, National Geographic, Chris Anderson and TED and many others for helping to promote and share sunfish stories with the masses.

References

Ahuir Bajara, A.E. 2020. Parasites of the ocean sunfishes. pp. 160–185. *In*: T.M. Thys, G.C. Hays and J.D.R. Houghton [eds.]. The Ocean Sunfishes: Evolution, Biology and Conservation, CRC Press. Boca Raton, FL, USA.

Akahora, F., T. Masaoka, F. Yamada, S. Arai and G. Kubo. 1990. Effects of liver extract from the ocean sunfish (*Mola mola*) on acute gastric lesions in the rat. Japan, J. Vet. Sci. 52(2): 419–421.

Andrews, A.C. 1948. Greek and Latin mouse-fishes and pig-fishes. Trans. Proc. Am. Philol. Assoc. 79: 232–253.

Baptista, M., J. Raimundo, C. Lopes, P.R. Costa, R. Rosa, J. Dutton et al. 2020. Biotoxins, trace elements and microplastics in the ocean sunfishes (Molidae). pp. 186–215. *In*: T.M. Thys, G.C. Hays and J.D.R. Houghton [eds.]. The Ocean Sunfishes: Evolution, Biology and Conservation, CRC Press. Boca Raton, FL, USA.

Barlow, Will. A paper concerning the *Mola salviani* and a glue made of it, in Philos. Transact. P. 1. 1740.456: 343–345.

Barnes, R.H. 1996. Sea Hunters of Indonesia: Fishers and Weavers of Lamalera, Clarendon Press, Oxford. pp. 467.

Bemis, K.E., J.C. Tyler, E.J. Hilton and W.E. Bemis. 2020. Overview of the anatomy of ocean sunfishes (Molidae Tetraodontiformes). pp. 55–71. *In*: T.M. Thys, G.C. Hays and J.D.R. Houghton [eds.]. The Ocean Sunfishes: Evolution, Biology and Conservation, CRC Press. Boca Raton, FL, USA.

Bennett, F.D. 1840. Narrative of a whaling voyage round the globe from the year 1833–1836, vol. 1-2, Richard Bentley, London UK.

Carnevale, G., L. Pellegrino and J.C. Tyler. 2020. Evolution and fossil record of the ocean sunfishes. pp. 1–17. *In*: T.M. Thys, G.C. Hays and J.D.R. Houghton [eds.]. The Ocean Sunfishes: Evolution, Biology and Conservation, CRC Press. Boca Raton, FL, USA.

Chabin, M. 2019. How to handle the worldwide jellyfish threat, Washington Post, 5 July 5 2019.

Chanet, B., C. Guintard, E. Betti, C. Gallut, A. Dettai and G. Lecointre. 2013. Evidence for a close phylogenetic relationship between the teleost orders Tetraodontiformes and Lophiiformes based on an analysis of soft anatomy. Cybium. 37(3): 179–198.

Chang, C.T., C.H. Chih, Y.C. Chang, W.C. Chiang, F.Y. Tsai, H.H. Hsu et al. 2018. Seasonal variations of species, abundance and size composition of Molidae in eastern Taiwan. J. Taiwan Fish. Res. 26: 27–42.

Clarey, C. 2014. 23 December. In ocean races, some of the biggest perils can't be seen, New York Times. Accessed March 20 2020.

Clua, E., N. Buray, P. Legendre, J. Mourier and S. Planes. 2011. Business partner or simple catch? The economic value of the sicklefin lemon shark in French Polynesia. Mar. Freshw. Res. 62(6): 764–770.

Cole, C.W. 2011. Running down a sunfish, Nov 6 1886. pp. 109. *In*: T. Bennett [ed.]. Japan and the Graphic: Complete Record of Reported Events 1870–1899. Global Oriental. Leiden, The Netherlands.

Dewar, H., T. Thys, S.L.H. Teo, C. Farwell, J. O'Sullivan, T. Tobayama et al. 2010. Satellite tracking the world's largest jelly predator, the ocean sunfish, *Mola mola*, in the Western Pacific. J. Exp. Mar. Biol. Ecol. (393)1-2: 32–42.

El Penguino. 2013. Gigantesco pez tropical hallado en el Canal Beagle. https://elpinguino.com/noticias/135797/Gigantesco-pez-tropical-hallado-en-el-Canal-Beagle. Accessed 9 March 2020.

Erlandson, J. 1994. Early hunter-gatherers of the California coast, Springer Science and Business Media New York, p. 336.

FAO. 2018. The State of World Fisheries and Aquaculture 2018—Meeting the sustainable development goals. Rome. Licence: CC BY-NC-SA 3.0 IGO.

Fishkeeping News. 2014. Sunfish auction raises over £12,500 for marine conservation. https://www.practicalfishkeeping.co.uk/fishkeeping-news/sunfish-auction-raises-over-12500-for-marine-conservation. Accessed on 16 March 2020.

Forsgren, K., R.S. McBride, T. Nakatsubo, T.M. Thys, C.D. Carson, E.K. Tholke et al. 2020. Reproductive biology of the ocean sunfishes. pp. 87–104. *In*: T.M. Thys, G.C. Hays and J.D.R. Houghton [eds.]. The Ocean Sunfishes: Evolution, Biology and Conservation, CRC Press. Boca Raton, FL, USA.

Fraser-Brunner, A. 1951. The ocean sunfishes (Family Molidae). Bull. Br. Mus. (Nat. Hist.) Zool. 1: 87–121.

Froese, R. and D. Pauly [eds.]. 2020. FishBase, version (02/2020). Online version, updated February 2020. https://www.fishbase.se/search.php (Accessed 02 February 2020).

Grasser, J.J. 1609. Newe und volkomme italianische, frantzösische, und englische Schatzkamer, Basel 6: 788–9. www.worldcat.org/title/newe-und-volkomme-italianische-frantzosische-und-englische-schatzkamer/oclc/715173775.

Green, E.H. 1899. The chemical composition of the sub-dermal connective tissue of the ocean sunfish. Contrib. Bio. Lab US Fish Comm Woods Hole MA 321.

Gudger, E.W. 1928. Capture of an Ocean Sunfish. The Scientific Monthly 26(3): 257–261.

Gudger, E.W. 1936. From atom to colossus. Nat. Hist. 38: 26–30.

Guinness World Records. 2019. Heaviest bony fish. Guinness World Records. http://www.guinnessworldrecords.com/world-records/heaviest-bony-fish (Accessed 12 May 2019).

Haas, A.R., T. Fedler and E.J. Brooks. 2017. The contemporary economic value of elasmobranchs in The Bahamas: Reaping the rewards of 25 years of stewardship and conservation. Biol. Conserv. 207: 55–63.

Hays, G.C., J.D.R. Houghton, T.M. Thys, D.H. Adams, A.E. Ahuir-Baraja, J. Alvarez et al. 2020. Unresolved questions about ocean sunfishes, molidae—a family comprising some of the world's largest teleosts. pp. 280–296. *In*: T.M. Thys, G.C. Hays and J.D.R. Houghton [eds.]. The Ocean Sunfishes: Evolution, Biology and Conservation, CRC Press. Boca Raton, FL, USA.

Hill, J. 1752. An history of animals containing descriptions of the birds, beasts, fishes and insects of the several parts of the world, and including accounts of the several classes of animalcules visible only by the assistance of microscopes. Thomas Osborn, London, pp. 584.

Hiroyuki, S., A. Toshihiro, U. Motohiko, K. Shigeru, N. Shimizu, M. Koichi et al. 2001. Analysis of the triacylglycerol species of ocean sunfish liver oil, Nihon Yuku Gakkai Nenkai Koen Yoshishu 40: 94 (In Japanese).

Houghton, J.D., T.K. Doyle, J. Davenport and G.C. Hays. 2006. The ocean sunfish *Mola mola*: insights into distribution, abundance and behaviour in the Irish and Celtic Seas. J. Mar. Biol. Assoc. U.K. 86: 1237–1243.

Howard, M.J., T. Nakatsubo, J.P. Correia, H. Batista, N. Baylina, C. Taura et al. 2020. Sunfish on display: husbandry of the ocean sunfish *Mola mola*. pp. 243–262. *In*: T.M. Thys, G.C. Hays and J.D.R. Houghton [eds.]. The Ocean Sunfishes: Evolution, Biology and Conservation, CRC Press. Boca Raton, FL, USA.

Huang, K., S. Liu, Y. Huang, K. Huang and D. Hwang. 2011. Food poisoning caused by sunfish *Masturus lanceolatus* in Taiwan. J. Food Drug Anal. 19: 191–196

Jarvis, O. 2016. Five of the best places to dive with *Mola mola* (www.UW360_Asia/best-places-dive-mola-mola-sunfish/ retrieved Jan 22 2020).

Jenkins. 1895. A description of a new species of Ranzania from the Hawaiian Islands, Proc. Cal. Acad. Sci. Ser. 2(5): 781–784.

Jingushicho. 1910 Koji ruien. Dobutsubu. (S.I.): Jingushicho. Tokyo, Japan, pp. 1698.

Johnson, G.D. and R. Britz. 2005. Leis' conundrum: Homology of the clavus of the ocean sunfishes. 2. Ontogeny of the median fins and axial skeleton of *Ranzania laevis* (Teleostei, Tetraodontiformes, Molidae). J. Morph. 266: 11–21.

Jones, T., D. Wood, J. Catlin and B. Norman. 2009. Expenditure and ecotourism: predictors of expenditure for whale shark tour participants. J. Ecotourism. 8: 32–50.

Jordan, D.S. and B.W. Evermann. 1898. The Fishes of North and Middle America, Part 2. Bull. US. Nat. Mus. No. 47.

Kondo, Y. 1986. Rearing and exhibit of ocean sunfish (*Mola mola*). Museum Chiba: Bull. Chiba Mus. Assoc. 17: 253–260.

Konow, N., R. Fitzpatrick and A. Barnett. 2006. Adult Emperor angelfish (*Pomacanthus imperator*) clean giant sunfishes (*Mola mola*) at Nusa Lembongan, Indonesia. Coral Reefs. 25: 208.

Kothe, R. 2012. Will it be Super-maxi or Sunfish? www.sail-world.com/Australia/Sydney-Hobart--Will-it-be-Supermaxi-or-Sunfish/64679?source=google.ch. Accessed on 6 March 2020.

Leachman, S. 2019. Rare hoodwinker sunfish, never before seen in the Northern Hemisphere, washes up at Coal Oil Point Reserve. UC Santa Barbara, updated 27 February 2019. https://www.news.ucsb.edu/2019/019361/hoodwinked (Accessed 09 March 2020).

Lee, D.S., H.A. Cho, N.Y. Yoon, Y.K. Kim, C.W. Lim and K.B. Shim. 2012. Biochemical composition of muscle from Tanaka's eelpout *Lycodes tanakae*, Magistrate Armhook squid *Berryteuthis magister*, and ocean sunfish *Mola mola* caught in the East Sea Korea. Fish. Aquat. Sci. 15: 2.

Montgomery, B. 2015. Rolex Sydney Hobart Yacht Race: Comanche fights back www.yachtsandyachting.com/news/187264/Comanche-fights-back Retrieved Jan 31 2020.

Mouton, P. 2004. Le thon del la Mediterranee. Edisud. Aix-en-Provence, pp. 140.

Nakamura, I., Y. Goto and K. Sato. 2015. Ocean sunfish rewarm at the surface after deep excursions to forage for siphonophores. J. Anim. Ecol. 84: 590–603.

Nakatsubo, T. and H. Hirose. 2007. Growth of captive ocean sunfish, *Mola mola*. Suisan Zoshoku 55: 403–7.

Nur, M. 2019. Ikan Mola-mola ditemukan terdampar di Perairan Pantai Barane, Majene, Sulawesi Barat. Masyarakat Iktiologi Indonesia. https://iktiologi-indonesia.org/ikan-mola-mola-mola-ditemukan-terdampar-di-perairan-pantai-barane-majene-sulawesi-barat/. Accessed on 12 March 2020.

Nyegaard, M. 2018. There be giants! The importance of taxonomic clarity of the large ocean sunfishes (genus Mola, Family Molidae) for assessing sunfish vulnerability to anthropogenic pressures (Doctoral dissertation, Murdoch University).

Nyegaard, M. and E. Sawai. 2018. Species identification of sunfish specimens (Genera *Mola* and *Masturus*, Family Molidae) from Australian and New Zealand natural history museum collections and other local sources, Data in Brief 19: 2404–2415.

Nyegaard, M., E. Sawai, N. Gemmell, J. Gillum, N.R. Loneragan, Y. Yamanoue et al. 2018. Hiding in broad daylight: molecular and morphological data reveal a new ocean sunfish species (Tetraodontiformes: Molidae) that has eluded recognition. Zool. J. Linn. Soc. 182: 631–658.

Nyegaard, M., S.G. Barcelona, N.D. Phillips and E. Sawai. 2020. Fisheries interactions, distribution modeling and conservation issues of the ocean sunfishes. pp. 216–242. *In*: T.M. Thys, G.C. Hays and J.D.R. Houghton [eds.]. The Ocean Sunfishes: Evolution, Biology and Conservation, CRC Press. Boca Raton, FL, USA.

O'Malley, M.P., K. Lee-Brooks and H.B. Medd. 2013. The global economic impact of manta ray watching tourism. PLoS ONE 8: e65051.

Packard, J. 2020. Foreword. pp. iii–iv. *In*: T.M. Thys, G.C. Hays and J.D.R. Houghton [eds.]. The Ocean Sunfishes: Evolution, Biology and Conservation, CRC Press. Boca Raton, FL, USA.

Pan, H.L., H. Yu, V. Ravi, C. Li, A.P. Lee, M.M. Lian et al. 2016. The genome of the largest bony fish, ocean sunfish (Mola mola), provides insights into its fast growth rate. Gigascience 5: doi: 10.1186/s13742-016-0144-3.

Parona, C. 1919. Il tonno e la sua pesca. Mem. R. Com. Talass. Ital. Venezia, pp. 265.

Phillips, N., N. Reid, T. Thys, C. Harrod, N.L. Payne, C.A. Morgan et al. 2017. Applying species distribution modelling to a data poor, pelagic fish complex: The ocean sunfishes. J. Biogeogr. 44(10): 2176–2187.

Pokja KKP NP. 2012. Rencana pengelolaan KKP Nusa Penida ["Nusa Penida MPA management plan"]. Kabupaten Klungkung, Propinis Bali. Pokja Kawasan Konservasi Perairan Nusa Penida, Kabupaten Klungkung, 6 pp [unpaginated] In Indonesian]. Available at: http://www.kkji.kp3k.kkp.go.id/index.php/dokumen/publikasi/buku/finish/2-buku/814-buku-1-rencana-pengelolaan-kkp-nusa-penida-v2. Accessed April 10 2018.

Porcasi, J.F. and S.L. Andrews. 2001. Evidence for a prehistoric *Mola mola* fishery on the southern California Coast. J. Calif. Gt. Basin. Anthropol. 23: 51–66.

Potter, I.F., B. Galuardi and W.H. Howell. 2010. Horizontal movement of ocean sunfish, *Mola mola*, in the northwest Atlantic. Mar. Biol. 158(3): 531–540.

Powell, D.C. 2001. A fascination for fish: adventures of an underwater pioneer. 2nd ed. Berkeley: University of California Press. USA.

Ranzani, C. 1819. Elementi di Zoologia, Bologna, Per le stampe di A. Nobili, pp. 1819–1825.

Ranzani, C. 1839. Dispositio familiae Molarum in genera et in species. Novi Comment. Acad. Sci. Inst. Bonon. 3: 63–82, pl 6.

Relini, L.O., G. Palandri and M. Relini. 2010. Medusivorous fishes of the Ligurian Sea 3. The young giant, *Mola mola* at the Camogli tuna trap. Rapports CIESM 39: 613.

Relini, L.O. and G. Vallarino. 2016. An artisanal fishery, a centuries-long experience for the future and intriguing relations between ocean sunfish and sharks. Biol. Mar. Mediterr. 23: 17–20.

RitaAmordicucina https://blog.giallozafferano.it/ritaamordicucina/ghiotta-di-mola-o-pesce-luna/ accessed March 20 2020.

Rondelet, G. 1558. La seconde partie de L'Histoire entiere des poissons/composée premierement en latin par maistre Guilaume Rondelet ...; maintenant traduite en françois [par Laurent Joubert] sans auoir rien omis estant necessaire à l'intelligence d'icelle; avec leurs pourtraits au naif. Lyon: Bonhome a la Masse D'Or.

Salviani, I. 1514–1572 Aquatilium animalium historiae, liber primus : cum eorumdem formis, aere excusis. Apud eundem Hippolytum Salvianum, Romae, 1558.

Sawai, E. 2017. The mystery of ocean sunfishes [Manbou no Himitsu]. Iwanami Shoten Publishers, Tokyo (In Japanese).

Sawai, E., Y. Yamanoue, L. Jawad, J. Al-Mamry and Y. Sakai. 2017. Molecular and morphological identification of *Mola* sunfish specimens (Actinopterygii: Tetraodontiformes: Molidae) from the Indian Ocean. Species Divers. 22: 99–104.

Sawai, E., M. Nyegaard and Y. Yamanoue. 2020. Phylogeny, taxonomy and size records of the ocean sunfishes. pp. 18–36. *In*: T.M. Thys, G.C. Hays and J.D.R. Houghton [eds.]. The Ocean Sunfishes: Evolution, Biology and Conservation, CRC Press. Boca Raton, FL, USA.

Schmidt, J. 1921. New studies of sun-fishes made during the "Dana" Expedition, 1920. Nature. 107: 76–79. Retrieved from https://www.nature.com/articles/107076a0.

Shufeldt, R.W. 1900. Chapters on the Natural History of the United States. Studer Brothers. New York USA. 472 p.

Smith, J.L.B. 1961. The sea fishes of Southern Africa, Central News Agency Ltd; 4th Edition.

Sousa, L., I. Nakamura and D.W. Sims. 2020. Movements and foraging behavior of ocean sunfishes. pp. 129–145. *In*: T.M. Thys, G.C. Hays and J.D.R. Houghton [eds.]. The Ocean Sunfishes: Evolution, Biology and Conservation, CRC Press. Boca Raton, FL, USA.

Sweeney, J.B. 1972. A Pictorial History of Sea Monsters and other Dangerous Marine Life, Crown Publishers NY, pp. 314.

Tanshu, K. 1825, Manboko, National Diet Library, Tokyo, Japan, 19pp. DOI: 10.11501/1286944.

The Daily Telegraph. 2019. Sydney to Hobart delivers big-time as Wild Oats XI holds on in supermaxi duel to win for a ninth time. www.dailytelegraph.com.au/sport/more-sports/sydney-to-hobart-delivers-bigtime-as-wild-oats-xi-holds-on-in-supermaxi-duel-to-win-for-a-ninth-time/news-story/cab059ba568c8c1b8c9efef4da4f02fb. Accessed 30 Oct 2019.

Thys, T. 1994. Swimming heads. Nat. Hist. 103: 36–39.

Thys, T.M., J.P. Ryan, H. Dewar, C.R. Perle, K. Lyons, J. O'Sullivan et al. 2015. Ecology of the ocean sunfish, *Mola mola,* in the southern California Current System. J. Exp. Mar. Biol. Ecol. 471: 64–76.

Thys, T., J.P. Ryan, K.W. Weng, M. Erdmann and J. Tresnati. 2016. Tracking a marine ecotourism star: movements of the short ocean sunfish *Mola ramsayi* in Nusa Penida, Bali, Indonesia. J. Mar. Biol., 6, Article ID 8750193.

Thys, T.M., A.R. Hearn, K.C. Weng, J.P. Ryan and C. Peñaherrera-Palma. 2017. Satellite tracking and site fidelity of short ocean sunfish, *Mola ramsayi,* in the Galápagos Islands. J. Mar. Biol. Article ID7097965.

Titcomb, M. 1977. Native Use of Fish in Hawai'i, Vol 29 of Memoirs of the Polynesian Society, University of Hawai'i Press Honolulu, pp. 175.

Uehara, M., Y.Z. Hosaka, H. Doi and H. Sakai. 2015. The shortened spinal cord in tetraodontiform fishes. J. Morph. 276(3): 290–300.

Vianna, G.M.S., J.J. Meeuwig, D. Pannell, H. Sykes and M.G. Meekan. 2011. The socioeconomic value of the shark-diving industry in Fiji. Perth: University of Western Australia. 26 p.

Vianna, G.M.S., M.G. Meekan, D.J. Pannell, S.P. Marsh and J.J. Meeuwig. 2012. Socio-economic value and community benefits from shark-diving tourism in Palau: a sustainable use of reef shark populations. Biol. Conserv. 145(1): 267–277.

Watanabe, Y. and K. Sato. 2008. Functional dorsoventral symmetry in relation to lift-based swimming in the ocean sunfish *Mola mola*. PLoS ONE 3(10): e3446.

Wantanabe, Y. and J. Davenport. 2020. Locomotory systems and biomechanics of ocean sunfish. pp. 72–86. *In*: T.M. Thys, G.C. Hays and J.D.R. Houghton [eds.]. The Ocean Sunfishes: Evolution, Biology and Conservation, CRC Press. Boca Raton, FL, USA.

Wegner, N.C., O.E. Snodgrass, H. Dewar and J.R. Hyde. 2015. Whole-body endothermy in a mesopelagic fish, the opah, *Lampris guttatus,* Science 348(6236): 786–789.

Yamanoue, Y., K. Mabuchi, E. Sawai, Y. Sakai, H. Hashimoto and M. Nishida. 2010. Multiplex PCR-based genotyping of mitochondrial DNA from two species of ocean sunfish from the genus Mola (Tetraodontiformes: Molidae) found in Japanese waters. Jpn. J. Ichthyol. 57: 27–34.

Yarrell, W. 1836. A history of British fishes. Vol.2. John van Voorst, London, UK.

Chapter 15

Unresolved Questions About Ocean Sunfishes, Molidae

A Family Comprising Some of the World's Largest Teleosts

#*Graeme C. Hays*,[1,*] *Jonathan D.R. Houghton*,[2] *Tierney M. Thys*,[3] *Douglas H. Adams*,[4] *Ana E. Ahuir-Baraja*,[5] *Jackie Alvarez*,[6] *Miguel Baptista*,[7] *Hugo Batista*,[8] *Núria Baylina*,[8] *Katherine E. Bemis*,[9] *William E. Bemis*,[10] *Eric J. Caldera*,[11] *Giorgio Carnevale*,[12] *Carol D. Carson*,[13] *João Pedro Correia*,[14] *Pedro Reis Costa*,[15] *Olivia Daly*,[16] *John Davenport*,[17] *Jessica Dutton*,[18] *Lawrence E. Eagling*,[16] *Cátia Figueiredo*,[19] *Kristy Forsgren*,[20] *Marko Freese*,[21] *Salvador García-Barcelona*,[22] *Chris Harrod*,[23] *Alex Hearn*,[24] *Lea Hellenbrecht*,[25] *Eric J. Hilton*,[26] *Michael J. Howard*,[27] *Rachel Kelly*,[28] *Lukas Kubicek*,[29] *Clara Lopes*,[19] *Tor Mowatt-Larssen*,[30] *Richard McBride*,[31] *Itsumi Nakamura*,[32] *Toshiyuki Nakatsubo*,[33] *Emily Nixon*,[34] *Marianne Nyegaard*,[35] *Enrique Ostalé-Valriberas*,[36] *Luca Pellegrino*,[37] *Natasha D. Phillips*,[16] *Edward C. Pope*,[38] *Inga Potter*,[39] *Joana Raimundo*,[19] *Martin Riis*,[40] *Rui Rosa*,[41] *John P. Ryan*,[42] *Etsuro Sawai*,[43] *Gento Shinohara*,[44] *David W. Sims*,[45] *Lara L. Sousa*,[46] *Carlos Taura*,[47] *Emilee Tholke*,[48] *Katsumi Tsukamoto*,[49] *James C. Tyler*,[50] *Yuuki Y. Watanabe*,[51] *Kevin C. Weng*,[52] *Jonathan L. Whitney*,[53] *Yusuke Yamanoue*[54] and *Kristina S. Ydesen*[55]

Introduction

The ocean sunfishes, family Molidae, currently consist of five species classified in three genera, including the largest of the teleostean fish, *Mola alexandrini* (Sawai et al. 2018), which can exceed 2,300 kg (see Sawai et al. 2020 [Chapter 2] and Caldera et al. 2020 [Chapter 3]). The other four species

For affiliations see end of the chapter

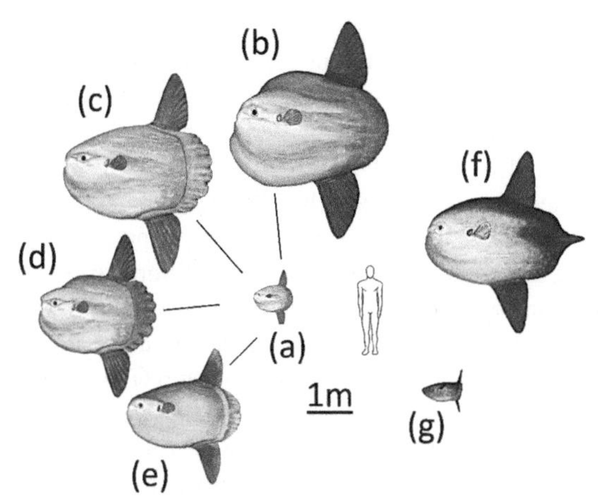

Figure 1. Key questions remain around the taxonomy of ocean sunfishes. Currently five species are recognized. a. Juvenile *Mola* spp., b. *Mola alexandrini*, c. *Mola mola* v1, d. *Mola mola* v2, e. *Mola tecta*, f. *Masturus lanceolatus* and g. *Ranzania laevis*. *M. mola* can have a variety of morphologies ranging from those with a head bump and wavy clavus to those with no head bump and a less slightly scalloped clavus. These different morphologies occur in both Atlantic and Pacific basins and can become more pronounced when individuals are in captivity. (See Caldera et al. 2020 [Chapter 3] for more details.) At small sizes, *Mola* spp. as seen in (a) *M. alexandrini* (b) *M. mola* (c,d) and *M. tecta* (e) are difficult to distinguish from each other. Illustration credit: Jamie Watts.

include *Mola mola*, *Mola tecta*, *Masturus lanceolatus* and *Ranzania laevis* (Fig. 1). As with studies of other marine megafauna, it is an exciting time for ocean sunfish research. Growing interest in the group, combined with emerging techniques, is driving new discoveries. For example, satellite tags have revealed the use of ocean depths to beyond 1000 m (Thys et al. 2017), animal-borne cameras and metabarcoding analyses are shedding light on diet and feeding behaviors (Nakamura et al. 2015, Sousa et al. 2016a), while genetic and morphological research have helped reveal the first new *Mola* species to be identified in 125 years (Nyegaard et al. 2017). However, many ocean sunfish mysteries still remain. For example, total fecundity remains unknown for any molid species despite Schmidt's (1921) often cited report (based on a single specimen) that *Mola mola* is the most fecund vertebrate on earth. It is therefore timely to take stock of our current knowledge of ocean sunfishes and triage important areas for future research. Following the theme of expert identification of key scientific research questions (e.g., Hays et al. 2016), we here synthesize recent findings, identify knowledge gaps and suggest tactics and techniques for rapidly advancing the field of ocean sunfish research.

Methods

Experts in the field of ocean sunfish biology were invited to contribute chapters to a book on ocean sunfishes (The Ocean Sunfishes: Evolution, Biology and Conservation) summarizing their fields of expertise and identifying remaining knowledge gaps. Key questions are summarized here along with potential methods and collaborations that could help advance the search for answers.

Results and Discussion

Diets of Ocean Sunfishes

The historic view that ocean sunfishes are obligatory jelly eaters has at last been overturned with the recent confirmation that *M. mola* diets change with age, shifting from benthic foraging to more pelagic prey as individuals grow larger (Phillips et al. 2020 [Chapter 9]). This realization is not new, as Fraser-Brunner (1951) reported that sunfish boast a diverse diet of fishes, squids and crustaceans as well as jellies. However, it took recent studies in the Mediterranean, NW Pacific (Japan) and NE Atlantic (Portugal) to finally consolidate our view that small molid individuals tend to occupy coastal areas and feed broadly on neritic invertebrates and fish, while larger specimens appear to live predominantly in the open ocean and have diets focused more on gelatinous zooplankton (Harrod et al. 2013, Nakamura and Sato 2014, Sousa et al. 2016a, Syväranta et al. 2012, Phillips et al. 2020). More studies of this type are needed to assess if size-related changes in habitat and diet occur more broadly across geographical areas and across all molid species. Here a range of techniques may help address questions of diet, including direct observations from animal-borne cameras (Nakamura et al. 2015), metabarcoding analysis of gut content or fecal samples (Sousa et al. 2016a), as well as stable isotope analysis (Syvaranta et al. 2012, Harrod et al. 2013, Nakamura and Sato 2014, Phillips et al. 2020).

For the latter, the use of compound specific stable isotopes (Phillips et al. 2020) warrants further attention given that this method, although considerably more expensive than bulk isotopes, can overcome issues with identifying correct nitrogen baselines for broad-ranging fishes (which vary markedly in space and time). Such research could reveal what drivers underlie ontogenetic dietary and habitat shifts including (i) the loss of agility in larger individuals to target small benthic or more maneuverable prey, (ii) biomechanic costs of transport linked to large body size, (iii) rapid growth and large size to reduce the pool of predators, and (iv) changing nutritional needs with body size.

Diet studies also need to consider the foraging ecology of sunfishes in the open ocean in addition to nearshore waters, where most research has focused (Fig. 2a). While information about the ecology

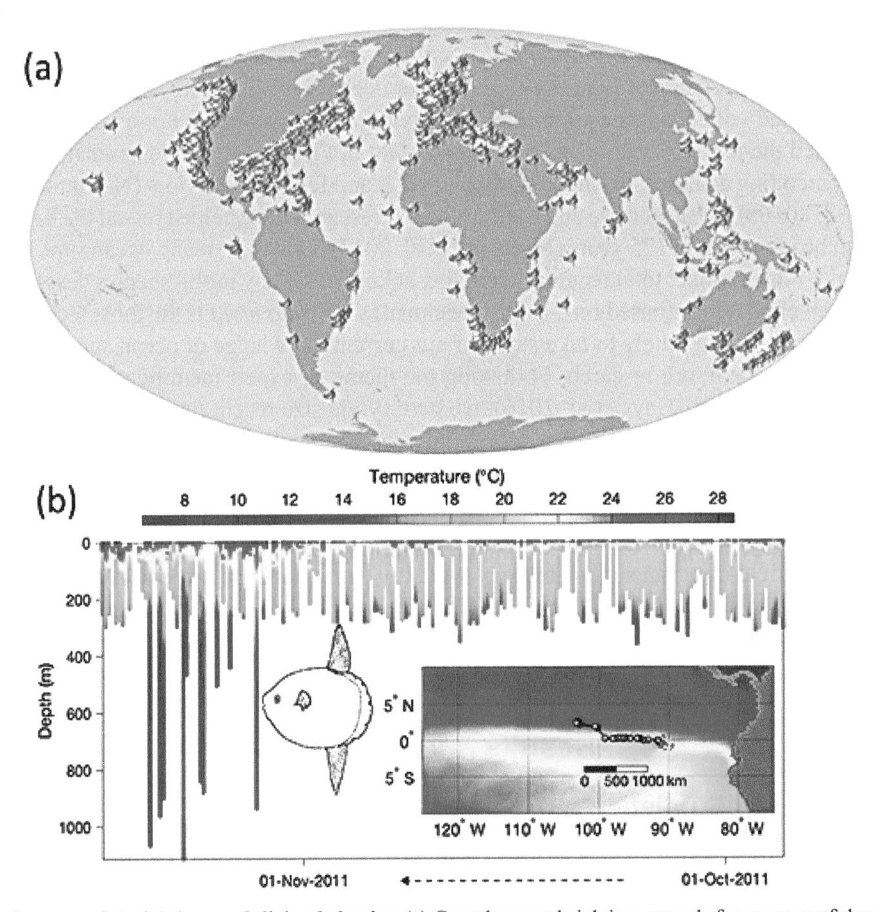

Figure 2. Ocean sunfish sightings and diving behavior. (a) Crowd-sourced sightings records for ocean sunfishes (www.oceansunfish.org) tend to be biased to near-shore areas where observer effort is greater. (b) Dive profile of a *M. alexandrini* tracked in the equatorial Pacific (modified with permission from Thys et al. 2017).

of ocean sunfishes in open-ocean areas is lacking, this may be a key habitat. Biologging studies and direct observation from submersibles have shown that sunfishes can on occasion dive deeper than 1000 m (Fig. 2b) and can spend long periods at 200–300 m (Sims et al. 2009a, Phillips et al. 2015, Thys et al. 2017), although their behavior at these depths is still poorly understood. One possibility is that sunfishes are feeding at depth on colonial gelatinous zooplankton such as siphonophores, salps and pyrosomes (Potter and Howell 2011, Nakamura et al. 2015, Phillips et al. 2020) or perhaps large deep-sea medusae such as *Stygiomedusa gigantea* (Benfield and Graham 2010). Interestingly, tracking data show that when in oceanic regions, their range of depths occupied is broadly similar to adult leatherback turtles that feed primarily on gelatinous zooplankton (Houghton et al. 2008).

In line with observations by Sims et al. (2009a), Nakamura et al. (2015) postulated that sunfishes may also feed upon bioluminescent prey given the evidence of their extensive nighttime excursions below 50 m. Similar strategies have been suggested for other gelativores (e.g., leatherback turtles, Davenport and Balazs 1991), but empirical data of sunfish feeding at depth are currently lacking. In the future, animal-borne cameras and other biologging techniques may resolve their feeding ecology below 200 m (Nakamura et al. 2015). Investigating the expression of visual opsin genes for both rod and cone photopigments could also help determine to which colors sunfishes are most attuned and if they have the visual sensitivity to see and target bioluminescent prey (Musilova et al. 2019). Such inquiry into sunfish visual acuity at low light levels will add insight into their prey foraging capacities during descent, ascent (e.g., silhouette hunting) and various phases of their deep dives.

Foraging Physiology and Ecology

We should also consider the biochemical composition of prey and how it relates to sunfish physiology (Hays et al. 2018). For example, gelatinous zooplankton may be rich in collagen (e.g., some species boast 69 percent collagen by dry mass; Khong et al. 2016), which may be important for the development of the thick subcutaneous layer known as the hypodermis (Davenport et al. 2018, Watanabe and Davenport 2020 [Chapter 5], Bemis et al. 2020 [Chapter 4]). The hypodermis has been suggested to play a central role in buoyancy (Arakawa and Masuda 1961, Watanabe and Sato 2008, Davenport et al. 2018), but it may be equally important in retaining heat through thermal inertia during deep and/or cold water feeding. Certainly, sunfishes appear to display a passive thermoregulatory strategy, which seems to reflect the physiological attributes associated with a large body mass, rather than physiological mechanisms. Specifically, larger sunfishes have lower heat-transfer coefficients, suggesting they also benefit from their large body masses to keep their body warm during foraging dives to deep, cold waters (Nakamura and Sato 2014).

Advances in understanding the foraging ecology of sunfish requires integration of animal tracking, environmental sensing, and ecosystem sampling. Satellite remote sensing is essential because of the large spatial scales over which sunfish migrate. Sunfish migratory paths have been associated with large scale open-ocean features such as the Pacific equatorial upwelling front (Figure 2b; Thys et al. 2017), as well as smaller scale, more ephemeral fronts in coastal upwelling habitat (Thys et al. 2015). While studies of sunfish migration patterns have effectively used relatively low accuracy (~ 100 km) position data from tags (Sousa et al. 2020 [Chapter 8]), understanding foraging ecology requires high accuracy tracking, particularly in ocean margin ecosystems having strong environmental gradients. In these systems high accuracy tracking provides certainty in defining relationships between sunfish and oceanographic features. For example, in the southern California Current System, GPS tracking of a sunfish and satellite sea surface temperature data revealed consistent habitat occupancy in coastal upwelling fronts throughout an 800 km migration, synthetic aperture radar images indicated convergent circulation in these fronts, and *in situ* net tow data revealed maximum concentrations of gelatinous prey at the warm side of these fronts (Thys et al. 2015). Improved information on the movements of sunfish will help identify key areas for targeted conservation, which has been an important goal in tracking studies with other marine megafauna (Hays et al. 2019, Queiroz et al. 2019).

Bioenergetics and Fasting Endurance

There is a need to develop robust bioenergetic models that explain how metabolic demands are met by feeding on different prey groups, particularly as body size increases. We need to better understand both the energetic content of different prey items and whether the energy density of different parts of prey is linked to selective feeding. For example, some low energy-density prey, such as scyphozoan jellyfish, may have body components (e.g., gonads) that are relatively energy rich (Doyle et al. 2007, Lucas et al. 2011) and therefore may be targeted by predators (Lucas et al. 2011, Hays et al. 2018). Techniques, such as the use of animal-borne cameras (Nakamura et al. 2015) and direct observations have started to reveal how some marine predators, e.g., penguins (Thiebot et al. 2017) can feed selectively on high energy density parts of prey (Milisenda et al. 2014, Nakamura et al. 2015). The use of this sort of technology is still in its infancy with ocean sunfishes, but shows great promise.

Bioenergetic models tend to focus on feeding rates and assimilation efficiency (i.e., energy intake) versus metabolic rate (Lawson et al. 2019). However, it is now well known that prey fields for ocean predators are not homogenous but rather prey is often patchily distributed. Consequently, there may be long intervals between encounters with rich prey patches (Sims et al. 2009b). With this in mind, it is important to understand how prey encounter rates relate to fasting endurance, for this relationship may be a key aspect of the foraging ecology of wide-ranging ocean predators (Hays et al. 2018). These links have been reported infrequently for ocean sunfishes, with the notable exception of Nakamura et al. (2015) who combined multichannel data loggers and HD cameras to determine prey selectivity and encounter rates for ocean sunfish (See Figure 6 in Nakamura et al. 2015).

We can also learn from developments in the ecology of taxa feeding on gelatinous zooplankton. For example, Thiebot et al. (2017) using similar technology to Nakamura et al. (2015), revealed that four species of penguins routinely feed on pelagic gelatinous organisms (gelata), and provided invaluable data on search time and encounter rates. Such empirical foraging data (encompassing prey encounter rate, selection, consumption, and handling time) can open the door to field-based studies of optimal foraging via functional response (e.g., Hays et al. 2018). This approach would drive a step change in our capacity to construct bioenergetics models. Likewise, movement data from leatherback turtles (*Dermochelys coriacea*) tracked for up to one year suggest individuals feed in jelly blooms only for about 30 percent of their time (Bailey et al. 2012). Thus, long fasting endurance may be the key requirement for a predator to feed only on jellies, with adult leatherback sea turtles likely having a particularly long fasting endurance of more than six months (Hays and Scott 2013).

One consequence of an ontogenetic shift from a broad diet to a more gelatinous zooplankton diet, is that prey may become increasingly patchily distributed (Houghton et al. 2006). From a theoretical viewpoint, fasting endurance is theorized to increase with body size, since metabolic rate typically scales with an exponent of less than one, while body reserves scale with an exponent close to one. In other words, as animals get larger, their mass-specific metabolic rate decreases, but their mass-specific energy stores stay broadly the same (Lindstedt and Boyce 1985). Work in this area has tended to focus on mammals, with the conclusion that the fasting endurance (i.e., the ratio of energy reserves to metabolic rate) increases in larger individuals (Lindstedt and Boyce 1985). However, the same key drivers likely hold true for fish. Typically metabolic rate is reported as $R = aM^b$, where R = metabolic rate and M = body mass. A great range of scaling exponents (b in the equation) have been reported across different fish species. For example, Killen et al. (2010) reported a mean scaling exponent of 0.698 for open-water pelagic fish, 0.776 for bentho-pelagic fish and 0.802 for benthic fish. This range of exponents across species broadly reflects that reported in other studies (e.g., Lawson et al. 2019) and means that larger fish will have a lower mass-specific metabolic rate. For example, using an exponent of 0.698, the mass-specific metabolic rate of a 1000 kg fish is predicted to be 20 percent of the metabolic rate of a 5 kg fish. Thus, we might expect that larger ocean sunfishes will have a longer fasting endurance than smaller individuals (Fig. 3). An increase in fasting endurance with body size may be key to the survival of large sunfishes feeding on gelatinous plankton, with individuals being able to survive long periods between encounters with rich prey patches. However,

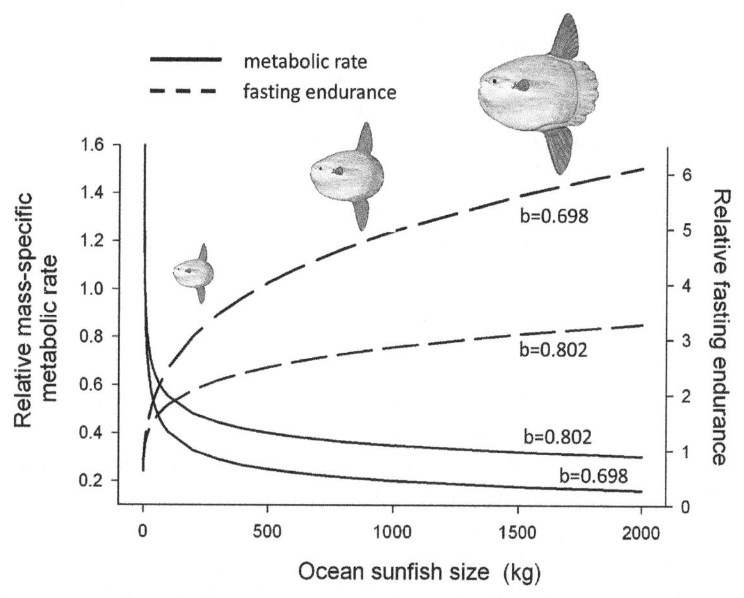

Figure 3. Is fasting endurance the key to a diet of patchily distributed gelatinous plankton? The mass-specific metabolic rate tends to decrease with animal size and hence fasting endurance increases. Typically metabolic rate is reported as $R = aM^b$, where R = metabolic rate and M = body mass. Solid lines represent the estimated mass-specific metabolic rates of different sized ocean sunfish (*Mola mola*) using two different scaling exponents between body size and metabolic rate (0.698 and 0.802), reflecting the range reported by Killen et al. (2010). Dashed lines represent the relative fasting endurance of different sized sunfish, which are derived from the mass-specific metabolic rates (assuming mass-specific energy reserves scale with body size to the power one). Values are expressed as a proportion of the values for a 5 kg fish. The plots show that fasting endurance likely increases in larger ocean sunfish. Improving estimates of this relationship requires data on the metabolic rate of ocean sunfish across a broad range of body sizes. Sunfish images credit: Jamie Watts.

aside from these general predictions based on metabolic rate measurements for other fish species, there remain key unresolved questions regarding the metabolic rate, energy stores and fasting endurance of different sized sunfishes.

Innovative biologging techniques that can record individual prey capture events may resolve how periods of feeding versus fasting change with body size in sunfishes. Furthermore, movement trajectories from tracking may reveal periods of feeding versus transit legs between prey patches (e.g., Sims et al. 2009b) as have been achieved for pelagic sea turtles. In time, measuring metabolic rates for ocean sunfishes (scaled for body mass and activity level) may become feasible. Although seemingly intractable for such large fishes (excluding *Ranzania* sp.), the recent development of the *in situ* mega-flume allowed Payne et al. (2015) to gather swimming metabolic rates for a 2.1 m, 36 kg zebra shark (*Stegostoma fasciatum*) under natural conditions. Similar data would help us refine scaled estimates of cost of travel (required in bioenergetic models) and complement recent advances in our understanding of sunfish biomechanics and locomotion (Watanabe and Sato 2008, Davenport et al. 2018).

Biomechanics and Breaching

A better understanding of how swimming behavior relates to foraging ecology of different sizes and species of wild ocean sunfishes (Sousa et al. 2020 [Chapter 8], Phillips et al. 2020 [Chapter 9]) is also needed. Studying the biomechanics of fast locomotion (i.e., sprints) in ocean sunfish may hold clues to various aspects of foraging and predator avoidance (*sensu* Soto et al. 2008, Wilson et al. 2018). Sunfishes under 2 m are also known to breach, with repeated reports of this behavior being made off California (see Fig. 2A in T. Thys et al. 2020 [Chapter 14]), New England (C. Carson personal observation), Italy (N. Phillips personal observation), Bali (Konow et al. 2006) and Angola (C. Weir personal observation). Various explanations for the adaptive significance of breaching have been put forth including dislodging ectoparasites, ditching overly zealous cleaner fishes (Konow et al. 2006), or avoiding predators. All, however, warrant formal testing.

Also of note are the major ontogenetic changes that occur in molid morphology including the shape, size and angle of their dorsal and anal fins and disproportionate thickening of their hypodermis (Watanabe and Davenport 2020 [Chapter 5]). How these dramatic morphological changes impact swimming ability and movement patterns, can be addressed through the use of animal-borne biologging with accelerometers, cameras and depth sensors.

Anatomy, Taxonomy and Evolution

Many questions remain with regards to molid anatomy, including the strange nature of their bones, which are spongy, fibrous and easily sliced with a knife. The unique gills, circulatory patterns, and large hearts of molids likely hold more clues to the group's evolutionary success and to their physiological abilities for diving. However, few specimens are properly preserved and available for study in natural history museums. In most cases, we simply lack sufficient specimens to study these organ systems in very large individuals (> 2.5 m, Bemis et al. 2020 [Chapter 4]). Also important is the need for well-fixed larvae and juvenile specimens for future histological study of anatomical systems including the nervous system, swim bladder and sensory organs.

Key questions in molid phylogenetic studies involve the number of valid, extant species of ocean sunfishes (Fig. 1). For example, are there different species of *Mola mola* in different ocean basins (Sawai et al. 2020 [Chapter 2])? Is the current monotypic status of the *Ranzania* and *Masturus* genera valid? More detailed information on morphological changes with growth across different species is crucial to establish unambiguous identification keys. Currently, there is no way to identify, from morphology alone, small individuals in the genus *Mola* to species. There is also considerable variation in the head, snout and chin bump features of *Mola* spp. that often, but not always, varies with size (Sawaii et al. 2020 [Chapter 2], Caldera et al. 2020 [Chapter 3]). Morphologic and genetic analyses of individuals possessing these possibly hybrid traits will be informative and essential to understanding the taxonomy of the group as a whole.

With only a fragmentary fossil record, the evolutionary history of ocean sunfishes remains poorly resolved as well. For example, most of the pre-Miocene history (> 23 million years ago) of ocean sunfishes is completely unknown, while the anatomy of now extinct species (*Mola pileata* and *Ranzania* spp.) that were relatively common in the Miocene (23 to 5 million years ago) is known only from isolated beaks and dermal plates (Carnevale et al. 2020 [Chapter 1]). Future fossil finds, particularly in the Oligocene (34 to 23 million years ago), will further illuminate the hidden pathway to modern molids.

Population Structure, Genetic Identity, eDNA and Trait Derivation

A host of questions remain concerning molid species' identity, population structure and size. To date, everything we have learned regarding phylogenetics and population genetics of molids has been based on Sanger sequencing of mitochondrial genes (reviewed in Caldera et al. 2020 [Chapter 3]). Remaining questions can now be addressed with emerging techniques benefiting from advances in next-generation sequencing technologies including genome-wide SNP (single nucleotide polymorphism) genotyping (e.g., RADseq). Environmental DNA (commonly referred to as eDNA), can also be used to assess the occurrence and seasonality of molids from seawater samples, as is already conducted for other fish taxa (e.g., sharks in Truelove et al. 2019). Genetic approaches can address fundamental questions including: (i) how many molid species exist, what are their geographic ranges, and do they hybridize? (ii) how and when did different molid species and populations emerge (e.g., molecular clock studies)?

(iii) how genetically connected are populations in different ocean basins? and (iv) what are the genetic underpinnings of their specialized traits like the reduced skeleton and thick hypodermis?

Growth Rates, Ageing and Reproductive Biology

While growth rates of ocean sunfishes in captivity are widely measured to help assess animal health and well-being, rates for free-living individuals are unknown. It is likely that captive molas receive much larger and more calorie-rich rations than wild molas, so growth rates may be artificially high in captivity. Key questions surrounding the links between diet, environmental conditions and growth remain poorly resolved. To improve long-term aquarium rearing conditions and husbandry management, wild growth rates are much needed. In addition, acquiring baseline data on the condition and histological status of internal organs for wild specimens would provide invaluable benchmark information for aquarists to assess organ condition and function over time for captive specimens (Howard et al. 2020 [Chapter 13]). Since it is unknown whether sunfishes are sexually dimorphic, developing blood assays or other methods to assess sex and level of maturity in sunfishes would also offer interesting insight.

Another major gap in our knowledge of sunfishes involves measuring their lifespan in the wild. Ageing sunfishes remains a challenge. Their strange ear bones (i.e., otoliths and octoconia; Nolf and Tyler 2006), are composed of vaterite rather than the more typical teleost aragonite (Gauldie 1990) and are not easily aged. Additional techniques such as measuring central corneal thickness and/or using radiocarbon dating to measure the carbon isotopes in their eyes, as done in Greenland sharks (Nielsen et al. 2016) may hold promise and are worthy of investigation.

Knowledge of the reproductive biology of sunfishes also remains limited (Forsgren et al. 2020 [Chapter 6]), with no observations of spawning in nature or captivity. Few large animals have been examined and total fecundity remains unknown for any molid species. For free-living individuals, spawning aggregations might be uncovered through citizen science or animal-borne cameras which could also help aquariums create suitable conditions for spawning. Likewise, drones are now widely used across other taxa to help assess abundance and interactions (Schofield et al. 2019) and may have merit in locating and assessing potential molid spawning aggregations. Long-term tracking of multiple mature individuals might also reveal spawning areas (Thys et al. 2020 [Chapter 7]), but further information on size at maturity (Nakatsubo et al. 2008, Forsgren et al. 2020 [Chapter 6]) is required from different sites.

Effects of Parasites on Sunfish Behavior and Overall Welfare

Understanding the pathological and ecological host consequences of infection is the next major goal for sunfish parasitology (Ahuir-Baraja 2020 [Chapter 10]). Specifically, an understanding of how parasite attachment locations and feeding modes affect growth, locomotion, buoyancy and survivorship is needed. Observational studies at cleaning stations are also recommended, given that individual sunfish allocate significant time to this behavior (Konow et al. 2006, Thys et al. 2017), which suggests parasites could be a major stressor. Where parasites are thought to be ingested together with prey, there is scope for collaboration with trophic ecologists to quantify at what size sunfish start to feed upon particular prey items. Where prey is spatially defined, it may also be possible to use parasites as indicators of broad-scale movements, either horizontally or throughout the water column. This logic extends to the use of parasites as biological tags, an approach that has been used with great success in other wide-ranging fishes (Marcogliese and Jacobson 2015). Furthermore, comparative studies of parasitic species abundance and diversity in juvenile and adult molids may shed more light on ontogenetic shifts in sunfish diet (e.g., Nakamura and Sato 2014, Sousa et al. 2016a, Phillips et al. 2020, Phillips et al. 2020 [Chapter 9]). It is also noteworthy that juvenile sunfish swim in schools (Abe et al. 2012) which may influence transmission of ectoparasites during early life history stages.

Established protocols for the collection of parasite data in areas of high bycatch need to be developed and disseminated widely so that such information becomes interwoven more routinely into overall studies of sunfish ecology. In turn, these data can help aquarists in refining their husbandry practice.

Elemental Pollution

While a baseline examination of trace element dynamics has been conducted mainly in Atlantic ocean sunfish populations (Baptista et al. 2020 [Chapter 11]) understanding bioaccumulation effects of these elements is unknown. Some elements are known to disrupt fish physiology, reproductive success and increase fish mortality at certain levels (Dunier 1996, Authman et al. 2015). Further ecotoxicological studies could explore physiological effects using a variety of elemental treatments administered in a controlled setting on small captive sunfish (or perhaps even their more accessible phylogenetic relatives such as pufferfish) (e.g., Holdway and Sprague 1979, Lushchak et al. 2009, Ricketts et al. 2015, Wang et al. 2016). Alternatively, measuring elemental concentrations and conducting histology studies of gill and liver tissues where deformities due to elemental accumulation can occur (Gaber 2007) may provide added insight into the threshold level at which elements impact fish physiology.

Plastic Pollution

The impacts of plastic debris, from micro ($\leq$ 5 mm) to meso (5–25 mm) to macro (> 25 mm), are pervasive concerns across all marine taxa, yet little is known about these impacts on ocean sunfishes. *Mola* have been known to ingest plastics as a result of eating monofilament attached to baited hooks and through ingested microfibers (Baptista et al. 2020 [Chapter 11]) wherein 80 percent of the individuals examined had at least one microfiber in their guts. The impact of these ubiquitous contaminants however is largely unknown and deserving of more attention.

Climate Change, Horizontal and Vertical Range Shifts

Many marine taxa (reviewed in Edwards 2016) are experiencing changes in their phenology, abundance and distribution due to climate change (Pinsky et al. 2013, Poloczanska et al. 2013, Halpern et al. 2019), yet how these influences affect sunfishes remain unclear. It is well documented that isotherms (lines of equal temperature) are generally moving polewards at a rate of > 100 km per decade in some places (e.g., McMahon and Hays 2006). Tracking studies have identified seasonal poleward movements of ocean sunfishes in the north Pacific (Dewar et al. 2010) and northeast Atlantic (Sims et al. 2009a, Sousa et al. 2016b). A shifting position of the summer isotherms will likely influence the extent of poleward migration of sunfishes. Changes in the trailing edge of animals distributions can be faster than the poleward edges (Robinson et al. 2015) so tracking studies of molids should also pay attention to possible loss of more equatorial habitats. The horizontal range of ocean sunfishes may be changing accordingly, a suggestion further supported by increased sightings in the northern latitudes of Iceland (Palsson and Astthorsson 2017) and Norway (Frafjord et al. 2017). Whether or not sunfishes are adjusting their vertical behaviors and shifting their preferred cleaning stations to be in deeper water based on ocean warming also remains a question. Further tracking studies, at-sea surveys of distribution (Breen et al. 2017) and *in situ* observations at cleaning stations, particularly in the mesophotic zones, will help clarify the extent of these range changes (Sousa et al. 2020 [Chapter 8]).

Crowd-Based Science, Non-Invasive Tracking, Skin and Color patterns

Harnessing and honing the power of crowd-based sunfish sightings (for example, www.oceansunfish.org, New England Coastal Wildlife Alliance (NECWA) tracking form at www.nebshark.org, molaphotos@outlook.com, MatchMyMola at (http://www.thebalisunfish.org/matchmymola/) offer a powerful means of collecting tissue samples for a wide array of studies including stable isotope and genetic analyses. These datasets can also offer a means of potentially tracking sunfish non-invasively. With their plethora of skin patterns and scarring, individual sunfish can theoretically be tracked visually given sufficient representative photographic material that can be supplied by SCUBA divers or ecotourism groups. The presence of sunfish on social media and the internet (Thys et al. 2020 [Chapter 14]) coupled with image recognition software and machine learning may help to improve both the quantity and quality of crowd-sourced observations, and make these datasets increasingly more valuable. Additional uses of well-curated crowd-sourced visual databases could be to: assess the variability of sunfish skin and color patterns during different behaviors, times of day and seasons; ascertain whether skin and color patterns are stable over time and; establish the basis for assessments of injury types and frequency.

Crowd-based sightings and reports can also reveal mass strandings and die-off events of *Mola mola* as witnessed between September and December 2019 on both the east (C. Carson personal communication) and west coast of the USA (T. Thys unpublished data). These observations can be coupled with environmental data to glean a better understanding of the factors underlying past stranding events noted for *M. mola* in Monterey Bay (Gotshall 1961) and for *Ranzania laevis* in Western Australia (Smith et al. 2010) and South Africa (L. Nupen personal communication).

Bycatch, Targeted Fisheries and Population Trends

Threats to ocean sunfishes are poorly documented although it is known that large numbers are caught as bycatch in commercial fisheries (Nyegaard et al. 2020 [Chapter 12]). While individuals accidentally caught may be released alive (Phillips et al. 2020 [Chapter 9]), post-release survival is unknown. For example, in the Moroccan driftnet fishery, the estimated bycatch was 36,450 in one year (Pope et al. 2010). Yet beyond these incidental 'hotspots' (Fig. 4), information on bycatch for all molid species is

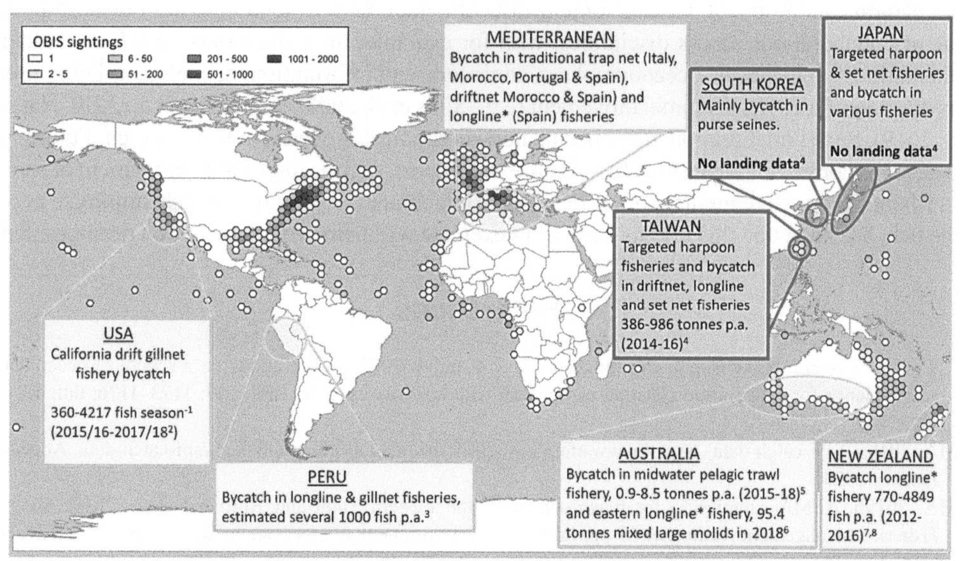

Figure 4. Ocean sunfish face fishing pressure across their broad range. World map showing locations of sightings for combined genera *Mola* and *Masturus* drawn from OBIS data (www.obis.org) accessed 10/1/2020 (total records 19,353). Commercial fisheries landing large molids are highlighted in red, those known to experience considerable levels of large molid bycatch are highlighted in gold. 1 Mason et al. (2019). 2 Common mola seasonal (1 May to 31 January) bycatch in the California drift gillnet swordfish fishery estimated from NOAA (2019): proportion of sunfish in observed nets and the number of total nets set. 3 Mangel et al. (2019). 4 FAO (2018). 5 AFMA (2019a). 6 AMFA (2019b). 7 MPI (2016). 8 MPI (2017).
* Large molids are widely reported as bycatch in pelagic longline fisheries worldwide but accurate estimates are not currently possible for most fisheries (see main text).

either lacking or sporadic, requiring further investigation and collation. Whether sunfish are captured via directed take or bycatch, the implications of removing large numbers of individuals from any system requires further attention. As we have moved beyond the notion of sunfishes being life-long gelatinous animal specialists, these implications will change in line with the size of individuals taken, and the complexity of the community from which they were removed. With some certainty we can say that large aggregations of small sunfish found in temperate and subtropical coastal seas exert a degree of top down control on both pelagic and benthic food webs (Grémillet et al. 2017), at least on a seasonal basis. Thus, their removal may have multi-faceted and profound effects.

More strikingly, we have yet to understand the connectivity and shared genetic heritage of seemingly independent aggregations of sunfish, whereby the extirpation from one region may have wide ranging ramifications for other sites. Filling these knowledge gaps will require a combination of trophic, molecular and tracking studies.

Additionally, we have a poor understanding of overall trends in abundance and effective populations sizes of any sunfish species. In the context of bycatch and emergent fisheries, such data are vital (Liu et al. 2015, Nyegaard et al. 2020 [Chapter 12]), especially where policy recommendations hinge on information relating to favorable conservation status. Again, a combination of approaches will be required starting with mark-recapture studies to assess whether areas of high bycatch are simply recapturing the same individuals. There is also an opportunity here to work with fisheries to gather much needed life history information from caught specimens including data on gonad maturity stage for assessing reproductive status and fecundity.

Concluding Remarks

We hope that this compilation of key questions for ocean sunfish research helps to convey the growing interest in and knowledge of the biology of this fascinating group of fishes. Emergent techniques will certainly help drive ocean sunfish research forward, and transform our capacity to study individuals under natural conditions. At the same time, work with other marine megafauna has revealed the huge value of collaborations across disciplines, with for example, the collaborative work of ecologists, mathematicians, physicists, oceanographers, engineers, and information technologists being used to identify general patterns in animal movement (e.g., Sims et al. 2008, Humphries et al. 2010, Harcourt et al. 2019). Such collaboration may open up new directions for ocean sunfish research. Future work also needs to consider ethical concerns of some techniques, such as long-term deployment of satellite tags, where continued refinement of tag design and attachment techniques can help minimize impacts to the fish. We hope that this horizon-scanning exercise will help promote work on ocean sunfishes.

References

Abe, T., K. Sekiguchi, H. Onishi, K. Muramatsu and T. Kamito. 2012. Observations on a school of ocean sunfish and evidence for a symbiotic cleaning association with albatrosses. Mar. Biol. 159: 1173–1176. doi: 10.1007/s00227-011-1873-6.

AFMA. 2019a. SPF catch data. https://www.afma.gov.au/fisheries/small-pelagic-fishery/spf-catch-data. Accessed 9 January 2020.

AFMA. 2019b. AFMA submission for reassessment of the Eastern Tuna and Billfish Fishery 2019. Australian Fisheries Management Authority. 67 pp. http://www.environment.gov.au/system/files/consultations/779c1883-0898-431c-9b85-25c796cea9ad/files/assessment-report-2019.docx.

Ahuir-Baraja, A.E. 2020. Parasites of the ocean sunfishes. pp. 160–185. *In*: T.M. Thys, G.C. Hays and J.D.R. Houghton [eds.]. The Ocean Sunfishes: Evolution, Biology and Conservation, CRC Press. Boca Raton, FL, USA.

Authman, M.M, M.S. Zaki, E.A. Khallaf and H.H. Abbas. 2015. Use of fish as bio-indicator of the effects of heavy metals pollution. Journal of Aquaculture Research and Development 6: 328. doi: 10.4172/2155-9546.1000328.

Arakawa, K.Y. and S. Masuda. 1961. Some observations on ocean sunfish *Mola mola* (Linnaeus) with reference to its swimming behaviour and feeding habits. J. Jap. Assoc. Zoos and Aquariums 34: 95–97.

Bailey, H., S. Fossette, S.J. Bograd, G.L. Shillinger, A.M. Swithenbank, J.Y. Georges et al. 2012. Movement patterns for a critically endangered species, the leatherback turtle (*Dermochelys coriacea*), linked to foraging success and population status. PLoS One 7(5): e36401. doi: 10.1371/journal.pone.0036401.

Baptista, M., C. Figueiredo, C. Lopes, P.R. Costa, J. Dutton, D.H. Adams et al. 2020. Biotoxins, trace elements and microplastics in the ocean sunfishes (Molidae). pp. 186–215. *In*: T.M. Thys, G.C. Hays and J.D.R. Houghton [eds.]. The Ocean Sunfishes: Evolution, Biology and Conservation, CRC Press. Boca Raton, FL, USA.

Bemis, K.E., E.J. Hilton, J.C. Tyler and W.E. Bemis. 2020. Overview of the anatomy of ocean sunfishes (Molidae: Tetraodontiformes). pp. 55–71. *In*: T.M. Thys, G.C. Hays and J.D.R. Houghton [eds.]. The Ocean Sunfishes: Evolution, Biology and Conservation, CRC Press. Boca Raton, FL, USA.

Benfield, M.C. and W.M. Graham. 2010. *In situ* observations of *Stygiomedusa gigantea* in the Gulf of Mexico with a review of its global distribution and habitat. J. Mar. Biol. Assoc. UK 90(6): 1079–1093. doi: 10.1017/S0025315410000536.

Breen, P., A. Cañadas, O.Ó. Cadhla, M. Mackey, M. Scheidat, S.C. Geelhoed et al. 2017. New insights into ocean sunfish (*Mola mola*) abundance and seasonal distribution in the northeast Atlantic. Sci. Rep. 7(1): 2025. doi: 10.1038/s41598-017-02103-6.

Caldera, E.J., J.L. Whitney, M. Nyegaard, E. Ostalé-Valriberas, L. Kubicek and T.M. Thys. 2020. Genetic Insights regarding the taxonomy, phylogeography and evolution of the ocean sunfishes (Molidae: Tetraodontiformes). pp. 37–54. *In*: T.M. Thys, G.C. Hays and J.D.R. Houghton [eds.]. The Ocean Sunfishes: Evolution, Biology and Conservation, CRC Press. Boca Raton, FL, USA.

Carnevale, G., L. Pellegrino and J.C. Tyler. 2020. Evolution and fossil record of the ocean sunfishes. pp. 1–17. *In*: T.M. Thys, G.C. Hays and J.D.R. Houghton [eds.]. The Ocean Sunfishes: Evolution, Biology and Conservation, CRC Press. Boca Raton, FL, USA.

Davenport, J. and G.H. Balazs. 1991. 'Fiery bodies'—are pyrosomas important items in the diet of leatherback turtles? Brit. Herp. Soc. Bull. 37: 33–38.

Davenport, J., N.D. Phillips, E. Cotter, L.E. Eagling and J.D. Houghton. 2018. The locomotor system of the ocean sunfish *Mola mola* (L.): role of gelatinous exoskeleton, horizontal septum, muscles and tendons. J. Anat. 223(3): 347–357. doi: 10.1111/joa.12842.

Dewar, H., T. Thys, S.L.H. Teo, C. Farwell, J. O'Sullivan, T. Tobayama et al. 2010. Satellite tracking the world's largest jelly predator, the ocean sunfish, *Mola mola*, in the Western Pacific. J. Exp. Mar. Biol. Ecol. 393(1-2): 32–42. doi: 10.1016/j.jembe.2010.06.023.

Doyle, T.K., J.D. Houghton, R. McDevitt, J. Davenport and G.C. Hays. 2007. The energy density of jellyfish: estimates from bomb-calorimetry and proximate-composition.J. Exp. Mar. Biol. Ecol. 343(2): 239–252. doi: 10.1016/j.jembe.2006.12.010.

Dunier, M. 1996. Water pollution and immunosuppression of freshwater fish. Italian J. Zool. 63(4): 303–309.

Edwards, M. 2016. Chapter 11—Sea life (Pelagic Ecosystems). pp. 167–182. *In*: T.M. Letcher [ed.]. Climate Change Observed Impacts on Planet Earth (2nd ed.). Elsevier. doi: 10.1016/B978-0-444-63524-2.00011-7.

FAO. 2018. Fishery and Aquaculture Statistics. Global capture production 1950–2016 (FishstatJ). *In*: FAO Fisheries and Aquaculture Department [online]. Rome. Updated 2018. www.fao.org/fishery/statistics/software/fishstatj/en.

Forsgren, K., R.S. McBride, T. Nakatsubo, T.M. Thys, C.D. Carson, E.K. Tholke et al. 2020. Reproductive biology of the ocean sunfishes. pp. 87–104. *In*: T.M. Thys, G.C. Hays and J.D.R. Houghton [eds.]. The Ocean Sunfishes: Evolution, Biology and Conservation, CRC Press. Boca Raton, FL, USA.

Frafjord, K., T. Bakken, L. Kubicek, A.-H. Rønning and P.O. Syvertsen. 2017. Records of ocean sunfish along the Norwegian coast spanning two centuries, 1801–2015. J. Fish Biol. 91(5): 1365–1377. doi: 10.1111/jfb.13456.

Fraser-Brunner, A. 1951. The ocean sunfishes (Family Molidae). Bulletin of the Natural History Museum, Zoology Series 1: 87–121.

Gaber, H. 2007. Impact of certain heavy metals on the gill and liver of the Nile tilapia (*Orechromis niloticus*). Egypt. J. Aquat. Biol. Fisher. 11(2): 79–100. doi: 10.21608/ejabf.2007.1936.

Gauldie, R.W. 1990. Vaterite otoliths from the opah, *Lampris immaculatus*, and two species of sunfish, *Mola mola* and *M. ramsayi*. Acta Zool. 71(4): 193–199. doi: 10.1111/j.1463-6395.1990.tb01077.x.

Gotshall, D.W. 1961. Observations on a die-off of molas (*Mola mola*) in Monterey Bay. Calif. Fish Game 47(4): 339–341.

Halpern, B.S., M. Frazier, J. Afflerbach, J.S. Lowndes, F. Micheli, C. O'hara et al. 2019. Recent pace of change in human impact on the world's ocean. Sci. Rep. 9: 11609. doi: 10.1038/s41598-019-47201-9.

Harcourt, R., A.M. Martins Sequeira, X. Zhang, F. Rouquet, K. Komatsu, M. Heupel et al. 2019. Animal-Borne Telemetry: an integral component of the ocean observing toolkit. Front. Mar. Sci. 6: 326. doi: 10.3389/fmars.2019.00326.

Harrod, C., J. Syväranta, L. Kubicek, V. Cappanera and J.D.R. Houghton. 2013. Reply to Logan and Dodge: 'Stable isotopes challenge the perception of ocean sunfish *Mola mola* as obligate jellyfish predators'. J. Fish Biol. 82(1): 10–16. doi: 10.1111/j.1095-8649.2012.03485.x.
Hays, G.C. and R. Scott. 2013. Global patterns for upper ceilings on migration distance in sea turtles and comparisons with fish, birds and mammals. Funct. Ecol. 27(3): 748–756. doi: 10.1111/1365-2435.12073.
Hays, G.C., L.C. Ferreira, A.M. Sequeira, M.G. Meekan, C.M. Duarte, H. Bailey et al. 2016. Key questions in marine megafauna movement ecology. Trends Ecol. Evol. 3(6): 463–475. doi: 10.1016/j.tree.2016.02.015.
Hays, G.C., T.K. Doyle and J.D.R. Houghton. 2018. A paradigm shift in the trophic importance of jellyfish? Trends Ecol. Evol. 33(11): 874–884. doi: 10.1016/j.tree.2018.09.001.
Hays, G.C., H. Bailey, S.J. Bograd, W.D. Bowen, C. Campagna, R.H. Carmichael et al. 2019. Translating marine animal tracking data into conservation policy and management. Trends Ecol. Evol. 34: 459–473. doi: 10.1016/j.tree.2019.01.009.
Holdway, D.A. and J.B. Sprague. 1979. Chronic toxicity of vanadium to flagfish. Water Res. 13(9): 905–910. doi: 10.1016/0043-1354(79)90226-4.
Houghton, J.D., T.K. Doyle, M.W. Wilson, J. Davenport and G.C. Hays. 2006. Jellyfish aggregations and leatherback turtle foraging patterns in a temperate coastal environment. Ecology 87(8): 1967–1972. doi: 10.1890/0012-9658(2006)87[1967:JAALTF]2.0.CO;2.
Houghton, J.D.R., T.K. Doyle, J. Davenport, R.P. Wilson and G.C. Hays. 2008. The role of infrequent and extraordinary deep dives in leatherback turtles (*Dermochelys coriacea*). J. Exp. Biol. 211(16): 2566–2575. doi: 10.1242/jeb.020065.
Howard, M.J., T. Nakatsubo, J.P. Correia, H. Batista, N. Baylina, C. Taura et al. 2020. Sunfish on display: husbandry of the ocean sunfish *Mola mola*. pp. 243–262. *In*: T.M. Thys, G.C. Hays and J.D.R. Houghton [eds.]. The Ocean Sunfishes: Evolution, Biology and Conservation, CRC Press. Boca Raton, FL, USA.
Humphries, N.E., N. Queiroz, J.M. Dyer, N.G. Pade, M.K. Musyl, K.M. Schaefer et al. 2010. Environmental context explains Lévy and Brownian movement patterns of marine predators. Nature 465: 1066–1069. doi: 10.1038/nature09116.
Khong, N.M., F.M. Yusoff, B. Jamilah, M. Basri, I. Maznah, K.W. Chan and J. Nishikawa. 2016. Nutritional composition and total collagen content of three commercially important edible jellyfish. Food Chem. 196: 953–960. doi: 10.1016/j.foodchem.2015.09.094.
Killen, S.S., D. Atkinson and D.S. Glazier. 2010. The intraspecific scaling of metabolic rate with body mass in fishes depends on lifestyle and temperature. Ecol. Lett. 13(2): 184–193. doi: 10.1111/j.1461-0248.2009.01415.x.
Konow, N., R. Fitzpatrick and A. Barnett. 2006. Adult emperor angelfish (*Pomacanthus imperator*) clean giant sunfishes (*Mola mola*) at Nusa Lembongan, Indonesia. Coral Reefs 25(2): 208. doi: 10.1007/s00338-006-0086-9.
Lawson, C.L., L.G. Halsey, G.C. Hays, C.L. Dudgeon, N.L. Payne, M.B. Bennett et al. 2019. Powering ocean giants: the energetics of shark and ray megafauna. Trends Ecol. Evol. 34: 1009–1021. doi: 10.1016/j.tree.2019.07.001.
Lindstedt, S.L. and M.C. Boyce. 1985. Seasonality, fasting endurance, and body size in mammals. Am. Nat. 125(6): 873–878. doi: 10.1086/284385.
Liu, J., G. Zapfe, K.-T. Shao, J.L. Leis, K. Matsuura, G. Hardy et al. 2015. *Mola mola* (errata version published in 2016). The IUCN Red List of Threatened Species 2015: e.T190422A97667070.
Lucas, C.H., K.A. Pitt, J.E. Purcell, M. Lebrato and R.H. Condon. 2011. What's in a jellyfish? Proximate and elemental composition and biometric relationships for use in biogeochemical studies. Ecology 92(8): 1704. doi: 10.1890/11-0302.1.
Lushchak, O.V., O.I. Kubrak, I.M. Torous, T.Y. Nazarchuk, K.B. Storey and V.I. Lushchak. 2009. Trivalent chromium induces oxidative stress in goldfish brain. Chemosphere 75(1): 56–62. doi: 10.1016/j.chemosphere.2008.11.052.
Mangel, J.C., M. Pajuelo, A. Pasara-Polack, G. Vela, E. Segura-Cobeña and J. Alfaro-Shigueto. 2019. The effect of Peruvian small-scale fisheries on sunfishes (Molidae). J. Fish. Biol. 94: 77–85. doi: 10.1111/jfb.13862.
Marcogliese, D.J. and K.C. Jacobson. 2015. Parasites as biological tags of marine, freshwater and anadromous fishes in North America from the tropics to the Arctic. Parasitology 142: 68–89. doi: 10.1017/S0031182014000110.
Mason, J.G., E.L. Hazen, S.J. Bograd, H. Dewar and L.B. Crowder. 2019. Community-level effects of spatial management in the California drift gillnet fishery. Fish. Res. 214: 175–182. doi: 10.1016/j.fishres.2019.02.010.
McMahon, C.R. and G.C. Hays. 2006. Thermal niche, large-scale movements and implications of climate change for a critically endangered marine vertebrate. Glob. Chang. Biol. 12(7): 1330–1338. doi: 10.1111/j.1365-2486.2006.01174.x.
Milisenda, G., S. Rosa, V.L. Fuentes, F. Boero, L. Guglielmo, J.E. Purcell et al. 2014. Jellyfish as prey: frequency of predation and selective foraging of *Boops boops* (Vertebrata, Actinopterygii) on the mauve stinger *Pelagia noctiluca* (Cnidaria, Scyphozoa). PLoS One 9(4): e94600. doi: 10.1371/journal.pone.0094600.

MPI. 2016. Fisheries Assessment Plenary November 2016: Stock Assessments and Stock Status. Compiled by the Fisheries Science Group, Ministry for Primary Industries. Fisheries Science Group. Ministry for Primary Industries, Wellington, New Zealand. 682p.

MPI. 2017. Fisheries Assessment Plenary November 2016: Stock Assessments and Stock Status. Compiled by the FIsheries Science Group, Ministry for Primary Industries. Fisheries Science Group. Ministry for Primary Industries, Wellington, New Zealand. 500p.

Musilova, Z.F. Cortesi and M. Matschiner. 2019. Vision using multiple distinct rod opsins in deep-sea fishes. Science 364(6440): 588–592. doi: 10.1126/science.aav4632.

Nakamura, I. and K. Sato. 2014. Ontogenetic shift in foraging habit of ocean sunfish *Mola mola* from dietary and behavioral studies. Mar. Biol. 161(6): 1263–1273. doi: 10.1007/s00227-014-2416-8.

Nakamura, I., Y. Goto and K. Sato. 2015. Ocean sunfish rewarm at the surface after deep excursions to forage for siphonophores. J. Anim. Ecol. 84(3): 590–603. doi: 10.1111/1365-2656.12346.

Nakatsubo, T. 2008. A study on the reproductive biology of ocean sunfish *Mola mola* (Unpublished doctoral dissertation). Nihon University, Tokyo, Japan.

Nielsen, J., R.B. Hedeholm, J. Heinemeier, P.G. Bushnell, J.S. Christiansen, J. Olsen et al. 2016. Eye lens radiocarbon reveals centuries of longevity in the Greenland shark (*Somniosus microcephalus*). Science 353(6300): 702–704. doi: 10.1126/science.aaf1703.

NOAA. 2019. West Coast Region Observer Program data summaries and reports. https://www.westcoast.fisheries.noaa.gov/fisheries/wc_observer_programs/sw_observer_program_info/data_summ_report_sw_observer_fish.html. Accessed 21 January 2020.

Nolf, D. and J.C. Tyler. 2006. Otolith evidence concerning interrelationships of caproid, zeiform and tetraodontiform fishes. Bulletin de L'Institut Royal des Sciences Naturelles de Belgique Biologie 76: 147–189.

Nyegaard, M., E. Sawai, N. Gemmell, J. Gillum, N.R. Loneragan, Y. Yamanoue et al. 2017. Hiding in broad daylight: molecular and morphological data reveal a new ocean sunfish species (Tetraodontiformes: Molidae) that has eluded recognition. Zool. J. Linn. Soc. 182(3): 631–658. doi: 10.1093/zoolinnean/zlx040.

Nyegaard, M., N. Loneragan, S. Hall, J. Andrew, E. Sawai and M. Nyegaard. 2018. Giant jelly eaters on the line: Species distribution and bycatch of three dominant sunfishes in the Southwest Pacific. Est. Coast. Shelf Sci. 207: 1–15. doi: 10.1016/j.ecss.2018.03.017.

Nyegaard, M., S. García-Barcelona, N.D. Phillips and E. Sawai. 2020. Fisheries interactions, distribution modeling and conservation issues of the ocean sunfishes. pp. 216–242. *In*: T.M. Thys, G.C. Hays and J.D.R. Houghton [eds.]. The Ocean Sunfishes: Evolution, Biology and Conservation, CRC Press. Boca Raton, FL, USA.

Palsson, J. and O.S. Astthorsson. 2017. New and historical records of the ocean sunfish *Mola mola* in Icelandic waters. J. Fish. Biol. 90(3): 1126–1132. doi: 10.1111/jfb.13237.

Payne, N.L., E.P. Snelling, R. Fitzpatrick, J. Seymour, R. Courtney, A. Barnett et al. 2015. A new method for resolving uncertainty of energy requirements in large water breathers: the 'mega-flume' seagoing swim-tunnel respirometer. Methods Ecol. Evol. 6: 668–677. doi: 10.1111/2041-210X.12358.

Phillips, N.D., C. Harrod, A.R. Gates, T.M. Thys and J.D.R. Houghton. 2015. Seeking the sun in deep, dark places: Mesopelagic sightings of ocean sunfishes (Molidae). J. Fish. Biol. 87(4): 1118–1126. doi: 10.1111/jfb.12769.

Phillips, N.D., E.C. Pope, C. Harrod and J.D.R. Houghton. 2020. The diet and trophic role of ocean sunfishes. pp. 146–159. *In*: T.M. Thys, G.C. Hays and J.D.R. Houghton [eds.]. The Ocean Sunfishes: Evolution, Biology and Conservation, CRC Press. Boca Raton, FL, USA.

Phillips, N.D., E.A.E. Smith, S.D. Newsome, J.D.R. Houghton, C.D. Carson, J.C. Mangel et al. 2020. Bulk tissue and amino acid stable isotope analyses reveal global ontogenetic patterns in ocean sunfish trophic ecology and habitat use. Mar. Ecol. Prog. Ser. 633: 127–140. doi: 10.3354/meps13166.

Pinsky, M.L., B. Worm, M.J. Fogarty, J.L. Sarmiento and S.A. Levin. 2013. Marine taxa track local climate velocities. Science 341: 1239–1242. doi: 10.1126/science.1239352.

Poloczanska, E.S., C.J. Brown, W.J. Sydeman, W. Kiessling, D.S. Schoeman, P.J. Moore et al. 2013. Global imprint of climate change on marine life. Nat. Clim. Chang. 3: 919–925. doi: 10.1038/nclimate1958.

Pope, E.C., G.C. Hays, T.M. Thys, T.K. Doyle, D.W. Sims, N. Queiroz et al. 2010. The biology and ecology of the ocean sunfish *Mola mola*: A review of current knowledge and future research perspectives. Rev. Fish Biol. Fisheries 20(4): 471–487. doi: 10.1007/s11160-009-9155-9.

Potter, I.F. and W.H. Howell. 2011. Vertical movement and behavior of the ocean sunfish, *Mola mola*, in the northwest Atlantic. J. Exp. Mar. Biol. Ecol. 396(2): 138–146. doi: 10.1016/j.jembe.2010.10.014.

Queiroz, N., N.E. Humphries, A. Couto, M. Vedor, I. Da Costa, A.M. Sequeira et al. 2019. Global spatial risk assessment of sharks under the footprint of fisheries. Nature 572(7770): 461–466. doi:10.1038/s41586-019-1444-4.

Ricketts, C.D., W.R. Bates and S.D. Reid. 2015. The effects of acute waterborne exposure to sublethal concentrations of molybdenum on the stress response in rainbow trout, *Oncorhynchus mykiss*. PLoS One 10(1): e0115334. ddoi: 10.1371/journal.pone.0115334.

Robinson, L., A. Hobday, H. Possingham and A. Richardson. 2015. Trailing edges projected to move faster than leading edges for large pelagic fish habitats under climate change. Deep Sea Res. Part II Top. Stud. Oceanogr. 113: 225–234. doi: 10.1016/j.dsr2.2014.04.007.

Sawai, E., Y. Yamanoue, M. Nyegaard and Y. Sakai. 2018. Redescription of the bump-head sunfish *Mola alexandrini* (Ranzani 1839), senior synonym of *Mola ramsayi* (Giglioli 1883), with designation of a neotype for *Mola mola* (Linnaeus 1758) (Tetraodontiformes: Molidae). Ichthyol. Res. 65(1): 142–160. doi:10.1007/s10228-017-0603-6.

Sawai, E., M. Nyegaard and Y. Yamanoue. 2020. Phyolgeny, taxonomy and size records of ocean sunfishes. pp. 18–36. *In*: T.M. Thys, G.C. Hays and J.D.R. Houghton [eds.]. The Ocean Sunfishes: Evolution, Biology and Conservation, CRC Press. Boca Raton, FL, USA.

Schmidt, J. 1921. New studies of sunfishes made during the 'Dana' Expedition, 1920. Nature 107: 76–79.

Schofield, G., N. Esteban, K.A. Katselidis and G.C. Hays. 2019. Drones for research on sea turtles and other marine vertebrates—A review. Biol. Cons. 238: 108214. doi: 10.1016/j.biocon.2019.108214.

Sims, D.W., E.J. Southall, N.E. Humphries, G.C. Hays, C.J. Bradshaw, J.W. Pitchford et al. 2008. Scaling laws of marine predator search behaviour. Nature 451(7182): 1098–1102. doi: 10.1038/nature06518.

Sims, D.W., N. Queiroz, T.K. Doyle, J.D. Houghton and G.C. Hays. 2009a. Satellite tracking of the world's largest bony fish, the ocean sunfish (*Mola mola* L.) in the North East Atlantic. J. Exp. Mar. Biol. Ecol. 370(1-2): 127–133. doi: 10.1016/j.jembe.2008.12.011.

Sims, D.W., N. Queiroz, N.E. Humphries, F.P. Lima and G.C. Hays. 2009b. Long-term GPS tracking of ocean sunfish *Mola mola* offers a new direction in fish monitoring. PLoS One 4(10): e7351. doi: 10.1371/journal.pone.0007351.

Smith, K.A., M. Hammond and P.G. Close. 2010. Aggregation and stranding of elongate sunfish (*Ranzania laevis*) (Pisces: Molidae) (Pennant, 1776) on the southern coast of Western Australia. J. R. Soc. West. Aust. 93: 181–188. doi: 10.1093/acprof:oso/9780199583652.003.0012.

Soto, N.A., M.P. Johnson, P.T. Madsen, F. Diaz, I. Dominguez, A. Brito et al. 2008. Cheetahs of the deep sea: deep foraging sprints in short-finned pilot whales off Tenerife (Canary Islands). J. Anim. Ecol. 77(5): 936–947. doi: 10.1111/j.1365-2656.2008.01393.x.

Sousa, L.L., R. Xavier, V. Costa, N.E. Humphries, C. Trueman, R. Rosa et al. 2016a. DNA barcoding identifies a cosmopolitan diet in the ocean sunfish. Sci. Rep. 6: 28762. doi: 10.1038/srep28762.

Sousa, L.L., N. Queiroz, G. Mucientes, N.E. Humphries and D.W. Sims. 2016b. Environmental influence on the seasonal movements of satellite-tracked ocean sunfish *Mola mola* in the north-east Atlantic. Anim. Biotelemetry 4(1): 7. doi: 10.1186/s40317-016-0099-2.

Sousa, L.L., I. Nakamura and D.W. Sims. 2020. Movements and foraging behavior of ocean sunfish. pp. 129–145. *In*: T.M. Thys, G.C. Hays and J.D.R. Houghton [eds.]. The Ocean Sunfishes: Evolution, Biology and Conservation, CRC Press. Boca Raton, FL, USA.

Syväranta, J., C. Harrod, L. Kubicek, V. Cappanera and J.D.R. Houghton. 2012. Stable isotopes challenge the perception of ocean sunfish *Mola mola* as obligate jellyfish predators. J. Fish. Biol. 80(1): 225–231. doi: 10.1111/j.1095-8649.2011.03163.x.

Thiebot, J.B., J.P. Arnould, A. Gómez-Laich, K. Ito, A. Kato, T. Mattern et al. 2017. Jellyfish and other gelata as food for four penguin species—insights from predator-borne videos. Front. Ecol. Environ. 15(8): 437–441. doi: 10.1002/fee.1529.

Thys, T.M., J.P. Ryan, H. Dewar, C.R. Perle, K. Lyons, J. O'Sullivan et al. 2015. Ecology of the ocean sunfish, *Mola mola*, in the southern California Current System. J. Exp. Mar. Biol. Ecol. 471: 64–76. doi: 10.1016/j.jembe.2015.05.005.

Thys, T.M., A.R. Hearn, K.C. Weng, J.P. Ryan and C. Peñaherrera-Palma. 2017. Satellite tracking and site fidelity of short ocean sunfish, *Mola ramsayi*, in the Galápagos Islands. J. Mar. Biol. 7097965. doi: 10.1155/2017/7097965.

Thys, T.M., M. Nyegaard, J.L. Whitney, J.P. Ryan, I. Potter, T. Nakatsubo et al. 2020. Ocean sunfish larvae: detection, identification and predation. pp. 105–128. *In*: T.M. Thys, G.C. Hays and J.D.R. Houghton [eds.]. The Ocean Sunfishes: Evolution, Biology and Conservation, CRC Press. Boca Raton, FL, USA.

Thys, T.M., M. Nyegaard and L. Kubicek. 2020. Ocean sunfishes and society. pp. 263–279. *In*: T.M. Thys, G.C. Hays and J.D.R. Houghton [eds.]. The Ocean Sunfishes: Evolution, Biology and Conservation, CRC Press. Boca Raton, FL, USA.

Truelove, N.K., EA. Adruszkiewicz and B.A. Block. 2019. A rapid environmental DNA method for detecting white sharks in the open ocean. Methods Ecol. Evol. 10: 1128–1135. doi: 10.1111/2041-210X.13201.

Wang, J., X. Zhu, X. Huang, L. Gu, Y. Chen and Z. Yang. 2016. Combined effects of cadmium and salinity on juvenile *Takifugu obscurus*: cadmium moderates salinity tolerance; salinity decreases the toxicity of cadmium. Sci. Rep. 6: 30968. doi: 10.1038/srep30968.

Watanabe, Y. and K. Sato. 2008. Functional dorsoventral symmetry in relation to lift-based swimming in the ocean sunfish *Mola mola*. PLoS One 3: e3446. doi: 10.1371/journal.pone.0003446.

Watanabe, Y.Y. and J. Davenport. 2020. Locomotory systems and biomechanics of ocean sunfish. pp. 72–86. *In*: T.M.Thys, G.C. Hays and J.D.R. Houghton [eds.]. The Ocean Sunfishes: Evolution, Biology and Conservation, CRC Press. Boca Raton, FL, USA.

Wilson, A.M., T.Y. Hubel, S.D. Wilshin, J.C. Lowe, M. Lorenc, O.P. Dewhirst et al. 2018. Biomechanics of predator-prey arms race in lion, zebra, cheetah and impala. Nature 554(7691): 183–188. doi: 10.1038/nature25479.

[1] Centre for Integrative Ecology, Deakin University, Geelong, Australia.

[2] Queen's University Belfast, School of Biological Sciences, Belfast, UK; Email: j.houghton@qub.ac.uk

[3] California Academy of Sciences, San Francisco, CA 94118, USA; Email: tierneythys@gmail.com

[4] Florida Fish and Wildlife Conservation Commission, Fish and Wildlife Research Institute, 1220 Prospect Avenue, No. 285, Melbourne, FL USA; Email: Doug.Adams@myfwc.com

[5] Department of Animal Production and Health, Public Veterinary Health and Food Science and Technology, Faculty of Veterinary Science, Universidad Cardenal Herrera-CEU, CEU Universities, Alfara del Patriarca, Valencia, Spain. Email: ana.ahuir@uchceu.es

[6] Colegio de Ciencias Biológicas y Ambientales (COCIBA), Universidad San Francisco de Quito, Ecuador, Diego de Robles sn y Pampite, Quito, Ecuador; Email: jackiealvarezc@gmail.com

[7] Marine and Environmental Sciences Centre (MARE), Laboratório Marítimo da Guia, Faculdade de Ciências da Universidade de Lisboa, Campo Grande 016, 1600-548 Lisboa, Portugal; Email: msbaptista@fc.ul.pt

[8] Oceanário de Lisboa, Esplanada D. Carlos I, 1990-005 Lisboa, Portugal; Email: hbatista@oceanario.pt; nbaylina@oceanario.pt

[9] Department of Fisheries Science, Virginia Institute of Marine Science, William and Mary, Gloucester Point, Virginia, USA; Department of Vertebrate Zoology, National Museum of Natural History, Smithsonian Institution, Washington DC, USA. Email: kebemis@vims.edu

[10] Department of Ecology and Evolutionary Biology and Cornell University Museum of Vertebrates, Cornell University, Ithaca, NY, USA; Email: web24@cornell.edu

[11] UCLA Department of Ecology and Evolutionary Biology, Los Angeles CA, USA; Email: ericcaldera@gmail.com

[12] Dipartimento di Scienze della Terra, Università degli Studi di Torino, Via Valperga Caluso, 35 I-10125 Torino, Italy. Email: giorgio.carnevale@unito.it

[13] The New England Coastal Wildlife Alliance, Middleboro, MA, USA; Email: krillcarson@mac.com

[14] Flying Sharks, Rua Farrobim do Sul 116, 9900-361, Horta, Portugal; MARE – Marine and Environmental Sciences Centre, ESTM, Instituto Politécnico de Leiria, 2520-641 Peniche, Portugal; Email: info@flyingsharks.eu

[15] IPMA—Portuguese Institute of the Sea and Atmosphere, Rua Alfredo Magalhães Ramalho, 6, 1495-006, Lisbon, Portugal; CCMAR—Centre of Marine Sciences, University of Algarve, Campus of, Gambelas, 8005-139 Faro, Portugal. Email: prcosta@ipma.pt

[16] Queen's Marine Laboratory, Portaferry, UK; Email: odaly04@qub.ac.uk; Leagling01@qub.ac.uk; nphillips01@qub.ac.uk

[17] School of Biological, Earth and Environmental Sciences and Environmental Research Institute, University College Cork, Ireland; Email: J.Davenport@ucc.ie

[18] Department of Biology, Texas State University, Aquatic Station, San Marcos, TX 78666, USA; Email: jdutton@txstate.edu

[19] Division of Oceanography and Marine Environment, IPMA - Portuguese Institute of the Sea and Atmosphere, Av. Brasília, 1449-006 Lisbon, Portugal; Email: catia.figueiredo@ipma.pt; clara.lopes@ipma.pt; jraimundo@ipma.pt

[20] College of Natural Science and Mathematics, Department of Biological Science, California State University, Fullerton CA, USA; Email: kforsgren@fullerton.edu

[21] Thuenen Institute of Fisheries Ecology, Bremerhaven, Germany; Email: Marko.Freese@thuenen.de

[22] Mediterranean Large Pelagics Group, Spanish Institute of Oceanography, Puerto pesquero s/n, 29640, Fuengirola, Malaga, Spain; Email: salvador.garcia@ieo.es

[23] Instituto de Ciencias Naturales Alexander von Humboldt & Instituto Antofagasta, University of Antofagasta, Antofagasta, Chile; Nucleo Milenio INVASAL, Concepción, Chile; Email: chris@harrodlab.net

[24] Colegio de Ciencias Biológicas y Ambientales (COCIBA), Universidad San Francisco de Quito, Ecuador, Diego de Robles sn y Pampite, Quito, Ecuador; Email: ahearn@usfq.edu.ec
[25] Department of Biological Sciences, University of Bergen, Bergen ,Norway; Email: Lea.Hellenbrecht@gmail.com
[26] Virginia Institute of Marine Science, William and Mary, Gloucester Point, Virginia, USA; Email: ehilton@vims.edu
[27] Monterey Bay Aquarium, 886 Cannery Row, Monterey, CA, USA; Email: mhoward@mbayaq.org
[28] Centre for Marine Socioecology, University of Tasmania, Hobart 7005, Tasmania, Australia; Email: r.kelly@utas.edu.au
[29] Aussermattweg 22, 3532, Zäziwil, Switzerland; Email: molamondfisch@gmail.com
[30] Scripps Institution of Oceanography, UC San Diego, La Jolla CA, USA; Email: tor.mowatt.larssen@gmail.com
[31] Northeast Fisheries Science Center, National Marine Fisheries Service, Woods Hole, MA USA; Email: richard.mcbride@noaa.gov
[32] Institute for East China Sea Research, Nagasaki University, 1551-7 Taira-machi, Nagasaki, Japan. Email: itsumi@nagasaki-u.ac.jp
[33] ORIX Aquarium Corporation, Nippon Life Hamamatsucho Crea Tower, 2-3-1 Hamamtsu-cho, Minato-ku, Tokyo 105-0013, Japan; Email: t.nakatsubo@aqua.email.ne.jp
[34] Adopt a Sunfish Project, oceansunfish.org, Carmel CA USA; Email: en.nix96@gmail.com
[35] Auckland War Memorial Museum Tamaki Paenga Hira, Natural Sciences, The Domain, Private Bag 92018, Victoria Street West, Auckland 1142, New Zealand; Email: mnyegaard@aucklandmuseum.com
[36] Laboratorio de Biología Marina/Estación de Biología Marina del Estrecho (Ceuta), Universidad de Sevilla, Avda. Reina Mercedes 6, Sevilla 41012, Spain.
[37] Dipartimento di Scienze della Terra, Università degli Studi di Torino, Via Valperga Caluso, 35 I-10125 Torino, Italy. Email: lu.pellegrino@unito.it
[38] Department of Biosciences, Swansea University, Singleton Park, Swansea, Wales, UK; Email: e.c.pope@swansea.ac.uk
[39] Affiliate Faculty, University of New Hampshire, Durham NC USA; Email: ingapotter@gmail.com
[40] Nordsøen Oceanarium, Willemoesvej 2, 9850 Hirtshals, Denmark; Email: mr@nordsoemail.dk
[41] MARE – Marine and Environmental Sciences Centre, Laboratório Marítimo da Guia, Faculdade de Ciências, Universidade de Lisboa, Av. Nossa Senhora do Cabo, 939, 2750-374 Cascais, Portugal; Email: rrosa@fc.ul.pt
[42] MBARI, 7700 Sandholdt Road, Moss Landing, CA 95039, USA; Email: ryjo@mbari.org
[43] Ocean Sunfishes Information Storage Museum (online), C-102 Plaisir Kazui APT, 13-6 Miho, Shimizu-ku, Shizuoka-shi, Shizuoka 424-0901, Japan; Email: sawaetsu2000@yahoo.co.jp
[44] Department of Zoology, National Museum of Nature and Science, 4-1-1 Amakubo, Tsukuba-shi, Ibaraki 305-0005, Japan. Email: s-gento@kahaku.go.jp
[45] Marine Biological Association of the United Kingdom, The Laboratory, Citadel Hill, Plymouth PL1 2PB, UK. Email: dws@mba.ac.uk
[46] WildCRU, Zoology, University of Oxford, The Recanati-Kaplan Centre, Tubney House, Abingdon Road, Tubney, Abingdon OX13 5QL, UK; Email: lara.l.sousa@gmail.com
[47] Oceanogràfic de València,, C/Eduardo Primo Yufera 1-B, 46013, València, Spain; Email: ctaura@oceanografic.org
[48] Integrated Statistics,US National Marine Fisheries Service, Northeast Fisheries Science Center, Woods Hole, MA, USA. Email: Emilee.Tholke@noaa.gov
[49] The University of Tokyo, 1-1-1 Yayoi, Bunkyo-ku, Tokyo 113-8657, Japan; Email: tsukamoto@marine.fs.a.u-tokyo.ac.jp
[50] National Museum of Natural History, Smithsonian Institution, Washington, DC USA; Email: tylerj@si.edu
[51] National Institute of Polar Research, Tachikawa, Tokyo 190-8518, Japan; Email: watanabe.yuuki@nipr.ac.jp
[52] Virginia Institute of Marine Science, William & Mary,Gloucester Point, VA, USA; Email: kevincmweng@gmail.com
[53] Joint Institute for Marine and Atmospheric Research, University of Hawai'i at Mānoa, Honolulu, HI 96822 USA; School of Ocean and Earth Science and Technology, Honolulu, HI USA; Email: jw2@hawaii.edu
[54] The University Museum, The University of Tokyo, Tokyo 113-0033, Japan; Email: yamanouey@yahoo.co.jp
[55] Nordsøen Oceanarium, Willemoesvej 2, 9850 Hirtshals, Denmark; Email: ky@nordsoemail.dk
* Corresponding author: g.hays@deakin.edu.au

Index

A

acanthocephalans 161, 171, 174
anatomy 3, 13, 55–67, 74, 83, 286
animal-borne camera 136
anthropogenic contaminant 186, 208
anthropogenic pressures 41, 146, 230–232, 234, 235
aquarium 162, 243-259, 287
areas of restricted search 133
art 272–274
aspect ratio 73, 74, 82, 201
atresia 88
Austromola angerhoferi 9, 10

B

back-fold 20, 21
bask 137, 175, 176, 246, 255
beak 3–9, 20, 21, 59, 60
behavioral thermoregulation 137
benthic diet 137, 147, 149, 151, 152, 194, 234, 253
bioaccumulation 188, 189
biotoxins 186–188, 204
brain 62, 63
branchiurans 174
buoyancy 80–86, 283, 287
burst swimming 83, 129, 132
bycatch 152–154, 217–219, 221–232, 236, 288–290

C

catch per unit effort 217, 228, 229
catchability 227
cestodes 161, 171, 173
clavus 56, 57, 61, 66
clavus margin 20–21, 24
cleaning behavior 175, 176
climate change 216, 234, 288
CO1 mtDNA 48
collection 244, 245, 247, 251, 258, 260
collision 235, 244, 252, 253, 270, 271
commercial landing 219, 224
community science 274, 275
conservation 281, 283, 290
conservation status 216, 217, 223, 236
contaminants 186–188, 199, 204
copepods 161, 162, 173–176

D

deaccession 258
deep excursion 133–139
definitive host 163
dentary beak 6–8
development 37, 44, 48, 50
diet 146–149, 151, 153, 154, 281, 282, 284, 285, 287
digestive system 63
distribution 24, 28, 29, 31, 32, 46–48, 109
diving 133, 134, 139, 282, 286
D-loop 37, 39, 40–44, 46, 47, 50, 51
drag 81, 82

E

ear 62, 63
ecosystem functioning 146
ecotourism 271, 276
ectoparasite 160–161
eggs 87, 88, 91, 92, 98, 99, 101, 102
Eocene 4, 5, 10
Eomola bimaxillaria 4, 5, 10
etymology 24, 28, 29
euthanasia 258
evolution 37, 286
evolutionary history 13
extant species 32
extinct 7, 10, 12
eye 62

F

fasting endurance 284, 285
fecundity 87–89, 92, 94, 98, 101, 102
feeding 244, 247, 249, 251–254, 257, 258
fin shape 73, 78
fisheries interactions 218–226
folklore 266
foraging 129, 133–139
fossil record 1, 2, 4, 5, 7, 12, 13
fossilization potential 3, 7
front 133–135, 227, 283

G

gelatinous zooplankton 147, 149, 151, 153, 154, 282–284
gelativore 147, 151

genome 37–39, 44, 47–51
gills 64–67
gonads 90–92, 94–97, 101

H

handling 245, 255–258
heart 65, 67
heat transfer 137, 138
histology 89–91
horizontal movements 129, 133
horizontal oval mouth 20
husbandry 243–245, 251, 255, 258, 259, 287, 288
hybrid 28
hypodermis 61, 62, 66, 74–83

I

intermediate host 161–163, 173, 174
interrelationships 10
intraspecific variation 23
isopods 161, 175

J

Japan 266–268, 270, 272
jaws 59, 66
jellyfish 284

L

landings 217–224, 226, 229
larva 107, 108, 112, 114
lateral ridge 20, 21
lift 73
liver 64
location 199, 200, 204–209
lore 266

M

marine plastic 235
Masturus 105–108, 114
Masturus lanceolatus 44, 45
maximum size 29–31, 56, 113
measurement method 24
Mediterranean 270
menu 266, 269
metabolic rate 284, 285
microplastics 186, 187, 201–204
midwater 253
migration 130–132, 134
Miocene diversification 12
mixed diet 133, 147, 253
Mola 105–112, 114, 115, 124–128
Mola alexandrini 18–23, 25, 27, 30–32, 37, 41, 43, 45
Mola mola 41, 45, 46, 48, 51
Mola pileata 5, 6, 13
Mola tecta 18–22, 27, 30–32, 45
molacanthus 106–109, 112, 114
molid 18–20, 23–26, 28–32
molidae 37–40, 48, 50, 186, 187, 201, 204
monogeneans 161, 172, 174–176
morphology 18–20, 23–29, 83, 84
movement 283–288, 290
musculature 74, 77, 82

N

nematodes 161, 171, 174

O

ocean sunfish 187, 188, 189, 192, 196, 198, 199, 201, 204, 243–247, 249, 252, 254–256, 258, 259
ocean sunfishes 37–39, 50
ontogenetic 56, 60–62, 64, 73, 82, 83, 106–109, 137, 148–152, 200
ontogenetic shift 146, 148, 149, 151, 153
ontogeny 152, 153
oocytes 87–93, 97, 99, 100–102
ossicle 3, 5–7, 10, 20–22, 56
osteodentine 3
ova 87, 88, 94, 99–101, 105, 114
ovaries 89, 92, 94, 102

P

parasites 160–171, 173–176, 287
phylogeny 18, 24, 42
Pliny 264
population genetics 47
post-hatching 106, 107
post-release mortality 218, 230, 231, 232, 236
pre-clavus band length 24
predation 105, 106, 111–114, 121
predator-prey 133, 146
predators 111–114
pre-juvenile 107
protists 160, 161, 163, 164
public aquarium iii

Q

quarantine 250–252

R

range shift 234
Ranzania 105–111, 113
Ranzania graham 7, 8
Ranzania laevis 43, 45, 46
Ranzania ogaii 7–9
Ranzania tenneyorum 7, 8
Ranzania zappai 7, 8
rearing 107, 162, 242, 256
red muscles 77, 78, 191, 197–199
reproduction 87–102
residency 131, 132
resilience 230, 236
reverence 267–270
rudder (see also clavus) 19, 72, 78, 79, 81

S

Sargasso 106, 108, 109–111, 122
satellite tracking 130, 131
scaling 284, 285
season 188, 194, 197
seasonality 88, 90, 96, 129
selective feeding 136, 137
sex size 188, 202
sexual dimorphism 19, 41, 56, 227
sexual maturity 87, 88, 90, 102
sharptail 18, 29, 38, 44, 130, 134, 204, 266, 268, 272
skeleton 56–66
skull 59, 60
space use 130
spawning areas 109, 110
spawning stock biomass 88, 102
speciation 13
species differences 188, 189, 195, 197, 202
species diversity 12, 44–48
SST 130, 133
stable isotope 147
stranding 32, 89, 102, 274, 289
swim bladder 64
swimming behaviour 72, 83

T

Taiwan 265, 268, 269, 270, 272, 273
target-training 250–252
taxonomy 18, 23, 24, 26, 31, 37–39, 281, 286
teeth 59, 60, 65
tendon 74–79, 81
thermal range 139–141
thermoregulation 137, 138
threats 289
thyroid gland 66
tissue distribution 189
top down control 152, 153
toxicity 204
trace elements 186–191, 193–197, 199, 200, 204–209
transportation 245–248, 250, 251, 258, 259
treatment 247, 250, 255, 257
trematodes 161, 163, 171
trends 217, 226–230, 236
trophic role 147–151
type specimen 25

V

valid species 18, 25
vertebrae 60, 62
vertical distribution 133–137
vertical movements 133–135, 138, 139
vertical oval mouth 20

W

white muscles 74–77, 191, 197–199
world record 31, 87, 99, 263